Numerical Methods in Electromagnetism

ACADEMIC PRESS SERIES IN ELECTROMAGNETISM
....

Electromagnetism is a classical area of physics and engineering that still plays a very important role in the development of new technology. Electromagnetism often serves as a link between electrical engineers, material scientists, and applied physicists. This series presents volumes on those aspects of applied and theoretical electromagnetism that are becoming increasingly important in modern and rapidly developing technology. Its objective is to meet the needs of researchers, students, and practicing engineers.

This is a volume in
ELECTROMAGNETISM
....

ISAAK MAYERGOYZ, SERIES EDITOR
UNIVERSITY OF MARYLAND, COLLEGE PARK, MARYLAND

Numerical Methods in Electromagnetism

M.V.K. Chari
S. J. Salon

Rensselaer Polytechnic Institute
Troy, New York

ACADEMIC PRESS
A Harcourt Science and Technology Company

San Diego   San Francisco   New York   Boston
London   Sydney   Tokyo

This book is printed on acid-free paper.

Copyright © 2000 by Academic Press

All rights reserved.
No part of this publication may be reproduced or transmitted in any form or by any means, electronic or mechanical, including photocopy, recording, or any information storage and retrieval system, without permission in writing from the publisher.

Requests for permission to make copies of any part of the work should be mailed to the following address: Permissions Department, Harcourt, Inc., 6277 Sea Harbor Drive, Orlando, Florida 32887-6777.

Explicit permission from Academic Press is not required to reproduce a maximum of two figures or tables from an Academic Press article in another scientific or research publication provided that the material has not been credited to another source and that full credit to the Academic Press article is given.

Academic Press
A Harcourt Science and Technology Company
525 B Street, Suite 1900, San Diego, California 92101-4495, USA
http://www.apnet.com

Academic Press
24-28 Oval Road, London NW1 7DX, UK
http://www.hbuk.co.uk/ap/

Library of Congress Catalog Card Number: 99-067217
International Standard Book Number: 0-12-615760-X

Printed in the United States of America
99 00 01 02 03 IP 9 8 7 6 5 4 3 2 1

CONTENTS

Foreword   xi

Preface   xiii

1   BASIC PRINCIPLES OF ELECTROMAGNETIC FIELDS   1
    1.1   Introduction   1
    1.2   Static Electric Fields   1
    1.3   The Electric Potential   3
    1.4   Electric Fields and Materials   11
    1.5   Interface Conditions on the Electric Field   13
    1.6   Laplace's and Poisson's Equations   15
    1.7   Static Magnetic Fields   24
    1.8   Energy in the Magnetic Field   36
    1.9   Quasi-statics: Eddy Currents and Diffusion   39
    1.10   The Wave Equation   42
    1.11   Discussion of Choice of Variables   44
    1.12   Classification of Differential Equations   60

2   OVERVIEW OF COMPUTATIONAL METHODS IN ELECTROMAGNETICS   63
    2.1   Introduction and Historical Background   63
    2.2   Graphical Methods   65
    2.3   Conformal Mapping   68
    2.4   Experimental Methods   68
    2.5   ElectroConducting Analog   69
    2.6   Resistive Analog   70

2.7	Closed Form Analytical Methods	71
2.8	Discrete Analytical Methods	74
2.9	Transformation Methods for Nonlinear Problems	75
2.10	Nonlinear Magnetic Circuit Analysis	79
2.11	Finite Difference Method	80
2.12	Integral Equation Method	84
2.13	The Finite Element Method	96

3 THE FINITE DIFFERENCE METHOD — 105

3.1	Introduction	105
3.2	Difference Equations	106
3.3	Laplace's and Poisson's Equations	108
3.4	Interfaces Between Materials	109
3.5	Neumann Boundary Conditions	111
3.6	Treatment of Irregular Boundaries	114
3.7	Equivalent Circuit Representation	115
3.8	Formulas For High-Order Schemes	117
3.9	Finite Differences With Symbolic Operators	122
3.10	Diffusion Equation	126
3.11	Conclusions	141

4 VARIATIONAL AND GALERKIN METHODS — 143

4.1	Introduction	143
4.2	The Variational Method	144
4.3	The Functional and its Extremum	145
4.4	Functional in more than one space variable and its extremum	153
4.5	Derivation of the Energy-Related Functional	157
4.6	Ritz's method	170
4.7	The Wave Equation	176
4.8	Variational Method for Integral Equations	179
4.9	Introduction to The Galerkin Method	182
4.10	Example of the Galerkin Method	183

Contents

5	**SHAPE FUNCTIONS**	189
	5.1 Introduction	189
	5.2 Polynomial Interpolation	201
	5.3 Deriving Shape Functions	207
	5.4 Lagrangian Interpolation	211
	5.5 Two-Dimensional Elements	214
	5.6 High-Order Triangular Interpolation Functions	222
	5.7 Rectangular Elements	227
	5.8 Derivation of Shape Functions for Serendipity Elements	234
	5.9 Three-Dimensional Finite Elements	240
	5.10 Orthogonal Basis Functions	278
6	**THE FINITE ELEMENT METHOD**	283
	6.1 Introduction	283
	6.2 Functional minimization and global assembly	295
	6.3 Solution to the nonlinear magnetostatic problem with first-order triangular finite elements	302
	6.4 Application of the Newton–Raphson Method to a First-Order Element	306
	6.5 Discretization of Time by the Finite Element Method	310
	6.6 Axisymmetric Formulation for the Eddy Current Problem Using Vector Potential	313
	6.7 Finite Difference and First-Order Finite Elements	320
	6.8 Galerkin Finite Elements	322
	6.9 Three-Element Magnetostatic Problem	326
	6.10 Permanent Magnets	338
	6.11 Numerical Example of Matrix Formation for Isoparametric Elements	342
	6.12 Edge Elements	353
7	**INTEGRAL EQUATIONS**	359
	7.1 Introduction	359
	7.2 Basic Integral Equations	359
	7.3 Method of Moments	362
	7.4 The Charge Simulation Method	370

	7.5	Boundary Element Equations for Poisson's Equation in Two Dimensions	374
	7.6	Example of BEM Solution of a Two-Dimensional Potential Problem	381
	7.7	Axisymmetric Integral Equations for Magnetic Vector Potential	389
	7.8	Two-Dimensional Eddy Currents With $T - \Omega$	393
	7.9	BEM Formulation of The Scalar Poisson Equation in Three Dimensions	403
	7.10	Green's functions for some typical electromagnetics applications	409
8	**OPEN BOUNDARY PROBLEMS**	413
	8.1	Introduction	413
	8.2	Hybrid Harmonic Finite Element Method	413
	8.3	Infinite Elements	417
	8.4	Ballooning	427
	8.5	Infinitesimal Scaling	433
	8.6	Hybrid Finite Element–Boundary Element Method	437
9	**HIGH-FREQUENCY PROBLEMS WITH FINITE ELEMENTS**	451
	9.1	Introduction	451
	9.2	Finite Element Formulation in Two Dimensions	452
	9.3	Boundary Element Formulation	459
	9.4	Implementation of the Hybrid Method (HEM)	467
	9.5	Evaluation of the Far-Field	471
	9.6	Scattering Problems	481
	9.7	Numerical Examples	488
	9.8	Three Dimensional FEM Formulation for the Electric Field	494
	9.9	Example	517
10	**LOW-FREQUENCY APPLICATIONS**	519
	10.1	Time Domain Modeling of Electromechanical Devices	519
	10.2	Modeling of Flow Electrification in Insulating Tubes	550
	10.3	Coupled Finite Element and Fourier Transform Method for Transient Scalar Field Problems	561
	10.4	Axiperiodic Analysis	570

11 SOLUTION OF EQUATIONS 591
11.1 Introduction 591
11.2 Direct Methods 595
11.3 *LU* Decomposition 598
11.4 Cholesky Decomposition 605
11.5 Sparse Matrix Techniques 607
11.6 The Preconditioned Conjugate Gradient Method 627
11.7 GMRES 645
11.8 Solution of Nonlinear Equations 658

A VECTOR OPERATORS 707

B TRIANGLE AREA IN TERMS OF VERTEX COORDINATES 709

C FOURIER TRANSFORM METHOD 711
C.1 Computation of Element Coefficient Matrices and Forcing Functions 714

D INTEGRALS OF AREA COORDINATES 719

E INTEGRALS OF VOLUME COORDINATES 721

F GAUSS–LEGENDRE QUADRATURE FORMULAE, ABSCISSAE, AND WEIGHT COEFFICIENTS 723

G SHAPE FUNCTIONS FOR 1D FINITE ELEMENTS 725

H SHAPE FUNCTIONS FOR 2D FINITE ELEMENTS 727

I SHAPE FUNCTIONS FOR 3D FINITE ELEMENTS 735

REFERENCES 749

INDEX 759

Academic Press Series in
ELECTROMAGNETISM

Edited by Isaak Mayergoyz, University of Maryland, College Park, Maryland

BOOKS PUBLISHED IN THE SERIES

Georgio Bertotti, *Hysteresis in Magnetism: For Physicists, Material Scientists, and Engineers*

Scipione Bobbio, *Electrodynamics of Materials: Forces, Stresses, and Energies in Solids and Fluids*

Alain Bossavit, *Computational Electromagnetism: Variational Formulations, Complementarity, Edge Elements*

Göran Engdahl, *Handbook of Giant Magnetostrictive Materials*

John C. Mallinson, *Magneto-Resistive Heads: Fundamentals and Applications*

Isaak Mayergoyz, *Nonlinear Diffusion of Electromagnetic Fields*

RELATED BOOKS

John C. Mallinson, *The Foundations of Magnetic Recording, Second Edition*

Reinaldo Perez, *Handbook of Electromagnetic Compatibility*

FOREWORD

This volume in the Academic Press Electromagnetism series presents a detailed and self-contained treatment of numerical methods in electromagnetics. It is written by two well-known researchers in the field. Dr. M.V.K. Chari pioneered (jointly with Professor P. Silvester) the use of finite element technique in electromagnetic field calculations, and for many years he was in charge of the electromagnetic computational program at General Electric Corporate Research and Development Center. Professor S. Salon is one of the most active researchers in the area of electromagnetic field analysis of electric power devices and is highly regarded in the applied electromagnetic community for his important contributions to the field.

This book reflects the unique expertise, extensive experience, and strong interests of the authors in the computational aspects of electromagnetism. The unique feature of this book is its broad scope and unbiased treatment of finite element, finite difference, and integral equation techniques. These three techniques are most dominant in modern electromagnetic field computations, and they are extensively implemented in various commercially available software packages. This book contains a large number of tutorial examples that are worked out in detail. This feature will help readers to follow all the computational steps involved in the realization of different numerical methods and will be very beneficial for graduate students and code developers. The book covers the numerical analysis of static fields, eddy currents, scattering and waveguide problems as well as the coupled problems where the electromagnetic field analysis is intertwined with mechanical and circuit aspects. In its style, scope, and emphasis, this book is complementary to *Computational Electromagnetism*, written by Dr. A. Bossavit, also published in the Academic Press Electromagnetism series.

I believe that this book will be a valuable reference for both experts and beginners in the field. Electrical and mechanical engineers, physicists, material scientists, code developers, and graduate students will find this book very informative.

ISAAK MAYERGOYZ, Series Editor

PREFACE

Since the formulation of electromagnetic theory, researchers and engineers have sought accurate solutions of the resulting boundary and initial value problems. Initially the focus of these efforts was to find closed form and analytical solutions. These were followed by analogue methods. Progress in digital computer hardware and software have now made numerical solutions, such as finite difference, finite element and integral equation formulations popular.

Great advances have been made in field computation for static, steady state sinusoidal, and transient problems, first in two dimensions and more recently in three dimensions. Both linear and nonlinear cases are now treated routinely. Recent work has resulted in new formulations involving mixed scalar and mixed vector variables, especially for three dimensional problems. There has also been considerable progress in the area of numerical computation resulting in fast equation solvers.

For finite difference, finite element and integral equations, the reader is presented here with a broad picture of the development and research in the area of numerical analysis applied to problems in electromagnetics. The book is suitable for a graduate course in the subject and this material has formed the basis of a graduate course presented at Rensselaer Polytechnic Institute. A valuable feature of this book is the large number of tutorial examples that are presented. These are small problems (a few finite elements or boundary elements) that are worked out in detail. With this feature the student can follow all of the steps involved in the different methods. This also allows the book to be used as a self-study guide. This work will also be of use to researchers and code developers in this area as it includes a great deal of reference material. Engineers and designers using these methods will also benefit as we present many examples and discuss the accuracy of the methods as well as their strong and weak points.

The chapters contain principle categories—namely, introductory, historical, and concept development; advanced computational techniques; and specific and illustrative examples. Chapter 1 reviews the basic equations of electromagnetic theory, the different equations and formulations used in the remaining chapters, and contains discussions of materials, boundary conditions, and constraints. Chapter 2 presents an historical perspective of the different approaches that have been applied to the solution of electromagnetic problems. These range from classical closed form methods to analogue methods to modern digital methods. A brief discussion of the relative merits of these methods is presented. In Chapter 3 we introduce the method of finite differences. General methods of approximating differential equations by finite differences and the accuracy of these approximations is presented. Boundary conditions and materials are discussed. In examples, the method is applied to Poisson's equation, the diffusion equation and the wave equation. In Chapter 4 we present two mathematical techniques that are important in the understanding of the finite element method and the integral equation method. These are the variational method and the Galerkin method. A general approach is taken and several lllustrative examples are used. In Chapter 5 the theory of shape functions is presented. This is a very complete treatment and is used in the development of the finite element and boundary element methods. In Chapter 6 the finite element method is presented. Examples given to illustrate the procedure include a magnetostatic problem, a problem with nonlinear materials, a problem with permanent magnets, and a problem using the isoparametric approach. A comparison of the system of equations obtained by finite elements and finite differences is also included. In Chapter 7 we present integral equation methods. These include the method of moments, the charge simulation method and the boundary element method. Examples illustrate the application of these methods. In Chapter 8 we discuss the use of finite elements in the solution of open boundary problems. The techniques discussed include analytic coupling, ballooning, infinitesimal scaling, infinite elements and the hybrid finite element boundary element method. In Chapters 9 and 10 we present applications of the finite element method in high frequency and low frequency problems. These include examples of waveguides and scattering and the coupling of low frequency fields to mechanical, circuit, and fluid problems. In chapter 11 we discuss methods of solution for large linear systems and the different formulations of nonlinear problems that are frequently used in numerical analysis. A number of appendices include reference material that elaborates some of the material presented in the chapters. For example, the reader will find shape functions and integration formulae for many finite elements.

The authors are indebted to many people; far too many to acknowledge here. These include our colleagues at General Electric, Westinghouse, RPI, and Magsoft Corporation. We are also indebted to many talented graduate students whose work is represented here. We would like to particularly acknowledge Mark DeBortoli, who worked out several of the examples presented, Laurent Nicolas, who provided some of the waveguide and scattering figures, and Kiruba Sivasubramanaim who contributed many of the figures and helped in many ways. We also acknowledge the late Peter Silvester and the many discussions we had during his visits to RPI. We thank our friend and colleague Isaak Mayergoyz for his support, advice, and encouragement during the writing of this book and appreciate the work of the helpful staff at Academic Press. Finally, we could not have written this book without the support and encouragement of our wives, Padma and Corine, to whom we dedicate this work.

1

BASIC PRINCIPLES OF ELECTROMAGNETIC FIELDS

1.1 INTRODUCTION

This chapter is intended to introduce and to discuss the basic principles and concepts of electromagnetic fields, which will be dealt with in detail in succeeding chapters. The concept of flux and potential is discussed and Maxwell's set of equations is presented. The description of electric and magnetic materials then follows. We then discuss the set of differential equations that are commonly used as a starting point for the numerical methods we will study. An explanation of some of the more popular choices of state variables (scalar and vector potential or direct field variables) is then presented.

1.2 STATIC ELECTRIC FIELDS

The electric charge is the source of all electromagnetic phenomena. Electrostatics is the study of electric fields due to charges at rest. In SI units the electric charge has the units of coulombs after Charles Augustin de Coulomb (1736–1806). The charge of one electron is 1.602×10^{-19} coulombs. In rationalized MKS units an electric charge of Q coulombs produces an electric flux of Q Coulombs. In Figure 1.1 we see the electric charge *emitting* flux uniformly in the radial direction in a spherical coordinate system with origin at the charge. If we consider a closed surface around the charge (for example, the sphere shown in Figure 1.1), then all of the flux must pass through the surface. This would be true for any number of charges in the volume enclosed by the surface. Using the principle of superposition, we conclude that *the total flux passing through any closed surface is equal to the total charge enclosed by that surface*. This is Gauss' law and one of Maxwell's equations. If we define the electric flux density $\vec{D}$ as the local value of the flux per unit area then we may write

$$\oint_S \vec{D} \cdot d\vec{S} = \int_v \rho \, dv \tag{1.1}$$

where ρ is the charge density. This is the integral form of Gauss' law. From the symmetry of Figure 1.1 we find immediately that for a point charge

$$\vec{D} = \frac{Q}{4\pi r^2} \hat{a}_r \tag{1.2}$$

where r is the distance from the charge (source point) to the location where we compute the field quantity (field point).

The inverse square law of equation (1.2) is one of the most important results of electromagnetism. There are two consequences of the inverse square law that we shall encounter in subsequent chapters. One is that there is a second-order singularity which occurs at $r = 0$. The second is that the electric flux density goes to zero as $r \to \infty$.

Coulomb, in a famous experiment, found that there is a force between two electric charges acting along the line connecting them that varies as $\frac{1}{r^2}$. This force is proportional to the magnitude of the charges and depends on the medium. Coulomb's law states that for two charges Q_1 and Q_2 in vacuum,

$$\vec{F}_{12} = \frac{Q_1 Q_2}{4\pi \epsilon_0 r^2} \hat{a}_r \tag{1.3}$$

Here ϵ_0 is the permittivity of free space, $\frac{1}{36\pi} \times 10^{-9}$ farads/meter. The force on a unit charge is called the electric field and is denoted as $\vec{E}$. The units of electric field are volts/meter. From (1.2) and (1.3) we see that

$$\vec{D} = \epsilon_0 \vec{E} \tag{1.4}$$

Basic Principles of Electromagnetic Fields

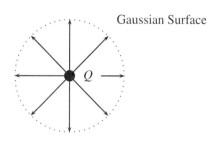

Figure 1.1 Point Charge and Flux

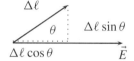

Figure 1.2 Work and Electric Field

1.3 THE ELECTRIC POTENTIAL

For computational purposes it is often difficult to compute the field quantities $\vec{D}$ and $\vec{E}$ directly. One of the reasons is that we need to find three components at each point and all of the algebra is vectorial. For these reasons we find it simpler to compute scalar variables and then compute the field from these values. In the case of the electric field we use the electric potential, V. Recalling that the electric field is the force on a unit charge, we define the electric potential at a point in space as the work expended in moving a unit charge from infinity to that point. Consider Figure 1.2, where we are moving a charge a distance $\Delta \ell$ at an angle θ with respect to the electric field. We see that we do work only when moving the charge in the direction of the electric field. The work done to move the charge in Figure 1.2 a distance $\Delta \ell$ is $E \Delta \ell \cos \theta$, because no work is required to move the distance $\Delta \ell \sin \theta$ perpendicular to the field.

We write

$$V = -\int_{\infty}^{P} \vec{E} \cdot d\vec{\ell} \tag{1.5}$$

The minus sign is a convention and means that we do work moving a positive charge toward a field produced by a positive charge. The work done moving a unit charge a distance $\Delta \ell$ in the presence of an electric field is $\Delta W = -\vec{E} \cdot \Delta \ell$ so that

$$\vec{E} = -\nabla V \tag{1.6}$$

Therefore, if we can find V, a scalar, we can compute the electric field by taking the gradient. An important property of the scalar potential is that the work done to move a charge between points P_1 and P_2 is independent of the path taken by the charge. Fields having this property are called *conservative*. This is typical of lossless systems in which we have thermodynamically reversible processes. Systems with losses do not have this property, and this will have implications later in our discussion of variational principles. A corollary of this principle is that it takes no work to move a charge around a closed path in a static electric field. This is a very important property of the potential. Without this, the potential would not be uniquely defined at a point and the path would have to be specified. We will see in the next section that the magnetostatic field with current sources does not have this property and the scalar potential is of more limited use. We have therefore

$$\oint \vec{E} \cdot d\vec{\ell} = 0 \tag{1.7}$$

another of Maxwell's equations.

A few examples of the electric field will be of interest to us.

Potential Due to a Point Charge

The electric field due to a point charge of magnitude Q is, from (1.3) and (1.4),

$$\vec{E} = \frac{Q}{4\pi \epsilon_0 r^2} \hat{a}_r \tag{1.8}$$

Basic Principles of Electromagnetic Fields

We find the potential by integrating equation (1.8) to obtain

$$V(P) = -\int_{\infty}^{P} \frac{Q}{4\pi\epsilon_0 r^2} \hat{a}_r \cdot dr = \frac{Q}{4\pi\epsilon_0 r} \quad (1.9)$$

We make note of some properties that will be of interest in later chapters. First, there is a singularity at $r = 0$, but it is of first order and not of second order as in the case of the field. Second, the potential goes to zero as r approaches infinity.

The Logarithmic Potential

If we consider a long line of charge density ρ_ℓ coulombs/meter, then the application of Gauss' law in Figure 1.3 gives a flux density of

$$\vec{D} = \frac{\rho_\ell}{2\pi r} \hat{a}_r \quad (1.10)$$

and an electric field of

$$\vec{E} = \frac{\rho_\ell}{2\pi \epsilon_0 r} \hat{a}_r \quad (1.11)$$

Integrating from infinity to a point P gives the potential

$$V(P) = \frac{\rho_\ell}{2\pi \epsilon_0} \ln r \quad (1.12)$$

This logarithmic potential is frequently found in two-dimensional problems and problems in cylindrical coordinates. The potential has a singularity at $r = 0$ but unfortunately does not go to zero as r goes to infinity. In fact, the potential slowly approaches infinity as r increases. The potential vanishes at $r = 1$.

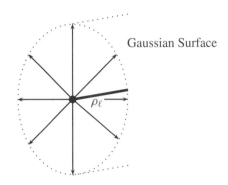

Figure 1.3 Logarithmic Potential

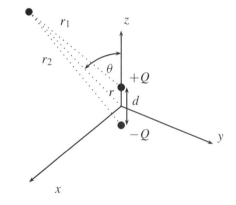

Figure 1.4 Dipole Field and Potential

Potential of a Dipole

The field and potential due to an electric dipole will be important in our discussion of integral equations. Consider two charges $+Q$ and $-Q$ separated by a distance d. We can find the potential by superposition. We will be interested in the potential at a

Basic Principles of Electromagnetic Fields

distance from the charges much greater than their separation. The potential due to a single point charge is

$$V(P) = \frac{Q}{4\pi\epsilon_0 r} \tag{1.13}$$

For both charges (see Figure 1.4) we have

$$V(P) = \frac{Q}{4\pi\epsilon_0 r_1} - \frac{Q}{4\pi\epsilon_0 r_2} = \frac{Q}{4\pi\epsilon_0} \frac{r_2 - r_1}{r_2 r_1} \tag{1.14}$$

We see that

$$r_1 = \sqrt{x^2 + y^2 + \left(z - \frac{d}{2}\right)^2}$$

$$r_2 = \sqrt{x^2 + y^2 + \left(z + \frac{d}{2}\right)^2} \tag{1.15}$$

We make the approximation that for $r \gg d$

$$r_2 - r_1 \approx d \cos\theta$$
$$r_1 r_2 \approx r^2 \tag{1.16}$$

so

$$V = \frac{Q}{4\pi\epsilon_0} \frac{d \cos\theta}{r^2} \tag{1.17}$$

This potential drops off faster than the potential of a point charge and the singularity at $r = 0$ is of second order.

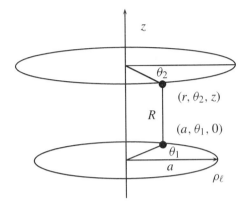

Figure 1.5 Potential of Ring of Charge

Potential Due to a Ring of Charge

Another configuration that will be useful is the potential due to a uniform ring of charge with charge per unit length of ρ_ℓ [1]. Consider the ring with center at the origin of a cylindrical coordinate system illustrated in Figure 1.5.

The potential is

$$\phi(r, z) = \frac{\rho_\ell}{4\pi \epsilon_0} \int_0^{2\pi} \frac{a\,d\theta}{\sqrt{(r - a\cos\theta)^2 + a^2 \sin^2\theta + z^2}} \quad (1.18)$$

where

$$R = \sqrt{r^2 + a^2 - 2ar\cos\theta + z^2} \quad (1.19)$$

and

$$\theta = \theta_1 - \theta_2 \quad (1.20)$$

Basic Principles of Electromagnetic Fields

This is equivalent to

$$\phi(r,z) = \frac{a\rho_\ell}{2\pi\epsilon_0} \frac{1}{(r+a)^2+z^2} \int_0^{\frac{\pi}{2}} \frac{d\alpha}{\sqrt{1-k^2\sin^2\alpha}} \quad (1.21)$$

where

$$k^2 = \frac{4ar}{(a+r)^2+z^2} \quad (1.22)$$

and

$$\alpha = \frac{\theta - \pi}{2} \quad (1.23)$$

Equation (1.21) is an elliptic integral of the first kind. Results for this are tabulated in many standard references [2].

1.3.1 Potential Energy and Energy Density

Using our definition of electric potential, the work required to move a charge of magnitude q_i from infinity to a point (x, y, z) is

$$w_i = q_i V(x, y, z) \quad (1.24)$$

For an ensemble of $N-1$ charges producing the potential, we have [3]

$$V(x', y', z') = \sum_{j=1}^{N-1} \frac{q_j}{4\pi\epsilon_0 r_{ij}} \quad (1.25)$$

where $r_{ij} = \sqrt{(x'-x)^2 + (y'-y)^2 + (z'-z)^2}$. The potential energy of the charge is then

$$w_i = q_i \sum_{j=1}^{N-1} \frac{q_j}{4\pi\epsilon_0 r_{ij}} \qquad (1.26)$$

The energy in all of the charges is

$$W_i = \sum_{i=1}^{N} \sum_{j<i} \frac{q_i q_j}{4\pi\epsilon_0 r_{ij}} \qquad (1.27)$$

The $j < i$ is required so that we do not count the contribution of each charge twice. We can also write

$$W_i = \frac{1}{2} \sum_{i=1}^{N} \sum_{j=1}^{N} \frac{q_i q_j}{4\pi\epsilon_0 r_{ij}} \qquad (1.28)$$

It is understood that $i \neq j$ in the summations. We can always express the charges as a distribution[1] to obtain

$$W_i = \frac{1}{2} \int_{\Omega'} \int_{\Omega} \frac{\rho(x',y',z')\rho(x,y,z)}{4\pi\epsilon_0 r_{ij}} d\Omega d\Omega' \qquad (1.29)$$

We now substitute the scalar potential into equation (1.29).

$$W_i = \frac{1}{2} \int_{\Omega} \rho(x,y,z) V(x,y,z) d\Omega \qquad (1.30)$$

---

[1] Even if we use point charges this can be done with the aid of Dirac delta functions.

Basic Principles of Electromagnetic Fields

From Poisson's equation (see Section 1.6), we express the charge density in terms of the Laplacian of the potential

$$W_i = \frac{1}{2} \int_\Omega \epsilon_0 \nabla^2 V(x, y, z) V(x, y, z) d\Omega \qquad (1.31)$$

We now integrate by parts over all space. The surface integral at infinity vanishes because both E and V go to zero, giving

$$W_i = \frac{1}{2} \int_\Omega \epsilon_0 |\nabla V(x, y, z)|^2 d\Omega = \frac{1}{2} \int_\Omega \vec{D} \cdot \vec{E} \, d\Omega \qquad (1.32)$$

1.4 ELECTRIC FIELDS AND MATERIALS

If we dealt only with charge distributions in free space, we would have no need for the numerical methods presented in this book. Given the charge distribution, we could use Coulomb's law and in principle find the electric field at any point in space by superposition. In reality, we must consider materials that affect and are affected by the electric field. There are two types of materials to consider, both idealized here, which we must deal with to solve realistic problems in electrostatics. They are conductors and insulators or dielectrics.

1.4.1 Conductors

An ideal conductor is a material with an unlimited number of free charges. When an electric field is applied to a conductor, the free charges move under the influence of the field. An equilibrium state occurs when the free charges adjust themselves in a distribution such that the net or total electric field is identically zero in the medium. If the field is not zero, the charges will continue to experience a force and to move. In materials that we consider good conductors (e.g., copper, aluminum, iron) this redistribution takes place very rapidly. In room temperature copper it is on the order of 10^{-18} seconds. Unless we are dealing with very high frequency effects, we can consider this redistribution of the charge to be instantaneous. In this case the net electric field in the conductor is zero, which means that at each point in the volume of the conductor the induced electric field due to the internal charges is exactly equal and opposite to the applied electric field. (See Figure 1.6.)

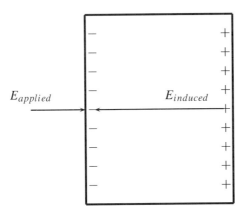

Figure 1.6 Electric Field Cancellation in a Conductor

1.4.2 Dielectrics

Real materials contain both free and bound charges. We define an ideal dielectric as a material with no free charges. Under the influence of an electric field the bound charges, which are originally electrically neutral, will distort into dipoles as shown in Figure 1.7.

These internal dipoles produce a *polarization*, which for linear materials is proportional to the electric field. The proportionality constant depends on the material. We can write

$$\vec{P} = \epsilon_0 \chi \vec{E} \tag{1.33}$$

where χ is the susceptibility of the material.

The net flux density is the sum of the applied flux density and the polarization flux density

$$\vec{D} = \epsilon_0 \vec{E} + \epsilon_0 \chi \vec{E} = \epsilon \vec{E} \tag{1.34}$$

Basic Principles of Electromagnetic Fields

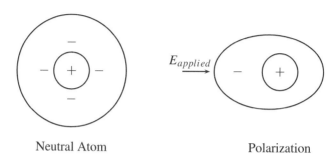

Figure 1.7 Polarization by an Electric Field

The product $\epsilon_0(1 + \chi)$ is called the permittivity of the material. The complex interactions related to the bound charges occurring in the dielectric are thus all included in this macroscopic view of the permittivity. Note that a material with a high permittivity begins to behave like a conductor. The greater the polarization, the smaller the internal electric field. In modeling applications it is therefore possible to represent a conductor as a dielectric with a very high value of ϵ.

1.5 INTERFACE CONDITIONS ON THE ELECTRIC FIELD

The electric fields and flux densities must satisfy certain interface conditions. A valid solution to the field equations will automatically result in these conditions being satisfied. We can find these conditions by considering the interface between two materials in Figure 1.8. Let the lower material have electric property ϵ_1 and the upper material have ϵ_2.

We first consider the incremental volume represented by the disk in Figure 1.8. We can let the height of the disk approach zero so that no flux leaves the disk through the cylindrical side. Gauss's law tells us that the total flux leaving the volume through the top and bottom caps on the cylinder is equal to the charge enclosed in the cylinder. The flux leaving the top surface is $\phi_t = \vec{D}_2 \cdot \vec{S}_2 = D_{n2}S$, where D_{n2} is the normal component

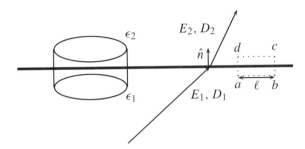

Figure 1.8 The Electric Field at a Boundary

of the flux density and S is the surface area. The flux leaving the bottom is similarly $\phi_b = \vec{D}_1 \cdot \vec{S}_1 = -D_{n1}S$. The total flux leaving the volume is $\phi = (D_{n2} - D_{n1})S$ and this equals the charge enclosed by the volume, which is $Q = \rho_s S$ where ρ_s is the surface charge density (C/m^2). So we have

$$D_{n2} - D_{n1} = \rho_s \tag{1.35}$$

If there is no surface charge density, then the normal component of the flux density is continuous. If there is a surface charge density, the normal components are discontinuous by the magnitude of the surface charge density.

Our second interface condition comes from considering the electric field around the closed path $abcda$ in Figure 1.8. We apply $\oint \vec{E} \cdot d\vec{\ell} = 0$. As the height of the rectangular path goes to zero, there will be no contribution from the vertical sides. The contribution from a to b is $\vec{E}_1 \cdot \vec{\ell}_{ab} = E_{t1}\ell_{ab}$, where E_{t1} is the tangential component of the electric field. Similarly, the contribution along cd is $\vec{E}_2 \cdot \vec{\ell}_{cd} = E_{t2}\ell_{cd}$. Because $|\ell_{ab}| = |\ell_{cd}|$,

$$E_{t2} - E_{t1} = 0 \tag{1.36}$$

Basic Principles of Electromagnetic Fields

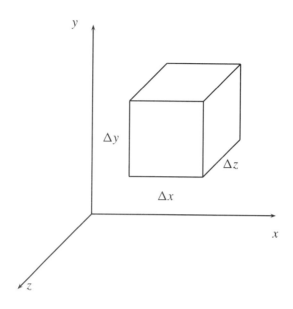

Figure 1.9 Differential Element for Laplace's Equation

So the tangential component of the electric field is continuous. This is true even if there is surface charge. These are our two interface conditions for electric fields. We have made no assumptions concerning the nature of the materials. They can be conductors or dielectrics and can have nonlinear, anisotropic properties. Note that these conditions are implied in Maxwell's equations and therefore usually need not be specifically enforced. We will see that in numerical solutions one of these is satisfied exactly (in a strong sense) and the other is only approximately satisfied (in a weak sense).

1.6 LAPLACE'S AND POISSON'S EQUATIONS

Two of the most important equations in physics are Laplace's and Poisson's equations. These equations describe the continuity of the field and apply not only to electric fields but also to magnetic fields, heat flow, gravitation, pressure, seepage, and so on. It is instructive to derive Laplace's equation from a differential point of view because we will be dealing with finite difference and finite element equations later on.

Consider the differential volume element in Figure 1.9, with sides Δx, Δy, and Δz. We will assume that the electric field in the center of the volume is $E(x, y, z)$. We first find the flux leaving the elemental volume. The flux leaving the volume on the right in the x direction, to a first-order approximation, is

$$\phi_{xr} = \epsilon \left(E_x + \frac{\partial E_x}{\partial x} \frac{\Delta x}{2} \right) \Delta y \Delta z \tag{1.37}$$

The flux entering on the left in the x direction is

$$\phi_{xl} = \epsilon \left(E_x - \frac{\partial E_x}{\partial x} \frac{\Delta x}{2} \right) \Delta y \Delta z \tag{1.38}$$

The total flux leaving in the x direction is the difference of (1.37) and (1.38) or

$$\phi_x = \epsilon \frac{\partial E_x}{\partial x} \Delta x \Delta y \Delta z \tag{1.39}$$

Similar formulae in the y and z directions give the total flux leaving the volume as

$$\phi = \epsilon \left(\frac{\partial E_x}{\partial x} + \frac{\partial E_y}{\partial y} + \frac{\partial E_z}{\partial z} \right) \Delta x \Delta y \Delta z \tag{1.40}$$

By Gauss' law this must be equal to the total charge in the volume or $\rho \Delta x \Delta y \Delta z$. Equation (1.40) becomes

$$\epsilon \left(\frac{\partial E_x}{\partial x} + \frac{\partial E_y}{\partial y} + \frac{\partial E_z}{\partial z} \right) = \rho \tag{1.41}$$

This is the differential form of Gauss' law, expressed as $\nabla \cdot D = \rho$. If we now substitute $E_x = -\frac{\partial V}{\partial x}$ etc. we get

$$\epsilon \left(\frac{\partial^2 V}{\partial x^2} + \frac{\partial^2 V}{\partial y^2} + \frac{\partial^2 V}{\partial z^2} \right) = -\rho \tag{1.42}$$

Basic Principles of Electromagnetic Fields

This is written in operator form as

$$\nabla \cdot \epsilon \nabla V = -\rho \tag{1.43}$$

or for homogeneous space,

$$\nabla^2 V = -\frac{\rho}{\epsilon} \tag{1.44}$$

Note that Poisson's equation is a second-order equation and therefore requires two boundary conditions. Generally speaking we must specify the potential and its normal derivative on the boundary. If we let $\rho = 0$ in equation (1.44), we have Laplace's equation.

The Uniqueness Theorem

If we have a set of conductors and specify either the potential or the total charge on each of these, we have completely specified the problem. Any solution to Laplace's equation that satisfies the boundary conditions is the unique solution. Consider the set of N conductors in Figure 1.10. Let ϕ_1 be a solution to Laplace's equation. We now expand

$$\nabla \cdot \phi_1 \nabla \phi_1 = \nabla \phi_1 \cdot \nabla \phi_1 + \phi_1 \nabla^2 \phi_1 = |\nabla \phi_1|^2 \tag{1.45}$$

because by hypothesis $\nabla^2 \phi_1 = 0$. Integrating both sides over a volume Ω and using the divergence theorem,

$$\int_\Omega \nabla \cdot \phi_1 \nabla \phi_1 \, d\Omega = \oint_S \phi_1 \nabla \phi_1 \cdot d\vec{S} = \oint_S \phi_1 \frac{\partial \phi_1}{\partial n} dS = \int_\Omega |\nabla \phi_1|^2 d\Omega \tag{1.46}$$

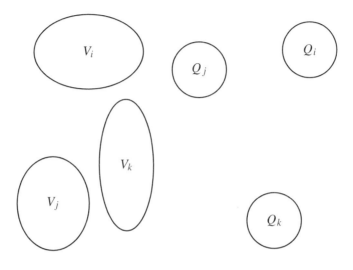

Figure 1.10 Laplace's Equation with Conducting Boundaries

If we let the radius of the surface enclosing the volume go to infinity, the surface integral term must vanish. This is because ϕ_1 goes to zero as $\frac{1}{r}$ or faster, $\nabla \phi_1$ goes to zero as $\frac{1}{r^2}$ or faster, and the surface area goes to infinity as r^2. We now have

$$\sum_{i=1}^{N} \oint_S \phi_1 \frac{\partial \phi_1}{\partial n} dS = \int_\Omega |\nabla \phi_1|^2 d\Omega \tag{1.47}$$

where the summation is over the N conductors. Now assume there are two solutions that satisfy Laplace's equation and the potential or total charge conditions on the conducting boundaries. Let us call the second solution ϕ_2. The Laplacian is a linear operator, so the difference of the two solutions must also be a solution. Let us define

$$\delta \phi = \phi_1 - \phi_2 \tag{1.48}$$

Then

$$\sum_{i=1}^{N} \oint_S (\phi_1 - \phi_2) \frac{\partial(\phi_1 - \phi_2)}{\partial n} dS = \int_\Omega |\nabla(\phi_1 - \phi_2)|^2 d\Omega \quad (1.49)$$

Now, on all of the conductors with specified potential, the term $(\phi_1 - \phi_2)$ is zero. On the remaining conducting surfaces the potential is constant and we have $\rho_{s1} = \epsilon_0 \frac{\partial \phi_1}{\partial n}$ and $\rho_{s2} = \epsilon_0 \frac{\partial \phi_2}{\partial n}$, where ρ_{s1} and ρ_{s2} are the charge densities on the conductor surfaces in the two solutions. We can write equation (1.49) as

$$\sum_{i=1}^{N} (\phi_1 - \phi_2) \oint_S \frac{(\rho_{s1} - \rho_{s2})}{\epsilon_0} dS = \int_\Omega |\nabla(\phi_1 - \phi_2)|^2 d\Omega \quad (1.50)$$

We know from Gauss' law that

$$\oint_S \rho_{s1} dS = \oint_S \rho_{s2} dS = Q \quad (1.51)$$

for each conductor, so

$$\int_\Omega |\nabla(\phi_1 - \phi_2)|^2 d\Omega = 0 \quad (1.52)$$

The integrand of equation (1.52) is nonnegative and we must have

$$\nabla(\phi_1 - \phi_2) = 0 \quad (1.53)$$

so ϕ_1 and ϕ_2 can differ by a constant. However, because they have the same value on the conductors which have potentials specified, they must be equal. There is one exception. If the potential is not specified on any conductor the solutions may differ by a constant because no reference is given. The electric fields will be identical. To ensure uniqueness, the potential must therefore be specified at least at one point in the domain.

1.6.1 The Time-Varying Electric Field

When sinusoidal time-varying fields are applied to material bodies, the polarization vector $\vec{P}$ may lag behind the applied field. In this case the permittivity will be complex and there will be a loss component due to the work done against "frictional" forces. We can write

$$\vec{P} = \epsilon_0 \alpha e^{-j\phi} \vec{E} \tag{1.54}$$

where ϕ is the phase angle between P and E. The susceptibility is then

$$\chi = \alpha e^{-j\phi} \tag{1.55}$$

and the permittivity is

$$\epsilon = \epsilon_0(1+\chi) = \epsilon_0(1 + \alpha(\cos\phi - j\sin\phi)) = \epsilon' - j\epsilon'' \tag{1.56}$$

An insulating material will have both conduction and displacement currents. From the sinusoidally time-varying Maxwell equation

$$\nabla \times \vec{H} = j\omega\vec{D} + \vec{J} = j\omega\epsilon'\vec{E} + (\omega\epsilon'' + \sigma)\vec{E} \tag{1.57}$$

we see that the conductivity causes an apparent increase in the imaginary part of the permittivity. The current density in phase with the electric field will produce a loss. We normally express the permittivity as a single complex number that includes both of these effects. So

$$\epsilon = \epsilon' - j\left(\epsilon'' + \frac{\sigma}{\omega}\right) \tag{1.58}$$

Basic Principles of Electromagnetic Fields

or we can express the permittivity as a magnitude and a loss tangent as

$$\epsilon = \epsilon_r \epsilon_0 (1 - j \tan \delta) \tag{1.59}$$

Sinusoidally Time-Varying Electric Fields

In the case in which we have both conduction and displacement current in the insulating structure, the complex potential distribution becomes capacitive at high frequency and resistive at low frequency. The formulation begins with the continuity equation

$$\nabla \cdot \vec{J} = 0 \tag{1.60}$$

From Maxwell's equation

$$\nabla \cdot (\sigma + j\omega\epsilon)\vec{E} = 0 \tag{1.61}$$

So

$$\nabla \cdot (\sigma + j\omega\epsilon)\nabla V = 0 \tag{1.62}$$

We now divide through by $j\omega$ to obtain

$$\nabla \cdot (\epsilon - j\frac{\sigma}{\omega})\nabla V = 0 \tag{1.63}$$

As an example of the application of these equations, consider the end region of a large electric generator as shown in Figure 1.11. In order to control the electric field at the end of the generator where the high-voltage conductors come out of the grounded core, a semiconducting layer is applied to the conductor's surface. The small conduction current that flows in this layer produces a nearly uniform potential gradient along the coil, which reduces the maximum electric field.

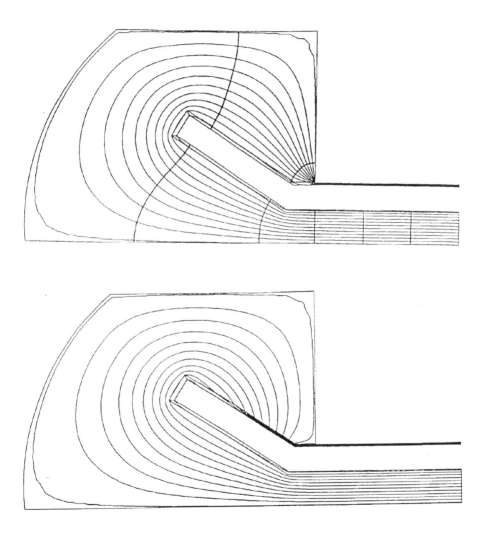

Figure 1.11 End Winding of Turbine Generator with (lower) and without Semiconducting Layer in a 60-Hertz Electric Field

Transient Electric Fields

Starting from

$$\vec{E} = -\nabla V \tag{1.64}$$

and

$$\vec{J} = \sigma \vec{E} + \frac{\partial \vec{D}}{\partial t} \tag{1.65}$$

we take the derivative of (1.65) with respect to time

$$\frac{\partial \vec{J}}{\partial t} = \sigma \frac{\partial \vec{E}}{\partial t} + \epsilon \frac{\partial^2 \vec{E}}{\partial t^2} \tag{1.66}$$

The continuity equation states that

$$\nabla \cdot \vec{J} = 0 \tag{1.67}$$

so

$$\nabla \cdot \left(\epsilon \frac{\partial}{\partial t} (\nabla V) + \sigma \nabla V \right) = 0 \tag{1.68}$$

In Cartesian coordinates, for homogeneous materials, we get

$$\left(\epsilon \frac{\partial}{\partial t} + \sigma \right) \left(\frac{\partial^2 V}{\partial x^2} + \frac{\partial^2 V}{\partial y^2} + \frac{\partial^2 V}{\partial z^2} \right) = 0 \tag{1.69}$$

For the axisymmetric case

$$\left(\epsilon\frac{\partial}{\partial t}+\sigma\right)\left(\frac{1}{r}\left(\frac{\partial}{\partial r}\left(r\frac{\partial V}{\partial r}\right)\right)+\frac{1}{r}\frac{\partial}{\partial z}\left(r\frac{\partial V}{\partial z}\right)\right)=0 \quad (1.70)$$

1.7 STATIC MAGNETIC FIELDS

The equations for the static magnetic field can be developed in the same manner as for the static electric field. Although so-called *monopoles*, single magnetic charges equivalent to electric charges, have not been observed in nature, it is often convenient from a computational standpoint to use this concept. If we assume a magnetic point charge of magnitude m, then this charge will produce m units of magnetic flux in the same way that an electric charge produces electric flux. In SI units the magnetic charge will have a magnitude of m webers, and this is the unit of magnetic flux. Following a similar development, we define the magnetic flux density B from a point charge as

$$\vec{B} = \frac{m}{4\pi r^2}\hat{a}_r \quad (1.71)$$

The equivalent of Gauss' law for magnetic fields is

$$\oint \vec{B} \cdot d\vec{S} = Q_m \quad (1.72)$$

where Q_m is the total magnetic charge enclosed. In differential form we have

$$\nabla \cdot \vec{B} = \rho_m \quad (1.73)$$

where ρ_m is the magnetic charge density. This quantity is always zero in nature but is often used as a source of magnetic fields in simulations. As in the electric field case,

Basic Principles of Electromagnetic Fields

there is a force on the magnetic charge that depends on the medium and varies inversely as r^2. $\vec{H}$, the magnetic field of a point charge in free space, is then

$$\vec{H} = \frac{Q_m}{4\pi \mu_0 r^2} \hat{a}_r \quad (1.74)$$

We have the constitutive equation

$$\vec{B} = \mu_0 \vec{H} \quad (1.75)$$

where μ_0 (henries/meter in SI units) is the permeability of free space.

If we consider the magnetic field due to charges alone, we again have a conservative system and it takes no work to move a magnetic charge around a closed path. So

$$\oint \vec{H} \cdot d\vec{\ell} = 0 \quad (1.76)$$

From this principle we can define the magnetic scalar potential[2] Ω in the same way we defined the electric scalar potential. In the SI system Ω has the units of amperes and is related to the magnetic field by

$$\vec{H} = -\nabla \Omega \quad (1.77)$$

The magnetic scalar potential is also described by Poisson's equation by combining equations (1.75) and (1.77).

$$\nabla^2 \Omega = -\rho_m / \mu_0 \quad (1.78)$$

The uniqueness properties found for the case of the static electric field apply here as well.

[2] Sometimes called total magnetic scalar potential.

26 CHAPTER 1

As already stated, free magnetic charges have not been observed. The source of the magnetic field is either permanent magnets (ie., magnetic material) or electric current. In the case of a field produced by current we have Ampère's law, which in integral form is

$$\oint \vec{H} \cdot d\vec{\ell} = i_{\text{enclosed}} \tag{1.79}$$

This tells us that the work required to move a magnetic charge around a closed path is equal to the current (ampere-turns) enclosed by the path. If the current is zero, of course we get equation (1.76). If the current is not zero, the field is not conservative and the scalar potential is not uniquely defined. The potential difference, or work required to move a charge between two points, depends on the path taken by the charge. In order to apply scalar potentials to these problems, cuts are sometimes introduced, or surfaces with discontinuous potentials are introduced. Great care must be taken in defining and interpreting these. We introduce the magnetic vector potential to overcome many of these difficulties in Section 1.7.3.

1.7.1 Magnetic Materials

If we place a magnetic material in a magnetic field, the magnetic flux increases. If we call the flux density due to the sources in free space B_s (for the source component) and the flux density due to the magnetic material B_i (for the induced component), then

$$\vec{B} = \vec{B}_s + \vec{B}_i \tag{1.80}$$

We now define the magnetization vector $\vec{M}$ by

$$\vec{B} = \mu_0(\vec{H} + \vec{M}) \tag{1.81}$$

This magnetization depends on the magnetic field and we define

$$\vec{M} = \xi_m \vec{H} \tag{1.82}$$

where ξ_m is the magnetic susceptibility. So from (1.81) and (1.82) we have

$$B = \mu_0(1 + \xi_m)H = \mu H \tag{1.83}$$

where

$$\mu = \mu_0(1 + \xi_m) = \mu_0 \mu_r \tag{1.84}$$

Here μ is defined as the permeability and μ_r is a dimensionless quantity called the relative permeability.

In real ferromagnetic materials the permeability is a nonlinear function of the field and depends on the past history of the material. For a detailed description of hysteresis see Mayergoyz [4]. Magnetic domains in the material line up with the applied field, increasing the magnetic flux. When most of these domains are lined up with the field, further increase of the field has little effect on the additional flux. We call this effect saturation. These materials are often represented by saturation curves.

Figure 1.12 shows a typical magnetization curve. Measurements of flux and current are converted to H and B in a controlled sample. The current is slowly varied from a positive maximum to a negative maximum and the corresponding flux density is found for each current value. At first the patterns do not repeat but after several cycles we reach a repeatable pattern, which is the major hysteresis loop. The procedure is repeated for a different peak value of current and so on until a series of these hysteresis loops are defined. A line connecting the tips of the loops as shown in the figure is called the saturation or magnetization curve. A *soft* magnetic material is one in which the hysteresis loops are narrow, as in Figure 1.13. For narrow loops, the normal magnetization curve is a good approximation and is often used to characterize the material. We note that the multivalued property of the hysteresis loops is lost in this case but nonlinearity is approximately represented.

Another type of magnetic material that is characterized by a wide sharp hysteresis loop is called hard magnetic material (Figure 1.14). These materials are used for permanent magnets and recording media. In these materials we often use the hysteretic property of the material and the normal magnetization curve is inappropriate. In the representation of permanent magnets we often use the second quadrant characteristic because that is where the magnet is usually designed to operate.

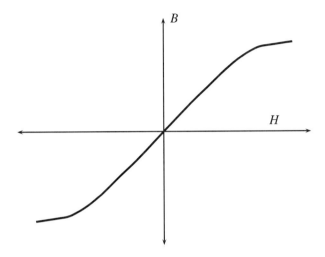

Figure 1.12 Saturation Curve of Magnetic Steel

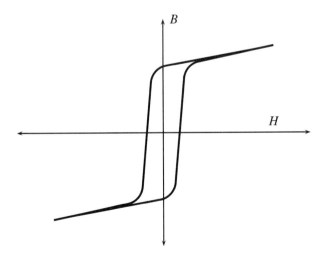

Figure 1.13 Soft Magnetic Material

Basic Principles of Electromagnetic Fields

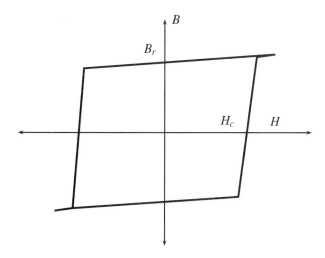

Figure 1.14 Hard Magnetic Material

Intrinsic and Normal Hysteresis Characteristics

We end our discussion of magnetic materials with a brief description of the hysteresis characteristics of magnet material [5]. We often assume that magnets are linear, yet the hysteresis loops look quite nonlinear in the second quadrant. We must consider that two different curves are used to represent the magnet characteristics. It is useful to consider the different information contained in each curve. As we begin to magnetize a material, first in one direction and then in the other, after a few cycles we obtain the normal hysteresis loop. In this case the flux density, B, on the ordinate is the flux density which would be measured. For the so-called *intrinsic* characteristic, the flux density is that portion due only to the magnetic material or equivalent sources in the magnetic material. This curve saturates at a point at which all of the magnetic dipoles are lined up in the direction of the magnetic field. The actual flux density can still be increased beyond this point by increasing the current in an external source. At the point where the magnetic field, H, is zero, the intrinsic and normal curves are the same. The flux density at this point is B_r, the remnant flux density. In second-quadrant operation, which is of most interest to us, the value of the magnetic field required to bring the normal flux density to zero is H_c, the coercive force. The value of H to bring the intrinsic flux density to zero is a larger (negative) value called H_{ci}, the intrinsic coercive force. For an ideal magnet, one with a square hysteresis loop, the intrinsic flux density is approximately a straight line over a wide range of magnetic excitation. This means that the magnetization vector, $\vec{M}$, inside the material is constant. Thus the ampere-turns available to magnetize an external magnetic circuit are constant, that is, independent of the external reluctance or external sources. It is the intrinsic curve that tells us the effect of the sources or equivalent current and it is the normal curve that tells how the magnet behaves in a magnetic circuit with external fields applied. Consider Figure 1.15. If we demagnetize the material to the point C and then increase the magnetic field, the material follows a new path to point D. If we repeat this process a few times between C and D, the characteristic settles down to a repeatable minor loop. For most magnet materials the loop is quite narrow and can be approximated by a straight line. The slope of this line is the *recoil* permeability. It is this permeability that determines the change in flux density if the air gap changes or the external field changes. The recoil permeability is fairly constant, that is, independent of the point C. In other words, the chords representing the minor loops are approximately parallel. This slope also turns out to be approximately equal to the slope of the major hysteresis loop at B_r. This value is widely used. The intrinsic and normal curves are plotted in Figure 1.16.

Basic Principles of Electromagnetic Fields 31

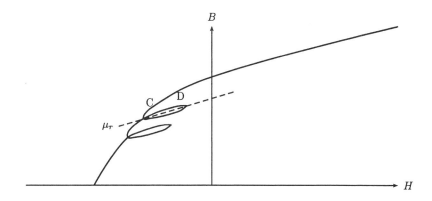

Figure 1.15 Recoil Line

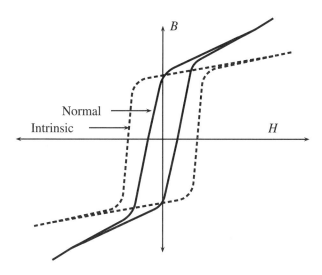

Figure 1.16 Intrinsic and Normal Demagnetization Curves

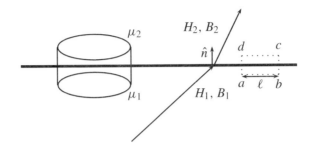

Figure 1.17 The Magnetic Field at a Boundary

1.7.2 Interface Conditions for the Magnetic Field

The magnetic field interface conditions are developed following the electric field case of Section 1.5. Gauss' law applied to the incremental disk in Figure 1.17 tell us that the normal component of flux density is continuous. Ampère's law around the rectangular path $abcda$ tells us that the tangential components of the magnetic field are continuous if no surface current exists. Otherwise there is a discontinuity in the tangential field equal to the surface current density. These are written as

$$B_{n1} = B_{n2} \tag{1.85}$$

and

$$H_{t2} - H_{t1} = J_s \tag{1.86}$$

where J_s is the surface current density (amperes/meter).

1.7.3 Magnetic Vector Potential (MVP)

We have seen that the magnetic scalar potential is not generally applicable in problems involving current, which are most of the problems we will face. We must therefore use a different variable. If we look at the integral form of Ampère's law

$$\oint \vec{H} \cdot d\vec{\ell} = i \tag{1.87}$$

we can think of the magnetic field as a variable that can be used to find the current through a surface. In a similar way, we define a variable called the magnetic vector potential such that the integral of the tangential component around a closed path gives the magnetic flux through the surface defined by that path.

$$\oint \vec{A} \cdot d\vec{\ell} = \psi_m \tag{1.88}$$

In vector form (compare to $\nabla \times \vec{H} = \vec{J}$) we have

$$\nabla \times \vec{A} = \vec{B} \tag{1.89}$$

Thus the curl of the magnetic vector potential gives the magnetic flux density. We also have by the vector identity that for any vector $\vec{F}$, $\nabla \cdot (\nabla \times \vec{F}) = 0$. Therefore the divergence of B is zero, as it must be. We note that while this is a necessary property, an inaccurate solution for $\vec{A}$ will produce an induction that is divergenceless ($\nabla \cdot B = 0$). Therefore this cannot be used as a validation of B computed from a magnetic vector potential solution. In the SI system the units of MVP are webers/meter.

The reader may wonder about the advisability of substituting one vector unknown for another. In the case of the scalar potential, we had the advantage of solving for one unknown at a point instead of three. Now we have the same number of unknowns and still must perform the curl operation to find the flux density. Although this is true, we shall now see that the MVP is described by Poisson's equation and that for two-dimensional and axisymmetric problems we have only a single component of MVP and for computational purposes can treat it as a scalar. We begin with Faraday's law

$$\nabla \times \vec{E} = -\frac{\partial \vec{B}}{\partial t} \tag{1.90}$$

Replacing B by $\nabla \times \vec{A}$

$$\nabla \times \vec{E} = -\frac{\partial}{\partial t}(\nabla \times \vec{A}) = -\nabla \times \frac{\partial \vec{A}}{\partial t} \qquad (1.91)$$

Rearranging terms, we get

$$\nabla \times \left(\vec{E} + \frac{\partial \vec{A}}{\partial t}\right) = 0 \qquad (1.92)$$

The quantity in parentheses can be replaced by the gradient of a scalar potential. Solving for E, this gives

$$\vec{E} = \nabla V - \frac{\partial \vec{A}}{\partial t} \qquad (1.93)$$

The first term is the conservative part of the electric field (due to charges) and the second term is the induced electric field or electromotive force (EMF) resulting from the rate of change of flux linkage.

We now consider Ampère's law

$$\nabla \times \vec{H} = \vec{J} + \frac{\partial \vec{D}}{\partial t} \qquad (1.94)$$

We can write this in terms of the scalar and vector potentials as

$$\nabla \times \nabla \times \vec{A} = \mu \vec{J} + \mu\epsilon \frac{\partial}{\partial t}\left(-\nabla V - \frac{\partial \vec{A}}{\partial t}\right) \qquad (1.95)$$

Basic Principles of Electromagnetic Fields

We can write Gauss' law as

$$\nabla \cdot \left(-\nabla V - \frac{\partial \vec{A}}{\partial t} \right) = \frac{\rho}{\epsilon} \qquad (1.96)$$

or

$$\nabla^2 V + \frac{\partial}{\partial t}(\nabla \cdot \vec{A}) = -\frac{\rho}{\epsilon} \qquad (1.97)$$

Using a vector identity, we can write equation (1.95) as

$$\nabla^2 \vec{A} - \nabla(\nabla \cdot \vec{A}) - \mu\epsilon \left(\frac{\partial \nabla V}{\partial t} + \frac{\partial^2 \vec{A}}{\partial t^2} \right) = -\mu \vec{J} \qquad (1.98)$$

Interface Conditions on A

Consider the interface of material 1 and material 2. We will develop interface conditions for the tangential component of the MVP. We postpone discussion of the normal component continuity until we discuss the gauge conditions. If we integrate

$$\oint \vec{A} \cdot d\vec{\ell} = \psi_m \qquad (1.99)$$

around the path *abcda* in Figure 1.18, then as the sides *bc* and *da* shrink to zero the flux enclosed by the path vanishes. Therefore we conclude that the tangential component of the magnetic vector potential is continuous. Then

$$\hat{n} \times (\vec{A}_1 - \vec{A}_2) = 0 \qquad (1.100)$$

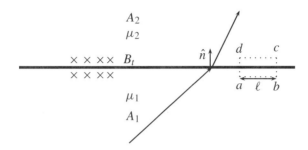

Figure 1.18 The Magnetic Vector Potential at a Boundary

1.8 ENERGY IN THE MAGNETIC FIELD

If we assume that magnetic charges are the source of magnetic fields, then the development of expressions for the energy stored in the magnetic field follows analogously to that of the electric field. If we consider the field due to currents, then we can find the energy required to establish a distribution of current and the associated fields. Consider a single circuit with constant current I. If we change the flux linkage of the circuit by an amount $\delta\lambda$, then we create an EMF in the circuit in a direction to oppose the change. In order to keep the current constant, work must be done by the source of magnitude

$$dw = I\delta\lambda \tag{1.101}$$

We now consider the work necessary to build up an arbitrary current distribution by breaking that distribution into a set of elementary current loops as shown in Figure 1.19.

Each loop encloses an area S. The incremental work resulting from the change in flux linkage of the loop is

$$dw = J\Delta \int_S \hat{n} \cdot \delta\vec{B}\, dS \tag{1.102}$$

Basic Principles of Electromagnetic Fields

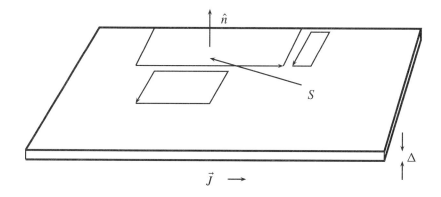

Figure 1.19 A Circuit Composed of Elemental Loops

We can write B in terms of the magnetic vector potential to get

$$dw = J\Delta \int_S \hat{n} \cdot \nabla \times \delta \vec{A} \, dS \qquad (1.103)$$

Using Stokes' theorem, we transform equation (1.103) to a line integral.

$$dw = J\Delta \oint_C \delta \vec{A} \cdot d\vec{c} \qquad (1.104)$$

We note that $J\Delta dc$ is the integral of the current density over the volume of the loop. Therefore the work done by external sources to change the flux linkage of the loop is the volume integral

$$dw = \int_v \vec{J} \cdot \delta \vec{A} \, dv \qquad (1.105)$$

Integrating over all of the current elements (i.e., all of space), we obtain

$$W = \int_v \vec{J} \cdot \vec{A} \, dv \tag{1.106}$$

We can also express the energy in terms of the field. Using Ampère's law $\nabla \times \vec{H} = \vec{J}$ in (1.105),

$$dw = \int_v (\nabla \times \vec{H} \cdot \delta \vec{A}) dv \tag{1.107}$$

Using the vector identity

$$\nabla \cdot (\vec{A} \times \vec{B}) = \vec{B} \cdot (\nabla \times \vec{A}) - \vec{A} \cdot (\nabla \times \vec{B}) \tag{1.108}$$

we get

$$dw = \int_v [\vec{H} \cdot (\nabla \times \delta \vec{A}) + \nabla \cdot (\vec{H} \times \delta \vec{A})] dv \tag{1.109}$$

The second term can be transformed to a surface integral and vanishes at infinity. In the first term we use the definition of magnetic vector potential to obtain

$$dw = \int_v [\vec{H} \cdot \delta \vec{B}] dv \tag{1.110}$$

This expression is analogous to the one obtained for electric fields and is valid for problems with magnetic materials.

1.9 QUASI-STATICS: EDDY CURRENTS AND DIFFUSION

Two-dimensional eddy current phenomena are described by the diffusion equation for the MVP. We obtain this equation from equations (1.93) and (1.98). We set the divergence of the vector potential to zero and then ignore the terms involving the displacement currents, which we can neglect at low frequency. For the steady-state time-harmonic case, this equation in terms of the magnetic vector potential is then

$$\frac{1}{\mu}\frac{\partial^2 A}{\partial x^2} + \frac{1}{\mu}\frac{\partial^2 A}{\partial y^2} = -J_o + j\omega\sigma A \tag{1.111}$$

where $A = A_z(x, y)$.

From Faraday's law, a time rate of change of flux linkage will produce an EMF. The direction of the EMF is to circulate a current that would oppose the change of flux linkage. In a conductor, the net field is altered due to this induced or eddy current. The eddy currents are affected by their own field, and thus currents and fields must be solved for simultaneously.

A one-dimensional example will give us some insight into the phenomenon [6]. Consider the semi-infinite slab of Figure 1.20. The material occupies the half-space $x \geq 0$. There is a field, H_o, applied in the y direction. The field varies sinusoidally with time at angular frequency ω. The material has constant properties μ henries/meter and σ mhos/meter. In this one-dimensional case B and H have only a y component, E and J have only a z component and these four quantities vary in the x direction. From Ampère's law

$$\frac{\partial H}{\partial x} = J \tag{1.112}$$

From Faraday's law

$$\frac{1}{\sigma}\frac{\partial J}{\partial x} = j\omega B \tag{1.113}$$

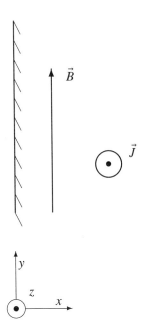

Figure 1.20 Semi-Infinite Slab

B and H are related by the constitutive equation

$$B = \mu H. \tag{1.114}$$

Using equation (1.113) we write B in terms of J as

$$B = -\frac{j}{\omega\sigma}\frac{\partial J}{\partial x} \tag{1.115}$$

Using equation (1.112) we get

$$\frac{\partial^2 B}{\partial x^2} - j\omega\mu\sigma B = 0 \tag{1.116}$$

Basic Principles of Electromagnetic Fields

This is a second-order homogeneous differential equation with constant coefficients. The general solution to this equation is

$$B = C_1 e^{\sqrt{j\omega\mu\sigma}\,x} + C_2 e^{-\sqrt{j\omega\mu\sigma}\,x} \quad (1.117)$$

This may be checked by substitution into equation (1.116). It is convenient to define

$$\delta = \sqrt{\frac{2}{\omega\mu\sigma}} \quad (1.118)$$

so that equation (1.117) becomes

$$B = C_1 e^{\sqrt{2j}\,x/\delta} + C_2 e^{-\sqrt{2j}\,x/\delta} \quad (1.119)$$

We can eliminate $\sqrt{j}$ by noting that

$$\sqrt{j} = \frac{1+j}{\sqrt{2}} \quad (1.120)$$

so

$$B = C_1 e^{(1+j)\frac{x}{\delta}} + C_2 e^{-(1+j)\frac{x}{\delta}} \quad (1.121)$$

From a physical standpoint C_1 must be zero. Otherwise, the first term would go to infinity as x approached infinity. If we call the flux density at the surface $B(0) = B_0$, then equation (1.121) becomes

$$B = B_0 e^{-\frac{x}{\delta}} e^{-\frac{jx}{\delta}} \quad (1.122)$$

Thus B is the product of two exponential terms. The first term indicates that the magnitude of the flux density is decreasing exponentially with x. At depth $x = \delta$, B is $1/e$ of its value at the surface. The second exponential term has a magnitude of one and describes the phase shift of the flux density. At a depth $x = \delta$, the flux density lags the surface flux density B_o by one radian. The parameter δ is called the depth of penetration. The depth of penetration, δ, becomes a useful parameter in eddy current analysis. The depth of penetration depends on the material properties (μ and σ) and on the frequency.

The current density can be found from the flux density. From equation (1.112)

$$J = \frac{1}{\mu} \frac{\partial B}{\partial x} \quad (1.123)$$

Taking the derivative of equation (1.122)

$$J = -\frac{B_0}{\mu} \frac{(1+j)}{\delta} e^{-(1+j)\frac{x}{\delta}} \quad (1.124)$$

or defining the surface current density as J_0

$$J = J_0 e^{-(1+j)\frac{x}{\delta}} \quad (1.125)$$

1.10 THE WAVE EQUATION

The wave equation (Helmholtz's equation) is often used with the field variable E or H directly. We can obtain the wave equation by beginning with Ampère's law.

$$\nabla \times \vec{H} = \vec{J} + \frac{\partial \vec{D}}{\partial t} \quad (1.126)$$

Basic Principles of Electromagnetic Fields

Taking the curl of both sides, we get

$$\nabla \times \nabla \times \vec{H} = \nabla \times \vec{J} + \frac{\partial}{\partial t} \nabla \times \vec{D} \tag{1.127}$$

For isotropic materials we can use the constitutive equations to find

$$\nabla \times \nabla \times \vec{H} = \sigma \nabla \times \vec{E} + \frac{\partial}{\partial t} \epsilon \nabla \times \vec{E} \tag{1.128}$$

We now use Faraday's law to replace the $\nabla \times \vec{E}$ by magnetic quantities.

$$\nabla \times \nabla \times \vec{H} = -\sigma \frac{\partial \vec{B}}{\partial t} - \epsilon \frac{\partial^2 \vec{B}}{\partial t^2} \tag{1.129}$$

We now have a second-order equation that we can write for the magnetic field vector.

$$\nabla \times \nabla \times \vec{H} = -\sigma \mu \frac{\partial \vec{H}}{\partial t} - \epsilon \mu \frac{\partial^2 \vec{H}}{\partial t^2} \tag{1.130}$$

Using a vector identity for the curl–curl operator, we rewrite equation (1.130) as

$$\nabla^2 \vec{H} - \sigma \mu \frac{\partial \vec{H}}{\partial t} - \epsilon \mu \frac{\partial^2 \vec{H}}{\partial t^2} = \nabla \nabla \cdot \vec{H} \tag{1.131}$$

Because the divergence of the magnetic field is zero in homogeneous material

$$\nabla^2 \vec{H} - \sigma \mu \frac{\partial \vec{H}}{\partial t} - \epsilon \mu \frac{\partial^2 \vec{H}}{\partial t^2} = 0 \tag{1.132}$$

We can obtain a similar equation for the electric field by starting out with Faraday's law, taking the curl of both sides, and substituting out the magnetic quantities. This results in

$$\nabla^2 \vec{E} - \sigma\mu \frac{\partial \vec{E}}{\partial t} - \epsilon\mu \frac{\partial^2 \vec{E}}{\partial t^2} = \frac{1}{\epsilon} \nabla \rho \qquad (1.133)$$

1.11 DISCUSSION OF CHOICE OF VARIABLES

If we start by considering the direct solution of Maxwell's equations, repeated in equation (1.134)

$$\nabla \times \vec{E} = -\frac{\partial \vec{B}}{\partial t}$$
$$\nabla \times \vec{H} = \vec{J} + \frac{\partial \vec{D}}{\partial t}$$
$$\nabla \cdot \vec{D} = \rho$$
$$\nabla \cdot \vec{B} = 0 \qquad (1.134)$$

We have five unknown quantities and four equations. In three dimensions, the unknowns are all vectors, so that there are 15 quantities to solve for, and since the first two equations are vector equations we have eight equations. We can solve the equations by using the constitutive relationships

$$\vec{D} = \epsilon \vec{E}$$
$$\vec{B} = \mu \vec{H}$$
$$\vec{J} = \sigma \vec{E} \qquad (1.135)$$

We can substitute out, say, D and B and solve for the six components of E and H.

Basic Principles of Electromagnetic Fields

In order to get the number of unknowns down to three we transform the first-order partial differential equations into second-order equations. Taking the curl of the second Maxwell equation

$$\nabla \times \nabla \times \vec{H} = \nabla \times \vec{J} + \frac{\partial}{\partial t}(\nabla \times \vec{D})$$

$$\nabla \times \nabla \times \vec{H} = \sigma \nabla \times \vec{E} + \frac{\partial}{\partial t}(\epsilon \nabla \times \vec{E})$$

$$\nabla \times \nabla \times \vec{H} = -\sigma \frac{\partial \vec{B}}{\partial t} - \epsilon \frac{\partial^2 \vec{B}}{\partial t^2} \quad (1.136)$$

We now use the vector identity $\nabla \times \nabla \times F = \nabla \nabla \cdot F - \nabla^2 F$ to obtain

$$\nabla^2 \vec{H} - \mu\sigma \frac{\partial \vec{H}}{\partial t} - \mu\epsilon \frac{\partial^2 H}{\partial t^2} = \nabla \nabla \cdot \vec{H} = 0 \quad (1.137)$$

Here we have assumed linear isotropic materials and that $\nabla \cdot \vec{H} = 0$. We now have a second-order equation with a single vector unknown.

Proof of the Uniqueness of the Fields

Any field that satisfies Maxwell's equations and tangential components on the boundaries is the only possible solution. Say that E_1 and H_1 are solutions. Then

$$\nabla \cdot \epsilon \vec{E}_1 = \rho$$
$$\nabla \cdot \vec{H}_1 = 0$$
$$\nabla \times \vec{E}_1 = -\mu \frac{\partial \vec{H}_1}{\partial t}$$
$$\nabla \times \vec{H}_1 = \vec{J}_0 + \sigma \vec{E}_1 + \epsilon \frac{\partial \vec{E}_1}{\partial t} \quad (1.138)$$

If we have another solution E_2 and H_2, then

$$\nabla \cdot \epsilon \vec{E}_2 = \rho$$
$$\nabla \cdot \vec{H}_2 = 0$$
$$\nabla \times \vec{E}_2 = -\mu \frac{\partial \vec{H}_2}{\partial t}$$
$$\nabla \times \vec{H}_2 = \vec{J}_0 + \sigma \vec{E}_2 + \epsilon \frac{\partial \vec{E}_2}{\partial t} \qquad (1.139)$$

Let the differences between the two solutions be $\delta E = E_2 - E_1$ and $\delta H = H_2 - H_1$. Then δE and δH must be solutions to the homogeneous Maxwell equations.

$$\nabla \cdot \epsilon \delta \vec{E} = 0$$
$$\nabla \cdot \delta \vec{H} = 0$$
$$\nabla \times \delta \vec{E} = -\mu \frac{\partial \delta \vec{H}}{\partial t}$$
$$\nabla \times \delta \vec{H} = \sigma \delta \vec{E} + \epsilon \frac{\partial \delta \vec{E}}{\partial t} \qquad (1.140)$$

We now take the dot product of the last of equations (1.140) with δE.

$$\delta \vec{E} \cdot \nabla \times \delta \vec{H} = \sigma |\delta \vec{E}|^2 + \epsilon \delta \vec{E} \cdot \frac{\partial \delta \vec{E}}{\partial t} \qquad (1.141)$$

Now use the vector identity

$$A \cdot \nabla \times B = B \cdot (\nabla \times A) - \nabla \cdot (A \times B) \qquad (1.142)$$

to get

$$\nabla \cdot (\delta \vec{E} \times \delta \vec{H}) = -\frac{1}{2} \frac{\partial}{\partial t} (\mu |\delta \vec{H}|^2 + \epsilon |\delta \vec{E}|^2) - \sigma |\delta \vec{E}|^2 \qquad (1.143)$$

Basic Principles of Electromagnetic Fields

We integrate this over the volume Ω and use the divergence theorem to obtain

$$\oint_S (\delta\vec{E} \times \delta\vec{H}) \cdot dS = -\frac{\partial}{\partial t} \int_\Omega [\frac{1}{2}(\mu|\delta\vec{H}|^2 + \epsilon|\delta\vec{E}|^2)]d\Omega - \int_\Omega \sigma|\delta\vec{E}|^2 \, d\Omega \quad (1.144)$$

If the tangential components of E and H are the same in both solutions on the boundary, then the left-hand side of equation (1.144) becomes zero. Each term on the right-hand side must be zero independently because they all involve squared quantities. Thus we have shown that by specifying the tangential component of the fields of the boundaries we have completely specified the problem and the solution is unique.

1.11.1 Total and Reduced Magnetic Scalar Potential

We begin with one of Maxwell's equations,

$$\nabla \times \vec{H} = J + \frac{\partial \vec{D}}{\partial t} \quad (1.145)$$

We have seen that in the case in which a region has no current so that

$$\nabla \times \vec{H} = 0 \quad (1.146)$$

we can write the field as the gradient of a scalar potential.

$$\vec{H} = -\nabla\Omega \quad (1.147)$$

where Ω has the units of amperes.

Because

$$\nabla \cdot \vec{B} = 0 \quad (1.148)$$

and $B = \mu H$ we obtain

$$\nabla \cdot (\mu \nabla \Omega) = 0 \quad (1.149)$$

This scalar potential is not unique in current-carrying regions, which severely reduces its applicability. Therefore, let us define a *reduced scalar potential* such that its gradient gives only part of the field. If we have a component of H due to currents, which we will call H_c then

$$\nabla \times \vec{H}_c = \vec{J} + \frac{\partial \vec{D}}{\partial t} \tag{1.150}$$

so

$$\nabla \times (\vec{H} - \vec{H}_c) = 0 \tag{1.151}$$

and

$$\vec{H} - \vec{H}_c = -\nabla \Omega_r \tag{1.152}$$

This reduced potential, Ω_r exists and can be uniquely defined in a region with current.

Now

$$\nabla \cdot \mu(\vec{H} - \vec{H}_c) = -\nabla \cdot \mu \nabla \Omega_r \tag{1.153}$$

Because $\nabla \cdot (\mu H) = 0$, we have

$$\nabla \cdot (\mu \nabla \Omega_r) = \nabla \cdot \mu \vec{H}_c \tag{1.154}$$

Mathematically, we can define H_c in a number of ways. For example, we can define H_c as the field that would exist due to the designated current sources in an open boundary region with no magnetic material. In this case the Biot–Savart law can be used to compute H_c as

$$\vec{H}_c = \frac{1}{4\pi} \int \frac{\vec{J} \times \vec{r}}{|r|^3} dv \tag{1.155}$$

This field has no divergence so

$$\begin{aligned}\nabla \cdot (\mu \nabla \Omega_r) &= \nabla \cdot (\mu \vec{H}_c) = \mu \nabla \cdot \vec{H}_c + \vec{H}_c \cdot \nabla \mu \\ \nabla \cdot (\mu \nabla \Omega_r) &= \vec{H}_c \cdot \nabla \mu\end{aligned} \tag{1.156}$$

Mixed Scalar Potential Formulation

We have seen that the scalar potential is not suitable in regions with current sources and that a reduced scalar potential could be used in these regions. It is also true that the reduced scalar potential is not often accurate in regions of high permeability. In these regions where $\mu \gg \mu_0$, the magnetic field is very small. If we use the reduced scalar potential formulation, the field is found as the sum of the source field (perhaps computed by the Biot–Savart law) and the negative gradient of the reduced scalar potential. Because $\vec{H} \approx 0$, these two fields must approximately cancel and small errors can result in large deviations in $\vec{H}$. For this reason a mixed formulation is often used. The reduced scalar potential is used in the current-carrying regions (usually

Basic Principles of Electromagnetic Fields

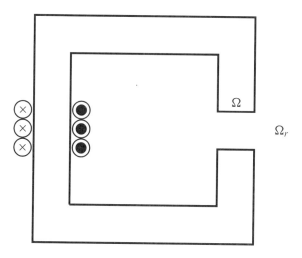

Figure 1.21 Correct Use of Mixed Scalar Potential Formulation

$\mu = \mu_0$) and the total scalar potential is used in the high-permeability regions (usually with no current sources). This is illustrated in Figures 1.21–1.24.

Of concern here is the problem of connectivity. As we have seen, the total scalar potential can be used only in curl-free regions. The formulation is not valid if we can find any path in a total scalar potential region that links a current. For example, in Figure 1.21 we are able to apply the mixed formulation because there is no path in the total scalar region that encloses current. For all paths in the total scalar region $\oint \vec{H} \cdot d\vec{\ell} = 0$. This is because of the break in the magnetic circuit where we apply the reduced potential formulation. In Figure 1.22 we have a continuous magnetic circuit with total scalar potential, but because this circuit is not linking the current we can apply the total scalar potential formulation. However, in Figure 1.23 we are not able to apply the mixed formulation because there is a path in the scalar potential region that links the current. There is a convenient *loophole* that may sometimes prove useful. In Figure 1.24 we have the same geometry as in Figure 1.23. In this case we have used symmetry to reduce the size of the problem. The result is that there is no longer a path in the iron region which links the current — at least in the problem domain. In this case the mixed scalar potential may be applied.

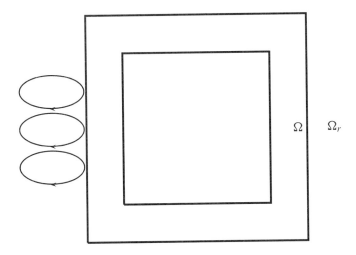

Figure 1.22   Closed Magnetic Circuit Linking No Current

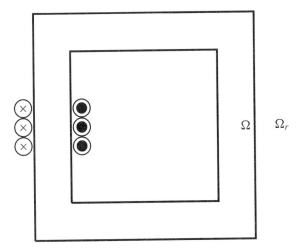

Figure 1.23   Incorrect Use of Mixed Scalar Potential Formulation

Basic Principles of Electromagnetic Fields

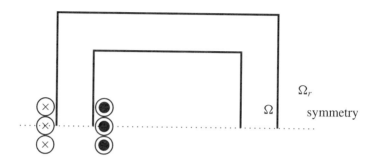

Figure 1.24 Scalar Potential is Valid With Symmetry Condition

As an illustration of the cancellation problem, consider the small transformer illustrated in Figures 1.25, 1.26, and 1.27. We see a cross section in the symmetry plane of a three-dimensional solution using tetrahedral elements. The mesh is the same for all cases. The permeability is $\mu_r = 1000$. The arrows represent the flux density in the magnetic material. Figure 1.25 is the result of using the reduced potential in the closed ferromagnetic region. Due to the high permeability, the magnetic field is very small, and this small value is found as the sum of the source field and the induced field. Small errors in either cause a large error in the result. The arrows show a poor result, especially near the conductor. We can improve the situation somewhat by using second-order elements as in Figure 1.26. The solution in the magnetic material is better but still not very good. In Figure 1.27 a small reduced potential region is added to the magnetic circuit, which is solved as almost entirely a total scalar potential region. Using the same mesh, we obtain a much more accurate solution in this case.

1.11.2 Magnetic Vector Potential, $A - V$ Formulation

We will now discuss the curl–curl equation for the vector potential. Beginning with

$$\nabla \times \vec{A} = \vec{B} \tag{1.157}$$

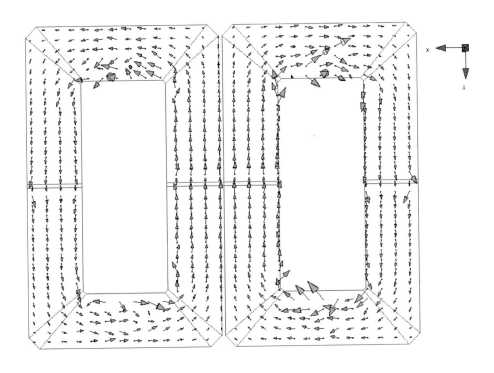

Figure 1.25 First-Order Mesh with All Reduced Scalar Potential

we take the curl of both sides and use $\nu\vec{B} = \vec{H}$ to get

$$\nabla \times \nu\nabla \times \vec{A} = \nabla \times \vec{H} \tag{1.158}$$

From Maxwell's equations

$$\nabla \times \nu\nabla \times \vec{A} = \vec{J} + \frac{\partial \vec{D}}{\partial t} \tag{1.159}$$

or

$$\nabla \times \nu\nabla \times \vec{A} = \sigma\vec{E} + \epsilon\frac{\partial \vec{E}}{\partial t} \tag{1.160}$$

Basic Principles of Electromagnetic Fields

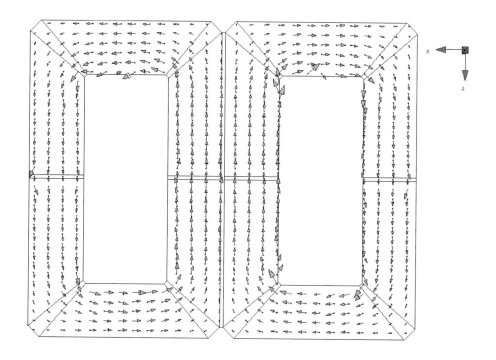

Figure 1.26 Second-Order Mesh with Reduced Scalar Potential

Substituting the vector and scalar potentials for the field quantities,

$$\nabla \times \nu \nabla \times \vec{A} = -\sigma \frac{\partial \vec{A}}{\partial t} - \epsilon \frac{\partial^2 \vec{A}}{\partial t^2} - \sigma \nabla V - \epsilon \nabla \frac{\partial V}{\partial t} \tag{1.161}$$

In a more conventional form the curl–curl operator is replaced by the Laplacian for homogeneous regions, although this is by no means necessary. Using the vector identity for a homogeneous region,

$$\nabla \times \nabla \times \vec{F} = \nabla \nabla \cdot \vec{F} - \nabla^2 \vec{F} \tag{1.162}$$

Equation (1.161) becomes

$$\nu \nabla^2 \vec{A} - \sigma \frac{\partial \vec{A}}{\partial t} - \epsilon \frac{\partial^2 \vec{A}}{\partial t^2} = \nabla(\nu \nabla \cdot \vec{A}) + \sigma \nabla V + \epsilon \nabla \frac{\partial V}{\partial t} \tag{1.163}$$

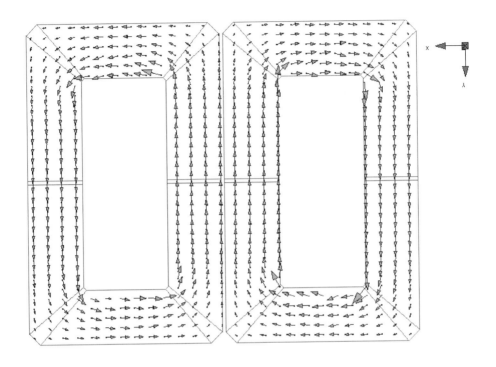

Figure 1.27 Total Scalar Potential in Cuts; Reduced Potential Elsewhere

If ν is not constant, then

$$\nabla \times \nu \nabla \times \vec{A} = -\nu \nabla^2 \vec{A} + \nu \nabla \nabla \cdot \vec{A} + \nabla \nu \times \nabla \times \vec{A} \qquad (1.164)$$

so

$$\nu \nabla^2 \vec{A} - \nabla \nu \times \nabla \times \vec{A} - \sigma \frac{\partial \vec{A}}{\partial t} - \epsilon \frac{\partial^2 \vec{A}}{\partial t^2} = \nu(\nabla \nabla \cdot \vec{A}) + \sigma \nabla V + \epsilon \nabla \frac{\partial V}{\partial t} \qquad (1.165)$$

We now consider some choices for the divergence.

Basic Principles of Electromagnetic Fields

The Choice of $\nabla \cdot A$

Consider the equation

$$\nabla^2 \vec{A} - \mu\sigma \frac{\partial \vec{A}}{\partial t} - \mu\epsilon \frac{\partial^2 \vec{A}}{\partial t^2} = \nabla \left[\nabla \cdot \vec{A} + \mu\sigma V + \mu\epsilon \frac{\partial V}{\partial t} \right] \tag{1.166}$$

One choice for the divergence of A is

$$\nabla \cdot \vec{A} = 0 \tag{1.167}$$

Because the divergence of $\vec{A}$ is zero, by the divergence theorem

$$\int_S \nabla \cdot \vec{A} \, dS = 0 \tag{1.168}$$

If we consider the material boundary of Figure 1.28 then the flux of the vector A which enters the disc from the bottom must leave the disc through the top as the height of the disc approaches zero. This means that

$$\hat{n} \cdot (A_1 - A_2) = 0 \tag{1.169}$$

We use the same argument that we used to show that the normal component of B was continuous at an interface. Therefore the normal component of A is continuous. This condition, the *Coulomb gauge*, is often chosen in static field problems, where $\frac{\partial A}{\partial t} = 0$. In this case we have

$$\nabla^2 \vec{A} = \nabla(\mu\sigma V) = -\mu \vec{J} \tag{1.170}$$

Going back to equation (1.166)

$$\nabla^2 \vec{A} - \mu\sigma \frac{\partial \vec{A}}{\partial t} - \mu\epsilon \frac{\partial^2 \vec{A}}{\partial t^2} = \nabla \left[\nabla \cdot A + \mu\sigma V + \mu\epsilon \frac{\partial V}{\partial t} \right] \tag{1.171}$$

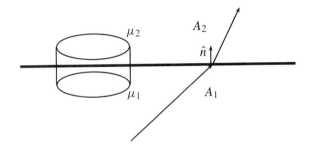

Figure 1.28 The Magnetic Vector Potential at a Boundary

Another possible choice for divergence is

$$\nabla \cdot \vec{A} = -\mu\epsilon \frac{\partial V}{\partial t} \tag{1.172}$$

In this case we have at the material boundary

$$\oint \vec{A} \cdot d\vec{S} = \vec{S} \cdot \hat{n}(\mu_2\epsilon_2 - \mu_1\epsilon_1)\frac{\partial V}{\partial t} \tag{1.173}$$

This gives

$$\hat{n} \cdot (A_1 - A_2) = (\mu_2\epsilon_2 - \mu_1\epsilon_1)\frac{\partial V}{\partial t} \tag{1.174}$$

In this case the normal component of A is not continuous.

Basic Principles of Electromagnetic Fields

The equation simplifies to

$$\nabla^2 \vec{A} - \mu\sigma \frac{\partial A}{\partial t} - \mu\epsilon \frac{\partial^2 \vec{A}}{\partial t^2} = \mu\sigma \nabla V \tag{1.175}$$

Another possible choice for divergence is

$$\nabla \cdot \vec{A} = -\mu\sigma V - \mu\epsilon \frac{\partial V}{\partial t} \tag{1.176}$$

We have from Maxwell's equation that

$$\nabla \cdot \vec{E} = \frac{\rho}{\epsilon} \tag{1.177}$$

Therefore

$$\nabla \cdot \vec{E} = \nabla \cdot \left(-\frac{\partial \vec{A}}{\partial t} - \nabla V \right) = -\frac{\partial}{\partial t} \nabla \cdot \vec{A} - \nabla^2 V \tag{1.178}$$

So we have the coupled set of equations

$$\begin{aligned}
\nabla^2 V - \mu\sigma \frac{\partial V}{\partial t} - \mu\epsilon \frac{\partial^2 V}{\partial t^2} &= -\frac{\rho}{\epsilon} \\
\nabla^2 \vec{A} - \mu\sigma \frac{\partial \vec{A}}{\partial t} - \mu\epsilon \frac{\partial^2 \vec{A}}{\partial t^2} &= 0
\end{aligned} \tag{1.179}$$

At the interface between two materials, we again have a discontinuity in the normal component of the vector potential, We see that

$$\hat{n} \cdot (A_1 - A_2) = -(\mu_1\sigma_1 - \mu_2\sigma_2)V - (\mu_1\epsilon_1 - \mu_2\epsilon_2)\frac{\partial V}{\partial t} \tag{1.180}$$

1.11.3 The $\vec{A}^*$ Formulation

Although the A–V formulation is quite general, it does require four unknowns at each point, and these typically involve double precision and sometimes complex numbers. In order to reduce the number of unknowns, the A^* formulation can be applied to certain problems. Let's consider the eddy current problem described by the diffusion equation,

$$\nabla \times \nu \nabla \times \vec{A} + \sigma \left(\frac{\partial \vec{A}}{\partial t} + \nabla V \right) = \vec{J}_0 \qquad (1.181)$$

The continuity equation ($\nabla \cdot \vec{J} = 0$) requires that

$$\nabla \cdot \sigma \left(\frac{\partial \vec{A}}{\partial t} + \nabla V \right) = 0 \qquad (1.182)$$

We now define a modified vector potential $\vec{A}^*$ such that

$$\frac{\partial \vec{A}^*}{\partial t} = \frac{\partial \vec{A}}{\partial t} + \nabla V = 0 \qquad (1.183)$$

This gives for equation (1.181)

$$\nabla \times \nu \nabla \times \vec{A}^* + \sigma \frac{\partial \vec{A}^*}{\partial t} = \vec{J}_0 \qquad (1.184)$$

Thus we have eliminated the scalar potential and need only compute three values at each point. The $\vec{A}^*$ formulation is less general and may be used only in problems with a single conductivity. The continuity condition $\nabla \cdot \vec{J} = \nabla \cdot \sigma \vec{A}^* = 0$ is satisfied in a weak sense as is $\vec{J} \cdot \hat{n}$ on the boundary. In static problems or cases in which $\sigma = 0$, the divergence is not specified and the solution is not unique.

1.11.4 Electric Vector Potential, T–Ω Formulation

Just as the continuity equation for the magnetic flux, $\nabla \cdot \vec{B} = 0$, allowed us to represent $\vec{B}$ by a vector potential such that $\nabla \times \vec{A} = \vec{B}$, the continuity of current $\nabla \cdot \vec{J} = 0$ allows us to define a current vector potential often referred to as the electric vector potential, $\vec{T}$

$$\nabla \times \vec{T} = \vec{J} \tag{1.185}$$

We note that $\nabla \times \vec{H} = \vec{J}$ as well, so $\vec{H}$ and $\vec{T}$ differ by the gradient of a scalar and have the same units. So

$$\vec{H} = \vec{T} - \nabla \Omega \tag{1.186}$$

Using $\vec{E} = \rho \vec{J}$, from Faraday's law

$$\nabla \times \vec{E} = -\frac{\partial \vec{B}}{\partial t} = \nabla \times \rho \nabla \times \vec{T} \tag{1.187}$$

Here ρ is the resistivity, the reciprocal of the conductivity. Substituting for the magnetic induction

$$\vec{B} = \mu \vec{H} = \mu (\vec{T} - \nabla \Omega) \tag{1.188}$$

we obtain

$$\nabla \times \rho \nabla \times \vec{T} + \mu \frac{\partial \vec{T}}{\partial t} - \mu \nabla \frac{\partial \Omega}{\partial t} = 0 \tag{1.189}$$

In current-free regions we can find the magnetic field from the scalar potential

$$\vec{H} = -\nabla \Omega \tag{1.190}$$

where Ω can be found from Laplace's equation

$$-\nabla \cdot \mu \nabla \Omega = 0 \tag{1.191}$$

The solution can be made unique by setting a gauge condition, for example the Coulomb gauge $\nabla \cdot \vec{T} = 0$. With this choice equation (1.189) becomes [7]

$$\nabla \times \rho \nabla \times \vec{T} - \nabla \rho \nabla \cdot \vec{T} + \mu \frac{\partial \vec{T}}{\partial t} - \mu \nabla \frac{\partial \Omega}{\partial t} = 0 \tag{1.192}$$

An equation relating the vector and scalar potential may be found by using $\nabla \cdot \vec{B} = 0$. From (1.188) we have

$$\nabla \cdot \mu(\vec{T} - \nabla \Omega) = 0 \tag{1.193}$$

1.12 CLASSIFICATION OF DIFFERENTIAL EQUATIONS

For completeness we define these equations here but the reader interested in more details should consult [8]. Considering the second-order partial differential equation (in two variables)

$$A \frac{\partial^2 \phi}{\partial x_1^2} + 2B \frac{\partial \phi}{\partial x_1} \frac{\partial \phi}{\partial x_2} + C \frac{\partial^2 \phi}{\partial x_2^2} = 0 \tag{1.194}$$

The equations are classified on the basis of the sign of the determinant

$$D = \begin{vmatrix} A & B \\ B & C \end{vmatrix} \tag{1.195}$$

- When $AC > B^2$ we have an elliptic equation. For $A = C$ and $B = 0$ we have Laplace's equation, which is elliptic.

- When $AC = B^2$ the equation is parabolic. This includes the class of initial value problems described by the diffusion equation.

- When $AC < B^2$ we have the hyperbolic equation. For example, if $A = 1$, $B = 0$, and $C = -1$, we obtain the one-dimensional wave equation.

In more than two variables this classification is not so simple. In many practical problems we may have some regions described by one type of equation and another region by a different type of partial differential equation.

2

OVERVIEW OF COMPUTATIONAL METHODS IN ELECTROMAGNETICS

2.1 INTRODUCTION AND HISTORICAL BACKGROUND

Since the early days of development of electrical devices, engineers and scientists have paid attention to effecting design improvements for economy and efficiency. The byproduct of this effort has been the development of new techniques to model devices and to predict their performance accurately at the design stage. In the early stages, these models consisted of hardware prototypes or scaled models and results obtained from tests conducted on these models were used to predict the performance of the actual devices and also to enable hardware changes to be made to meet specifications. However, these proved to be expensive, cumbersome, and time consuming. Further, often only terminal quantities could be measured owing to the nonavailability of sophisticated sensors and instrumentation. It was neither practical nor always possible to determine from these tests the changes to device design required for improving performance and reducing costs.

In the wake of the difficulties of modeling with prototypes, field plotting techniques were developed, such as conformal mapping and the method of curvilinear squares. This was soon followed by experiments carried out with electrolytic tanks and conducting paper analogs. Owing to temperature and humidity variations and their effects on measurements, these experimental methods could not be relied upon for accuracy and consistency of results.

Therefore, analytical methods were pursued, although at the time, these were limited to techniques applicable to linear isotropic media, the entire half-space, or semi-infinite regions. Only constant values of permeability and conductivity for magnetostatic and eddy current problems and constant dielectric permittivity for electrostatic problems could be used. Different analytical techniques were required depending upon the system of coordinates chosen. For example, harmonic series solutions of space and time could be considered if the boundary value problem was formulated in Cartesian coordinates. Bessel functions were used for axi-symmetric and eddy current problems and spherical harmonics were required when the problem was posed in spherical coordinates. Several simplifications of the formulation and associated boundary conditions were required in view of the limitations on available solution techniques.

Further complexities became evident depending upon whether the boundary value problem was formulated in differential form or in terms of integral equations. Even in the case of the latter, only linear problems could be solved and limitations due to the choice of coordinate system still applied. Modeling of nonlinearity was not possible except for simple problems where transformation methods could be used.

With the advent of calculating machines, analog and digital computers, new techniques were developed, which are generally considered numerical methods. The forerunners of these were the resistance analogs based on the so-called *magnetic Ohm's law*. This was soon followed by divided difference schemes generally called finite differences. Several variations of these were developed based on five-point and nine-point operators, cell schemes, and others. The underlying principle of these methods is to convert the partial differential equations used in describing the boundary value problem into difference equations and solving these by iterative methods. Finite difference schemes were initially applied to linear problems formulated in Cartesian or polar coordinates and later were extended to nonlinear problems.

First-order finite difference schemes that were used by many researchers and engineers suffered from the fact that a large number of nodes were required, leading to large computer memory and execution times. This was to some degree alleviated by the use of iterative solution techniques, although the latter posed serious problems of numerical stability and lack of convergence.

Integral equation methods gained importance in accelerator magnet design and scattering problems for high-frequency applications. Boundary integral methods are the most popular ones used in engineering applications such as metal forming, induction furnace design, and others.

Finite element methods based on variational schemes or the Galerkin weighted residual technique were first used in structural and continuum mechanics in the 1960s and later found a wide variety of applications in electromagnetic field problems. These schemes are either used in their own right or in combination with other numerical methods. Coupling of finite element methods with closed form solutions has found applications in solving open boundary problems. A fuller discussion of the various numerical methods of solving boundary value problems is provided in succeeding chapters of this book. In this introductory chapter, we shall review the different experimental and analytical methods of field analysis for linear and nonlinear problems that have been developed over the past few decades.

2.2 GRAPHICAL METHODS

This method was first developed by Richardson [9], and was advanced by Lehman [10], Kuhlman [11], Moore [12], Stevenson and Park [13]. The procedure is fully described by Hague [14], Bewley [15], Binns and Lawrenson [16], and Moon and Spencer [17]. A summary is presented by Wright and Deutsch [18] and the following description of the method is based on their summary.

This method is called flux plotting and is based on the principle, in the electrostatic case, that the equipotential lines and electrostatic flux lines are mutually orthogonal: they cross at right angles at all points in space. Because the potential satisfies Laplace's equation in free space, we can choose analytic functions for the solution. By analytic we mean that the chosen functions satisfy the Cauchy Riemann conditions.

Let us consider a potential function ϕ that satisfies Laplace's equation. Therefore

$$\frac{\partial^2 \phi}{\partial x^2} + \frac{\partial^2 \phi}{\partial y^2} = 0 \tag{2.1}$$

Any function z that is analytic may be chosen in terms of complex variables that satisfies equation (2.1). Therefore, we can write w as a function of z such that

$$w = F(z) = u(x, y) + jv(x, y) \tag{2.2}$$

where u and v are real functions of x and y. They also satisfy the Cauchy-Riemann conditions because the function z is analytic, so that

$$\frac{\partial u}{\partial x} = \frac{\partial v}{\partial y} \tag{2.3}$$

$$\frac{\partial u}{\partial y} = -\frac{\partial v}{\partial x} \tag{2.4}$$

To test whether u and v satisfy Laplace's equation, we differentiate (2.3) with respect to x and (2.4) with respect to y and add the resulting equations, so that

$$\frac{\partial^2 u}{\partial x^2} + \frac{\partial^2 u}{\partial y^2} = \frac{\partial^2 v}{\partial x \partial y} - \frac{\partial^2 v}{\partial x \partial y} = 0 \tag{2.5}$$

Similarly, if we differentiate (2.3) with respect to y and 2.4 with respect to x and subtract the latter from the former result, we obtain

$$\frac{\partial^2 u}{\partial x \partial y} - \frac{\partial^2 u}{\partial x \partial y} = \frac{\partial^2 v}{\partial x^2} + \frac{\partial^2 v}{\partial y^2} = 0 \tag{2.6}$$

Therefore, u and v are conjugates of each other. Bewley [15] has shown that the families of curves $u(x, y)$ = constant and $v(x, y)$ = constant intersect at right angles and, therefore, can be chosen to represent equipotential lines and flux lines, respectively. To obtain a graphical solution to Laplace's equation, it is necessary to subdivide the preceding orthogonal contours into equal increments Δu and Δv, respectively. The field plot can then be obtained by setting up a system of curvilinear squares or rectangles between the boundary surfaces. To illustrate this procedure, let us consider the annulus between a circular conductor at potential V and a concentric ground plane, as shown in Figure 2.1. Owing to symmetry along the x and y axes, we can consider only a quarter of the cross section.

In Figure 2.2, the potential of the conductor is V and that of the ground plane is zero. It is also evident that concentric circles in the annulus and radial lines from the center of the conductor will intersect orthogonally. In this simple case, extremely accurate solutions may be obtained because of symmetry in both directions. This may not always be the case; therefore a trial-and-error method of field plotting may be necessary. Let us now consider a small region, which may be subdivided into curvilinear rectangles as shown in Figure 2.3.

Assuming δV is an equal subdivision of the equipotential surface or line and $\delta \psi$ of the flux line, we can construct curvilinear rectangles as shown in Figure 2.3. The only requirement is that in each subregion of δV and $\delta \psi$, the equipotential and flux lines at

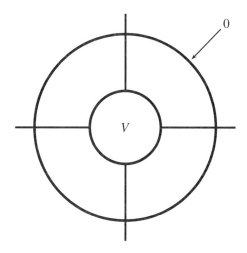

Figure 2.1 Geometry of the Annulus

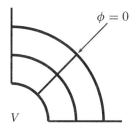

Figure 2.2 Quarter-Section of the Annulus

each corner of the deformed rectangle A–B–C–D intersect at right angles. We may start with this rectangle and extend the procedure to other neighboring rectangles by trial and error until orthogonality is obtained at any corner. This is no doubt a cumbersome and tedious procedure, but many useful solutions were obtained in the early days of field plotting by this method. Further, this method also provided an insight into the physical meaning of the solution to Laplace's equation.

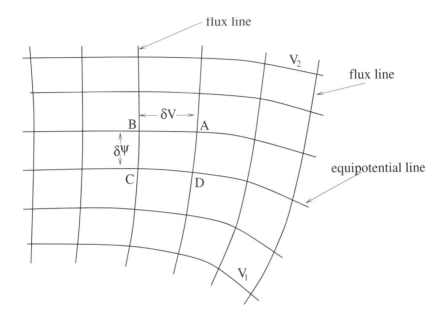

Figure 2.3 Laplacian Region

2.3 CONFORMAL MAPPING

In this method, the boundary value problem is transformed to the so-called *z-plane* such that the actual geometry of the field region is replaced by a semi-infinite continuous region. Then Lapalce's equation is solved in this continuous region and the solution is transformed back into the original plane. The most useful of these methods is the Schwartz-Christoffel transformation, by means of which discontinuities of slots and teeth in electrical machinery problems could be handled effectively. The method is strictly applicable to linear problems with constant material characteristics.

2.4 EXPERIMENTAL METHODS

An important experimental method is related to the electrolytic tank attributed to Kirchhoff [19] as modified by Adams [20]. The method was implemented by Kennelly and Whiting [21]. The apparatus consists of a metallic tank containing an electrolyte with electrodes immersed in it. Surface equipotentials are measured by a probe mounted on a moving carriage over the tank. A fuller description of the arrangement is given in ref-

Overview of Computational Methods in Electromagnetics 69

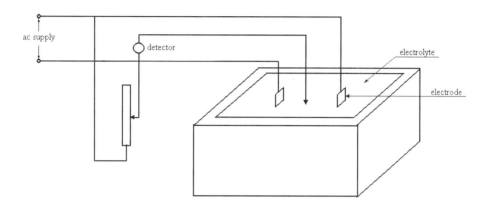

Figure 2.4 Electrolytic Tank

erence [22] and is summarized here. Figure 2.4 shows the arrangement of the tank and the probe with the external circuitry for measurements. A variety of two-dimensional problems requiring mapping of Laplacian fields have been investigated by this method. Insulating sheets of glass cut to the required shape are immersed in the electrolyte so as to change its bulk resistivity. With certain modifications and a change over to AC supply to eliminate polarization at the electrodes and duly compensate for capacitance effects, three-dimensional cases with axial symmetry have also been studied. Sander and Yates [23] have reported accuracy of results up to a tenth of a percent by this technique.

2.5 ELECTROCONDUCTING ANALOG

Important among conducting analog methods is the use of *Teledeltos* paper. This is a graphite-based material with a surface resistivity between 1000 and 5000 ohms per square for different samples but homogeneous to 1% within the same roll. The resistivity is also somewhat anisotropic, with a resistivity ratio in the orthogonal directions varying between 1.03 and 1.15 as reported by Raby [24]. This anisotropy can be ignored in some cases and compensated for in others. Electrodes are applied by means of silver paint or aluminum foil stuck with conducting paint at appropriate places. Change in surface resistivity to model changes in the medium is achieved by providing holes or adding layers of colloidal graphite or Teledeltos sheets as required. Equipotential plots

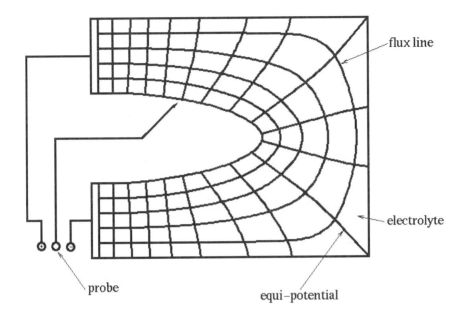

Figure 2.5 Teledeltos Paper Setup and Flux Plotting

are made directly on the paper surface using a DC potential applied across the electrodes, because there is no polarization in the absence of an electrolyte. The method is inexpensive, quick, and easily implemented. However, it is susceptible to change in temperature and humidity. The technique has been used extensively for engineering applications, and an accuracy up to 2% has been reported in the literature. Figure 2.5 shows the setup of the analog as described in Raby [25].

2.6 RESISTIVE ANALOG

In place of the electrolytic tank and conducting paper analogs, resistive elements have been used between electrodes [26]. Potential measurements made at the nodes yield sufficient information to map the field provided a large number of resistive elements are used. This method also requires the use of a large number of resistors where the field gradient is large and fewer resistors where the field gradient is low. Random errors in

Overview of Computational Methods in Electromagnetics

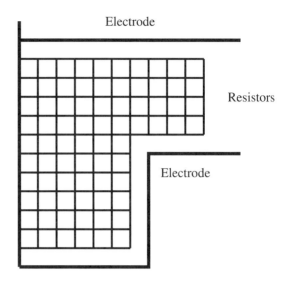

Figure 2.6 Resistance Analog Setup

resistance values are cancelled on a statistical basis and yield a high degree of accuracy. Figure 2.6 shows a typical arrangement.

2.7 CLOSED FORM ANALYTICAL METHODS

To illustrate these methods, we shall consider Laplace's and Poisson's equations. From Ampére's law we can derive the following:

$$\oint \vec{H} \cdot d\vec{\ell} = \iint \vec{J} \cdot d\vec{S} = I \qquad (2.7)$$

where $\vec{H}$ is the magnetic field, I the current, $\vec{J}$ the current density, $d\vec{\ell}$ the elemental length of the trajectory of $\vec{H}$, and $d\vec{S}$ the elemental cross section of the current-carrying conductor.

Applying Stokes' theorem to (2.7), the line integral on the left hand-side is transformed as a surface integral such that

$$\oint \vec{H} \cdot d\vec{\ell} = \iint \nabla \times \vec{H} \cdot d\vec{S} = \iint \vec{J} \cdot d\vec{S} = I \qquad (2.8)$$

Equating the integrands of the surface integrals in (2.8), there is

$$\nabla \times \vec{H} = \vec{J} \qquad (2.9)$$

Equation (2.9) is Ampére's law for the magnetostatic problem. In source-free regions, $J = 0$, so that

$$\nabla \times \vec{H} = 0 \qquad (2.10)$$

We now take the curl on both sides of (2.10) and expand the result by the vector identity

$$\nabla \times \nabla \times \vec{H} = \nabla(\nabla \cdot \vec{H}) - \nabla^2 \vec{H} = 0 \qquad (2.11)$$

Because the divergence of the magnetic field in a homogeneous region is zero, equation (2.11) reduces to Laplace's equation

$$\nabla^2 \vec{H} = 0 \qquad (2.12)$$

The general solution of equation 2.12 can be obtained for a two-dimensional problem by the separation of variables method [27] as a summation of harmonic terms, so that

$$H = \sum_{n=1}^{N} (C_n \cos m_n x + D_n \sin m_n x)(E_n e^{m_n y} + F_n e^{-m_n y}) \qquad (2.13)$$

where C_n, D_n, E_n, and F_n are integration constants that are determined by interface and boundary conditions and n is the order of the harmonic in the series solution.

Solution in the current region can also be obtained by modeling the current source as a sheet or line current so that it can be expressed as a discontinuity in the tangential component of the magnetic field above and below the current region. It is evident that equation (2.13) applies only to periodic structures where the field region is assumed to be semi-infinite in the x direction.

A solution similar to (2.13) for Laplace's equation in polar coordinates is

$$H = \sum_{n=1}^{N}(C_n r^n + D_n r^{-n})(E_n \cos n\theta + F_n \sin n\theta) \qquad (2.14)$$

For axisymmetric and three-dimensional problems, one may be required to use Bessel functions and spherical harmonic functions, respectively. These are by no means simple, and the different regions such as material and air regions need to be represented as extending to infinity. As has already been stated, currents can be represented only as current sheets, and their actual geometrical configuration cannot be modeled in these methods. For example, in a rotating electrical machine, the current-carrying conductors in stator iron slots would be required to be modeled as a thin current sheet at the inner bore of the machine below the slots and teeth. This may not be adequate because the slots and other discontinuities cannot be represented in the model. Further, nonlinearity due to iron saturation cannot be taken into account, as the formulation permits only a constant value of permeability or free space permeability. Solutions for the flux density and the electric field in electrostatic field problems can be obtained by a similar procedure.

Closed form methods have also been successfully applied to steady-state time-periodic eddy current problems with modifications to equation (2.13) for the conducting region such that

$$H = \sum_{n=1}^{N}(C_n \cos m_n x + D_n \sin m_n x)(E_n e^{\beta_n y} + F_n e^{-\beta_n y}) \qquad (2.15)$$

where $\beta_n = \sqrt{m_n^2 + j\omega\mu\sigma}$, and σ is the conductivity.

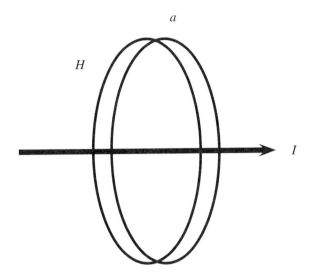

Figure 2.7 Current and Magnetic Field

2.8 DISCRETE ANALYTICAL METHODS

The foremost of these is based on simulating the resistive analog analytically. This is best illustrated by determining the relationship between the applied potential or source magnetomotive force and the flux. Let us consider Ampére's law, which relates the current in a conductor and the magnetic field associated with it as shown in Figure 2.7.

The following relationship between the current and magnetic field holds:

$$\oint \vec{H} \cdot d\vec{\ell} = I \qquad (2.16)$$

where $\vec{H}$ is the magnetic field, I is the current, and $d\vec{\ell}$ is an elemental length of the magnetic field trajectory. We assume that the flux follows a well-defined path and that the flux density and magnetic field are constant along the path.

The magnetic flux through the cross-sectional area S of an iron bar of permeability μ is given by

$$\psi = \vec{B} \cdot \vec{S} = \mu \vec{H} \cdot \vec{S} \qquad (2.17)$$

Overview of Computational Methods in Electromagnetics

If the path length of $\vec{H}$ is denoted by $\vec{\ell}$, then

$$\vec{H} = \frac{I}{\vec{\ell}} \tag{2.18}$$

Substituting for $\vec{H}$ from (2.18) into equation (2.17) for the flux, we have

$$\psi = \frac{\mu S I}{\ell} = \frac{I}{\ell/(S\mu)} \tag{2.19}$$

If we designate $\frac{\ell}{\mu S} = R$, the reluctance, and replace the current I by the excitation ampere-turns or the so-called magnetomotive force, then

$$\text{Flux} = \frac{\text{MMF}}{\text{Reluctance}} \tag{2.20}$$

This relationship is analogous to the voltage–current relationship in a circuit described by Ohm's law. Here MMF takes the place of voltage, the reluctance becomes the resistance or impedance, and magnetic flux is the current in the circuit of Figure 2.8. We can calculate the reluctances of elements of the magnetic circuit *a priori* for linear problems with constant permeability and the network of reluctances can be assembled in the same way as the resistive analog. The MMF is then applied across the terminals of the network, and the fluxes through the branches are calculated. By increasing the number of reluctance elements, a good approximation to the field distribution can be obtained.

2.9 TRANSFORMATION METHODS FOR NONLINEAR PROBLEMS

So far we have discussed various experimental, analog, and analytical methods, which are applicable to linear problems and where the principle of superposition applies. However, these methods cannot be applied to nonlinear problems in a rigorous mathematical sense. Nonlinear equations must be linearized or solved in some other way. Methods used for this purpose are known as transformations, which are powerful analytical tools for solving nonlinear equations in general. Typically these techniques

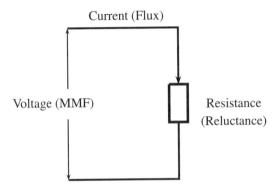

Figure 2.8 Magnetic Circuit

linearize the system of equations (e.g., the Kirchhoff and Hodograph transformations), reduce the partial differential equations to nonlinear ordinary differential equations (e.g., the similarity transformation), transform the system to one already solved or perform some other reduction of complexity.

In general, these transformations can be classified into three groups:

1. Those that change the dependent variables

2. Those that change the independent variable

3. Those that change both the dependent and independent variables

In the Kirchhoff transformation, a new dependent variable is introduced so as to linearize the nonlinear equations.

Let us consider the pseudo-Laplacian, which occurs in diffusion heat conduction and magnetic field problems.

$$\nabla \cdot [f(\phi)\nabla\phi] = 0 \qquad (2.21)$$

Introducing a new dependent variable such that

$$\begin{aligned} \psi &= \psi(\phi) \\ \frac{d\psi}{d\phi} &= f(\phi) \end{aligned} \qquad (2.22)$$

equation (2.21) reduces to the linear Laplace equation

$$\nabla^2 \psi = 0 \qquad (2.23)$$

The boundary conditions are changed similarly and it has been shown in reference [28] that for the Dirichlet problem (i.e., where boundary conditions are specified as potentials), the boundary conditions transform to yet another Dirichlet form. For the Neumann-type boundary conditions, however, the Kirchhoff transformation introduces nonlinearities, resulting in complicated boundary conditions. This transformation on the dependent variable has the feature that the physical range of the independent variable is unchanged, but the method is limited in its application to very simple geometries and boundaries.

An illustration of the transformation of the independent variable is the *similarity transformation* due to Boltzmann [29], which transforms the independent variable such that the partial differential equation is changed into an ordinary differential equation. The technique can be applied to a one-dimensional diffusion equation of the form

$$\frac{\partial C}{\partial t} = \frac{\partial}{\partial x}\left[D(C)\frac{\partial C}{\partial x}\right] \qquad (2.24)$$

We choose a function of the independent variables x and t given by

$$\eta = x^\alpha t^\beta \qquad (2.25)$$

where α and β are to be determined. Equation (2.25) is modified as an ordinary differential equation in η free of x and t so that

$$\frac{d}{d\eta}\left[D(C)\frac{dC}{d\eta}\right] + \frac{\eta}{2}\frac{dC}{d\eta} = 0 \qquad (2.26)$$

This transformation can be used effectively only if the boundary and initial conditions are consolidated, the medium is homogeneous, and the geometry of the region and boundary is a simple one.

The *Hodograph transformation,* so named by Hamilton in 1869 [28], is a typical example of the mixed method and it permits a certain amount of flexibility in the geometry. A set of quasi-linear equations of the form

$$F_1\frac{\partial u}{\partial x} + F_2\frac{\partial u}{\partial y} + F_3\frac{\partial v}{\partial x} + F_4\frac{\partial v}{\partial y} = F(u, v, x, y)$$
$$G_1\frac{\partial u}{\partial x} + G_2\frac{\partial u}{\partial y} + G_3\frac{\partial v}{\partial x} + G_4\frac{\partial v}{\partial y} = G(u, v, x, y) \qquad (2.27)$$

where F_i and G_i are functions of $u, v, x,$ and y, together representing second-order equations, are transformed by changing the independent variables x and y as functions of u and v, so that

$$\begin{aligned} F_1 y_v - F_2 x_v - F_3 y_u + F_4 x_u &= 0 \\ G_1 y_v - G_2 x_v - G_3 y_u + G_4 x_u &= 0 \end{aligned} \qquad (2.28)$$

Hence the solution of the modified set of equations (2.28) leads to the solution of (2.27) provided the Jacobian

$$J = x_u y_v - x_v y_u \neq 0 \qquad (2.29)$$

Overview of Computational Methods in Electromagnetics 79

This transformation has been successfully applied in fluid mechanics problems where the geometry of the region of interest may not be simple or regular. The advantage of linearity gained by this hodograph transformation is, however, offset by complicated boundary conditions.

The preceding examples of transformation are but a few of a large number of such techniques that have been used and illustrate their usefulness for solving quasi-linear partial differential equations. There is, however, no general way to obtain the required transformation, and imagination, ingenuity, and good fortune play a major role in their choice. The chief limitations of the methods are that, in general, they lead to complex boundary conditions and are really suitable only for cases where the materials are homogeneous and the boundaries and geometry are simple.

2.10 NONLINEAR MAGNETIC CIRCUIT ANALYSIS

This is a sequel to the discrete analysis method described earlier and is a forerunner of numerical methods. In this method, a relaxation technique is used for solving the nonlinear field problem. A magnetic circuit is developed as before, with lumped reluctances representing various parts of the field region. The flux densities and MMF drops are determined for an initial estimate of the flux in the iron. With these values of flux density, the appropriate permeabilities are determined from the $B - H$ curve and the new reluctances are estimated. The iterative cycle is continued until the total MMF drop around the circuit reaches an acceptable value.

The merits of the method are that it is an advance over nomographic techniques of field plotting [30], and with the aid of a digital computer, the field region can be faithfully represented by an equivalent magnetic circuit. This technique was employed by Binns [16] for the estimation of the open-circuit saturation curve of a turboalternator. He reported that only six to eight iterations were required to obtain a solution of acceptable accuracy. Since then, the method has been used for a number of applications of magnetic field computation by various researchers.

The chief limitation of the method, however, is that the circuit representation of the region is specific for each problem and cannot, therefore, be generalized. Further, because the flux paths are restricted to the branches containing lumped reluctances, the network representation must be sufficiently fine to obtain a useful flux plot. Thus, the computational advantage gained by the small number of iterations is offset by the large number of branches of the magnetic circuit and the corresponding number of equations to be solved.

2.11 FINITE DIFFERENCE METHOD

Except for the cases of simple and geometrically well-defined problems, closed form analytical solutions are not available for solving partial differential equations representing boundary value problems. For complex geometrical shapes, with varying material characteristics and often mixed boundary conditions, numerical methods offer the best and often the most economical solution.

Finite difference methods are perhaps the oldest numerical techniques and can be traced back to Gauss [31]. One of the oldest iteration methods dates back to 1873 [32]. The word relaxation was introduced by Southwell [33], who described a method of solving stresses in jointed frames by the systematic relaxation of the strains.

The simple finite difference method (or five-point star method) is the most elementary form of the point-value techniques, that is, methods of finding solutions at discrete points in the field region. The principle underlying this technique is one of replacing the partial differential equations of the field problem by a number of difference approximations, so that difference expressions approximate the derivatives. The derivative of a function ϕ in the x direction is defined as

$$\frac{\partial \phi}{\partial x} = \lim_{\Delta x \to 0} \frac{\phi(x + \Delta x, y, z) - \phi(x, y, z)}{\Delta x} \tag{2.30}$$

If the separation distance between points is sufficiently small, we may approximate the first and second derivatives of the function (Figure 2.9) as

$$\frac{\partial \phi}{\partial x} \approx \frac{\phi(x + \Delta x, y, z) - \phi(x, y, z)}{\Delta x} \tag{2.31}$$

$$\frac{\partial^2 \phi}{\partial x^2}\bigg|_p \approx \frac{\phi_w + \phi_e - 2\phi_p}{h_1^2}$$

$$\frac{\partial^2 \phi}{\partial y^2}\bigg|_p \approx \frac{\phi_n + \phi_s - 2\phi_p}{h_2^2} \tag{2.32}$$

This procedure is strictly applicable to the interior points of the region R. Points on the boundary C must be dealt with separately. This can be best illustrated by a simple problem shown in Figure 2.10.

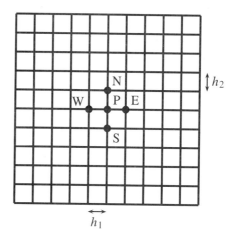

Figure 2.9 Rectangular Mesh and Five-Point Finite Difference Operator

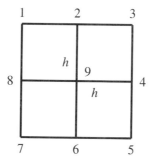

Figure 2.10 Rectangular Region with a Single Five Point Finite Difference Operator

Let us assume that the potential at each of the nodes (1) through (7) is equal to 1. Also let the gradient at node 8 be zero, so that

$$\frac{\partial V}{\partial x} = \frac{V(9) - V(8)}{h} = 0 \qquad (2.33)$$

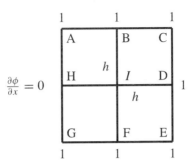

Figure 2.11 Example of Relaxation

For the interior point (9), the following equation applies:

$$\nabla^2 \phi = \frac{V(2) + V(4) + V(6) + V(8) - 4V(9)}{h^2} = -\frac{\rho}{\epsilon} \quad (2.34)$$

Equations (2.33) and (2.34) can be put in matrix form, yielding a (9 × 9) square matrix. The right hand forcing function will be a column vector of (9 × 1). It will be noted that the matrices formed by the finite difference method are sparse with a large number of zero entries. Therefore, it is often convenient to solve the resulting set of linear equations iteratively, instead of using the straightforward but expensive inversion technique. One such scheme, known as *alternating relaxation,* was used by Erdelyi and Ahmed [34] for solving the electromagnetic field problem in a DC machine under load.

We shall illustrate a relaxation method for solving Laplace's equation, which may also be applied to a digital computer. Consider a region shown in Figure 2.11 bounded by a potential of unity along A–B–C–D–E–F–G and with $\frac{\partial \phi}{\partial x} = 0$ across A–H and H–G and along H–I. Also, the volumetric charge density $\rho = 0$ everywhere.

For the interior point I, the potential is given by

$$\nabla^2 \phi = 0 = \frac{\phi_B + \phi_F + \phi_H + \phi_D - 4\phi_I}{h^2} \quad (2.35)$$

Rearranging terms in (2.35) and assuming $h = 1$

$$\phi_I = \frac{\phi_B + \phi_F + \phi_H + \phi_D}{4} \qquad (2.36)$$

For the first pass, if we assume

$$\phi_H = 0.5$$
$$\phi_I = \frac{1+1+0.5+1}{4} = 0.875$$

The deviation is $0.875 - 0.5 = 0.375$.

Because $\frac{\partial \phi}{\partial x} = 0$ along H–I, we have $\frac{\phi_I - \phi_H}{h} = 0$ or $\phi_I = \phi_H$.

Hence, for the second pass $\phi_H = 0.875$. Again calculating ϕ in terms of ϕ_B, ϕ_D, ϕ_F and ϕ_H, we have $\phi_I = \frac{1+1+1+0.875}{4} = 0.96875$ and the deviation is $0.96875 - 0.875 = 0.09325$. As before $\phi_H = \phi_I$.

After the sixth pass of the iteration procedure, we obtain $\phi_I = 0.9998779296875$ and the deviation is 0.0003660.

This process can be accelerated by adding a multiple of the deviation to the potential. Therefore, we may write

$$\phi_{new} = \phi_{old} + K \cdot (\text{deviation})$$

where K is the acceleration factor. Intuitively, we had set $K = 1$ in the previous calculation. An optimal value of K lies between 1 and 2. The correct choice of this factor, however, is largely dependent on the nature of the problem and experience.

There are many sophisticated finite difference schemes such as those that employ irregular rectangular meshes, polycentric grids, a nine-point star, and hexagonal meshes. A fuller description of the finite difference method is included in Chapter 3.

2.12 INTEGRAL EQUATION METHOD

Many field problems can be formulated with equal ease as boundary value problems of differential equations, or alternatively as integral equations. In certain instances, the attempt to define a well-posed problem by differential equations fails altogether. As an example, we may consider the skin effect problem in a long straight wire. This may be formulated as a combination of Helmholtz's or the eddy current diffusion equation within the wire, Laplace's equation outside, and a continuity condition to link the two at the surface. It will be apparent that there is not a single useful boundary condition anywhere in the problem, and therefore all attempts to solve the problem by differential equations have not been successful. However, the problem may be easily posed as an integral equation and solved.

Almost without exception, integral equations arising in electromagnetic theory are of the Fredholm type.

$$\nu y(x) = f(x) + \lambda \int_a^b k(x, \xi) y(\xi) d\xi \qquad (2.37)$$

If $\nu = 0$, the equation is said to be a Fredholm equation of the first kind. If $\nu = 1$, the equation is said to be a Fredholm equation of the second kind. Of course, ν could have some other value, in which case the equation could be scaled to yield an equation of the second kind. In Fredholm equations the variable x does not appear in the limits of integration.

The function $k(x, \xi)$ is termed the kernel function. If this function is not continuous throughout the region of integration, or if the limits are infinite, the equation is said to be singular; otherwise, it is called regular. Singular kernels are the rule rather than the exception in electromagnetic theory. These are usually not convenient to handle. However, equations with finite limits of integration, but with a singular kernel, can often be converted into equations with a regular kernel but infinite limits and vice versa.

Equation(2.37) is solvable if $f(x)$, $k(x, \xi)$, and λ are given. Unlike differential equations, integral equations as a rule have no need for added conditions for their solution. This implies that the boundary conditions are part of the kernel function. A particular kernel is related to particular boundary conditions and does not have universal validity. Solution of electromagnetics problems by means of integral equations consists of (1) formulating the kernel to suit the given boundary conditions and (2) solving of the integral equation itself.

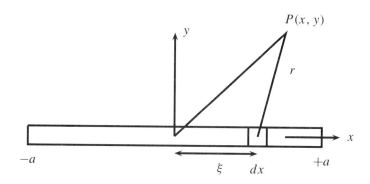

Figure 2.12 Electric Field of a Long Conductor

Equation(2.37) can be extended to higher dimensionality. In two dimensions

$$v\phi(x,y) = f(x,y) + \lambda \iint k(x,y,\xi,\eta)\phi(\xi,\eta)\,d\xi\,d\eta \qquad (2.38)$$

the integration being extended over some prescribed region of the $\xi - \eta$ plane. A further advantage of the integral equation representation is that often a 2D problem can be stated as an integral equation in only one variable.

2.12.1 Example of Electric Field of a Long Conductor

Figure 2.12 shows a charge distribution on a strip along the x axis of density $\sigma(x)$.

The charge associated with an element of the strip of length dx will be σdx. The field due to a point charge $\sigma(\xi)$ at a point $P(x,y)$ is given by

$$D = \frac{Q}{2\pi r} = \frac{\sigma(\xi)}{2\pi r} = \epsilon E = \epsilon \frac{dV'}{dr} \qquad (2.39)$$

Integrating throughout equation (2.39), one obtains

$$V' = \frac{\sigma(\xi)}{2\pi\epsilon} \int \frac{dr}{r} = \frac{\sigma(\xi)}{2\pi\epsilon} \ln\sqrt{(x-\xi)^2 + y^2} \qquad (2.40)$$

Therefore, the field at $P(x, y)$ due to the charge σdx is given by

$$dV(x, y) = \frac{\sigma(\xi)dx}{2\pi\epsilon} \ln\sqrt{(x-\xi)^2 + y^2} \qquad (2.41)$$

By the principle of superposition, the field at $P(x, y)$ due to all the charge on the total length of the conductor strip is obtained as the sum of the effects of the individual charges so that

$$V(x, y) = \frac{1}{2\pi\epsilon} \int_{-a}^{a} \sigma(\xi) \ln\sqrt{(x-\xi)^2 + y^2} d\xi \qquad (2.42)$$

2.12.2 Boundary Conditions

We shall assume that the potential on the conducting strip is V_0, so that

$$V(x, 0) = V_0, \quad -a \leq x \leq a \qquad (2.43)$$

From equations (2.42) and (2.43),

$$V(x, 0) = V_0 = \frac{1}{2\pi\epsilon} \int_{-a}^{a} \sigma(\xi) ln(x-\xi) d\xi \qquad (2.44)$$

or

$$\int_{-a}^{a} \sigma(\xi) ln(x-\xi) d\xi - 2\pi\epsilon V_0 = 0 \qquad (2.45)$$

Comparing equation (2.45) with (2.37), we see that (2.45) is a Fredholm integral equation of the first kind, with singular kernel function

$$k(x, \xi) = \ln |x - \xi| \qquad (2.46)$$

and $f(x) = -2\pi \epsilon V_0$.

It should be noted that this simple problem is very hard to formulate in terms of a differential equation, because of the lack of an outer boundary. Moreover, the integral equation is one dimensional. This frequently occurs with boundary value problems, the PDE. being replaced by an integral equation valid over the boundaries, which of course is of one order lower dimensionality.

Solution of equation (2.45) yields the charge density distribution. Substituting this value of $\sigma(\xi)$ in equation (2.42), one obtains the required potential $V(x, y)$ by straightforward numerical integration.

A more detailed discussion of integral equations is presented in Chapter 7.

2.12.3 Electrostatic Integral Formulation

Let us consider a volume V bounded by the surface S as shown in Figure 2.13 [35]. Let a charge of volumetric charge density ρ be uniformly distributed. Let x, y, and z be the coordinates of the source point of the charge and x', y', and z' be the coordinates of the observation point.

The potential at the observation point due to the entire charge distribution is given by $\phi(x', y', z')$. Let us also assume that ϕ and ψ are scalar functions of position that are continuous throughout the volume V and on the surface S. We know that for a spherically symmetric solution of Laplace's equation, we can obtain the function ψ such that

$$\psi(x, y, z; x', y', z') = \frac{1}{r} \qquad (2.47)$$

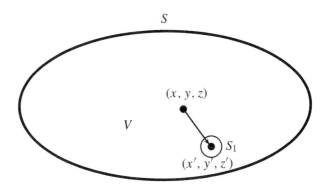

Figure 2.13 Volume and Surface with Field and Source Points

where r is the distance between the source and observation points and is of magnitude

$$r = \sqrt{(x'-x)^2 + (y'-y)^2 + (z'-z)^2} \tag{2.48}$$

This function, however, is undefined at $r = 0$ (the so-called $1/r$ singularity). To exclude this singularity, a small sphere of radius r_1 is circumscribed about x', y', z' as center. The volume V is then bound externally by S and internally by S_1.

Within V, both ϕ and ψ satisfy Green's second identity (Green's theorem) and also Laplace's equation

$$\nabla^2 \psi = 0 \tag{2.49}$$

We shall now consider Green's theorem or Green's identity

$$\int_V (\psi \nabla^2 \phi - \phi \nabla^2 \psi) dV = \int_S (\psi \frac{\partial \phi}{\partial n} - \phi \frac{\partial \psi}{\partial n}) dS \tag{2.50}$$

Setting $\psi = \frac{1}{r}$ in equation (2.50), and noting that ψ satisfies Laplace's equation, we get

$$\int_V \frac{\nabla^2 \phi}{r} dV = \int_{S+S_1} \left(\frac{1}{r} \frac{\partial \phi}{\partial n} - \phi \frac{\partial}{\partial n} \left(\frac{1}{r} \right) \right) dS \qquad (2.51)$$

Over the sphere S_1, the positive normal is directed radially toward the center (x', y', z'), because this is pointing out of the volume V. Over S_1 therefore,

$$\frac{\partial \phi}{\partial n} = -\frac{\partial \phi}{\partial r} \qquad (2.52)$$

and

$$\frac{\partial}{\partial n}\left(\frac{1}{r}\right) = -\frac{\partial}{\partial r}\left(\frac{1}{r}\right) = \frac{1}{r_1^2} \qquad (2.53)$$

Because r_1 is a constant, the contributions of the sphere S_1 to the right-hand side of equation (2.51) will be

$$-\frac{1}{r_1}\int_{S_1} \frac{\partial \phi}{\partial r} dS - \frac{1}{r_1^2}\int_{S_1} \phi \, dS \qquad (2.54)$$

If ϕ_0 and $\frac{\partial \phi_0}{\partial r}$ represent the mean values of ϕ and $\frac{\partial \phi}{\partial r}$, then equation (2.54) becomes

$$-\frac{1}{r_1} 4\pi r_1^2 \frac{\partial \phi_0}{\partial r} - \frac{1}{r_1^2} 4\pi r_1^2 \phi_0 \qquad (2.55)$$

In the limit, when r_1 tends to zero, (2.55) reduces to $-4\pi \phi_0$ or $-4\pi \phi(x', y', z')$.

Let us now return to equation (2.51), which may be expanded as

$$\int_V \frac{\nabla^2 \phi}{r} dV = \int_S \left(\frac{1}{r} \frac{\partial \phi}{\partial n} - \phi \frac{\partial}{\partial n}\left(\frac{1}{r}\right) \right) dS + \int_{S_1} \left(\frac{1}{r} \frac{\partial \phi}{\partial n} - \phi \frac{\partial}{\partial n}\left(\frac{1}{r}\right) \right) dS \qquad (2.56)$$

The second term of (2.56) equals $-4\pi\phi$. Therefore,

$$\int_V \frac{\nabla^2\phi}{r}\,dV = \int_S \left(\frac{1}{r}\frac{\partial\phi}{\partial n} - \phi\frac{\partial}{\partial n}\left(\frac{1}{r}\right)\right)dS - 4\pi\phi \tag{2.57}$$

or

$$\phi = -\frac{1}{4\pi}\int_V \frac{\nabla^2\phi}{r}\,dV + \frac{1}{4\pi}\int_S \left(\frac{1}{r}\frac{\partial\phi}{\partial n} - \phi\frac{\partial}{\partial n}\left(\frac{1}{r}\right)\right)dS \tag{2.58}$$

Because $\nabla^2\phi = -\frac{\rho}{\epsilon}$ for a homogeneous medium

$$\phi(x', y', z') = \frac{1}{4\pi\epsilon}\int_V \frac{\rho}{r}\,dV + \frac{1}{4\pi}\int_S \left(\frac{1}{r}\frac{\partial\phi}{\partial n} - \phi\frac{\partial}{\partial n}\left(\frac{1}{r}\right)\right)dS \tag{2.59}$$

If there are no charges external to S, the surface integral vanishes and

$$\phi(x', y', z') = \frac{1}{4\pi\epsilon}\int_V \frac{\rho}{r}\,dV \tag{2.60}$$

2.12.4 Magnetostatic Integral Formulation

As in the electrostatic case, equation (2.60) is transformed for the magnetic vector potential as

$$A(x', y', z') = \frac{\mu}{4\pi}\int_V \frac{J(x, y, z)}{r}\,dV \tag{2.61}$$

because $\nabla^2 A = -\mu J$.

The magnetic field is obtained by taking the curl of A and dividing the result by μ so that

$$H(x', y', z') = \frac{1}{4\pi} \int_V \nabla' \times \left(\frac{J(x, y, z)}{r}\right) dV \qquad (2.62)$$

Expansion of the integrand on the right-hand side of (2.62) yields

$$\nabla' \times \left(\frac{1}{r} J(x, y, z)\right) = \frac{1}{r} \nabla' \times J(x, y, z) + \nabla' \frac{1}{r} \times J(x, y, z) \qquad (2.63)$$

Because the current density J is a function of x, y, z, the curl with respect to x', y', z' is zero. Therefore (2.63) reduces to

$$\nabla' \times \left(\frac{1}{r} J(x, y, z)\right) = \nabla' \frac{1}{r} \times J(x, y, z) \qquad (2.64)$$

and

$$H(x', y', z') = \frac{1}{4\pi} \int_V \nabla' \frac{1}{r} \times J(x, y, z) dV \qquad (2.65)$$

or

$$H(x', y', z') = \frac{1}{4\pi} \int_V J(x, y, z) \times \nabla \frac{1}{r} dV \qquad (2.66)$$

If we assume that the conductor cross section is small compared with r, then (2.66) can be written in terms of the conductor current I as

$$H(x', y', z') = \frac{I}{4\pi} \int_S \hat{s} \times \nabla \frac{1}{r} dS \qquad (2.67)$$

where $\hat{s}$ is the unit vector in the direction of the current element $I\,ds$. Further simplification of (2.67) is carried out as follows:

$$\nabla \frac{1}{r} = \frac{\partial}{\partial x}\left(\frac{1}{r}\right)\hat{i} + \frac{\partial}{\partial y}\left(\frac{1}{r}\right)\hat{j} + \frac{\partial}{\partial y}\left(\frac{1}{r}\right)\hat{k} \qquad (2.68)$$

$$\begin{aligned}\frac{\partial}{\partial x}\left(\frac{1}{r}\right) &= \frac{\partial}{\partial x}\frac{1}{\sqrt{(x'-x)^2+(y'-y)^2+(z'-z)^2}} \\ &= \frac{(x'-x)\hat{i}}{\left((x'-x)^2+(y'-y)^2+(z'-z)^2\right)^{3/2}}\end{aligned} \qquad (2.69)$$

$$\begin{aligned}\frac{\partial}{\partial y}\left(\frac{1}{r}\right) &= \frac{\partial}{\partial y}\frac{1}{\sqrt{(x'-x)^2+(y'-y)^2+(z'-z)^2}} \\ &= \frac{(y'-y)\hat{j}}{\left((x'-x)^2+(y'-y)^2+(z'-z)^2\right)^{3/2}}\end{aligned} \qquad (2.70)$$

$$\begin{aligned}\frac{\partial}{\partial z}\left(\frac{1}{r}\right) &= \frac{\partial}{\partial z}\frac{1}{\sqrt{(x'-x)^2+(y'-y)^2+(z'-z)^2}} \\ &= \frac{(z'-z)\hat{k}}{\left((x'-x)^2+(y'-y)^2+(z'-z)^2\right)^{3/2}}\end{aligned} \qquad (2.71)$$

Substituting (2.69) through (2.71) into (2.67), we get

$$H(x', y', z') = \frac{I}{4\pi}\int_S \frac{\hat{s}\times\vec{r}}{r^2}dS \qquad (2.72)$$

where $\vec{r} = \dfrac{(x'-x)\hat{i} + (y'-y)\hat{j} + (z'-z)\hat{k}}{\sqrt{(x'-x)^2+(y'-y)^2+(z'-z)^2}}$. Equation (2.72) is known as the Biot–Savaart law.

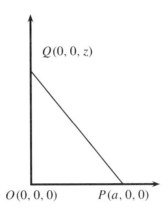

Figure 2.14 Field of a Line Current

Example 1

It is required to find the the magnetic field at a point $P(a, 0, 0)$ due to a line current directed along OQ as shown in Figure 2.14. The coordinates of the point Q are $(0, 0, z)$ and the origin is at point O at $(0, 0, 0)$.

We shall use the formula for finding the magnetic field given by equation (2.67) repeated here.

$$H(x', y', z') = \frac{I}{4\pi} \int_S \hat{s} \times \nabla \frac{1}{r} dS$$

where $\hat{s} = \hat{k}$ and $\vec{r} = \dfrac{(a-0)\hat{i} + (0-0)\hat{j} + (0-z)\hat{k}}{\sqrt{a^2 + z^2}}$.

Because

$$\nabla \left(\frac{1}{r}\right) = \frac{(x'-x)\hat{i} + (y'-y)\hat{j} + (z'-z)\hat{k}}{((x'-x)^2 + (y'-y)^2 + (z'-z)^2)^{3/2}} \qquad (2.73)$$

we have

$$H(x', y', z') = \frac{I}{4\pi} \int_{-\infty}^{\infty} \frac{\hat{k} \times (a\hat{i} - z\hat{k})}{(a^2 + z^2)^{3/2}} dz = \frac{I}{4\pi} \int_{-\infty}^{\infty} \frac{a\hat{j}}{(a^2 + z^2)^{3/2}} dz \quad (2.74)$$

or

$$H(x', y', z') = \frac{I}{4\pi} 2 \int_0^{\infty} \frac{a\, dz}{(a^2 + z^2)^{3/2}} \quad (2.75)$$

The preceding integral is in standard form, and from tables of integrals, one obtains the solution as

$$H(x', y', z') = H_y \hat{j} = \frac{I}{2\pi} \frac{az}{a^2\sqrt{a^2 + z^2}} \Big|_0^{\infty}$$

$$H_y \hat{j} = \frac{I}{2\pi a} \frac{1}{\sqrt{1 + \left(\frac{a}{z}\right)^2}} = \frac{I}{2\pi a} \quad (2.76)$$

Example 2

It is required to find the field at a point $P(0, 0, h)$ due to a current loop as shown in Figure 2.15,

$$H(\rho', \phi', z') = \frac{I}{4\pi} \int_S \hat{s} \times \nabla\left(\frac{1}{r}\right) dS \quad (2.77)$$

where $r = \sqrt{\rho^2 + \rho'^2 - 2\rho\rho' \cos(\phi' - \phi) + (z' - z)^2}$. In cylindrical coordinates the gradient is given by

$$\nabla\left(\frac{1}{r}\right) = \frac{\partial}{\partial \rho}\left(\frac{1}{r}\right)\hat{a}_\rho + \frac{1}{\rho}\frac{\partial}{\partial \phi}\left(\frac{1}{r}\right)\hat{a}_\phi + \frac{\partial}{\partial z}\left(\frac{1}{r}\right)\hat{a}_z \quad (2.78)$$

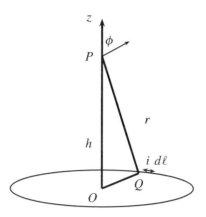

Figure 2.15 Field Due to a Current Loop

We also have $\hat{s} = \hat{a}_\phi$ so that

$$\begin{aligned} H &= \frac{I}{4\pi} \int \hat{a}_\phi \times \left(\frac{\partial}{\partial \rho}\left(\frac{1}{r}\right)\hat{a}_\rho + \frac{1}{\rho}\frac{\partial}{\partial \phi}\left(\frac{1}{r}\right)\hat{a}_\phi + \frac{\partial}{\partial z}\left(\frac{1}{r}\right)\hat{a}_z \right) ds \\ &= \frac{I}{4\pi} \int \left(\frac{\partial}{\partial z}\left(\frac{1}{r}\right)\hat{a}_\rho - \frac{\partial}{\partial \rho}\left(\frac{1}{r}\right)\hat{a}_z \right) ds \end{aligned} \qquad (2.79)$$

Because $\hat{a}_\rho \times \hat{a}_\phi = \hat{a}_z$, $\hat{a}_\phi \times \hat{a}_z = \hat{a}_\rho$, and $\hat{a}_z \times \hat{a}_\rho = \hat{a}_\phi$.

We also have

$$\begin{aligned} \frac{\partial}{\partial z}\left(\frac{1}{r}\right) &= \frac{z' - z}{(\rho^2 + \rho'^2 - 2\rho\rho'\cos(\phi' - \phi) + (z' - z)^2)^{3/2}} \\ \frac{\partial}{\partial \rho}\left(\frac{1}{r}\right) &= -\frac{\rho - \rho'\cos(\phi' - \phi)}{(\rho^2 + \rho'^2 - 2\rho\rho'\cos(\phi' - \phi) + (z' - z)^2)^{3/2}} \end{aligned} \qquad (2.80)$$

Here $\rho' = 0$, $\phi' = 0$, $z = 0$, $z' = h$, and $\rho = a$.

Therefore

$$\frac{\partial}{\partial z}\left(\frac{1}{r}\right) = \frac{(h-0)}{(a^2 + 0 - 0 + (h-0)^2)^{3/2}} = \frac{h}{(a^2 + h^2)^{3/2}} \quad (2.81)$$

and

$$\frac{\partial}{\partial \rho}\left(\frac{1}{r}\right) = \frac{-a}{(a^2 + h^2)^{3/2}} \quad (2.82)$$

The ρ component of H will be zero, as this component due to a current element $i\,d\ell$ will be opposed by a component due to a diametrically opposite current element.

The remaining z component of H will be given by the following expression, because $dS = a\,d\phi$

$$H_z = \frac{I}{4\pi} \int_0^{2\pi} \frac{a \cdot a\,d\phi}{(a^2 + h^2)^{3/2}} = \frac{I}{4\pi} \frac{2\pi a^2}{(a^2 + h^2)^{3/2}} = \frac{Ia^2}{2(a^2 + h^2)^{3/2}} \quad (2.83)$$

2.13 THE FINITE ELEMENT METHOD

The finite element method based on a variational approach [36] or the Galerkin weighted residual technique [37] is perhaps the most practical one for solving boundary value problems. In this method based on a variational approach, unlike finite differences and integral equations, the partial differential equation representing the field problem is formulated as an energy-related expression called a functional. Minimization of this functional yields the solution to the PDE. Alternatively, the PDE could be multiplied by a weighting function and integrated over the domain of the field region to give the required solution.

The procedure for implementing the finite element method based on a variational approach is briefly summarized in this introductory chapter as follows. A more detailed description of the method is provided in later chapters.

- Define the boundary value problem by a partial differential equation.
- Obtain a variational formulation of the PDE in terms of an energy-related functional
- Subdivide the field region into discrete subregions such as triangles and quadrilaterals as shown in Figure 2.16.
- Choose a trial solution defined in terms of nodal values of the solution yet to be determined and interpolatory functions.
- Minimize the functional with respect to the nodal values of the unknown solution.
- Solve the resulting set of algebraic equations and obtain the required solution to the field problem.

Advantages of the finite element method are that

1. It is versatile for application to engineering problems.
2. It has the flexibility for modeling irregular field geometry and boundaries.
3. It handles material nonlinearity and eddy currents well.
4. It yields a stable solution of required accuracy.

Natural boundary conditions are implicit in the method and, therefore, do not need to be imposed explicitly.

2.13.1 Examples

The finite element method is best illustrated by a numerical example in electrostatics as follows.

Parallel Plate Capacitor

Figure 2.17 shows a parallel plate capacitor with voltage V_0 applied at A and C is a ground plane. It is required to determine the potential at point B.

For the sake of simplicity, we shall consider this as a one-dimensional field problem with applied potentials.

Figure 2.16 Subdivision of a 2D Field Region into Triangular and Quadrilateral Elements

Overview of Computational Methods in Electromagnetics

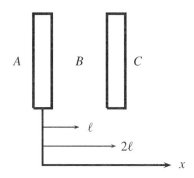

Figure 2.17 Parallel Plate Capacitor Problem

The partial differential equation describing the potential is

$$\frac{\partial^2 V}{\partial x^2} = 0 \tag{2.84}$$

with boundary conditions $V = V_0$ at A and $V = 0$ at C.

The functional is

$$\mathcal{F} = \frac{\epsilon}{2} \int \left(\frac{\partial V}{\partial x}\right)^2 dx \tag{2.85}$$

The field region is modeled as a straight line A–B–C, (Figure 2.18) over which the solution is defined in terms of its nodal values and second-order interpolation polynomials.

The solution can be expressed as

$$V = \sum C_k V_k = C_A V_A + C_B V_B + C_C V_C \tag{2.86}$$

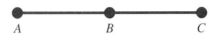

Figure 2.18 Second Order Line Element

where

$$C_A = \left(1 - \frac{x}{\ell}\right)\left(1 - \frac{x}{2\ell}\right)$$
$$C_B = \frac{x}{\ell}\left(2 - \frac{x}{\ell}\right)$$
$$C_C = -\frac{x}{2\ell}\left(1 - \frac{x}{\ell}\right) \qquad (2.87)$$

These coefficients are second-order interpolation polynomials as they contain second-degree terms in x. The suitability of these functions can be verified by substituting the values of x at the nodes along the line A–B–C, so that at $x = 0$, $V = V_A$; at $x = \ell$, $V = V_B$; and at $x = 2\ell$, $V = V_C$.

We now minimize the functional $\mathcal{F}$ in equation (2.85) with respect to the nodal potentials by setting its first derivative to zero, a necessary condition for a minimum. Therefore,

$$\frac{\partial \mathcal{F}}{\partial V_K}\bigg|_{K=A,B,C} = 0 = \frac{\epsilon}{2}\frac{\partial}{\partial V_K}\int \left(\frac{\partial V}{\partial x}\right)^2 dx = \epsilon \int \frac{\partial V}{\partial x}\frac{\partial}{\partial V_K}\frac{\partial V}{\partial x} dx \qquad (2.88)$$

Substituting for C_A, C_B, and C_C from (2.87) into (2.86) and differentiating with respect to x, we get

$$\frac{\partial V}{\partial x} = \left(-\frac{3}{2\ell} - \frac{x}{\ell^2}\right) V_A + \left(\frac{2}{\ell} - \frac{2x}{\ell^2}\right) V_B - \left(\frac{1}{2\ell} - \frac{x}{\ell^2}\right) V_C \qquad (2.89)$$

If we now substitute (2.89) into (2.88), perform the necessary differentiation with respect to the nodal potentials one at a time, integrate the result over the length of the element, and collect terms, we obtain the following matrix equation.

$$\frac{\epsilon}{6\ell} \begin{pmatrix} 7 & -8 & 1 \\ -8 & 16 & -8 \\ 1 & -8 & 7 \end{pmatrix} \begin{pmatrix} V_A \\ V_B \\ V_C \end{pmatrix} = 0 \qquad (2.90)$$

Applying the boundary condition $V = V_0$ at A and $V = 0$ at C, we obtain

$$\frac{16\epsilon V_B}{6\ell} = \frac{8\epsilon V_0}{6\ell} \qquad (2.91)$$

from which $V_B = \frac{V_0}{2}$.

2.13.2 Electrical Conduction Problem

The finite element method will be further illustrated by a one-dimensional electrical conduction example. The differential equation can be solved either by minimizing an energy-related functional or by a weighted residual procedure. The details of these processes are presented in Chapter 4. Returning to the electrical conduction problem, Laplace's equation will be solved subject to boundary conditions as follows:

$$\sigma \nabla^2 V = 0 \qquad (2.92)$$

where σ is the conductivity and V the potential. The functional expression for equation (2.92) is

$$\mathcal{F} = \int_{x_A}^{x_C} \frac{\sigma}{2} \left(\frac{\partial V}{\partial x} \right)^2 dx \qquad (2.93)$$

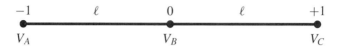

Figure 2.19 One-Dimensional Field Region

Referring to Figure 2.19, the region is represented by a straight line in which the potential is described by a second order interpolation function. Potentials are applied at nodes A and C, such that the potential at A is V_0 and the potential at node C is 0.

We are required to determine the potential at B and at any point within A, B, C by interpolation. The potential at any point can be expressed as a linear combination of nodal potentials weighted by interpolatory functions. These weighting functions in terms of nondimensional coordinates $-1, 0, +1$ are given by

$$\zeta_A = \frac{\xi}{2}(\xi - 1); \quad \zeta_B = (1 - \xi^2); \quad \zeta_C = \frac{\xi}{2}(\xi + 1) \tag{2.94}$$

where $\xi = x/\ell$.

Thus the potential at any point in the region A, B, C can be defined as an interpolate of the vertex values, so that

$$V = \zeta_A V_A + \zeta_B V_B + \zeta_C V_C \tag{2.95}$$

$$\frac{\partial V}{\partial x} = \frac{\partial \zeta_A}{\partial x} V_A + \frac{\partial \zeta_B}{\partial x} V_B + \frac{\partial \zeta_C}{\partial x} V_C \tag{2.96}$$

or

$$\frac{\partial V}{\partial x} = \frac{\partial \zeta_A}{\partial \xi}\frac{\partial \xi}{\partial x} V_A + \frac{\partial \zeta_B}{\partial \xi}\frac{\partial \xi}{\partial x} V_B + \frac{\partial \zeta_C}{\partial \xi}\frac{\partial \xi}{\partial x} V_C \tag{2.97}$$

Overview of Computational Methods in Electromagnetics

Substituting for ζ_A, ζ_B, and ζ_C from (2.94) into (2.97) and after some algebraic manipulation,

$$\frac{\partial V}{\partial x} = \left(\left(\xi - \frac{1}{2}\right) V_A - 2\xi V_B + \left(\xi + \frac{1}{2}\right) V_C\right) / \ell \tag{2.98}$$

After substituting equation (2.98) into (2.93), minimizing $\mathcal{F}$ with respect to V_A, V_B, and V_C yields

$$\begin{pmatrix} \frac{\partial \mathcal{F}}{\partial V_A} \\ \frac{\partial \mathcal{F}}{\partial V_B} \\ \frac{\partial \mathcal{F}}{\partial V_C} \end{pmatrix} = 0$$

$$= \int_{-1}^{1} \frac{1}{\ell} \begin{pmatrix} (\xi - 1/2)^2 & -2\xi(\xi - 1/2) & (\xi^2 - 1/4) \\ -2\xi(\xi - 1/2) & 4\xi^2 & -2\xi(\xi + 1/2) \\ (\xi^2 - 1/4) & -2\xi(\xi + 1/2) & (\xi + 1/2)^2 \end{pmatrix} \times \begin{pmatrix} V_A \\ V_B \\ V_C \end{pmatrix} d\xi \tag{2.99}$$

Equation (2.99), after performing the integrations term by term, becomes

$$\begin{pmatrix} \frac{\partial \mathcal{F}}{\partial V_A} \\ \frac{\partial \mathcal{F}}{\partial V_B} \\ \frac{\partial \mathcal{F}}{\partial V_C} \end{pmatrix} = 0 = \frac{1}{6\ell} \begin{pmatrix} 7 & -8 & 1 \\ -8 & 16 & -8 \\ 1 & -8 & 7 \end{pmatrix} \begin{pmatrix} V_a \\ V_B \\ V_c \end{pmatrix} d\xi \tag{2.100}$$

Applying the boundary conditions $V_A = V_0$ and $V_C = 0$ and eliminating rows and columns corresponding to the applied potentials, we obtain

$$\frac{16 V_B}{6\ell} = \frac{8 V_0}{6\ell} \qquad (2.101)$$

$$V_B = \frac{V_0}{2} \qquad (2.102)$$

Because $V_A = V_0$, $V_B = V_0/2$, and $V_C = 0$, we can determine the potential anywhere in the region from (2.95).

Also, the current density is given by

$$J = -\sigma \frac{\partial V}{\partial x} \qquad (2.103)$$

and from (2.98)

$$J = -\sigma[(\xi - 1/2)V_0 - V_0 \xi] = \frac{\sigma V_0}{2} \qquad (2.104)$$

3

THE FINITE DIFFERENCE METHOD

3.1 INTRODUCTION

In Chapter 1 we found that electromagnetic phenomena are described by partial differential equations, with the dependent variables being either potentials or field values. One of the earliest attempts to solve these problems was by the finite difference method. In the finite difference method, the partial derivatives are replaced by difference equations. Space is divided into finite difference cells and the equations are written for the unknown potential or field values at nodes. The partial differential equations are replaced by a set of simultaneous equations. The unknowns are computed on a finite set of points (the nodes). If we need the values at other locations, we must interpolate. A brief summary of the method is given in Chapter 2.

In this chapter we introduce the concept of difference equations in detail and apply them to the Laplacian operator. We look at the application of boundary conditions and the treatment of nonhomogeneous materials and irregular boundaries. We then show that the difference equations can be represented by an equivalent circuit. Although second-order finite difference methods are the most common, higher accuracy can be achieved by using high-order finite difference formulas. We look at some of these high-order finite difference methods and means of generating any high order finite difference expansion. We proceed from the solution of Poisson's equation to the finite difference representation of the diffusion equation. In these time domain problems, explicit and implicit integration methods are explored. We find that there is also an equivalent circuit for the finite difference equations related to these problems. We end the chapter with the finite difference time domain formulas for the wave equation.

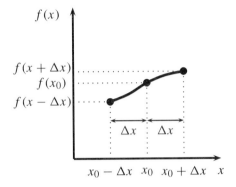

Figure 3.1 Forward, Backward, and Central Differences

3.2 DIFFERENCE EQUATIONS

The basic approximations to derivatives in finite differences are

$$\begin{aligned}
f'(x_0) &\approx \frac{f(x_0 + \Delta x) - f(x_0)}{\Delta x}, &\text{forward difference} \\
f'(x_0) &\approx \frac{f(x_0) - f(x_0 - \Delta x)}{\Delta x}, &\text{backward difference} \\
f'(x_0) &\approx \frac{f(x_0 + \Delta x) - f(x_0 - \Delta x)}{2\Delta x}, &\text{central difference}
\end{aligned} \quad (3.1)$$

These values are illustrated in Figure 3.1.

The second derivatives are found by using a central difference formula on the first derivatives, where the first derivatives are estimated at the intermediate points $(x + \Delta x/2)$ and $(x - \Delta x/2)$.

The Finite Difference Method

$$f''(x_0) = \frac{f'(x_0 + \Delta x/2) - f'(x_0 - \Delta x/2)}{\Delta x}$$

$$f''(x_0) = \frac{\frac{f(x_0 + \Delta x) - f(x_0)}{\Delta x} - \frac{f(x_0) - f(x_0 - \Delta x)}{\Delta x}}{\Delta x} \quad (3.2)$$

$$f''(x_0) = \frac{f(x_0 + \Delta x) - 2f(x_0) + f(x_0 - \Delta x)}{(\Delta x)^2}$$

We can find the truncation errors by expanding the function $f(x)$ around $x_0 + \Delta x$ and $x_0 - \Delta x$ in a Taylor series.

$$f(x_0 + \Delta x) = f(x_0) + \Delta x f'(x_0) + \frac{(\Delta x)^2}{2!} f''(x_0) + \cdots \quad (3.3)$$

$$f(x_0 - \Delta x) = f(x_0) - \Delta x f'(x_0) + \frac{(\Delta x)^2}{2!} f''(x_0) - \cdots \quad (3.4)$$

If we subtract equation (3.4) from (3.3) we obtain

$$f(x_0 + \Delta x) - f(x_0 - \Delta x) = 2\Delta x f'(x_0) + O(\Delta x)^3 \quad (3.5)$$

where the symbol O means *order of*. Dividing equation (3.5) by $2\Delta x$, we obtain the central difference formula (3.1). We see that the truncation error is of the order $(\Delta x)^2$.

Solving for the first derivatives in equations (3.3) and (3.4) separately, we obtain the forward and backward difference formulas of equation (3.1). We see from the Taylor formula that the truncation error associated with each of these is of the order of Δx.

If we add equations (3.3) and (3.4) we obtain

$$f(x_0 + \Delta x) + f(x_0 - \Delta x) = 2f(x_0) + (\Delta x)^2 f''(x_0) + O(\Delta x)^4 \quad (3.6)$$

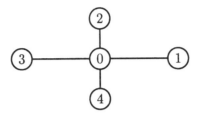

Figure 3.2 Finite Difference Cell and Diagram

Solving equation (3.6) for the second derivative, we obtain equation (3.2) and see that the truncation error is of the order of $(\Delta x)^2$.

3.3 LAPLACE'S AND POISSON'S EQUATIONS

Let us consider Poisson's equation for homogeneous materials in two-dimensional Cartesian coordinates.

$$k\nabla^2 \phi = k \left(\frac{\partial^2 \phi}{\partial x^2} + \frac{\partial^2 \phi}{\partial y^2} \right) = f(x, y) \tag{3.7}$$

where k is a material property (permittivity, for example). Using the expansion for the second derivative in equation (3.2) we obtain (see Figure 3.2)

$$\frac{\partial^2 \phi}{\partial x^2} = \frac{\phi_{i+1,j} - 2\phi_{i,j} + \phi_{i-1,j}}{(\Delta x)^2} + O(\Delta x)^2 \tag{3.8}$$

and

$$\frac{\partial^2 \phi}{\partial y^2} = \frac{\phi_{i,j+1} - 2\phi_{i,j} + \phi_{i,j-1}}{(\Delta y)^2} + O(\Delta y)^2 \tag{3.9}$$

The Finite Difference Method

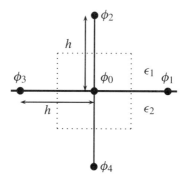

Figure 3.3 Interface Between Two Dielectrics

Substituting into (3.7), multiplying through by $\Delta x \Delta y$, and adding, we obtain

$$\Delta y \frac{\phi_{i+1,j} - 2\phi_{i,j} + \phi_{i-1,j}}{\Delta x} + \Delta x \frac{\phi_{i,j+1} - 2\phi_{i,j} + \phi_{i,j-1}}{\Delta y} = \frac{f(x,y)}{k} \Delta x \Delta y \quad (3.10)$$

This equation is written for each node in the problem to obtain a set of simultaneous equations. The set of equations is singular until the potential of at least one node is specified. We see that for Poisson's equation, the right-hand side is the source function times the area of the finite difference cell. This is (for uniform sources) the total source in the cell. If the source function is not constant, then the right-hand side becomes an approximation to the source in the cell and this approximation gets better as the cell size decreases.

3.4 INTERFACES BETWEEN MATERIALS

One of the reasons why we use numerical methods is the ease with which they handle materials and complicated interfaces. Consider Figure 3.3, in which we have a five-point finite difference scheme and a boundary between two materials. We choose the case of an electrostatic problem here for illustration, but the method can be applied to

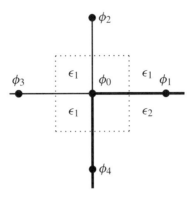

Figure 3.4 Corner of a Dielectric Box

any phenomenon described by Poisson's equation. If we assume that the interface has no charge, then the application of Gauss' law will give us an equation for the potential at node 0 that includes the effect of the material interface. Recall that the well-known boundary condition of $D_{n1} = D_{n2}$ comes directly from application of Gauss' law. Using the dotted line of the finite difference cell as the Gaussian surface, we evaluate $\iint D \cdot dS = 0$.

$$\epsilon_1 \frac{\phi_0 - \phi_1}{h} \frac{h}{2} + \epsilon_2 \frac{\phi_0 - \phi_1}{h} \frac{h}{2} + \epsilon_1 \frac{\phi_0 - \phi_2}{h} h + \epsilon_1 \frac{\phi_0 - \phi_3}{h} \frac{h}{2}$$
$$+ \epsilon_2 \frac{\phi_0 - \phi_3}{h} \frac{h}{2} + \epsilon_2 \frac{\phi_0 - \phi_4}{h} h = 0 \quad (3.11)$$

Rearranging (3.11) we get

$$2(\epsilon_1 + \epsilon_2)\phi_0 = \frac{\epsilon_1 + \epsilon_2}{2}\phi_1 + \epsilon_1 \phi_2 + \frac{\epsilon_1 + \epsilon_2}{2}\phi_3 + \epsilon_2 \phi_4 \quad (3.12)$$

Note that for the case $\epsilon_1 = \epsilon_2$, and with $\Delta x = \Delta y$ and no sources (right-hand side), we have equation (3.10). The special case of the corners can be dealt with in the same manner. We will illustrate by considering the upper left corner of the dielectric box shown in Figure 3.4. By summing the flux leaving the finite difference cell, we have

The Finite Difference Method

$$\frac{\epsilon_1}{2}(\phi_0-\phi_1)+\frac{\epsilon_2}{2}(\phi_0-\phi_1)+\epsilon_1(\phi_0-\phi_2)+\epsilon_1(\phi_0-\phi_3)+\frac{\epsilon_1}{2}(\phi_0-\phi_4)+\frac{\epsilon_2}{2}(\phi_0-\phi_4) = 0 \quad (3.13)$$

Combining terms, we have

$$(3\epsilon_1 + \epsilon_2)\phi_0 = \frac{\epsilon_1 + \epsilon_2}{2}\phi_1 + \epsilon_1\phi_2 + \epsilon_1\phi_3 + \frac{\epsilon_1 + \epsilon_2}{2}\phi_4 \quad (3.14)$$

Formulas for the other corners are found in the same manner.

Example

As an example of the solution of Laplace's equation by the finite difference method, consider the electrostatic problem illustrated in Figure 3.5. The outer box is set to zero potential; the rectangular electrode is set to 100 volts. Two material bodies are included. One is a dielectric body with $\epsilon_r = 5$ and the other is a floating conductor. On the floating conductor, the potential is constant but unknown. The problem was solved using a standard spreadsheet program. The values of the cells for the Dirichlet boundaries were set to their known values (i.e., 100 or 0). All other cells had a formula. For most of the cells this formula was that the value of the cell was 1/4 of the sum of the top, bottom, right, and left cells. This formula corresponds to the finite difference formula for a uniform grid and homogeneous materials. The formulas of the cells on the boundaries of the dielectric and floating conductor were modified as described in equation (3.12). The floating conductor was represented as a dielectric body with $\epsilon_r = 1000$. The spreadsheet program then iterated until the solution converged. The spreadsheet solution is illustrated in Figure 3.6. Here we see that the floating conductor is an equipotential and that the gradient of the potential is reduced in the dielectric region.

3.5 NEUMANN BOUNDARY CONDITIONS

For Poisson's equation to have a unique solution, either the potential or the normal derivative of the potential must be specified at each point on the boundary. If the potential is specified, this is a Dirichlet condition and the unknown nodal potential is eliminated. If the normal derivative is specified, then we can proceed as follows. Consider Figure 3.7. In this case the potential at node (i, j) is unknown. If we consider the finite difference expansion for the potential at node (i, j), then we see that it involves

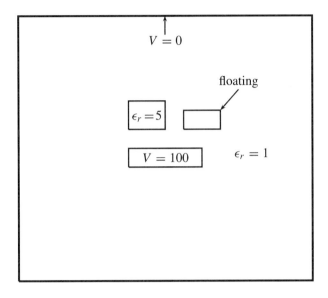

Figure 3.5 Laplacian Problem with Materials

the potential at point $(i+1, j)$ which is outside the domain of interest. We express the normal derivative at (i, j), as

$$\frac{\partial \phi}{\partial n} = \frac{\partial \phi}{\partial x} = \frac{\phi_{i+1,j} - \phi_{i-1,j}}{2\Delta x} \tag{3.15}$$

From (3.15) we solve for the potential at $(i+1, j)$.

$$\phi_{i+1,j} = 2\Delta x \frac{\partial \phi}{\partial n} + \phi_{i-1,j} \tag{3.16}$$

The Finite Difference Method

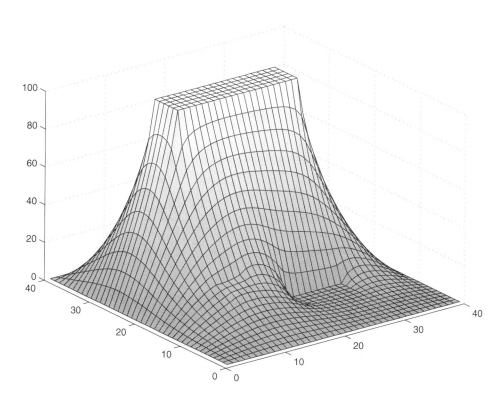

Figure 3.6 Spreadsheet Solution to Laplace's Equation

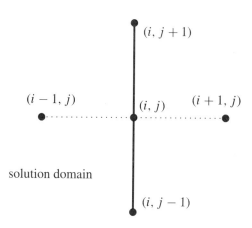

Figure 3.7 Boundary with Normal Derivative Specified

The finite difference expansion for the potential at node (i, j) now becomes (using equation 3.10),

$$k_x \frac{2\Delta x \frac{\partial \phi}{\partial n} - 2\phi_{i,j} + 2\phi_{i-1,j}}{(\Delta x)^2} + k_y \frac{\phi_{i,j+1} - 2\phi_{i,j} + \phi_{i,j-1}}{(\Delta y)^2} \quad (3.17)$$

where k_x and k_y are the material properties in the x and y directions, respectively. A special but very common case is the homogeneous Neumann boundary condition where $\frac{\partial \phi}{\partial n} = 0$. In this case the equipotential lines are perpendicular to the surface. This is the case if, for example, we have a plane of symmetry where the potential on one side of the surface is the same as the potential on the other side of the surface. From Figure 3.7 we have $\phi_{i-1,j} = \phi_{i+1,j}$, and setting $\frac{\partial \phi}{\partial n} = 0$ in equation (3.16) we see that this condition is satisfied.

3.6 TREATMENT OF IRREGULAR BOUNDARIES

One of the difficulties with the finite difference method is its reliance on a regular grid. When the boundaries are irregular it is often difficult or expensive to model the contour accurately with finite difference cells. Many approximations have been developed and the following describes a method that has been found to be effective by the authors. Consider the irregular boundary and finite difference nodes of Figure 3.8. The potential at node 0 will be written as a weighted average of the potentials at the four surrounding nodes. Let us assume that the normal spacing of the grid is a in the x direction and h in the y direction. Let the distance between nodes 0 and 1 be $k_1 a$ and the distance between nodes 0 and 2 be $k_2 h$. Here k_1 and k_2 are constants between zero and one. Let us now assume that $\phi(x)$ is a quadratic function in the vicinity of the boundary. So

$$\phi(x) = \alpha_0 + \alpha_1 x + \alpha_2 x^2 \quad (3.18)$$

Laplace's equation then gives

$$\frac{\partial^2 \phi}{\partial x^2} = 2\alpha_2 \quad (3.19)$$

At the node points we have

$$\begin{aligned} \phi_0 &= \alpha_0 \\ \phi_1 &= \alpha_0 + \alpha_1 k_1 a + \alpha_2 k_1^2 a^2 \\ \phi_3 &= \alpha_0 - \alpha_1 a + \alpha_2 a^2 \end{aligned} \quad (3.20)$$

We have three equations and three unknowns and can solve for α_2 as

$$\alpha_2 = \frac{\phi_1 - (1 + k_1)\phi_0 + k_1 \phi_3}{k_1 a^2 + k_1^2 a^2} \quad (3.21)$$

Using equation (3.19) we obtain

$$\frac{\partial^2 \phi}{\partial x^2} = 2\frac{\phi_1 - (1 + k_1)\phi_0 + k_1 \phi_3}{k_1 a^2 + k_1^2 a^2} \quad (3.22)$$

As we see, if $k_1 = 1$ we have the standard finite difference equation.

We do the same in the y direction to find

$$\frac{\partial^2 \phi}{\partial y^2} = 2\frac{k_2 \phi_4 - (1 + k_2)\phi_0 + \phi_2}{k_2 h^2 + k_2^2 h^2} \quad (3.23)$$

We add equations (3.22) and (3.23) to obtain the final equation.

3.7 EQUIVALENT CIRCUIT REPRESENTATION

We will now show that the finite difference expression has an equivalent circuit representation [38] [39]. This is often useful, and in fact circuit simulators have been used to solve field problems by the use of these equivalent networks. As an example we will consider the two-dimensional expression for the magnetic vector potential

$$\frac{1}{\mu}\nabla^2 A = -J_0 \quad (3.24)$$

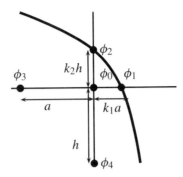

Figure 3.8 Irregular Boundary with Finite Difference Nodes

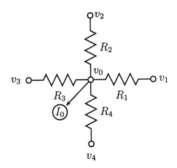

Figure 3.9 Resistive Circuit for the Vector Potential

Consider the resistive network of Figure 3.9. Writing Kirchhoff's current law for node 0, we find

$$\frac{v_0 - v_1}{R_1} + \frac{v_0 - v_2}{R_2} + \frac{v_0 - v_3}{R_3} + \frac{v_0 - v_4}{R_4} = I_0 \quad (3.25)$$

We note that equation (3.25) has the same form as equation 3.10. We can make these the same with the following definitions.

$$R_1 = R_3 = \frac{\mu \Delta x}{\Delta y}$$
$$R_2 = R_4 = \frac{\mu \Delta y}{\Delta x}$$
$$v_i = A_i$$
$$I_0 = -J_0(\Delta x \Delta y) \tag{3.26}$$

Thus in this equivalent circuit representation for the magnetic vector potential, the nodal voltages are the z components of the vector potential, the resistances are permeances normal to the direction of the current, the source current is the total input current to the finite difference cell, and the currents in the resistors are the ampere-turns magnetizing that section of the cell.

3.8 FORMULAS FOR HIGH-ORDER SCHEMES

The five-point scheme for Laplace's equation is the most popular because of its ease of programming and because it has a simple and physically meaningful equivalent circuit representation. There are, however, high-order finite difference schemes involving a greater number of nodes than the four nearest neighbors in two dimensions. These schemes give higher accuracy but at the cost of larger matrix bandwidth. These high-order schemes can be found directly from Taylor series representations of the functions where more terms are kept. A more compact method involving operators will be presented in the next section.

3.8.1 Formulas Involving Any Number of Points

We have seen that the forward and backward difference formulas using two points give first-order accuracy. Consider the one-dimensional case of Figure 3.10. We may write the first derivative as the linear combination of the values at the three nodes.

$$\left.\frac{\partial \phi}{\partial x}\right|_i = \frac{a\phi_i + b\phi_{i-1} + c\phi_{i-2}}{\Delta x} + O(\Delta x)^2 + \cdots \tag{3.27}$$

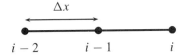

Figure 3.10 Second-Order Backward Difference at Node i

Expanding the values at the nodes around node i,

$$\phi_{i-2} = \phi_i - 2\Delta x \left.\frac{\partial \phi}{\partial x}\right|_i + (2\Delta x)^2 \frac{\partial^2 \phi}{\partial x^2} - \frac{(2\Delta x)^3}{6}\frac{\partial^3 \phi}{\partial x^3} + \cdots \qquad (3.28)$$

$$\phi_{i-1} = \phi_i - \Delta x \left.\frac{\partial \phi}{\partial x}\right|_i + (\Delta x)^2 \frac{\partial^2 \phi}{\partial x^2} - \frac{(\Delta x)^3}{6}\frac{\partial^3 \phi}{\partial x^3} + \cdots \qquad (3.29)$$

We now multiply equation (3.28) by c, multiply equation (3.29) by b and add them along with $a\phi$.

$$a\phi + b\phi + c\phi - \Delta x(2c+b)\frac{\partial \phi}{\partial x} + (\Delta x)^2(4c+b)\frac{\partial^2 \phi}{\partial x^2} + O(\Delta x)^3 \qquad (3.30)$$

Multiplying through equation (3.27) and equating the coefficients we have

$$\begin{aligned} a+b+c &= 0 \\ 2c+b &= -1 \\ 4c+b &= 0 \end{aligned} \qquad (3.31)$$

The Finite Difference Method

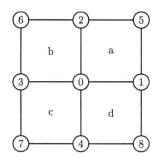

Figure 3.11 High-Order Finite Difference Cell

This gives $a = 3/2$, $b = -2$, and $c = 1/2$ and a backward difference formula involving three points of higher order. We have

$$\left.\frac{\partial \phi}{\partial x}\right|_i = \frac{3\phi_i - 4\phi_{i-1} + \phi_{i-2}}{2\Delta x} \tag{3.32}$$

We will now look at a simple extension of the second-order method to include more nodes in two dimensions. Referring to Figure 3.11, we can obtain a high-order scheme for Laplace's equation by including the eight surrounding nodes. First, write the second-order formulas for the dummy nodes $a, b, c,$ and d. For example, at node a we have

$$\phi_a = \frac{1}{4}(\phi_0 + \phi_1 + \phi_5 + \phi_2) \tag{3.33}$$

We obtain similar expressions for nodes $b, c,$ and d. We then use these in the formula for node 0. So

$$\phi_0 = \frac{1}{4}(\phi_a + \phi_b + \phi_c + \phi_d) \tag{3.34}$$

This gives the formula

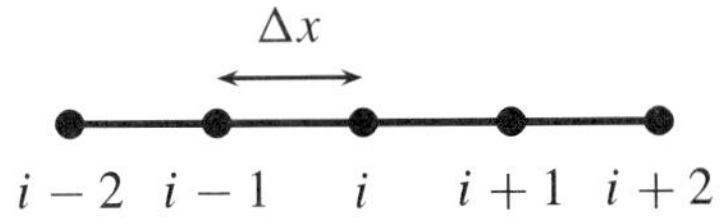

```
         Δx
         ↔
●────●───●───●────●
i−2  i−1  i  i+1  i+2
```

Figure 3.12 Five-Node Scheme to Find Derivative at Node i

$$12\phi_0 = 2\phi_1 + 2\phi_2 + 2\phi_3 + 2\phi_4 + \phi_5 + \phi_6 + \phi_7 + \phi_8 \tag{3.35}$$

We shall now show that this equation is fourth order.

3.8.2 Fourth-Order Formula from Taylor Series

A high-order finite difference formula can be obtained directly from a Taylor series expansion of the derivatives around the node of interest. As an example consider the one-dimensional mesh in Figure 3.12. We have an equal node spacing of Δx and we will find an approximation to the first derivative at node i using not only the adjacent points $(i+1)$ and $(i-1)$ but also contributions from $(i+2)$ and $(i-2)$. The method can be used in general to find formulas with high accuracy but which include more nodes. We begin by assuming that the derivatives can be written as a linear combination of the values at the node points.

$$\frac{d\phi}{dx}\Big|_i = \alpha_1 \phi_{i+2} + \alpha_2 \phi_{i+1} + \alpha_3 \phi_i + \alpha_4 \phi_{i-1} + \alpha_5 \phi_{i-2} \tag{3.36}$$

We now expand ϕ_i about the five node points using the notation that $\frac{d\phi}{dx} = \phi_x$.

$$\begin{aligned}
\phi_{i+2} = &\ \phi_i + 2\Delta x \phi_x + \frac{(2\Delta x)^2}{2!}\phi_{xx} + \frac{(2\Delta x)^3}{3!}\phi_{xxx} \\
&+ \frac{(2\Delta x)^4}{4!}\phi_{xxxx} + \frac{(2\Delta x)^5}{5!}\phi_{xxxxx} + \cdots
\end{aligned}$$

The Finite Difference Method

$$\begin{aligned}
\phi_{i+1} &= \phi_i + \Delta x \phi_x + \frac{(\Delta x)^2}{2!}\phi_{xx} + \frac{(\Delta x)^3}{3!}\phi_{xxx} \\
&\quad + \frac{(\Delta x)^4}{4!}\phi_{xxxx} + \frac{(\Delta x)^5}{5!}\phi_{xxxxx} + \cdots \\
\phi_i &= \phi_i \\
\phi_{i-1} &= \phi_i - \Delta x \phi_x + \frac{(\Delta x)^2}{2!}\phi_{xx} - \frac{(\Delta x)^3}{3!}\phi_{xxx} \\
&\quad + \frac{(\Delta x)^4}{4!}\phi_{xxxx} - \frac{(\Delta x)^5}{5!}\phi_{xxxxx} + \cdots \\
\phi_{i-2} &= \phi_i - 2\Delta x \phi_x + \frac{(2\Delta x)^2}{2!}\phi_{xx} - \frac{(2\Delta x)^3}{3!}\phi_{xxx} \\
&\quad + \frac{(2\Delta x)^4}{4!}\phi_{xxxx} - \frac{(2\Delta x)^5}{5!}\phi_{xxxxx} + \cdots
\end{aligned} \qquad (3.37)$$

Multiplying the first of these equations by α_1, the second by α_2, and so on, we equate coefficients of the derivatives. For example, the coefficient of ϕ_i is $(\alpha_1 + \alpha_2 + \alpha_3 + \alpha_4 + \alpha_5)$, which must be zero. The coefficient of ϕ_x is $2\alpha_1 \Delta x + \alpha_2 \Delta x - \alpha_4 \Delta x - 2\alpha_5 \Delta x$, which must equal $(1)\Delta x$. We obtain the following system of simultaneous equations for α.

$$\begin{pmatrix} 1 & 1 & 1 & 1 & 1 \\ 2 & 1 & 0 & -1 & -2 \\ 2 & \frac{1}{2} & 0 & \frac{1}{2} & 2 \\ \frac{4}{3} & \frac{1}{6} & 0 & -\frac{1}{6} & -\frac{4}{3} \\ \frac{2}{3} & \frac{1}{24} & 0 & \frac{1}{24} & \frac{2}{3} \end{pmatrix} \begin{pmatrix} \alpha_1 \\ \alpha_2 \\ \alpha_3 \\ \alpha_4 \\ \alpha_5 \end{pmatrix} = \begin{pmatrix} 0 \\ 1 \\ 0 \\ 0 \\ 0 \end{pmatrix} \qquad (3.38)$$

This gives

$$\begin{pmatrix} \alpha_1 \\ \alpha_2 \\ \alpha_3 \\ \alpha_4 \\ \alpha_5 \end{pmatrix} = \begin{pmatrix} -0.0833 \\ 0.6667 \\ 0 \\ -0.6667 \\ 0.0833 \end{pmatrix} \qquad (3.39)$$

The equation for the derivative is now

$$\frac{d\phi}{dx}\bigg|_i = \frac{-\phi_{i+2} + 8\phi_{i+1} - 8\phi_{i-1} + \phi_{i-2}}{12\Delta x} + \frac{(\Delta x)^4}{30}\frac{d^5\phi}{dx^5} + \cdots \quad (3.40)$$

From equation (3.40) we see that the truncation error in the derivative is fourth order. The reader may check by using the computed values of α in equation (3.37) that the second- and third-order terms vanish.

3.9 FINITE DIFFERENCES WITH SYMBOLIC OPERATORS

Perhaps the most powerful and certainly the most elegant method of obtaining finite difference formulas for high-order derivatives and high accuracy is by the use of operators [40][41][36][42]. We define the following symbolic operators:

$$E\phi_n = \phi_{n+1} \quad (3.41)$$

$$\mu\phi_n = \frac{1}{2}(\phi_{n+\frac{1}{2}} + \phi_{n-\frac{1}{2}}) \quad (3.42)$$

$$\delta\phi_n = \phi_{n+\frac{1}{2}} - \phi_{n-\frac{1}{2}} \quad (3.43)$$

$$\delta^+\phi_n = \phi_{n+1} - \phi_n \quad (3.44)$$

$$\delta^-\phi_n = \phi_n - \phi_{n-1} \quad (3.45)$$

$$D\phi = \frac{\partial\phi}{\partial x} \quad (3.46)$$

Here E is the shift operator, μ is the averaging operator, δ is the central difference operator, δ^+ is the forward difference operator, δ^- is the backward difference operator, and D is the differential operator.

The reader can verify from these definitions that

$$\delta^+ = E - 1 \quad (3.47)$$

$$\delta^- = 1 - E^{-1} \quad (3.48)$$

The Finite Difference Method

$$\delta^- = E^{-1}\delta^+ \tag{3.49}$$
$$\delta = E^{\frac{1}{2}} - E^{-\frac{1}{2}} \tag{3.50}$$
$$(\delta^+)^2 = E^2 - 2E + 1 \tag{3.51}$$
$$(\delta^+)^3 = E^3 - 3E^2 + 3E - 1 \tag{3.52}$$
$$etc.$$

We use E^{-1} to indicate a negative shift.

We now obtain formulas from the derivatives by expanding the function of interest in a Taylor series. In one dimension this becomes

$$\phi(x+\Delta x) = \phi(x) + \Delta x \phi_x + \frac{(\Delta x)^2}{2!}\phi_{xx} + \cdots$$

$$\phi(x+\Delta x) = \phi(x) + \Delta x D\phi(x) + \frac{(\Delta x)^2}{2!}D^2\phi(x) + \cdots$$

$$\phi(x+\Delta x) = (1 + \Delta x D + \frac{\Delta x^2 D^2}{2!} + \cdots)\phi(x)$$

$$\phi(x+\Delta x) = e^{\Delta x D}\phi(x) \tag{3.53}$$

Therefore

$$E\phi(x) = e^{\Delta x D}\phi(x) \tag{3.54}$$

or

$$E = e^{\Delta x D} \tag{3.55}$$

From this result we can write the derivative operator as

$$D = \frac{\ln E}{\Delta x} \tag{3.56}$$

Expansion of equation (3.56) now yields the finite difference formulas. As an example, the forward difference formula is found by using $E = 1 + \delta^+$. Expanding the natural logarithm, we obtain

$$\ln E = \ln(1 + \delta^+) = \delta^+ - \frac{(\delta^+)^2}{2} + \frac{(\delta^+)^3}{3} - \cdots \tag{3.57}$$

Keeping the first term only gives the first-order formula

$$D\phi = \frac{\phi_{i+1} - \phi_i}{\Delta x} \tag{3.58}$$

Keeping the first two terms gives the second-order formula.

$$\Delta x D = \delta^+ - \frac{\delta^{+2}}{2} = \phi_{i+1} - \phi_i - \frac{\delta^{+2}}{2} \tag{3.59}$$

We now replace the δ^2 term as follows:

$$\delta^{+2} = (E - 1)^2 = E^2 - 2E + 1 \tag{3.60}$$

Substituting into equation (3.59) and solving for D,

$$D = \frac{-3\phi_i + 4\phi_{i+1} - \phi_{i+2}}{2\Delta x} \tag{3.61}$$

This is a second-order equation involving three points. Note that the coefficients are the same as we found in equation (3.32).

The Finite Difference Method

Similarly, for the backward difference formula we use

$$\ln E = -\ln(1 - \delta^-) = \delta^- + \frac{(\delta^-)^2}{2} + \frac{(\delta^-)^3}{3} + \cdots \tag{3.62}$$

For the central difference formula

$$\delta \phi_i = (E^{\frac{1}{2}} - E^{-\frac{1}{2}}) \phi_i \tag{3.63}$$

In this case

$$\delta = e^{\frac{\Delta x D}{2}} - e^{\frac{\Delta x D}{2}} = 2 \sinh \frac{\Delta x D}{2} \tag{3.64}$$

So

$$\begin{aligned}
\Delta x D &= 2 \sinh^{-1} \frac{\delta}{2} = 2(\frac{\delta}{2}) - \frac{1}{2 \cdot 3}(\frac{\delta}{2})^3 + \frac{1 \cdot 3}{2 \cdot 4 \cdot 5}(\frac{\delta}{2})^5 - \frac{1 \cdot 3 \cdot 5}{2 \cdot 4 \cdot 6 \cdot 7}(\frac{\delta}{2})^7 + \cdots \\
&= \delta - \frac{\delta^3}{24} + \frac{3\delta^5}{640} - \frac{5\delta^7}{7168} + \cdots
\end{aligned} \tag{3.65}$$

Keeping only the first term gives the second-order formula

$$\left. \frac{\partial \phi}{\partial x} \right|_i = \frac{\phi_{i+1/2} - \phi_{i-1/2}}{\Delta x} - \frac{(\Delta x)^2}{24} \frac{\partial^3 \phi}{\partial x^3} + \cdots \tag{3.66}$$

Keeping the first two terms gives the fourth-order formula

$$\left. \frac{\partial \phi}{\partial x} \right|_i = \frac{-\phi_{i+3/2} + 27\phi_{i+1/2} - 27\phi_{i-1/2} + \phi_{i-3/2}}{24 \Delta x} + \frac{3}{640}(\Delta x)^4 \frac{\partial^5 \phi}{\partial x^5} + \cdots \tag{3.67}$$

We see that the operator method is a very general method of obtaining formulas of arbitrary order.

3.10 DIFFUSION EQUATION

We now consider the one-dimensional diffusion equation in homogeneous material.

$$\nabla^2 A - \mu\sigma \frac{dA}{dt} = -\mu J_0 \qquad (3.68)$$

Notice that this is the same as Poisson's equation except for the extra term involving the time derivative of the vector potential. The finite difference treatment of the Laplacian and of the right hand-side is the same. We now look at the finite difference representation of the time derivative. As before, we have our choice of forward, backward, or central difference forms. This differential equation requires not only Dirichlet and/or Neumann boundary conditions but initial conditions as well. From these initial conditions (at all of the nodes) we project forward in time using a finite difference estimate of the derivatives. To study the behavior of the time derivative and to explore different integration schemes, we turn to the one-dimensional (space) form of the diffusion equation.

Explicit Scheme

Using a central difference for the Laplacian and a backward difference formula for the time derivative, the finite difference expansion for this equation is

$$\frac{A_{i-1}(t - \Delta t) - 2A_i(t - \Delta t) + A_{i+1}(t - \Delta t)}{(\Delta x)^2} \\ -\mu\sigma \frac{A_i(t) - A_i(t - \Delta t)}{\Delta t} = \mu J \qquad (3.69)$$

Rearranging terms, we can solve for $A_i(t)$ directly.

$$A_i(t) = \frac{\Delta t}{\mu\sigma(\Delta x)^2}(A_{i-1}(t - \Delta t) - 2A_i(t - \Delta t) + A_{i+1}(t - \Delta t)) \\ + A_i(t - \Delta t) - J\frac{\Delta t}{\sigma} \qquad (3.70)$$

The Finite Difference Method

If we define

$$\alpha = \frac{\Delta t}{\mu \sigma (\Delta x)^2} \tag{3.71}$$

we obtain a more convenient form

$$A_i(t) = \alpha A_{i-1}(t - \Delta t) + (1 - 2\alpha) A_i(t - \Delta t)$$
$$+ \alpha A_{i+1}(t - \Delta t) - \alpha \mu (\Delta x)^2 J \tag{3.72}$$

This is known as an explicit scheme. The advantage of this representation is that the solution at time t can be found directly by using the (known) values at time $t - \Delta t$. In this way we avoid having to solve a large system of algebraic equations. We simply *sweep* through the set of equations and obtain the solution for the next time step. Although this is a desirable feature, in practice the backward difference scheme is not commonly used. Recall that the error associated with the backward difference is $O(\Delta t)$. Therefore very small time steps are required. Furthermore, the solution may be unstable, depending on the time step, as will be illustrated in the following section.

Stability of the Solution

To study the stability of the solutions let us consider the one-dimensional problem in Figure 3.13[43]. We have a regular grid and the boundary conditions are homogeneous Dirichlet[1] (i.e., $A = 0$ at both ends). In this case we can write a matrix representation of equation (3.72).

$$\begin{pmatrix} A_1(t) \\ A_2(t) \\ \vdots \\ A_N(t) \end{pmatrix} = \begin{pmatrix} 1 - 2\alpha & \alpha & 0 & \cdots & 0 \\ \alpha & 1 - 2\alpha & \alpha & \cdots & 0 \\ \vdots & & & & \vdots \\ \cdots & & & \alpha & 1 - 2\alpha \end{pmatrix} \begin{pmatrix} A_1(t - \Delta t) \\ A_2(t - \Delta t) \\ \vdots \\ A_N(t - \Delta t) \end{pmatrix} \tag{3.73}$$

We can write this more simply as

$$(A(t)) = (S)(A(t - \Delta t)) \tag{3.74}$$

---
[1] We can do this without loss of generality because for the linear problems discussed here the stability will not depend on the reference value of the potential.

$$A(1) = 0 \quad \bullet\!\!-\!\!\bullet\!\!-\!\!\bullet\!\!-\!\!\bullet\!\!-\!\!\bullet\!\!-\!\!\bullet\!\!-\!\!\bullet\!\!-\!\!\bullet\!\!-\!\!\bullet\!\!-\!\!\bullet \quad A(N) = 0$$
$$A(2) \;\cdots\cdots$$

Figure 3.13 One-Dimensional Illustration of Explicit Scheme

Similarly,

$$(A(t + \Delta t)) = (S)(A(t)) = (S)^2(A(t - \Delta t)) \tag{3.75}$$

and so on.

We can directly find the solution at any time step from the initial conditions, without having to evaluate the solution at intermediate steps.

$$(A(t + M\Delta t)) = (S)(A(t)) = (S)^{M-1}(A(t - \Delta t)) \tag{3.76}$$

We can write the matrix S as the product of three matrices.

$$S = T^{-1}\lambda T \tag{3.77}$$

where T is a matrix whose columns are the eigenvectors of S and λ is a diagonal matrix of the eigenvalues. We can evaluate S to the power P as

$$S^P = T^{-1}\lambda^P T \tag{3.78}$$

The Finite Difference Method

We note that for this matrix, the eigenvalues are known in closed form and are given by the expression [43]

$$\lambda_k = 1 - 4\alpha \sin^2\left(\frac{k\pi}{2M+2}\right) \tag{3.79}$$

Let us now discuss the error propagation associated with the explicit method. If we call the exact solution $\bar{A}$ and the error ϵ, then the value we compute is

$$A(t) = \bar{A}(t) + \epsilon(t) \tag{3.80}$$

Substituting (3.80) into (3.74) we get

$$\hat{A}(t + \Delta t) = S\hat{A}(t)$$
$$\epsilon(t + \Delta t) = S\epsilon(t) \tag{3.81}$$

At the kth time step the error can be written in terms of the error at the first time step as

$$\epsilon(t + k\Delta t) = T^{-1}\Lambda^k T \epsilon(t) \tag{3.82}$$

In order for this error not to grow without bounds, we require that the magnitude of the eigenvalues be less than or equal to 1. So

$$-1 \leq 1 - 4\alpha \sin^2\left(\frac{k\pi}{2M+2}\right) \leq 1 \tag{3.83}$$

Because α is always positive and $\sin^2$ ranges from zero to one, we find that

$$0 < \alpha \leq \frac{1}{2} \tag{3.84}$$

In general, the time step must be very small in order to get good results from the explicit scheme. Further, if we refine the mesh we must also reduce the time step, so in order to achieve high accuracy, the computation becomes very expensive.

Optimum Time Step

We have shown that the explicit method is stable for any $\alpha \leq \frac{1}{2}$, but it is not necessarily true that the smaller the time step, the more accurate the solution. If we look at the finite difference expansion of the one-dimensional diffusion equation and include the first two truncation terms, we have

$$\phi_t - \alpha^2 \phi_{xx} = \frac{\phi(t + \Delta t) - \phi(t)}{\Delta t}$$

$$-\alpha^2 \frac{\phi(x + \Delta x, t) - 2\phi(x, t) + \phi(x - \Delta x, t)}{(\Delta x)^2} - \frac{\Delta t}{2} \phi_{tt} + \alpha^2 \frac{(\Delta x)^2}{12} \phi_{xxxx} + \cdots (3.85)$$

This reminds us that the central difference formulas we used for the space discretization are second order and the forward difference formula for the time discretization is first order. If we now just consider the first two terms of the truncation

$$-\frac{\Delta t}{2} \phi_{tt} + \alpha^2 \frac{(\Delta x)^2}{12} \phi_{xxxx}$$

we know that $\phi_t = \alpha^2 \phi_{xx}$ so

$$\phi_{tt} = \alpha^2 \phi_{xxt} = \alpha^2 (\phi_t)_{xx} \qquad (3.86)$$

so that

$$\phi_{tt} = \alpha^2 (\alpha^2 \phi_{xx})_{xx} = \alpha^4 \phi_{xxxx} \qquad (3.87)$$

Here ϕ_t is the derivative of ϕ with respect to t, and ϕ_{tt} and ϕ_{xx} are the second derivatives of ϕ with respect to t and x, respectively. The truncation term can be written as

$$\left(\frac{\Delta t}{2} \alpha^4 - \alpha^2 \frac{(\Delta x)^2}{12} \right) \phi_{xxxx} \qquad (3.88)$$

We can make this term vanish by choosing $\Delta t = \frac{(\Delta x)^2}{6\alpha^2}$

3.10.1 Implicit Scheme

To avoid the very short time steps required in the explicit integration scheme, we now consider implicit schemes, which express the values of the unknown at a particular time in terms of the potentials of the previous time step and the current time step. Clearly, the disadvantage of this type of method is that we must solve a system of simultaneous equations instead of sweeping through the nodes. Although this is more time consuming and uses considerably more memory, we can use much larger time steps, so that these methods are often preferable.

We begin by again considering the one-dimensional diffusion equation for the magnetic vector potential. In the implicit schemes we replace the Laplacian by the weighted average of the central difference expansions at time t and time Δt. This weighting factor is between 0 and 1 and is often given the symbol θ. These implicit methods are sometimes called θ methods, where the value of θ determines the exact method. We have for the expansion of the Laplacian

$$\frac{\partial^2 A}{\partial x^2} \approx \frac{A(x - \Delta x, t) - 2A(x, t) + A(x + \Delta x, t)}{(\Delta x)^2}(1 - \theta)$$
$$+ \frac{A(x - \Delta x, t + \Delta t) - 2A(x, t + \Delta t) + A(x + \Delta x, t + \Delta t)}{(\Delta x)^2}\theta \qquad (3.89)$$

The derivative of A with respect to time is approximated by a forward difference

$$\frac{\partial A}{\partial t} \approx \frac{A(x, t + \Delta t) - A(x, t)}{\Delta t} \qquad (3.90)$$

Writing the equation in terms of α

$$\alpha(1 - \theta)A(x - \Delta x, t) + (1 - 2\alpha + 2\alpha\theta)A(x, t)$$
$$+ \alpha(1 - \theta)A(x + \Delta x, t) + \alpha\theta A(x - \Delta x, t + \Delta t)$$
$$- (2\alpha\theta + 1)A(x, t + \Delta t) + \alpha\theta A(x + \Delta x, t + \Delta t) = 0 \qquad (3.91)$$

We see that for $\theta = 0$ we have a backward difference and the equations are exactly the same as in the explicit method. For $\theta = 1$ we have a forward difference equation. For $\theta = 0.5$ we have the Crank–Nicholson method.

Stability Criterion for the Implicit Scheme

To determine the stability of the method we again look at the eigenvalues of the matrix. For the implicit scheme the matrix equation that applies for our one-dimensional problem, using the Crank–Nicholson method, is

$$\begin{pmatrix} 2+2\alpha & -\alpha & 0 & \cdots & 0 \\ -\alpha & 2+2\alpha & -\alpha & \cdots & 0 \\ \vdots & & \vdots & & \\ 0 & \cdots & & -\alpha & 2+2\alpha \end{pmatrix} \begin{pmatrix} A_1(t+\Delta t) \\ \vdots \\ A_n(t+\Delta t) \end{pmatrix} = \begin{pmatrix} 2-2\alpha & \alpha & 0 & \cdots & 0 \\ \alpha & 2-2\alpha & \alpha & \cdots & 0 \\ \vdots & & \vdots & & \\ 0 & \cdots & & \alpha & 2-2\alpha \end{pmatrix} \begin{pmatrix} A_1(t) \\ \vdots \\ A_n(t) \end{pmatrix} \quad (3.92)$$

This equation is of the form

$$(C)\{A(t+\Delta t)\} = (S)\{A(t)\} \quad (3.93)$$

The matrix C is nonsingular and we can solve for the vector potential at the end of the time step as

$$\{A(t+\Delta t)\} = (C)^{-1}(S)\{A(t)\} = (T)\{A(t)\} \quad (3.94)$$

As before, we can find the vector potential at any time step from the initial conditions and a power of the matrix T, so

$$\{A(t+k\Delta t)\} = (T)^{k-1}\{A(t)\} \quad (3.95)$$

The Finite Difference Method

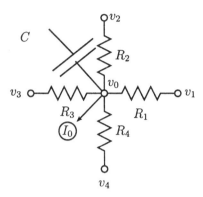

Figure 3.14 R–C Circuit for the Vector Potential Eddy Current Problem

Therefore in order to ensure stability the eigenvalues of the T matrix must all have magnitude less than 1. These eigenvalues are given by [43]

$$\lambda_k = \frac{2 - 4\alpha \sin^2(\frac{k\pi}{2N} - 2)}{2 + 4\alpha \sin^2(\frac{k\pi}{2N} - 2)} \tag{3.96}$$

where N is the order of the matrix.

This expression lies between -1 and 1 for any values of α, so the implicit method is stable (although not necessarily accurate, for any choice of time step).

Equivalent Circuit Representation

Let us now consider the electric circuit shown in Figure 3.14 [44]

$$\frac{v_0 - v_1}{R_1} + \frac{v_0 - v_2}{R_2} + \frac{v_0 - v_3}{R_3} + \frac{v_0 - v_4}{R_4} + C\frac{dv_0}{dt} = I_0 \tag{3.97}$$

As before, note that equation (3.25) has the same form as equation 3.97. We can make these the same with the following definitions.

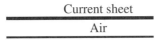

Conducting Region μ, σ

Figure 3.15 One-Dimensional Diffusion Problem

$$R_1 = R_3 = \frac{\mu a}{h}$$
$$R_2 = R_4 = \frac{\mu h}{a}$$
$$v_i = A_i \tag{3.98}$$
$$C = \sigma a h I_0 = J_0(ah)$$

This circuit is similar to the circuit for Poisson's equation (Figure 3.9) except for the addition of the capacitor to ground. This capacitor is the eddy current path and current into the capacitor is the eddy current in the cell. The total (z directed) current in the cell is the sum of the capacitor current and the current source. Note that the ground in the circuit now means that the zero of vector potential is no longer arbitrary. Care must be taken in setting the boundary potentials.

To see that this circuit reflects the observed behavior of skin effect problems consider the one-dimensional problem of a current sheet and a linear conducting material as shown in Figure 3.15. The finite difference equivalent circuit is shown in Figure 3.16. As the frequency of the source increases, the skin depth decreases and we would expect that the current is crowded near the surface. This behavior is represented in the equivalent circuit. As the frequency increases the reactance of the capacitors is decreased and more current (the actual eddy current) goes to ground at the capacitors near the surface. This is due to the smaller skin depth. As the conductivity increases, the capacitors

The Finite Difference Method

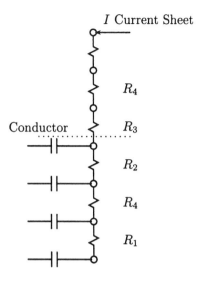

Figure 3.16 Equivalent Circuit for the One-Dimensional Diffusion Problem

get larger and again the current crowds more toward the surface. As the permeability increases, the resistors in the circuit get larger. This also forces the current to flow to ground through the capacitors near the surface. By a similar argument, if we decrease the frequency, decrease the conductivity, or decrease the permeability, the current will penetrate farther into the material.

3.10.2 FDTD for the Wave Equation

The vector Maxwell's equations for the magnetic and electric field are

$$\nabla \times H = \sigma E + \epsilon \frac{\partial E}{\partial t} \qquad (3.99)$$

and

$$\nabla \times E = -\mu \frac{\partial H}{\partial t} \qquad (3.100)$$

The most popular method of finite difference expansion for these equations was introduced by Yee [45] [46] and makes use of an explicit representation in time. Let us

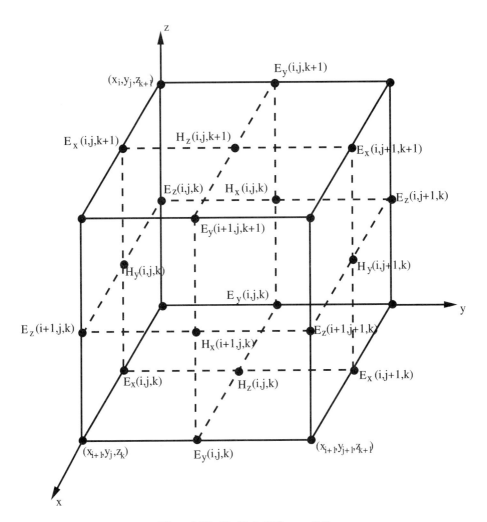

Figure 3.17 Yee Finite Difference Cell

use a uniform grid in Cartesian coordinates with $\Delta x = \Delta y = \Delta z = \Delta$ as shown in Figure 3.17. Then the indices i, j, and k are the numbering indices in the x, y, and z directions, respectively. The time step is Δt and the time is $n\Delta t$. A point is space–time

The Finite Difference Method

is therefore $(i\Delta, j\Delta, k\Delta, n\Delta t)$. Choosing one component of E as an example, the central difference formulas in space have the form

$$\frac{\partial E_y(x, y, z, t)}{\partial x} \approx \frac{E_y^n(i + \frac{1}{2}, j, k) - E_y^n(i - \frac{1}{2}, j, k)}{\Delta x} \qquad (3.101)$$

and in time

$$\frac{\partial E_y(x, y, z, t)}{\partial t} \approx \frac{E_y^{n+\frac{1}{2}}(i, j, k) - E_y^{n-\frac{1}{2}}(i, j, k)}{\Delta t} \qquad (3.102)$$

We evaluate the three components of E and H at spacings of $\frac{\Delta}{2}$.

The electric field is then

$$E_x^{n+1}\left(i + \frac{1}{2}, j, k\right) = \left(1 - \frac{\sigma(i + \frac{1}{2}, j, k)\Delta t}{\epsilon(i + \frac{1}{2}, j, k)}\right) E_x^n\left(i + \frac{1}{2}, j, k\right)$$
$$+ \frac{\Delta t}{\epsilon(i + \frac{1}{2}, j, k)\Delta} \left(H_z^{n+\frac{1}{2}}\left(i + \frac{1}{2}, j + \frac{1}{2}, k\right) - H_z^{n+\frac{1}{2}}\left(i + \frac{1}{2}, j - \frac{1}{2}, k\right) \right. \quad (3.103)$$
$$\left. + H_y^{n+\frac{1}{2}}\left(i + \frac{1}{2}, j, k - \frac{1}{2}\right) - H_y^{n+\frac{1}{2}}\left(i + \frac{1}{2}, j, k + \frac{1}{2}\right) \right)$$

$$E_y^{n+1}\left(i, j + \frac{1}{2}, k\right) = \left(1 - \frac{\sigma(i, j + \frac{1}{2}, k)\Delta t}{\epsilon(i, j + \frac{1}{2}, k)}\right) E_y^n\left(i, j + \frac{1}{2}, k\right)$$
$$+ \frac{\Delta t}{\epsilon(i, j + \frac{1}{2}, k)\Delta} \left(H_x^{n+\frac{1}{2}}\left(i, j + \frac{1}{2}, k + \frac{1}{2}\right) - H_x^{n+\frac{1}{2}}\left(i, j + \frac{1}{2}, k - \frac{1}{2}\right) \right. \quad (3.104)$$
$$\left. + H_z^{n+\frac{1}{2}}\left(i - \frac{1}{2}, j + \frac{1}{2}, k\right) - H_z^{n+\frac{1}{2}}\left(i + \frac{1}{2}, j + \frac{1}{2}, k\right) \right)$$

$$E_z^{n+1}\left(i, j, k+\frac{1}{2}\right) = \left(1 - \frac{\sigma(i, j, k+\frac{1}{2})\Delta t}{\epsilon(i, j, k+\frac{1}{2})}\right) E_z^n\left(i, j, k+\frac{1}{2}\right)$$
$$+ \frac{\Delta t}{\epsilon(i, j, k+\frac{1}{2})\Delta} \left(H_y^{n+\frac{1}{2}}\left(i+\frac{1}{2}, j, k+\frac{1}{2}\right) - H_y^{n+\frac{1}{2}}\left(i-\frac{1}{2}, j, k+\frac{1}{2}\right) \right. \quad (3.105)$$
$$\left. + H_x^{n+\frac{1}{2}}\left(i, j-\frac{1}{2}, k+\frac{1}{2}\right) - H_x^{n+\frac{1}{2}}\left(i, j+\frac{1}{2}, k+\frac{1}{2}\right) \right)$$

$$H_x^{n+\frac{1}{2}}\left(i, j+\frac{1}{2}, k+\frac{1}{2}\right) = H_x^{n-\frac{1}{2}}\left(i, j+\frac{1}{2}, k+\frac{1}{2}\right)$$
$$+ \frac{\Delta t}{\mu(i, j+\frac{1}{2}, k+\frac{1}{2})\Delta} \left(E_y^n\left(i, j+\frac{1}{2}, k+1\right) - E_y^n\left(i, j+\frac{1}{2}, k\right) \right. \quad (3.106)$$
$$\left. + E_z^n\left(i, j, k+\frac{1}{2}\right) - E_z^n\left(i, j+1, k+\frac{1}{2}\right) \right)$$

$$H_y^{n+\frac{1}{2}}\left(i+\frac{1}{2}, j, k+\frac{1}{2}\right) = H_y^{n-\frac{1}{2}}\left(i+\frac{1}{2}, j, k+\frac{1}{2}\right)$$
$$+ \frac{\Delta t}{\mu(i+\frac{1}{2}, j, k+\frac{1}{2})\Delta} \left(E_z^n\left(i+1, j, k+\frac{1}{2}\right) - E_z^n\left(i, j, k+\frac{1}{2}\right) \right. \quad (3.107)$$
$$\left. + E_x^n\left(i+\frac{1}{2}, j, k\right) - E_x^n\left(i+\frac{1}{2}, j, k+1\right) \right)$$

$$H_z^{n+\frac{1}{2}}\left(i+\frac{1}{2}, j+\frac{1}{2}, k\right) = H_z^{n-\frac{1}{2}}\left(i+\frac{1}{2}, j+\frac{1}{2}, k\right)$$
$$+ \frac{\Delta t}{\mu(i+\frac{1}{2}, j+\frac{1}{2}, k)\Delta} \left(E_x^n\left(i+\frac{1}{2}, j+1, k\right) - E_x^n\left(i+\frac{1}{2}, j, k\right) \right. \quad (3.108)$$
$$\left. + E_y^n\left(i, j+\frac{1}{2}, k\right) - E_y^n\left(i+1, j+\frac{1}{2}, k\right) \right)$$

The Finite Difference Method

Time Step and Stability

As in the case of the parabolic equations, we have a maximum time step requirement for the hyperbolic equations. Applying an intuitive approach, the time step must be small enough so that the wave front will not cross more than one node in the finite difference cell in one time step. If the wave front skipped over a node, then using our sweeping method to update the field values at the node, we would end up skipping the update of the field values at that node. The situation is illustrated in Figure 3.18. We have here a one-dimensional problem. Assume that we have a wave propagating in the z direction. At node i the values are updated based on the values of nodes $i+1$ and $i-1$ of the previous time step. If the wave front is allowed to travel, say from node i to node $i+2$ in one time step, then the field values at node $i+2$ would not be correctly updated because they depend on the field values at node $i+1$ at the previous time step.

In the case of two dimensions we use the same criterion. The wave front must not be allowed to skip over nodes so that a particular node would miss the updating process. In Figure 3.19 we see that if the wave was coming in at a 45°, angle then its plane would cross nodes b and c if the time step was greater than $\frac{\Delta}{\sqrt{2}u}$. For the three-dimensional case the reader can confirm that the criterion is

$$\Delta t \leq \frac{\Delta}{\sqrt{3}u} \qquad (3.109)$$

This assumes that the nodal spacing in all directions is equal to Δ. If not, the criterion must be modified to

$$\Delta t \leq \frac{1}{u\sqrt{\frac{1}{(\Delta x)^2} + \frac{1}{(\Delta y)^2} + \frac{1}{(\Delta z)^2}}} \qquad (3.110)$$

Boundary Conditions

One of the difficulties with the solution of the wave equation by the finite difference (and finite element) methods is that most problems are open boundary problems. We consider open boundary problems in Chapter 8 and absorbing boundary conditions and coupling to integral equations for the wave equations in Chapter 9. However, a simple method of terminating the finite difference grid for these explicit methods was developed by Taflove *et. al.* [47] To illustrate this method, let's consider a one-dimensional wave propagation problem. Assume that we have a wave propagating

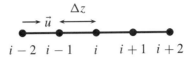

Figure 3.18 Stability for the 1D Wave Equation

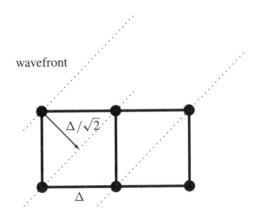

Figure 3.19 Stability Criterion for Finite Difference Time Domain

in the z direction with field components E_x and H_y and a velocity of propagation u. We create a uniform finite difference grid with spacing Δz. We now select the time step to be $\Delta t = \frac{\Delta z}{u}$. This is the largest possible time step allowable for a stable finite difference solution. Any larger time step will result in the wave traveling more than one finite difference cell in one time step. If we have a problem domain extending from $z = 0$ to $z = L$, then by requiring that

$$E_x(0, t) = E_x(\Delta z, t - \Delta t) \tag{3.111}$$

we have eliminated reflections at the $z = 0$ boundary. Similarly, by requiring that

$$E_x(L, t) = E_x(L - \Delta z, t - \Delta t) \tag{3.112}$$

we have eliminated the reflections from the $z = L$ boundary. These ideas have been extended to problems of two and three dimensions in which the fields can have all three components and the waves can travel in arbitrary directions.

3.11 CONCLUSIONS

The finite difference method is one of the oldest and one of the most reliable methods of solving electromagnetics problems. It is a generalization of the well-known magnetic circuit. The method is both rigorous and flexible. It remains the method of choice for many classes of problems, such as time domain wave problems. The method is easily adaptable to problems with nonhomogeneous materials and sources. Convenient equivalent circuits exist for a number of 2D and 3D problems. The method results in a simple and easy to program set of equations that are generally sparse, symmetric, and well behaved. The references at the end of the book contain a number of good examples of finite difference applications and several books specializing in the subject.

4

VARIATIONAL AND GALERKIN METHODS

4.1 INTRODUCTION

A good number of practical engineering problems can be formulated in terms of partial differential equations subject to specified boundary conditions. All static field problems involve either the Laplacian or closely related partial differential operators. A great many time-varying problems, particularly where periodic processes are involved, can be reduced to elliptic form by separation of the time variable. It is, therefore, of great interest in engineering applications to examine this class of boundary value problems in some detail.

Except in the case of simple and geometrically well-defined problems, closed form analytical solutions are impossible. For complex geometrical shapes, with varying material characteristics and often mixed boundary conditions, numerical methods offer the best and often the most economical solution. With the advent of modern digital computers and sophisticated software, it is possible to represent the field region fairly accurately without the need for excessive computer storage or execution time.

In this section, we shall discuss the variational representation of the partial differential equations, modeling the field problem by energy-related expressions called functionals, and derive the condition that the solutions to the PDEs also minimize the functionals. We shall briefly state the formulation of the partial differential equations for magnetostatic, eddy current diffusion, and high-frequency field problems and discuss their solution by the variational method by minimizing appropriate energy-related functionals. We shall also discuss the application of the variational principle to solution techniques, namely Ritz's method and the finite element method. A brief description of the variational formulation for integral equations and their solution will also be included. A closely related but a more general method called the weighted residual method or Galerkin method is discussed in Section 4.9.

4.2 THE VARIATIONAL METHOD

The variational method is one of the principal mathematical techniques and perhaps the most practical and accurate method for solving boundary value problems. In this method, unlike finite differences and classical integral equation methods, the partial differential equation of the field problem is formulated as an energy-related expression called a functional. The solution to the PDE is then obtained by minimizing the functional.

In this section, we shall derive the functional and its variation from first principles. We shall also show that the necessary condition for functional minimization yields the partial differential equation.

From this, we can generalize that in order to solve a boundary value problem, we must first formulate the PDE in variational terms as a functional and minimize this functional with respect to a trial solution or set of trial solutions. The necessary condition for functional minimum is the so-called *Euler–Lagrange* equation, which yields the partial differential equation of the field problem. The foregoing procedure is schematically illustrated in Figure 4.1.

Although this procedure seems cumbersome and somewhat indirect, it has the following advantages over traditional closed form methods, finite differences, and integral equation techniques:

1. The variational method is based on energy minimization and, as with structural and continuum mechanics, the solution yields the minimum potential energy.
2. The technique is applicable to homogeneous isotropic media as well as inhomogeneous, anisotropic nonlinear media. The method lends itself easily to discretization of the field region into subregions. Because the energy is a scalar quantity, the summation of energies computed in the discretized subregions can be added to yield the total energy in the field region.
3. Minimization of the global energy yields accurate potential solutions from which the field quantities can be evaluated.
4. Unlike most finite difference schemes, in which addition of nodes is essential to obtain a convergent and accurate solution, in the variational technique high-order elements can be used to improve accuracy.
5. The boundary conditions represented by the surface integral associated with the functional has physical meaning. Also, setting the surface integral to zero automatically satisfies natural boundary conditions (Dirichlet or homogeneous Neumann conditions).

Variational and Galerkin Methods

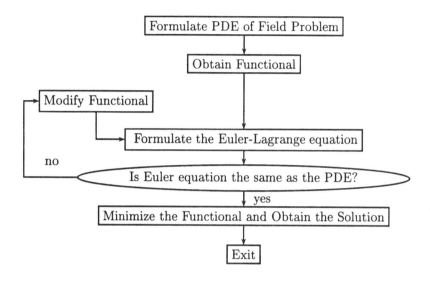

Figure 4.1 Schematic of Solving Boundary Value Problems by the Variational Method

4.3 THE FUNCTIONAL AND ITS EXTREMUM

A functional is an operation that assigns a number to each of a certain class of functions. If we define ϕ to be a set of functions of two variables (x, y) subject to the condition $\phi = f(s)$ on S (where S is the surface bounding the region R and is continuously differentiable), then any quantity such as $\mathcal{F}$ that takes a specific numerical value corresponding to each function in the set is said to be a functional on the set ϕ. It is a function of a function.

As a first step, we shall consider a functional in a single independent variable and extend the same to two or more space variables. Let us consider the maximization or minimization of an integral of the form

$$\mathcal{F} = \int_a^b f(x, u, u') dx \qquad (4.1)$$

where $u' = \frac{\partial u}{\partial x}$, and the boundary conditions are

$$u(a) = A, \quad u(b) = B \tag{4.2}$$

where a, b, A, and B are constants. We shall assume f has continuous second-order derivatives with respect to its three arguments and require that the unknown function $u(x)$ possesses two derivatives everywhere in (a, b). We shall consider a one-parameter family of admissible functions that includes $u(x)$ and is of the form

$$u(x) + \epsilon \eta(x) \tag{4.3}$$

where $\eta(x)$ is an arbitrary twice-differentiable function that vanishes at the end points (a, b). Thus

$$\eta(a) = \eta(b) = 0 \tag{4.4}$$

Also, ϵ is a parameter that is a constant in any one function in the set but varies from one function to another.

We seek all admissible functions $\eta(x)$ in the neighborhood of $u(x)$. Admissibility means that $\eta(x)$ vanishes identically on the boundary and it is at least once differentiable. $\epsilon(x)$ represents the increment or the difference between the varied function and the actual function. It is therefore called a variation of $u(x)$.

If $u(x)$ is replaced by $u(x) + \epsilon \eta(x)$, then the integral $\mathcal{F}$ assumes the form (after dropping the terms in the parentheses for convenience)

$$\mathcal{F} = \int_a^b f(x, u + \epsilon \eta, u' + \epsilon \eta') \, dx \tag{4.5}$$

It is apparent that the maximum or minimum value (extremum) of $\mathcal{F}(\epsilon)$ occurs when $\epsilon = 0$, that is, when the variation of u is zero. This is analogous to differentiation of

Variational and Galerkin Methods

a function of a single variable. However, the functional is a function of a set of variables, namely x, u, u'. Therefore, the term "variation" is used instead of "differential coefficient." Hence it follows that

$$\frac{d\mathcal{F}}{d\epsilon} = 0 \qquad (4.6)$$

when $\epsilon = 0$.

Substituting for $\mathcal{F}(\epsilon)$ from (4.5) into (4.6) and using the chain rule of differentiation

$$\frac{d\mathcal{F}}{d\epsilon} = \int_a^b \left(\frac{\partial f(x, u+\epsilon\eta, u'+\epsilon\eta')}{\partial(u+\epsilon\eta)} \frac{\partial(u+\epsilon\eta)}{\partial\epsilon} \right.$$
$$\left. + \frac{\partial f(x, u+\epsilon\eta, u'+\epsilon\eta')}{\partial(u'+\epsilon\eta')} \frac{\partial(u'+\epsilon\eta')}{\partial\epsilon} \right) dx \qquad (4.7)$$

or

$$\frac{d\mathcal{F}}{d\epsilon} = \int_a^b \left(\frac{\partial f(x, u+\epsilon\eta, u'+\epsilon\eta')}{\partial(u+\epsilon\eta)} \eta + \frac{\partial f(x, u+\epsilon\eta, u'+\epsilon\eta')}{\partial(u'+\epsilon\eta')} \eta' \right) dx \qquad (4.8)$$

Setting $\epsilon = 0$, we have

$$\frac{d\mathcal{F}}{d\epsilon} = \int_a^b \left(\frac{\partial f}{\partial u} \eta(x) + \frac{\partial f}{\partial u'} \eta'(x) \right) dx = \mathcal{F}'(0) = 0 \qquad (4.9)$$

The second term of the definite integral (4.9) can be transformed by integration by parts as follows:

$$\int_a^b \left(\frac{\partial f}{\partial u'} \eta'(x) \right) dx = \left[\frac{\partial f}{\partial u'} \eta(x) \right]_a^b - \int_a^b \left[\frac{\partial}{\partial x} \frac{\partial f}{\partial u'} \eta(x) \right] dx \qquad (4.10)$$

Because $\eta(a) = \eta(b) = 0$, equation (4.10) reduces to the form

$$\int_a^b \left(\frac{\partial f}{\partial u'}\eta'(x)\right) dx = -\int_a^b \left[\frac{\partial}{\partial x}\frac{\partial f}{\partial u'}\eta(x)\right] dx \qquad (4.11)$$

Substituting (4.11) into (4.9),

$$\mathcal{F}'(0) = -\int_a^b \left[\frac{\partial}{\partial x}\frac{\partial f}{\partial u'}\eta(x)\right] dx + \int_a^b \left(\frac{\partial f}{\partial u}\eta(x)\right) dx \qquad (4.12)$$

or

$$\mathcal{F}'(0) = -\int_a^b \left[\frac{\partial}{\partial x}\frac{\partial f}{\partial u'} - \frac{\partial f}{\partial u}\right]\eta(x)\, dx \qquad (4.13)$$

We can deduce from (4.13) that its integrand must be zero for any arbitrary function $\eta(x)$ in the neighborhood of $u(x)$, so that

$$\frac{\partial}{\partial x}\left(\frac{\partial f}{\partial u'}\right) - \left(\frac{\partial f}{\partial u}\right) = 0 \qquad (4.14)$$

The expression of equation (4.14) is the so-called *Euler–Lagrange* equation or, for short, the Euler equation of the functional (4.1), subject to boundary conditions (4.2).

Example

A. Find the first variation of the integral

$$\mathcal{F} = \int \left[\left(\frac{\partial V}{\partial x}\right)^2 - \frac{2\rho V}{\epsilon_0}\right] dx$$

subject to boundary conditions $V = V_0$ or $\frac{\partial V}{\partial x} = 0$. Show that the condition for an extremum of the functional $\mathcal{F}$ yields the scalar Poisson equation for the electrostatic problem.

Variational and Galerkin Methods

B. Apply the general Euler–Lagrange equation to the integral $\mathcal{F}$ to verify the answer to part A.

Solution:

A. Let $\epsilon\eta(x)$ be the variation on V such that

$$\mathcal{F} + \delta\mathcal{F} = \int \left[\left(\frac{\partial}{\partial x}(V + \epsilon\eta) \right)^2 - \frac{2\rho}{\epsilon_0}(V + \epsilon\eta) \right] dx \qquad (4.15)$$

and

$$\mathcal{F} = \int \left[\left(\frac{\partial V}{\partial x} \right)^2 - \frac{2\rho}{\epsilon_0}(V) \right] dx \qquad (4.16)$$

Subtracting (4.16) from (4.15), the first variation in $\mathcal{F}$ will be

$$\delta\mathcal{F} = \int \left(\left(\frac{\partial}{\partial x}(V + \epsilon\eta) \right)^2 - \frac{2\rho}{\epsilon_0}(V + \epsilon\eta) - \left(\frac{\partial V}{\partial x} \right)^2 + \frac{2\rho V}{\epsilon_0} \right) dx \qquad (4.17)$$

Rearranging terms in (4.17), we have

$$\delta\mathcal{F} = \int 2 \left(\epsilon \frac{\partial V}{\partial x} \left(\frac{\partial \eta}{\partial x} \right) - \frac{\rho \epsilon \eta}{\epsilon_0} \right) dx + \int \epsilon^2 \left(\frac{\partial \eta}{\partial x} \right)^2 dx \qquad (4.18)$$

The first term on the right-hand side of the first integral can be transformed by integration by parts as follows:

$$\int \epsilon \frac{\partial V}{\partial x} \frac{\partial \eta}{\partial x} dx = \frac{\partial V}{\partial x} \epsilon \eta - \int \frac{\partial}{\partial x} \frac{\partial V}{\partial x} \epsilon \eta \, dx \qquad (4.19)$$

Substituting (4.19) into (4.18)

$$\delta \mathcal{F} = 2\frac{\partial V}{\partial x}\epsilon\eta - 2\int \left(\frac{\partial}{\partial x}\frac{\partial V}{\partial x} + \frac{\rho}{\epsilon_0}\right)\epsilon\eta \, dx + \int \epsilon^2 (\frac{\partial \eta}{\partial x})^2 \, dx \quad (4.20)$$

As the parameter $\epsilon \to 0$, the approximate function $V + \epsilon\eta$ approaches the true solution V, and the last integral in (4.20) vanishes. Also, the first term vanishes because $\eta = 0$ everywhere on the boundary. Thus

$$\delta \mathcal{F} = \int \left(\frac{\partial}{\partial x}\frac{\partial V}{\partial x} + \frac{\rho}{\epsilon_0}\right)\epsilon\eta \, dx = 0 \quad (4.21)$$

For any arbitrary function η, equation (4.21) holds only if the integrand is identically zero. Therefore,

$$\left(\frac{\partial}{\partial x}\frac{\partial V}{\partial x} + \frac{\rho}{\epsilon_0}\right) = 0 \quad (4.22)$$

or

$$\frac{\partial^2 V}{\partial x^2} = -\frac{\rho}{\epsilon_0} \quad (4.23)$$

Equation (4.23) is the scalar Poisson equation for the one-dimensional electrostatic field problem, where V is the potential and ρ the volumetric charge density.

B. Applying the Euler–Lagrange equation (4.14) to the functional directly, we have

$$\frac{\partial}{\partial x}\frac{\partial f}{\partial u'} - \frac{\partial f}{\partial u} = 0 \quad (4.24)$$

Variational and Galerkin Methods

where the integrand

$$f = \left(\frac{\partial V}{\partial x}\right)^2 - \frac{2\rho V}{\epsilon_0} \qquad (4.25)$$

We define $u = V$ and $u' = \frac{\partial V}{\partial x}$

Substituting (4.25) into (4.24)

$$\frac{d}{dx}\left(\frac{\partial}{\partial \frac{dV}{dx}}\left(\frac{dV}{dx}\right)^2\right) - \frac{\partial}{\partial V}\left(\frac{-2\rho V}{\epsilon_0}\right) = 0 \qquad (4.26)$$

Therefore

$$\frac{d^2 V}{dx^2} = \frac{-\rho}{\epsilon_0} \qquad (4.27)$$

4.3.1 Accuracy of the Variational Method

An elegant discussion of the accuracy of the variational method is given in reference [48] and is summarized here for completeness. Let us consider Laplace's equation for the electrostatic problem in terms of the potential V, so that

$$\nabla^2 V = 0 \qquad (4.28)$$

The energy-related functional will be

$$\mathcal{F} = \frac{1}{2} \int |\nabla V|^2 \, dS \qquad (4.29)$$

Let us suppose that V is the true solution to the problem, while η is the neighborhood admissible function having exactly zero value on the boundary. The variation on $\mathcal{F}$ can then be written as $\epsilon \eta$, where ϵ is a scalar parameter. Therefore

$$\mathcal{F}(V + \epsilon \eta) = \mathcal{F}(V) + \epsilon \int \nabla V \cdot \nabla \eta \, dS + \frac{\epsilon^2}{2} \int |\nabla \eta|^2 \, dS \qquad (4.30)$$

From vector identities (Green's theorem)

$$\nabla \cdot (\eta \nabla V) = \nabla \eta \cdot \nabla V + \eta \nabla^2 V \qquad (4.31)$$

Using (4.31) in (4.30)

$$\mathcal{F}(V + \epsilon \eta) = \mathcal{F}(V) + \epsilon \oint \eta \frac{\partial V}{\partial n} \, d\Gamma - \epsilon \int \eta \nabla^2 V \, dS + \epsilon^2 \mathcal{F}(\eta) \qquad (4.32)$$

Because V satisfies Laplace's equation

$$\int \epsilon \eta \nabla^2 V \, dS = 0 \qquad (4.33)$$

Also the surface integral in (4.32) vanishes because on the boundary either $\eta = 0$ or $\frac{\partial V}{\partial n} = 0$.

Thus

$$\mathcal{F}(V + \epsilon \eta) = \mathcal{F}(V) + \epsilon^2 \mathcal{F}(\eta) \qquad (4.34)$$

The second term on the right-hand side is always positive. Therefore, $F(V)$ is the minimum value of the energy attained when $\epsilon = 0$ for any function η defined such that η must vanish on the boundary where V is prescribed and η must also be once differentiable.

From equation (4.34), one can observe that the error in the estimate of the energy depends on ϵ. If ϵ is sufficiently small, the error in energy is much smaller than the error in the potential. In many engineering applications, energy-related quantities are more important than the potential. Some examples are inductance, capacitance (ratio of energy to the square of the current or voltage), and power loss. These quantities can be found very accurately.

4.4 FUNCTIONAL IN MORE THAN ONE SPACE VARIABLE AND ITS EXTREMUM

Defining a functional $\mathcal{F}$ in two space variables, we have

$$\mathcal{F} = f(x, y, \phi, \phi_x, \phi_y) \, dx \, dy \tag{4.35}$$

where

$$\phi_x = \frac{\partial \phi}{\partial x} \quad \text{and} \quad \phi_y = \frac{\partial \phi}{\partial y} \tag{4.36}$$

As in the one-dimensional case, we shall perturb the function $\phi(x, y)$ to its new value as

$$\phi(x, y) + \epsilon \eta \tag{4.37}$$

The change $\epsilon \eta(x, y)$ in $\phi(x, y)$ is called the variation of ϕ and is defined by

$$\delta \phi = \epsilon \eta(x, y) \tag{4.38}$$

Here $\eta(x, y)$ is also a function that is continuous and differentiable and is subject to the same boundary conditions as ϕ. Also, ϵ is a parameter associated with the function. Corresponding to this change in ϕ, the functional will assume the value

$$\mathcal{F} = \iint_R (f(x, y, \phi + \epsilon \eta, \phi_x + \epsilon \eta_x, \phi_y + \epsilon \eta_y)) \, dx \, dy \tag{4.39}$$

Differentiating $\mathcal{F}(\epsilon)$ with respect to ϵ, we obtain as before

$$\frac{d\mathcal{F}}{d\epsilon} = 0 = \iint_R \left(\frac{\partial f(x, y, \phi + \epsilon\eta, \phi_x + \epsilon\eta_x, \phi_y + \epsilon\eta_y)\eta}{\partial(\phi + \epsilon\eta)} \right.$$
$$+ \frac{\partial f(x, y, \phi + \epsilon\eta, \phi_x + \epsilon\eta_x, \phi_y + \epsilon\eta_y)\eta_x}{\partial(\phi_x + \epsilon\eta_x)} \quad (4.40)$$
$$\left. + \frac{\partial f(x, y, \phi + \epsilon\eta, \phi_x + \epsilon\eta_x, \phi_y + \epsilon\eta_y)\eta_y}{\partial(\phi_y + \epsilon\eta_y)} \right) dx\, dy$$

Setting $\epsilon = 0$, there is

$$\frac{d\mathcal{F}(\epsilon)}{d\epsilon} = \mathcal{F}'(0) = \iint_R \left(\frac{\partial f}{\partial \phi}\eta + \frac{\partial f}{\partial \phi_x}\eta_x + \frac{\partial f}{\partial \phi_y}\eta_y \right) dx\, dy = 0 \quad (4.41)$$

The second term on the right-hand side of (4.41) can be expanded by integration by parts as follows.

$$\iint_R \frac{\partial f}{\partial \phi_x}\eta_x\, dx\, dy = \oint_S \frac{\partial f}{\partial \phi_x}\eta\, dy - \iint \frac{\partial}{\partial x}\frac{\partial f}{\partial \phi_x}\eta\, dx\, dy \quad (4.42)$$

Similarly,

$$\iint_R \frac{\partial f}{\partial \phi_y}\eta_y\, dx\, dy = \oint_S \frac{\partial f}{\partial \phi_y}\eta\, dx - \iint \frac{\partial}{\partial x}\frac{\partial f}{\partial \phi_y}\eta\, dx\, dy \quad (4.43)$$

Substituting (4.42) and (4.43) into equation (4.41) yields

$$\frac{d\mathcal{F}}{d\epsilon} = \iint_R \left(\frac{\partial}{\partial x}\frac{\partial f}{\partial \phi_x} + \frac{\partial}{\partial y}\frac{\partial f}{\partial \phi_y} - \frac{\partial f}{\partial \phi} \right) \eta\, dx\, dy$$
$$- \oint_S \left(\frac{\partial f}{\partial \phi_x}dy + \frac{\partial f}{\partial \phi_y}dx \right) \eta(x, y) = 0 \quad (4.44)$$

Variational and Galerkin Methods

Removing $\eta(x, y)$ from both terms of equation (4.44), there is

$$\iint_R \left(\frac{\partial}{\partial x} \frac{\partial f}{\partial \phi_x} + \frac{\partial}{\partial y} \frac{\partial f}{\partial \phi_y} - \frac{\partial f}{\partial \phi} \right) dx\, dy$$
$$- \oint_S \left(\frac{\partial f}{\partial \phi_x} dy + \frac{\partial f}{\partial \phi_y} dx \right) = 0 \tag{4.45}$$

We have thus split the original area integral of $\frac{d\mathcal{F}}{d\epsilon}$ over the domain of integration R including the boundary S into two integrals, one of which is for the region R excluding the boundary and the other a boundary integral on S.

Equation (4.45) represents the condition for the extremum (i.e., maximum or minimum) of the functional $\mathcal{F}$. In order to show that the functional is a minimum, we must establish the relation

$$\frac{\partial^2 \mathcal{F}}{\partial \epsilon^2} > 0 \tag{4.46}$$

In most cases condition (4.46) is met when the first variation of the functional is zero. This is true for quadratic functionals and quasi-linear functionals. Thus the necessary condition for an extremum is also the sufficient condition for a minimum in the preceding cases. Therefore, we shall concern ourselves with only the first variation of the functional henceforth.

Returning to equation (4.44), if we set the boundary integral to zero for any arbitrary variation in the function $\phi(x, y)$ given by $\delta\phi = \epsilon\eta(x, y)$, the condition for a functional minimum becomes

$$\iint_R \left(\frac{\partial}{\partial x} \frac{\partial f}{\partial \phi_x} + \frac{\partial}{\partial y} \frac{\partial f}{\partial \phi_y} - \frac{\partial f}{\partial \phi} \right) \eta\, dx\, dy = 0 \tag{4.47}$$

Hence the integrand of equation (4.47) must vanish, yielding the result

$$\frac{\partial}{\partial x} \frac{\partial f}{\partial \phi_x} + \frac{\partial}{\partial y} \frac{\partial f}{\partial \phi_y} - \frac{\partial f}{\partial \phi} = 0 \tag{4.48}$$

which is the well-known Euler equation.

The effect of setting the boundary integral to zero will be examined in a later section. For the present we can conclude that the functional is a minimum (extremum more generally) if the Euler equation is satisfied. In general, this condition is met if the Euler equation of the functional is the same as the partial differential equation of the field problem.

Example

Show that the Euler equation of the functional

$$\iint \nu \left[\left(\frac{\partial A}{\partial x} \right)^2 + \left(\frac{\partial A}{\partial y} \right)^2 \right) - 2A \cdot J \right] dx\, dy \qquad (4.49)$$

yields the Poisson's equation for the two-dimensional magnetostatic problem.

Solution

Let

$$\mathcal{F} = \iint f\, dx\, dy \qquad (4.50)$$

where f, the integrand, is

$$\nu \left[\left(\frac{\partial A}{\partial x} \right)^2 + \left(\frac{\partial A}{\partial y} \right)^2 \right) - 2A \cdot J \right] \qquad (4.51)$$

A is the z component of the magnetic vector potential, J is the current density, and ν is the reciprocal permeability (or reluctivity).

Substituting for f from (4.51) into (4.48),

$$\frac{\partial}{\partial x} \left(\frac{\partial}{\partial A_x} \left(\nu \left(\left(\frac{\partial A}{\partial x} \right)^2 + \left(\frac{\partial A}{\partial y} \right)^2 \right) \right) - 2A \cdot J \right)$$

$$+ \frac{\partial}{\partial y} \left(\frac{\partial}{\partial A_y} \left(\nu \left(\left(\frac{\partial A}{\partial x} \right)^2 + \left(\frac{\partial A}{\partial y} \right)^2 \right) \right) - 2A \cdot J \right)$$

Variational and Galerkin Methods

$$-\frac{\partial}{\partial A}\left(\nu\left(\left(\frac{\partial A}{\partial x}\right)^2 + \left(\frac{\partial A}{\partial y}\right)^2\right) - 2A \cdot J\right) = 0 \quad (4.52)$$

where $A_x = \frac{\partial A}{\partial x}$ and $A_y = \frac{\partial A}{\partial y}$. Performing the necessary differentiation in (4.52) and transposing terms

$$\frac{\partial}{\partial x}\nu\frac{\partial A}{\partial x} + \frac{\partial}{\partial y}\nu\frac{\partial A}{\partial y} + J = 0 \quad (4.53)$$

It is evident that (4.53) is the two-dimensional Poisson equation for the magnetostatic problem with the vector potential $\vec{A} = A_z(x, y)$.

4.5 DERIVATION OF THE ENERGY-RELATED FUNCTIONAL

In the previous section, the variational method was discussed and the necessary condition for functional minimization was derived. It was shown that the function that minimizes the energy-related functional yields the true solution to the field problem. In this section, we shall derive the functional for a linear Poisson equation and then extend the same to nonlinear and time-varying fields. Two methods of deriving the functional — the residual excitation method and the Poynting vector method — will be discussed and the functional formulation will be generalized.

4.5.1 Residual Excitation Method

Consider the linear Poisson equation

$$\nu\nabla^2\phi_0 = f_0 \quad (4.54)$$

where ϕ_0 is the true potential solution, f_0 is the forcing function, and ν is the material parameter, a constant scalar quantity in this case.

If ϕ is the approximate solution, then a residue will be obtained such that

$$\nu \nabla^2 \phi = f_0 - R \tag{4.55}$$

Expressing the right-hand side of (4.55) as an equivalent source density f_1, we have from (4.55)

$$R = f_0 - f_1 = \nu \nabla^2 \phi_0 - f_1 \tag{4.56}$$

Premultiplying both sides of (4.56) by ϕ_0 and integrating the result over the entire region, we obtain

$$\mathcal{F} = \int \phi_0 R \, dV = \int (\phi_0 \nu \nabla^2 \phi_0 - \phi_0 f_1) \, dV \tag{4.57}$$

The integral on the right-hand side of (4.57) yields the residual energy in the system and shall be termed the functional $\mathcal{F}$. It is then required to show that minimization of this energy-related functional yields the true solution to the field problem defined by (4.54). The integral on the right-hand side of equation (4.57) can be transformed by vector identities as follows:

$$\int \nu \phi_0 \nabla^2 \phi_0 \, dV = \int \nabla \cdot (\nu \phi_0 \nabla \phi_0) \, dV - \int \nu |\nabla \phi_0|^2 \, dV \tag{4.58}$$

Substituting (4.58) into (4.57),

$$\mathcal{F} = -\int \phi_0 f_1 \, dV + \int \nabla \cdot (\nu \phi_0 \nabla \phi_0) \, dV - \int \nu |\nabla \phi_0|^2 \, dV \tag{4.59}$$

By the divergence theorem, the second integral of (4.59) is transformed into a surface integral so that

$$\mathcal{F} = -\int \phi_0 f_1 \, dV + \oint (\nu \phi_0 \nabla \phi_0) \, dS - \int \nu |\nabla \phi_0|^2 \, dV \tag{4.60}$$

Variational and Galerkin Methods

For the functional of (4.60) to yield the true solution to the field problem when minimized, its Euler equation must result in the partial differential equation (4.54). This can be verified, for example, in two dimensions, so that

$$\frac{\partial}{\partial x}\frac{\partial f}{\partial \phi_x} + \frac{\partial}{\partial y}\frac{\partial f}{\partial \phi_y} - \frac{\partial f}{\partial \phi} = 0 \tag{4.61}$$

Substituting the integrand of (4.60) for f in (4.61) and setting the surface integral to zero, we have

$$-\frac{\partial}{\partial x}\frac{\partial}{\partial \phi_{0x}}\nu|\nabla\phi_0|^2 - \frac{\partial}{\partial y}\frac{\partial}{\partial \phi_{0y}}\nu|\nabla\phi_0|^2 + \frac{\partial(\phi_0 f_1)}{\partial \phi_0} = 0 \tag{4.62}$$

where $\phi_{0x} = \frac{\partial \phi_0}{\partial x}$ and $\phi_{0y} = \frac{\partial \phi_0}{\partial y}$.

Performing the required differentiation in (4.62)

$$2\nu\nabla^2\phi_0 - f_1 = 0 \tag{4.63}$$

However, the Poisson equation for the field problem is given by

$$\nu\nabla^2\phi_0 - f_0 = 0 \tag{4.64}$$

Comparing (4.63) and (4.64), we conclude that

$$f_1 = 2 f_0 \tag{4.65}$$

Substituting the value of f_1 in the expression for the functional (4.60), we have

$$\mathcal{F} = \oint_S (\nu\phi_0 \nabla\phi_0) \cdot dS - \int_V (\nu|\nabla\phi_0|^2 + 2 f_0\phi_0)\, dV \tag{4.66}$$

4.5.2 Generalization of the Functional Formulation for Poisson's Equation

Referring to the previous section, we obtained the functional as

$$\mathcal{F} = \oint_S (\nu\phi_0\nabla\phi_0) \cdot dS - \int_V (\nu|\nabla\phi_0|^2 \, dV - \int_V 2f_0\phi_0 \, dV \tag{4.67}$$

But

$$\nabla \cdot (\phi_0\nabla\phi_0) = |\nabla\phi_0|^2 + \phi_0\nabla^2\phi_0 \tag{4.68}$$

Multiplying equation (4.68) throughout by ν and integrating over the volume, using the divergence theorem and transposing terms, we have

$$\int_V \nu\phi_0\nabla^2\phi_0 \, dV = -\int_V \nu|\nabla\phi_0|^2 \, dV + \oint_S (\nu\phi_0\nabla\phi_0) \cdot dS \tag{4.69}$$

Substituting (4.69) into (4.67)

$$\mathcal{F} = \int_V (\nu\phi_0\nabla^2\phi_0 - 2f_0\phi_0) \, dV \tag{4.70}$$

If we replace the term for the operator ∇^2 by $\mathcal{L}$, ϕ_0 by ψ, f_0 by f, and use the inner product notation we can write

$$\mathcal{F} = -2\langle\psi^T|f\rangle + \langle\psi^T|\mathcal{L}\psi\rangle \tag{4.71}$$

where

$$2\langle\psi^T|f\rangle = \int 2\psi^T f \, dV \tag{4.72}$$

Variational and Galerkin Methods

$$\langle \psi^T | \mathcal{L}\psi \rangle = \int \psi^T \mathcal{L}\psi \, dV \tag{4.73}$$

Here ψ^T is the transpose of ψ.

Effect of Setting the Surface Integral to Zero

The surface integral in (4.60) reduces to the form

$$\oint_S (v\psi_0 \nabla \psi_0) \cdot d\vec{S} = \oint_S v\psi_0 \frac{\partial \psi_0}{\partial n} \, dS \tag{4.74}$$

where n is the outward unit normal to the surface S.

- Case (a), $\frac{\partial \phi}{\partial n} = 0$; then the surface integral in (4.74) assumes a zero value, the so-called *homogeneous Neumann condition*
- Case (b), ψ_0 is specified; then there will be no variation in ψ_0 on the boundary and, therefore, the variation in the surface integral makes no contribution to the functional and its minimization.

4.5.3 Functional Formulation from Poynting's Vector for Magnetic Field Problems

The functional is a measure of the net energy from the volume domain flowing across the bounding surface. It is a scalar quantity, and its minimum value is the precondition for obtaining the true solution to scalar and vector potential field problems. For electromagnetic fields, Poynting's vector suggests itself as a possible approach to the functional formulation.

$$-\oint_S (E \times H) \cdot dS = \int_V J \cdot E \, dV + \frac{\partial}{\partial t} \int_V \left(\frac{\epsilon}{2} E^2 + \frac{\mu}{2} H^2 \right) dV \tag{4.75}$$

We shall neglect the energy term associated with the electrostatic field and also assume that displacement currents are absent. Therefore,

$$-\oint_S (E \times H) \cdot dS = \int_V J \cdot E \, dV + \frac{\partial}{\partial t} \int_V \frac{\mu}{2} H^2 \, dV \qquad (4.76)$$

Defining a vector potential A such that

$$B = \nabla \times A \qquad (4.77)$$

and using the relationship

$$\nabla \times E = -\frac{\partial B}{\partial t} \qquad (4.78)$$

we obtain

$$\nabla \times E = -\frac{\partial}{\partial t}(\nabla \times A) \qquad (4.79)$$

or

$$E = -\frac{\partial A}{\partial t} \qquad (4.80)$$

If we assume that the current density J does not vary with time as in magnetostatic problems or as in a DC current source,

$$\frac{\partial}{\partial t}(J \cdot A) = J \cdot \frac{\partial A}{\partial t} \qquad (4.81)$$

Variational and Galerkin Methods

Thus from equations (4.80) and (4.81)

$$\int_V J \cdot E \, dV = -\frac{\partial}{\partial t} \int_V J \cdot A \, dV \tag{4.82}$$

Substituting equation (4.82) into (4.76)

$$\oint_S (E \times H) \cdot dS = -\frac{\partial}{\partial t} \left(\int_V J \cdot A \, dV - \int_V \frac{\mu}{2} H^2 \, dV \right) \tag{4.83}$$

Equation (4.83) can be rewritten in terms of the reluctivity ν ($\nu = \frac{1}{\mu}$) so that

$$H = \nu B = \nu \nabla \times A \tag{4.84}$$

and

$$-\oint_S (E \times H) \cdot dS = -\frac{\partial}{\partial t} \left(\int_V J \cdot A \, dV - \int \frac{\nu}{2} (\nabla \times A)^2 \, dV \right) \tag{4.85}$$

Assuming that A has only a z-directed component A_z, we have for the two-dimensional problem

$$(\nabla \times A)^2 = |\nabla A|^2 = \left(\frac{\partial A}{\partial x}\right)^2 + \left(\frac{\partial A}{\partial y}\right)^2 \tag{4.86}$$

Substituting for $(\nabla \times A)^2$ from (4.86) into (4.85) and integrating both sides of equation (4.85) with respect to time, we have

$$\int \left(-\oint_S (E \times H) \cdot dS \right) dt = -\int_V J \cdot A \, dV + \int_V \frac{\nu}{2} \left(\left(\frac{\partial A}{\partial x}\right)^2 + \left(\frac{\partial A}{\partial y}\right)^2 \right) dV \tag{4.87}$$

which is the energy flowing out of the surface enclosing the volume of the field region. The true solution A for the field problem will be such as to minimize the energy flowing out of the volume. Therefore, the expression of equation (4.87) is the required linear functional that is identical to equation (4.49).

4.5.4 The Nonlinear Energy-Related Functional for the Magnetostatic Field Problem

In this section we shall derive the nonlinear energy-related functional for the magnetic field problem based on the procedure described in the previous section. Green's identities for vector and scalar quantities are used for separating the volume integrals for the respective operators into volume and surface integrals. A discussion of the principal and natural boundary conditions associated with the surface integrals is presented. The procedure for attaining stationarity of the energy-related functional by means of an approximation to the field solution is described. A discussion of the gauge conventions used with magnetic field problems is included.

The magnetostatic field problem including magnetic saturation of the iron parts was modeled by the nonlinear curl–curl equation described in Chapter 1 as

$$\nabla \times \nu \nabla \times A = J \quad (4.88)$$

In addition to equation (4.88), the vector potential satisfies the Coulomb gauge. The resulting equation is

$$\nabla \cdot A = 0 \quad (4.89)$$

By the method of variational calculus, as has already been shown, the true solution to the field problem represented by partial differential equations is obtained by minimizing the associated energy-related functional.

The expression for the energy-related functional is given by the following expression similar to equation (4.71).

Substituting for ϕ^T, the vector $[A, \nabla \cdot A]$, and for $\mathcal{L}\phi$, the system of equations (4.88) and (4.89) into (4.71), and expanding the inner products in the manner described in equations (4.72) and (4.73), one obtains

$$\mathcal{F} = \int_V (A \cdot \nabla \times \nu \nabla \times A + \nu(\nabla \cdot A)^2 - 2A \cdot Js) \, dV \quad (4.90)$$

Variational and Galerkin Methods

where A is the vector potential function, ν is the reluctivity, and Js is the source current density.

The term ν is associated with $\nabla \cdot A$ to make the term $\nu(\nabla \cdot A)^2$ of the same dimensionality as the rest of equation (4.90). It is also advantageous to use a multiplier λ for $(\nabla \cdot A)^2$, much in the same way as a penalty function.

From vector identities the first term in the integral of (4.90) may be transformed as

$$A \cdot \nabla \times \nu \nabla \times A = \nu(\nabla \times A)^2 - \nabla \cdot (A \times \nu \nabla \times A) \tag{4.91}$$

Integrating both sides of (4.91) over the volume of the field region,

$$\int_V A \cdot \nabla \times \nu \nabla \times A \, dV = \int_V \nu(\nabla \times A)^2 \, dV - \int_V \nabla \cdot (A \times \nu \nabla \times A) \, dV \tag{4.92}$$

Using the divergence theorem, the second volume integral in (4.92) is transformed as a surface integral, so that

$$\int_V \nabla \cdot (A \times \nu \nabla \times A) \, dV = \oint_S A \cdot (n \times \nu \nabla \times A) \, dS \tag{4.93}$$

Substituting equations (4.92) and (4.93) for the first integral in equation 4.90, we obtain

$$\mathcal{F} = \int_V (\nu(\nabla \times A)^2 + \lambda\nu(\nabla \cdot A)^2 - 2A \cdot J) \, dV - \oint_S A \cdot (n \times \nu \nabla \times A) \, dS \tag{4.94}$$

The surface integral, when set to zero, imposes the homogeneous Neumann boundary condition automatically. This ensures tangential continuity of the H field and normal component of the B field, which are natural boundary and interface conditions for the field problem. On the other hand, if A is specified (Dirichlet boundary condition), this integral identically vanishes during the functional minimization process. Thus

$$\mathcal{F} = \int_V (\nu(\nabla \times A)^2 + \lambda \nu (\nabla \cdot a)^2 - 2A \cdot J) \, dV \qquad (4.95)$$

The first volume integral in equation (4.95) may be recognized as twice the energy stored in the nonlinear magnetic field. Therefore,

$$\frac{1}{2} \int_V \nu (\nabla \times A)^2 \, dV = \int_V \left(\int_0^B \nu B \cdot dB \right) dV \qquad (4.96)$$

Equations (4.95) and (4.96) yield

$$\mathcal{F} = \int_V 2 \left(\left(\int_0^B \nu B \cdot dB \right) - J \cdot A \right) dV + \int_V \lambda \nu (\nabla \cdot A)^2 \, dV \qquad (4.97)$$

If the material has orthotropic characteristics, equation (4.97) becomes, for the 3D case,

$$\mathcal{F} = \int_V 2((\nu_x B_x^2 + \nu_y B_y^2 + \nu_z B_z^2) - J \cdot A) \, dV + \int_V \lambda \nu (\nabla \cdot A)^2 \, dV \qquad (4.98)$$

4.5.5 Energy-Related Functional for the Diffusion Equation

The electromagnetic field problem including eddy current effects and magnetic saturation is modeled by the nonlinear diffusion equation in terms of a magnetic vector potential A, source current density J_s, and scalar potential function, ϕ as

$$\nabla \times \nu \nabla \times A + \sigma \frac{\partial A}{\partial t} - \sigma \nabla \phi = J_s \qquad (4.99)$$

Variational and Galerkin Methods

In addition to the equation (4.99), the vector and scalar potentials are related by a zero divergence condition on the current density vector. The resulting equation is

$$\nabla \cdot \sigma \frac{\partial A}{\partial t} - \nabla \cdot \sigma \nabla \phi = 0 \qquad (4.100)$$

Integrating equation (4.100) over time, assuming σ is constant with position,

$$\sigma \nabla \cdot A - \sigma \int \nabla^2 \phi \, dt = 0 \qquad (4.101)$$

Equations (4.99) and (4.101) fully define the time-dependent eddy current problem. For two-dimensional problems with a single component of vector potential, the divergence of A is automatically zero, and therefore there is no need for a scalar potential function, ϕ.

As for the magnetostatic problem, the functional for the generalized nonlinear diffusion equation (4.99) and gauge condition (4.101) are obtained from equation (4.71) by substituting for ψ, f, and $\mathcal{L}\psi$ from equations 4.99 and 4.101, with the result

$$\begin{aligned}\mathcal{F} = &\int_V \left(A \cdot \nabla \times \nu \nabla \times A + \sigma A \cdot \frac{\partial A}{\partial t} \right.\\ &\left. - \sigma A \cdot \nabla \phi + \sigma \phi \nabla \cdot A - \int \sigma \phi \nabla^2 \phi \, dt - 2 A \cdot J_s \right) dV \end{aligned} \qquad (4.102)$$

Using vector identities (Green's theorem and the divergence theorem), substituting ψ for the term containing the integral over time in (4.102), replacing the integrals containing $\nabla \times \nu \nabla \times A$ and Laplace's equation in ϕ by volume and surface integrals, and rearranging terms after considerable algebra, we obtain the energy-related functional for the nonlinear time-dependent diffusion equation as

$$\begin{aligned}\mathcal{F} = &\int_V \left(\nu (\nabla \times A)^2 + \sigma A \cdot \frac{\partial A}{\partial t} - \sigma A \cdot \nabla \phi + \sigma \phi \nabla \cdot A \right.\\ &\left. + \sigma \nabla \phi \cdot \nabla \phi - 2 A \cdot J_s \right) dV - \oint_S A \cdot (n \times \nu \nabla \times A) \, dS \\ &- \oint \phi \frac{\partial \phi}{\partial n} \, dS \end{aligned} \qquad (4.103)$$

Setting the surface integral terms to zero ensures, as for the magnetostatic case, that natural boundary (Dirichlet and Neumann) conditions hold on the exterior boundary.

4.5.6 Functional for the Steady-State Linear Time-Harmonic Diffusion Equation

The steady-state diffusion equation and the related functional are obtained by setting the vector potential A, the scalar potential ϕ, and the source current density J_s as harmonic functions of time. Thus

$$\begin{aligned} A &= |A|e^{j\omega t} \\ \phi &= |\phi|e^{j\omega t} \\ J_s &= |J_s|e^{j\omega t} \end{aligned} \tag{4.104}$$

Substituting for A, ϕ, and J_s from equation (4.104) into equations (4.99) and (4.101), one obtains the following set of equations for the steady-state diffusion equation and the divergence condition on A.

$$\nabla \times \nu \nabla \times A + j\omega \sigma A - \sigma \nabla \phi = J_s \tag{4.105}$$

$$\sigma \nabla \cdot A - \frac{\sigma}{j\omega} \nabla \cdot \nabla \phi = 0 \tag{4.106}$$

From the generalized expression for the functional of equation (4.71), we have

$$\mathcal{F} = \langle \psi^T | \mathcal{L} \psi \rangle - 2 \langle \psi^T | f \rangle \tag{4.107}$$

Making the following substitutions from equations (4.105) and (4.106) in equation (4.107)

Variational and Galerkin Methods

$$\psi = \begin{pmatrix} A \\ \phi \end{pmatrix}$$
$$f = J_s \qquad (4.108)$$

$$\mathcal{L}\psi = \begin{pmatrix} \nabla \times \nu \nabla \times A + j\omega\sigma A - \sigma\nabla\phi \\ \sigma\nabla \cdot A - \frac{\sigma}{j\omega}\nabla \cdot \nabla\phi \end{pmatrix} \qquad (4.109)$$

the functional for the steady-state diffusion equation is obtained as

$$\mathcal{F} = \int_V \left(A \cdot \nabla \times \nu \nabla \times A + j\omega\sigma A^2 - \sigma A \cdot \nabla\phi \right.$$
$$\left. + \sigma\phi\nabla \cdot A - \frac{\sigma\phi}{j\omega}\nabla \cdot \nabla\phi - 2A \cdot J_s \right) dV = 0 \qquad (4.110)$$

Using the vector identities

$$A \cdot \nabla \times \nu \nabla \times A = \nu(\nabla \times A)^2 - \nabla \cdot [A \times \nu\nabla \times A] \qquad (4.111)$$

$$-\frac{j\sigma}{\omega}\phi\nabla \cdot \nabla\phi = \frac{j\sigma}{\omega}(\nabla\phi)^2 - \frac{j\sigma}{\omega}\nabla \cdot (\phi\nabla\phi) \qquad (4.112)$$

and substituting them in the expression for the functional of equation (4.110), we obtain

$$\mathcal{F} = \int_V \left(\nu(\nabla \times A)^2 + j\omega\sigma A^2 - \sigma A \cdot \nabla\phi - \sigma\phi\nabla \cdot A + \frac{j\sigma}{\omega}(\nabla\phi)^2 - 2A \cdot J_s \right) dV$$
$$- \int_V \nabla \cdot (A \times \nu\nabla \times A) \, dV - \int_V \frac{j\sigma}{\omega}\nabla \cdot (\phi\nabla\phi) \, dV \qquad (4.113)$$

The last two volume integrals on the right-hand side of equation (4.113) can be transformed as surface integrals by the divergence theorem and, therefore,

$$\mathcal{F} = \int_V \left(\nu(\nabla \times A)^2 + j\omega\sigma A^2 - \sigma A \cdot \nabla\phi - \sigma\phi\nabla \cdot A + \frac{j\sigma}{\omega}(\nabla\phi)^2 - 2A \cdot J_s \right) dV$$
$$- \oint_S A \cdot (n \times \nu\nabla \times A \, dS - \oint_S \frac{j\sigma}{\omega}\phi\nabla\phi \cdot dS \qquad (4.114)$$

Equation (4.114) represents the functional for the generalized steady-state diffusion equation. Specialization for the curl–curl and diffusion equations for two-dimensional problems will be discussed in applications to be presented in later chapters.

We shall, in the following section, discuss the application of the variational method just described for solution techniques such as Ritz's method and two-dimensional finite elements.

4.6 RITZ'S METHOD

This solution technique is based on the variational principle of minimizing or making stationary an energy-related functional appropriate to the partial differential equation to be solved. It consists of assuming the form of the unknown solution in terms of trial functions and unknown adjustable parameters. Then, from among a family of trial functions, those functions are chosen that make the functional stationary. As has already been stated, this procedure is implemented by substituting trial functions into the functional and setting the first variation of the latter to zero with respect to each of the adjustable parameters.

This process results in a set of simultaneous equations that are required to be solved for evaluating the parameters. This method gives the best solution from the family of trial functions, and the accuracy depends on the choice of the latter. The trial solutions themselves are defined over the entire solution domain, and they usually satisfy all the boundary conditions, or at least a few of them. The accuracy of the method can be improved by increasing the number of trial solutions and the adjustable parameters [49].

Ritz's method is the forerunner of the finite element method and it is limited to homogeneous isotropic media. Because the trial functions are defined over the entire domain, only simple geometrical shapes can be modeled by this technique.

Variational and Galerkin Methods

We shall illustrate the implementation of the procedure with reference to Poisson's equation as follows.

$$k\nabla^2 \psi = -g \tag{4.115}$$

where ψ is a vector or scalar potential function, k a material constant, and g the forcing function.

We are required to solve equation (4.115) satisfying principal and natural boundary conditions (principal means applied or Dirichlet boundary conditions, and natural refers to Neumann boundary conditions). We shall assume that g is a continuous function in the domain.

The functional may be defined as

$$\mathcal{F} = \int_V (k(\nabla \psi)^2 - 2g\psi)\, dV - \oint_\Gamma \psi \frac{\partial \psi}{\partial n}\, d\Gamma \tag{4.116}$$

We must find $\psi(x)$ that renders the functional of equation (4.116) stationary.

The desired solution may be assumed to be a linear combination of trial functions such that

$$\psi = c_1\theta_1 + c_2\theta_2 + \cdots + c_n\theta_n \tag{4.117}$$

where $c_1, c_2, ..., c_n$ are the unknown parameters to be determined, which render the functional stationary, and $\theta_1, \theta_2, ..., \theta_n$ are the trial functions. We then substitute the expansion for ψ from equation (4.117) in the functional (4.116), and set the first variation of the latter to zero with respect to each of the parameters.

Therefore, we set

$$\frac{\partial \mathcal{F}}{\partial c_1} = 0, \quad \frac{\partial \mathcal{F}}{\partial c_2} = 0, \quad ..., \quad \frac{\partial \mathcal{F}}{\partial c_n} = 0 \tag{4.118}$$

This procedure yields n simultaneous equations that must be solved to determine the n parameters $c_1, c_2, ..., c_n$

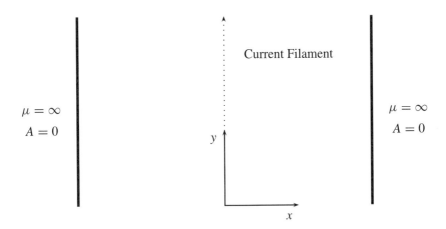

Figure 4.2   Current Filament with Line Current Density J

A One-Dimensional Example

Figure 4.2 shows a semi-infinite line current source (in the y-direction) placed between an infinitely permeable and infinitesimally thin magnetic shield (also semi-infinite in the y direction).

We are required to solve for A that satisfies the one-dimensional Poisson's equation

$$\nu \frac{d^2 A}{dx^2} = -J \qquad (4.119)$$

subject to the boundary conditions $A(-d) = 0$, $A(+d) = 0$.

The functional that has to be minimized to find A is of the form

$$\mathcal{F} = \int_{-d}^{d} \left(\frac{\nu}{2} \left(\frac{dA}{dx} \right)^2 - A \cdot J \right) dx \qquad (4.120)$$

Variational and Galerkin Methods

Because the field problem is symmetric, we need to solve for A only over half of the domain. Therefore, the functional becomes

$$\mathcal{F} = \int_0^d \left(\frac{\nu}{2} \left(\frac{dA}{dx} \right)^2 - A \cdot J \right) dx \qquad (4.121)$$

subject to $A(d) = 0$

We shall assume the solution to be of the form

$$A = c_1 \theta_1 + c_2 \theta_2 + \cdots + c_n \theta_n \qquad (4.122)$$

satisfying the boundary condition $A(x = d) = 0$ regardless of the choice of the c's.

Using one-dimensional polynomial trial functions, we can write

$$A = (x - d)(c_1 + c_2 x + c_3 x^2 + \cdots + c_n x^{n-1}) \qquad (4.123)$$

It is evident that the boundary condition $A(d) = 0$ is satisfied in equation (4.123) irrespective of the values of $c_1, c_2, \ldots, c_n$. We may truncate the polynomial up to only two terms such that

$$A = (x - d)(c_1 + c_2 x) \qquad (4.124)$$

Differentiating A with respect to x

$$\frac{dA}{dx} = c_1 + c_2(2x - d) \qquad (4.125)$$

Substituting equations (4.124) and (4.125) in the functional of (4.121).

$$\mathcal{F} = \int_0^d \left(\frac{v}{2}(c_1 + c_2(2x - d))^2 - ((x - d)(c_1 + c_2 x))J \right) dx \qquad (4.126)$$

Differentiating (4.126) with respect to c_1 and c_2 successively and setting the result to zero, we obtain

$$\frac{\partial \mathcal{F}}{\partial c_1} = \int_0^d (v(c_1 + c_2(2x - d)) - (x - d)J) \, dx = 0 \qquad (4.127)$$

$$\frac{\partial \mathcal{F}}{\partial c_2} = \int_0^d (v(c_1 + c_2(2x - d))(2x - d) - x(x - d)J) \, dx = 0 \qquad (4.128)$$

This set of equations reduces to the matrix form

$$\begin{pmatrix} \frac{\partial \mathcal{F}}{\partial c_1} \\ \frac{\partial \mathcal{F}}{\partial c_2} \end{pmatrix} = v \begin{pmatrix} d & 0 \\ 0 & \frac{d^3}{3} \end{pmatrix} - J \begin{pmatrix} -\frac{d^2}{2} \\ -\frac{d^3}{6} \end{pmatrix} = 0 \qquad (4.129)$$

Solving for c_1 and c_2, we have

$$c_1 = -\frac{\mu J d}{2}$$
$$c_2 = -\frac{\mu J}{2} \qquad (4.130)$$

Therefore the required solution is of the form

$$A = -\frac{\mu J}{2}(d + x)(x - d) = \frac{\mu J}{2}(d^2 - x^2) \qquad (4.131)$$

Variational and Galerkin Methods

Exact Solution of the Differential Equation

$$\nu \frac{d^2 A}{dx^2} = -J \quad (4.132)$$

or

$$\frac{d^2 A}{dx^2} = -\mu J \quad (4.133)$$

Integrating equation (4.133) with respect to x yields

$$\frac{dA}{dx} = -\mu J x + C \quad (4.134)$$

Once again integrating with respect to x,

$$A = -\mu J \frac{x^2}{2} + Cx + D \quad (4.135)$$

At $x = d$, $A = 0$. Therefore

$$D = \frac{\mu J d^2}{2} - Cd \quad (4.136)$$

Further

$$\frac{dA}{dx} = 0 \quad (4.137)$$

at $x = 0$ and hence $C = 0$ so

$$A = \frac{\mu J}{2}(d^2 - x^2) \quad (4.138)$$

Equation (4.138) for the vector potential is the same as (4.131) obtained by Ritz's method.

4.7 THE WAVE EQUATION

In antenna and wave propagation, the following Maxwell's equations hold. Although these are covered in a separate chapter, we shall specialize them to certain applications to illustrate the variational formulation and solution.

$$\nabla \times H = J + \frac{\partial D}{\partial t} \qquad (4.139)$$

$$\nabla \times E = -\frac{\partial B}{\partial t} \qquad (4.140)$$

$$\nabla \cdot D = \rho \qquad (4.141)$$

$$\nabla \cdot B = 0 \qquad (4.142)$$

The associated constitutive relations are

$$J = \sigma E \qquad (4.143)$$

$$D = \epsilon E \qquad (4.144)$$

$$B = \mu H \qquad (4.145)$$

For the present let us assume that H has only a z-directed component H_z as required for modeling the TE or transverse electric modes in a waveguide.

Taking curls on both sides of the first of the set of equations (4.139) through (4.142) and using the constitutive relationships of equations (4.143) through (4.145),

$$\nabla \times \nabla \times H = \sigma \nabla \times E + \epsilon \frac{\partial}{\partial t} \nabla \times E \qquad (4.146)$$

Variational and Galerkin Methods

Substituting the second equation of the set (4.139) — (4.142) in the first, we have

$$\nabla \times \nabla \times H = -\sigma\mu \frac{\partial H}{\partial t} - \mu\epsilon \frac{\partial^2 H}{\partial t^2} \qquad (4.147)$$

Expanding the left-hand side of (4.147) and setting the divergence of H to zero (because $H = H_z$ and is invariant along z)

$$\nabla \cdot \nabla H_z = \sigma\mu \frac{\partial H_z}{\partial t} + \mu\epsilon \frac{\partial^2 H_z}{\partial t^2} \qquad (4.148)$$

Assuming H_z to be time periodic (i.e., $H_z = H_z e^{j\omega t}$) and the conductivity σ to be zero, one obtains the homogeneous wave equation

$$\nabla \cdot \nabla H_z + \omega^2 \mu\epsilon H_z = 0 \qquad (4.149)$$

or

$$\nabla \cdot \nabla H_z + \lambda^2 H_z = 0 \qquad (4.150)$$

where $\lambda = \sqrt{\omega^2 \mu\epsilon}$ is called the cutoff wave number and is determined by the problem and the applicable boundary conditions. The quantity $\frac{1}{\lambda}$ has the dimension of length and represents the wavelength of a plane electromagnetic wave in an unbounded dielectric medium at the cutoff frequency of the waveguide. The quantity λ is also related to the linear dimension of the waveguide cross section.

The variational expression for this problem will be of the form

$$\mathcal{F}(\phi) = \int_V \frac{1}{2}((\nabla\phi)^2 - \lambda^2 \phi^2)\, dV - \oint_\Gamma \phi \frac{\partial \phi}{\partial n}\, d\Gamma \qquad (4.151)$$

where $\phi = H_z(x, y, z)$

Minimization of $\mathcal{F}$ in equation 4.151 with respect to unknown nodal potentials in the discretized field region yields the solution to the problem subject to the applied boundary conditions. It must also be noted that omitting the surface integral in (4.151) ensures homogeneous Neumann boundary conditions and also interelement continuity.

4.7.1 Vector Helmholtz Equation

In a number of applications, the scalar potential function ϕ may not fully describe the field problem. It is, therefore, necessary to express the field quantities by the electric field E or the magnetic field H. Alternatively, the electrostatic flux density D or the magnetic induction B may be used to describe the field problem adequately. In all these cases, one must solve a vector wave equation (Helmholtz equation) of the form

$$\nabla \cdot \nabla R + \lambda^2 R = -Q \qquad (4.152)$$

where R is the solution vector to be determined and Q is the known vector representing sources.

For a homogeneous medium, if we use the vector identity

$$\nabla \times \nabla \times R = \nabla \nabla \cdot R - \nabla \cdot \nabla R \qquad (4.153)$$

we can replace the first term in equation (4.152) such that

$$\nabla \nabla \cdot R - \nabla \times \nabla \times R + \lambda^2 R = -Q \qquad (4.154)$$

The variational expression for (4.154) is then given by the functional

$$\mathcal{F}(R) = \int_V \left((\nabla \times R) \cdot (\nabla \times R) - R \nabla \nabla \cdot R - \lambda^2 R \cdot R - 2Q \cdot R \right) dV - \oint_\Gamma R \cdot (n \times \nabla \times R) \, d\Gamma \qquad (4.155)$$

The functional is then minimized in the discretized space with respect to unknown nodal potentials after setting

$$\nabla \nabla \cdot R = 0 \qquad (4.156)$$

Variational and Galerkin Methods

As in the scalar case, the boundary conditions cause the surface integral term in (4.155) to vanish.

Alternative variational formulations of the scalar and vector Helmholtz equation have been described at length by Hammond [50] using the virtual work method.

4.8 VARIATIONAL METHOD FOR INTEGRAL EQUATIONS

In many engineering problems, an alternative to solving partial differential equations is the method of integral equations. In the latter method an elemental solution known as Green's function is made use of and the solution to the field problem is directly sought by integration of the source distribution that is known. In cases in which only boundary conditions are stated, the field is obtained by first solving the integral equation for the source distribution required to sustain the given boundary conditions.

Let us consider the partial differential equation for the electrostatic field produced by a charge.

$$-\nabla \cdot \epsilon \nabla \phi = \rho \tag{4.157}$$

where ϕ is the scalar potential (volts), ϵ is the permittivity as a function of position, and ρ is the volumetric charge density.

Because the differential operator is both self-adjoint and positive definite, we may apply the following procedure for the functional (which has already been discussed in great detail in the earlier part of this chapter).

$$\mathcal{F} = -\int_V (\phi \nabla \cdot \epsilon \nabla \phi - 2\phi \rho) \, dV \tag{4.158}$$

By Green's identity

$$\nabla \cdot \phi \epsilon \nabla \phi = \epsilon (\nabla \phi)^2 + \phi \nabla \cdot \epsilon \nabla \phi \tag{4.159}$$

Substituting (4.159) into (4.158), we have

$$\mathcal{F} = \int_V (\epsilon(\nabla\phi)^2 - \phi\rho)\, dV - 2\oint_\Gamma \epsilon\phi \frac{\partial\phi}{\partial n}\, d\Gamma \tag{4.160}$$

We may specify ϕ or $\frac{\partial\phi}{\partial n}$ on the boundary or each on part of the boundary. If ϕ is not specified, but the surface integral is set to zero, homogeneous Neumann boundary conditions are satisfied in a least squares sense.

Because the field solution ϕ may be described by a charge distribution over a conductor, the integral operator in place of the differential operator must also be positive definite. The integral form of the Poisson's equation (4.157) with a spatial charge distribution $\sigma(r')$ is given by the following expression.

$$I\sigma(r') = \int \frac{\sigma(r)}{\epsilon_0} G(r', r)\, dr = \phi(r') \tag{4.161}$$

where σ is the surface charge density, $G(r', r)$ is the free space Green's function, and r and r' are the source and observation points, respectively.

The integration in (4.161) is carried out over the entire free space region. For two- and three-dimensional problems, the free space Green's functions are given respectively as

$$G(r', r) = \frac{1}{2\pi} \ln(|r' - r|) \tag{4.162}$$

$$G(r', r) = \frac{1}{4\pi |r' - r|} \tag{4.163}$$

where $|r' - r|$ is the distance between the observation and source points r' and r, respectively, as shown in Figure 4.3. The preceding Green's functions include singularities and care must be taken in numerical computation of the same.

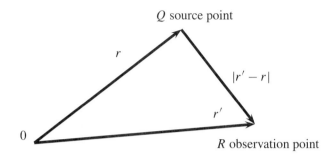

Figure 4.3 Source and Field Points for the Electrostatic Problem

Now we can write the functional using the operator equation in variational form

$$\mathcal{F} = \langle I\sigma, \sigma \rangle - 2\langle \sigma, f \rangle \quad (4.164)$$

Substituting for $I\sigma$ from (4.161) into (4.164) and expanding the inner products

$$\mathcal{F} = \iint \left(\frac{\sigma(r)}{\epsilon_0} G(r', r) \, dr \right) \sigma(r') \, dr' - 2 \int \sigma(r') f(r') \, dr' \quad (4.165)$$

If Dirichlet conditions are specified on the boundary, then (4.165) becomes

$$\mathcal{F} = \iint_s \left(\frac{\sigma(r)}{\epsilon_0} G(s', s) \, ds \right) \sigma(s') \, ds' - 2 \int_s \sigma(s') f(s') \, ds' \quad (4.166)$$

4.9 INTRODUCTION TO THE GALERKIN METHOD

The Galerkin method, like the variational method, is based on minimization of an integral expression [51] [52]. In the variational method we minimize a functional, which usually has an energy interpretation. In the Galerkin method, we minimize a weighted error on the domain. In this respect the Galerkin method is a special case of the more general method of weighted residuals (MWR). The weighted residual method is a more general and more universally applicable method than the variational approach because to apply it the variational principle need not be known. In fact, there are a number of practical cases in which the variational expression does not exist and the weighted residual method can be applied.

The method of weighted residuals can be stated simply as follows. Our objective is to approximate the solution to an operator equation

$$\mathcal{L}(x) = 0 \tag{4.167}$$

Equation (4.167) can be an ordinary differential equation or a partial differential equation of the elliptic, parabolic, or hyperbolic type. In electromagnetics, of course, we are most interested in these partial differential equations.

In the Galerkin method we first define a trial function

$$\hat{\phi}(x, t) = \hat{\phi}_0(x, t) + \sum_{j=1}^{N} \alpha_j(t) \zeta_j(x) \tag{4.168}$$

The following requirements are placed on the trial functions:

- The trial functions must satisfy the boundary conditions.
- The set of functions ζ_j are linearly independent and complete. In this case as $N \to \infty$ all possible solutions can be represented by equation (4.168).
- The ζ_j's are analytic. Typically, we choose polynomials.

We first substitute an approximate solution $\hat{x}$. Because $x \neq \hat{x}$ we obtain a *residual*

$$\mathcal{L}(\hat{x}) = R \tag{4.169}$$

Variational and Galerkin Methods

Of course, if the trial solution is the exact solution, then $R = 0$. The MWR now requires that the projection of the residual on a specified weighting function is zero over the domain of interest. The weighting functions depend on the same set of independent variables as the trial functions, ζ. The choice of the weighting function determines the type of MWR. Some popular choices are as follows.

- If we divide the region Ω into N subdomains and define the weighting functions such that $W_j = 1$ in Ω and $W_j = 0$ outside the domain then we have the *subdomain* method.

- If we choose the weighting function to be the Dirac delta function, $w_j(x) = \delta(x - x_j)$, then we have the *collocation* method.

- If we minimize the expression $\int R^2 dx$, then we have the least squares method. In this case $W_j(x) = \frac{\partial R}{\partial \alpha_j}$.

- If
$$W_j = \sum_{i=0}^{J} a_i x^j$$
we have the *method of moments*.

- If we choose the weighting function to have the same form as the shape function, we have the Galerkin method.

4.10 EXAMPLE OF THE GALERKIN METHOD

Consider the two-dimensional space shown in Figure 4.4.

We will solve Poisson's equation

$$\nabla^2 \phi = -\frac{\rho}{\epsilon} \qquad (4.170)$$

in the rectangular region subject to Dirichlet boundary conditions, $\phi = 0$ on the contour Γ. For this example we will consider a uniform charge density $\rho = 1.0$ and assume that the permittivity is that of free space.

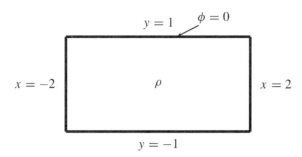

Figure 4.4 Rectangle with Uniform Charge Density

To apply the Galerkin method we will first approximate the solution with a set of trial functions that have the proper symmetry and satisfy the boundary conditions. This is, of course, not a unique set. In this case we choose

$$\hat{\phi} = \sum_{j=1}^{4} \alpha_j (4 - x^2)^j (1 - y^2)^j \qquad (4.171)$$

We see that this set of functions satisfies the boundary conditions at $x = \pm 2$ and $y = \pm 1$ and exhibits the necessary symmetry so that $\phi(x) = \phi(-x)$ and $\phi(y) = \phi(-y)$. We also note that the functions are twice differentiable. The choice of four terms is arbitrary in this case, and we would expect that more terms would yield a better result. The coefficients α_j will be found and equation (4.171) can then be used to find the potential at any point in the problem space.

We now substitute equation (4.171) into equation (4.170) and compute a residual R as

$$R = \frac{1}{\epsilon_0} + \sum_{j=1}^{4} \alpha_j f_j \qquad (4.172)$$

Variational and Galerkin Methods

where

$$\begin{align}
f_1 &= -2(4-x^2) - 2(1-y^2) \\
f_2 &= (12x^2 - 16)(y^4 - 2y^2 + 1) + (12y^2 - 4)(x^4 - 8x^2 + 16) \\
f_3 &= 3(1-y^2)^3(-10x^4 + 48x^2 - 32) \\
&\quad + 3(4-x^2)^3(-10y^4 + 12y^2 - 2) \\
f_4 &= 4(1-y^2)^4(-2(4-x^2)^3 + 12x^2(4-x^2)^2) \\
&\quad + 4(4-x^2)^4(-2(1-y^2)^3 + 12y^2(1-y^2)^2)
\end{align} \tag{4.173}$$

The values f_j are found by differentiating the polynomials in equation (4.171).

We now integrate the residual times a weighting function over the problem domain and set the integral to zero.

$$\int_{-1}^{1}\int_{-2}^{2} R \cdot w \, dx \, dy = 0 \tag{4.174}$$

In the Galerkin method the weighting function is chosen to have the same form as the test function so

$$w_i = (4-x^2)^i (1-y^2)^i \tag{4.175}$$

We now use this equation to find the α_i's. We obtain four equations with four unknowns. For $i = 1, ...4$,

$$\sum_{j=1}^{4} \alpha_j \int_{-1}^{1}\int_{-2}^{2} f_j(x,y) w_i(x,y) \, dx \, dy = \int_{-1}^{1}\int_{-2}^{2} -\frac{1}{\epsilon_0} w_i(x,y) \, dx \, dy \tag{4.176}$$

This results in a matrix equation

$$(S)(\alpha) = (b) \tag{4.177}$$

where the elements of S are given by

$$s_{ij} = \int_{-1}^{1}\int_{-2}^{2} f_j w_i \, dx \, dy \tag{4.178}$$

and the vector b has elements

$$s_{ij} = \int_{-1}^{1}\int_{-2}^{2} -\frac{1}{\epsilon_0} w_i \, dx \, dy \tag{4.179}$$

where the factor $-\frac{1}{\epsilon_0}$ is the right-hand side of equation (4.170). As an example, consider the (1, 2) term of the S matrix. To find this we evaluate

$$s_{12} = \int_{-1}^{1}\int_{-2}^{2} -2((4 - x^2) + (1 - y^2))(4 - x^2)^2(1 - y^2)^2 \, dx \, dy \tag{4.180}$$

This gives $s_{12} = -312.076$.[1]

By applying equation (4.178) we obtain

$$S = -\begin{pmatrix} 113.778 & 312.076 & 951.089 & 3074. \\ 312.076 & 1268. & 4611. & 16510 \\ 951.089 & 4611. & 18570. & 71120 \\ 3074. & 16510. & 71120. & 285600 \end{pmatrix} \tag{4.181}$$

Note that the matrix is symmetric. This will be the case when applying the Galerkin method to problems with self-adjoint operators, such as the Laplacian.

The b vector is

$$b = -\begin{pmatrix} 14.22 \\ 36.4 \\ 106.99 \\ 338.165 \end{pmatrix} \tag{4.182}$$

---

[1] The integral can be evaluated in closed form, but in this example numerical integration was used.

Variational and Galerkin Methods

Table 4.1 Comparison of Potential from Galerkin and Finite Element Results

x	y	Galerkin	Finite element
0	0	5.16E10	5.15E10
0.5	0	4.92E10	4.99E10
1.0	0	4.11E10	4.39E10
1.5	0	2.56E10	2.99E10
2.0	0	0.0	0.0

The solution is

$$\alpha = \begin{pmatrix} 0.148 \\ -0.013 \\ 0.002 \\ -0.000216 \end{pmatrix} \qquad (4.183)$$

This result was compared with a finite element solution of the problem at a number of points. These results are given in Table 4.1.

5
SHAPE FUNCTIONS

5.1 INTRODUCTION

The finite element method requires subdivision of the field region into subregions or subdomains and approximation of the solution in each subregion by a trial function. Then functional minimization or a weighted residual procedure is employed to determine the parameters of the interpolation function satisfying boundary conditions for the field problem. In this respect the finite element method is similar to Ritz's method discussed in Chapter 4. However, unlike Ritz's method, where a continuous approximating function is defined over the whole domain, in the finite element method, only a piecewise continuous function in each subregion or element is described. This trial function can be of any form that best fits the expected solution.

In closed form methods, one chooses an approximating function according to the geometry, 1D, 2D or 3D, whether the region is described in Cartesian, cylindrical, polar, or spherical coordinates; whether the problem solved is electrostatic, magnetostatic, eddy current or a wave problem; and finally whether the solution is sought in the frequency domain or in the time domain.

Examples of such trial functions are trigonometric (sine, cosine) functions, Bessel functions, Legendre polynomials, complex Bessel functions (Ber, Bei, Ker, Kei functions), and so on. In addition, boundary conditions have a significant impact on the choice of the approximating function. In contrast to closed form methods, the finite element method requires a relatively simple but general approximating function for wide and easy application.

The attractive feature of the finite element method is that even from the early days of its development, piecewise polynomial functions were chosen to satisfy the twin requirements of (1) simplicity and generality, and (2) the need to define the approximate solution as an interpolate of the nodal values of the state variable. There are, no doubt, other requirements such as completeness and compatibility conditions that these functions must satisfy, which will be discussed later on. Also, the simple but elegant choice of interpolation functions has been compounded by the complexity of engineering requirements such as potential or derivative continuity at nodes and interfaces, the need to couple finite element solutions with analytical and other numerical methods, and others. Nevertheless, polynomial approximating functions are by far the most used in finite element analysis. We shall, therefore, initiate our discussion of shape functions with simple first-order interpolatory functions and gradually increase the degree of complexity of the choice of these functions to high-order interpolation and for application to 2D and 3D problems, time-varying solutions, coupled problems, and so on.

In this chapter, we shall cover the following topics:

- Interpolatory functions and their choice
- Geometrical or shape functions and their properties
- High-order interpolation for regular and curved boundaries
- Isoparametric mapping
- C^1 continuous elements
- Singular elements
- Orthogonal polynomial basis functions

5.1.1 Interpolatory Functions

These are functions, as the name implies, that interpolate the value of the function within a region defined by end points or nodes. A simple and minimal requirement for the choice of an interpolatory function is that it tracks the behavior of the solution over the domain within acceptable accuracy. This provides a variety of choices for the interpolatory functions, such as first- or high-order functions, and the number of piecewise functions used to discretize (or model) the region of interest. We shall illustrate this process by the following example.

Shape Functions

Let us consider the following first-order differential equation for which the solution is required in the interval $(0, \frac{\pi}{2})$ for the state variable x. The boundary conditions are $y = 0$ at $x = 0$ and $y = 1$ at $x = \frac{\pi}{2}$.

$$\frac{d^2 y}{dx^2} = \sin x \quad \text{for } 0 < x < \frac{\pi}{2} \tag{5.1}$$

The analytical solution obtained by direct integration is

$$y = \frac{4x}{\pi} - \sin x \tag{5.2}$$

Suppose we choose a first-degree polynomial in x defined over a single element for approximating the function; then the solution will be of the form

$$y = a_0 + a_1 x \tag{5.3}$$

A simple collocation of the function values at the two end points of x yields

$$\begin{aligned} a_0 &= 0 \\ a_1 &= \frac{2}{\pi} \\ y &= \frac{2x}{\pi} \end{aligned} \tag{5.4}$$

The maximum error is obtained by differentiation of e, finding the difference of (5.2) and (5.4) with respect to x, and setting the result to zero, so that

$$\frac{de}{dx} = \frac{d}{dx}\left[\frac{4x}{\pi} - \sin x - \frac{2x}{\pi}\right] = \frac{2}{\pi} - \cos x = 0 \tag{5.5}$$

Thus

$$x = \cos^{-1} \frac{2}{\pi} \qquad (5.6)$$

and the maximum error $e = \frac{4x}{\pi} - \sin x - \frac{2x}{\pi}$

or

$$|e| = 0.21051 \qquad (5.7)$$

Next we shall approximate the function y in the specified region by a second-degree polynomial in x defined over a single element. Then

$$y = a_0 + a_1 x + a_2 x^2 \qquad (5.8)$$

By collocating at the end points as before, we have

$$a_0 = 0 \text{ and } a_1 + a_2 \frac{\pi}{2} = \frac{2}{\pi} \qquad (5.9)$$

We may determine a_1 and a_2 by one of several methods:

- Point collocation
- Least squares method
- Functional minimization (or weighted residual method)

In the case of point collocation, we need to choose an intermediate value of x in the interval $(0, \frac{\pi}{2})$ rather arbitrarily. For simplicity, we may select a point midway at $x = \frac{\pi}{4}$ with the exact solution at the midpoint obtained from equation (5.2) as 0.293. Then

$$a_1 \frac{\pi}{4} + a_2 \frac{\pi^2}{16} = 0.293 \qquad (5.10)$$

Shape Functions

Table 5.1 Error Computation

x	y from (5.2)	y from (5.13)	Error
0	0	0	0
$\pi/6$	0.167	0.1493	0.0177
$\pi/4$	0.293	0.293	0
$\pi/3$	0.4673	0.4827	0.0154
$\pi/2$	1	1	0

Also, from equation (5.9),

$$a_1 + a_2 \frac{\pi}{2} = \frac{2}{\pi} \qquad (5.11)$$

Solving (5.10) and (5.11) simultaneously, we obtain

$$a_1 = 0.1094986 \quad \text{and} \quad a_2 = 0.33557576 \qquad (5.12)$$

and

$$y = 0.1094986x + 0.33557576x^2 \qquad (5.13)$$

Table 5.1 shows the values of y obtained using equations (5.2) and (5.13).

So far we have used a single element over the whole domain using point collocation in each case. We shall now consider piecewise discretization of the region, say by two first-order elements. Here the order of the element refers to the highest degree of the polynomial. Again we shall use three points of collocation, 0, $\frac{\pi}{4}$, and $\frac{\pi}{2}$, and function values of 0.0, 0.293, and 1.0 as for the quadratic element. This is shown in Figure 5.1.

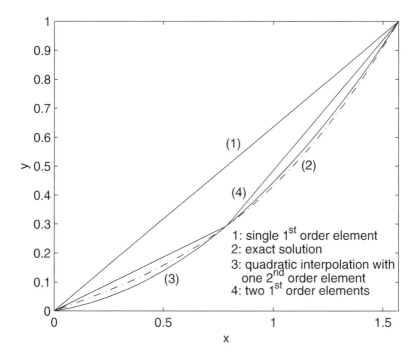

Figure 5.1 Comparison of Function Values in a Linear and a Quadratic Element with Those of the Analytical Solution

Using the procedure as for the single first-order element of equation (5.4), we can obtain the function approximation by two first-order elements as

$$y = 0.373x, \quad \text{for} \quad 0 < x < \frac{\pi}{4} \tag{5.14}$$

$$y = -0.414 + 0.9x, \quad \text{for} \quad \frac{\pi}{4} < x < \frac{\pi}{2} \tag{5.15}$$

The error at $x = \frac{\pi}{3}$, in this case from equations (5.2) and (5.15), is 0.061, which is much less than the error for the single-element case of 0.21051. We also notice from equations (5.14) and (5.15) that the function y is defined in the two regions 0 to $\frac{\pi}{4}$ and

Shape Functions

$\frac{\pi}{4}$ to $\frac{\pi}{2}$ by two different first-order polynomials. We can interpolate the approximate value of y at any point in the two intervals correspondingly.

So far we have considered point collocation to determine the parameters a_0, a_1, and a_2. Except in the case of the single first-order element, we were required to choose an intermediate value of x arbitrarily and make the approximation coincide exactly with the value of y given by the analytical solution (5.2). In order to bring some rationale into this choice of the intermediate value of x and that of the function, we shall use the finite element procedure and find the value of y by functional minimization.

Let us define a functional $\mathcal{F}$ that when minimized, will yield an acceptably accurate solution to the function. This functional is an energy-related integral expression and we shall assume it for the present to be of the form

$$\mathcal{F} = \int_0^{\frac{\pi}{2}} \left[\left(\frac{dy}{dx} \right)^2 + 2y \sin x \right] dx \tag{5.16}$$

An expression similar to (5.16) could also be obtained directly by the weighted residual procedure.

Let us now consider the one-element quadratic approximation for y in the interval $0 < x < \frac{\pi}{2}$. Therefore as in equation (5.8),

$$y = a_0 + a_1 x + a_2 x^2 \tag{5.17}$$

Applying the end conditions $y = 0$ at $x = 0$, and $y = 1$ at $x = \frac{\pi}{2}$, equation (5.17) yields $a_0 = 0$, $a_1 + a_2 \frac{\pi}{2} = \frac{2}{\pi}$, from which one obtains

$$a_2 = \frac{4}{\pi^2} - \frac{2}{\pi} a_1 \tag{5.18}$$

Substituting the value of a_2 from equation (5.18) into equation (5.17), we have

$$y = \frac{4}{\pi^2} x^2 + a_1 \left[x - \left(2 \frac{x^2}{\pi} \right) \right] \tag{5.19}$$

The only unknown here is a_1, which we shall determine by the finite element procedure. We can evaluate $\frac{dy}{dx}$ from equation (5.19) by differentiation, so that

$$\frac{dy}{dx} = \frac{8x}{\pi^2} + a_1\left(1 - 4\frac{x}{\pi}\right) \tag{5.20}$$

After substituting (5.20) into (5.16), minimizing $\mathcal{F}$ with respect to a_1 by setting the first derivative of $\mathcal{F}$ to zero, and performing the necessary algebraic steps and integrations, one obtains

$$\frac{d\mathcal{F}}{da_1} = \left(\frac{8x^2}{\pi^2}\right) - \left(\frac{64x^3}{3\pi^3}\right) + 2a_1\left(x - \left(\frac{4x^2}{\pi}\right) + \left(\frac{16x^3}{3\pi^2}\right)\right)$$
$$+ 2(\sin x - x\cos x) - \frac{4}{\pi}(2x\sin x + 2\cos x - x^2\cos x) \tag{5.21}$$

in the interval $[0, \frac{\pi}{2}]$.

Equation (5.21) yields

$$a_1 = \frac{8}{\pi} - \frac{24}{\pi^2}$$
$$a_2 = \frac{4}{\pi^2} - \frac{2}{\pi}a_1 = \frac{48}{\pi^3} - \frac{12}{\pi^2} \tag{5.22}$$

and

$$y = \left(\frac{8}{\pi} - \frac{24}{\pi^2}\right)x + \left(\frac{48}{\pi^3} - \frac{12}{\pi^2}\right)x^2 \tag{5.23}$$

or

$$y = 0.114770682x + 0.332219449x^2$$

Shape Functions

Table 5.2 Comparison of Computed Values of y

x	y from (5.2)	y from interpolation and FE minimization
0	0	0
$\pi/12$	0.0745	0.0528168
$\pi/6$	0.167	0.1511736
$\pi/4$	0.293	0.2950703
$\pi/3$	0.4673	0.4845069
$5\pi/12$	0.7007	0.7194835
$\pi/2$	1	1

The values of the function y obtained from equation (5.23) for different values of x are given in Table 5.2.

It can be observed from Table 5.2 that the finite element (FE) method used for determining the coefficients of the quadratic interpolation function yields a small error. Furthermore, unlike the collocation method, for this application the FE technique does not require the choice of an intermediate point in the domain.

Next we shall consider two first-order elements and use the finite element process to determine the coefficients. Here we require an intermediate value of x but not the exact value of y at this point. Choosing an intermediate value of $x = \frac{\pi}{4}$, we shall define the interpolation functions of the two elements as

$$y = a_0 + a_1 x, \quad \text{for } 0 < x < \frac{\pi}{4} \tag{5.24}$$

and

$$y = a_2 + a_3 x, \quad \text{for } \frac{\pi}{4} < x < \frac{\pi}{2} \tag{5.25}$$

From the end conditions of y, namely $y = 0$ at $x = 0$, and $y = 1$ at $x = \frac{\pi}{2}$, we have

$$a_0 = 0, \quad a_2 = \left(a_1 \frac{\pi}{2} - 1\right), \quad a_3 = \left(\frac{4}{\pi} - a_1\right) \tag{5.26}$$

Substituting (5.26) into equations (5.24) and (5.25)

$$y = a_1 x, \quad \text{for } 0 < x < \frac{\pi}{4} \tag{5.27}$$

$$y = \left(a_1 \frac{\pi}{2} - 1\right) + \left(\frac{4}{\pi} - a_1\right) x, \quad \text{for } \frac{\pi}{4} < x < \frac{\pi}{2} \tag{5.28}$$

The derivatives with respect to x of equations (5.27) and (5.28) are

$$\frac{dy}{dx} = a_1, \quad \text{for } 0 < x < \frac{\pi}{4} \tag{5.29}$$

$$\frac{dy}{dx} = \left(\frac{4}{\pi} - a_1\right), \quad \text{for } \frac{\pi}{4} < x < \frac{\pi}{2} \tag{5.30}$$

The total functional for the whole domain will be the sum of the functionals for the individual elements so that

$$\mathcal{F} = \int_0^{\frac{\pi}{4}} \left[\left(\frac{dy}{dx}\right)^2 + 2y \sin x\right] dx + \int_{\frac{\pi}{4}}^{\frac{\pi}{2}} \left[\left(\frac{dy}{dx}\right)^2 + 2y \sin x\right] dx \tag{5.31}$$

(element 1)   (element 2)

Substituting for $\frac{dy}{dx}$ from (5.29) and (5.30) and for y from (5.27) and (5.28) respectively in equation (5.31),

Shape Functions

$$\mathcal{F} = \int_0^{\frac{\pi}{4}} (a_1^2 + 2a_1 x \sin x)\, dx + \int_{\frac{\pi}{4}}^{\frac{\pi}{2}} \left[\left(\frac{4}{\pi} - a_1\right)^2 + 2\left(\left(a_1 \frac{\pi}{2} - 1\right)\right. \right.$$
$$\left.\left. + \left(\frac{4}{\pi} - a_1\right) x\right) \sin x \right] dx \quad (5.32)$$

Minimization of (5.32) is accomplished by setting its derivative with respect to a_1 to zero, so that

$$\frac{d\mathcal{F}}{da_1} = \int_0^{\frac{\pi}{4}} 2(a_1 + x \sin x)\, dx + \int_{\frac{\pi}{4}}^{\frac{\pi}{2}} \left[2\left(a_1 - \frac{4}{\pi}\right) + \pi \sin x - 2x \sin x \right] dx = 0$$
$$(5.33)$$

Performing, as before, the required integrations in (5.33) and rearranging terms, we obtain

$$\begin{aligned} a_1 &= \frac{4 - 2\sqrt{2}}{\pi} \\ a_2 &= 1 - \sqrt{2} \\ a_3 &= \frac{2\sqrt{2}}{\pi} \end{aligned} \quad (5.34)$$

Substituting (5.34) into (5.26) through (5.28)

$$y = (4 - 2\sqrt{2})\frac{x}{\pi}, \quad \text{for } 0 < x < \frac{\pi}{4} \quad (5.35)$$

$$y = (1 - \sqrt{2}) + 2\sqrt{2}\frac{x}{\pi}, \quad \text{for } \frac{\pi}{4} < x < \frac{\pi}{2} \quad (5.36)$$

Table 5.3   Error Computation

x	y from (5.35) and (5.36)	y from (5.2)	Error
0	0	0	0
$\pi/12$	0.0976	0.0745	0.0231
$\pi/6$	0.1953	0.1667	0.0286
$\pi/4$	0.2929	0.293	0.0001
$\pi/3$	0.5286	0.4673	0.0613
$5\pi/12$	0.7643	0.7007	0.0636
$\pi/2$	1.0	1.0	0

Table 5.3 shows a comparison of values of y for different values of x obtained from equations (5.35) and (5.36) with the exact solution from (5.2).

The maximum error occurs at $x = \frac{5\pi}{12}$ and has a value of 0.0636, which is a great deal less than the error for a single first-order element of 0.21051.

Once again, it must be emphasized that in this application of the finite element method to two first-order elements spanning the domain, we needed to select only an intermediate value of x, but not in any way restricted to the midpoint of the domain, which we have selected for convenience. However, with this method, unlike the collocation method, we did not have to fix the value of y *a priori* corresponding to this intermediate value of x. We then determined the values of the coefficients by minimizing the total functional in the domain made up of the sum of the piecewise functionals in the two first-order elements. In turn, these coefficients were used to find the respective interpolation functions from which the value of y at any value of x could be found easily.

Another interesting feature of the finite element approximation has been highlighted by this example. The coefficients of the second element interpolation function were determined in terms of the coefficients of the first by equating the values of y interpolated at the intermediate value of x, in the two elements. This ensures the continuity of the function between the two elements at the point of their connection, an essential requirement for the choice of interpolatory functions. We shall discuss this in more detail later on.

Shape Functions

To summarize this section, we introduced the concept of interpolation and illustrated the choice of interpolatory functions for first- and second-order approximations of the solutions to a differential equation. Point collocation procedure and end conditions were used to determine the coefficients of the interpolatory functions. When more than one element was employed, the collocation procedure required the *a priori* selection of the function value at an intermediate point such as the value obtained from a closed form solution. The use of the finite element method in place of the collocation procedure was illustrated as a means of obtaining the best approximation with an acceptable level of accuracy. The functional corresponding to the differential equation was minimized to determine the coefficients of the interpolatory functions. The choice of an interpolation function to obtain the least error is an iterative process that is best illustrated by the flowchart of Figure 5.2.

In the following section, we shall generalize the choice of interpolatory functions and discuss their properties and requirements.

5.2 POLYNOMIAL INTERPOLATION

Polynomial interpolation functions offer a suitable means of describing the complex behavior of the unknown solution and its approximation by the finite element method. In addition, they easily lend themselves to the process of integration and differentiation.

We may write the approximation of the unknown solution in one dimension by the general polynomial expansion

$$f(x) = a_0 + a_1 x + a_2 x^2 \cdots + a_n x^n \tag{5.37}$$

where $a_0, a_1, a_2, \cdots, a_n$ denote the unknown coefficients or parameters of the expansion and x is the independent variable. The total number of coefficients will equal $(n+1)$ and the order of the polynomial, which is the highest degree of the independent variable x, will equal n. It may also be noted that (5.37) represents the complete nth order polynomial in one dimension; that is, the expansion has all the terms including the constant term and powers of x up to the highest degree corresponding to the order of the polynomial.

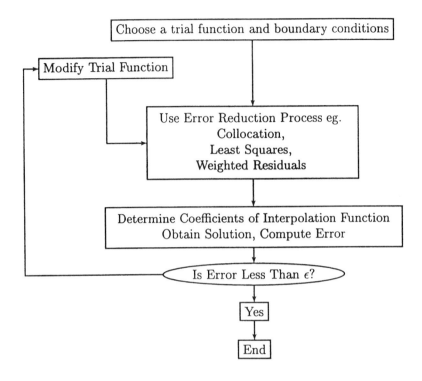

Figure 5.2 Flowchart Illustrating the Choice of Interpolation Functions to Minimize Error

The polynomial expansion may also be represented by the summation convention

$$f = \sum_{i=0}^{n} a_i x^i \tag{5.38}$$

Extending to two dimensions, a complete nth order polynomial may be written as

$$f = a_0 + a_1 x + a_2 y + a_3 xy + a_4 x^2 + a_5 y^2 \cdots + a_m y^n \tag{5.39}$$

Shape Functions

where n is the order of the polynomial, and the total number of coefficients in the expansion equals $m + 1$, where

$$m = \frac{(n+1)(n+2)}{2} \tag{5.40}$$

By the summation convention, (5.39) may be expressed as

$$f = \sum_{i=0}^{m} a_i x^j y^k, \quad j + k < n \tag{5.41}$$

where j and k are integer exponents whose values are related to the integer subscript i as follows.

$$i = \frac{(k+1)(j+k+2) + j(j+k) - 2}{2} \tag{5.42}$$

Here all combinations of j and k should be used in ascending order from 0 to m. We shall illustrate (5.41) and (5.42) by the following example for a two-dimensional first-order polynomial.

$$f = a_{i1} x^0 y^0 + a_{i2} x^1 y^0 + a_{i3} x^0 y^1 \tag{5.43}$$

Now from (5.42)

$$i1 = \frac{(0+1)(0+0+2) + 0(0+0)}{2} - 1 = 0 \tag{5.44}$$

$$i2 = \frac{(0+1)(1+0+2) + 1(1+0)}{2} - 1 = 1 \tag{5.45}$$

$$
\begin{array}{c}
a_0 \\
a_1 x \quad a_2 y \\
a_3 x^2 \quad a_4 xy \quad a_5 y^2 \\
a_6 x^3 \quad a_7 x^2 y \quad a_8 xy^2 \quad a_9 y^3 \\
a_{10} x^4 \quad a_{11} x^3 y \quad a_{12} x^2 y^2 \quad a_{13} xy^3 \quad a_{14} y^4 \\
a_{15} x^5 \quad a_{16} x^4 y \quad a_{17} x^3 y^2 \quad a_{18} x^2 y^3 \quad a_{19} xy^4 \quad a_{20} y^5
\end{array}
$$

Figure 5.3 Pascal Triangle of Polynomial Expansion

$$i3 = \frac{(1+1)(0+1+2) + 0(0+1)}{2} - 1 = 2 \tag{5.46}$$

As we stated, the order of the polynomial is the highest power to which the independent variable is raised in the chosen expansion, and the latter is complete to a given order if it contains all the terms of that order. The number of terms N in a polynomial of order n in two dimensions is calculated from equation (5.40). For the preceding case of first-order interpolation

$$N = \frac{(1+1)(1+2)}{2} = 3 \tag{5.47}$$

We may conveniently determine the terms in a complete two-dimensional polynomial by referring to the Pascal triangle as suggested in reference [53], wherein the terms of the expansion are arranged as shown in Figure 5.3.

Thus procedure given up to the fifth order polynomial can easily be extended to any order. It shows the number of terms in a complete polynomial of a given order and the number of terms associated with a particular order. For example, the complete quintic polynomial has 21 terms and 6 terms of fifth order. The coefficients a_i in the preceding polynomial expansion for representing the approximating function f are called generalized coordinates [48]. These generalized coordinates are independent parameters that determine the distribution of the function f, while the shape of the distribution is determined by the order of the polynomial selected. These generalized coordinates have no particular definition but can be shown to be linear combinations of the nodal degrees of freedom in the domain. Also, they are not identified with particular nodes in any form or manner.

We need to define some essential requirements for the choice of the interpolation functions. These are predicated by the need to ensure (1) that the piecewise functionals associated with finite elements may be added together to yield the overall functional for the entire region or domain and (2) that the solution converges to the correct solution when an increased number of small elements are used to represent the finite element mesh.

In order to ensure monotonic convergence to the true solution as described before, the interpolation polynomials must also meet the following further requirements.

- At element interfaces and along the edges, the function or the field variable and any of its derivatives up to one order less than the highest order derivative appearing in the functional must be continuous.
- The value of the field variable and its partial derivatives up to the highest order in the functional must be represented in the approximating function, when in the limit the element size shrinks to zero.

These requirements are described in references [54] and [55] as compatibility and completeness requirements, respectively. The first one ensures continuity of the field variable and the second ensures convergence to the correct solution as the element size is progressively reduced. Yet another requirement the interpolation functions must satisfy is that the field variable and the polynomial expansion for the element remain unchanged under a linear transformation of the coordinate system such as translation of the origin or change in orientation. This requirement is called geometric isotropy.

Complete polynomials, which contain all the terms to the required degree corresponding to the order, possess geometric isotropy. Even if the polynomials are not complete but contain the appropriate terms to preserve symmetry, they possess geometric isotropy.

In the light of this, complete polynomials in two dimensions described by equation (5.41) may also be rewritten as

$$f(x, y) = \sum_{i=0}^{m} a_i x^j y^k, \quad \text{for } j + k \leq n, \quad m = \frac{(n+1)(n+2)}{2} \quad (5.48)$$

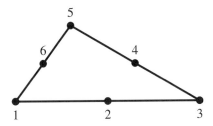

Figure 5.4 Second-Order Two-Dimensional Triangular Element

When these are used as interpolation polynomials, they will remain invariant under linear coordinate transformation.

The property of symmetry may be illustrated by the following example. A cubic polynomial has 10 terms and the corresponding number of coefficients. If we desire to truncate the polynomial, how should we do so without impairing its geometric isotropy? We need to drop symmetric pairs, in this case (x^2, y^2) or $(x^2 y, xy^2)$, resulting in a cubic polynomial of eight terms that exhibits geometric isotropy. We can easily extend this procedure to construct other incomplete polynomials from the array of terms in Figure 5.3. In reference [56], it is pointed out that meeting the criterion of geometric isotropy ensures that the interpolation function along the edge of a finite element becomes a complete polynomial in one dimension of the same order as the interpolation function within the two-dimensional region of the element. This can be illustrated by the following example of a two-dimensional second-order triangular element of Figure 5.4. These ideas hold for three-dimensional problems also.

Gallagher [53] shows that for a two-dimensional nth order polynomial function described in an R-sided polygon, the number of nodes on the edge or boundary segment must satisfy the relation

$$\frac{(n+1)(n+2)}{2} = Rn \qquad (5.49)$$

Shape Functions

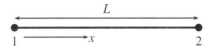

Figure 5.5 One-Dimensional Linear Element

This condition is met by a three-noded (first-order) or a six-noded (second-order) triangular element. Even for incomplete polynomials defined in rectilinear elements, the number of points on each edge equals $(n + 1)$. Thus, for a first-order rectangle, the number of points on the edge is two, which defines a linear polynomial of first order and so on. Heubner [49] discusses the completeness property of generally used interpolatory functions in problems that require potential continuity and presents guidelines for their selection for rectangular elements of any order.

5.3 DERIVING SHAPE FUNCTIONS

In the preceding section, we discussed the concept of interpolation and interpolatory functions and their requirements of compatibility, completeness, and continuity along edges with adjacent elements. The foregoing procedure was described in terms of generalized coordinates or parameters. However, in practice, the solution of the physical boundary value problem requires discretization of the field region into elements and description of the potential or field variables in terms of their nodal values of each element. In this section, we shall examine the method of defining interpolation functions in each of the elements of the discretized space.

We shall define the nodal values of the physical field variables as degrees of freedom. The nodal values can be potentials or field quantities or their derivatives of different order. The use of nodal values as derivatives of the potential function may yield certain computational advantages but may entail complexities of implementation. The advantages and limitations of derivatives as nodal values are fully discussed in reference [49]. In this section, we shall limit our attention to the field variables as nodal degrees of freedom and investigate how the interpolation functions are derived. These interpolation functions emerge from the basic procedure of expressing the generalized coordinates in terms of nodal degrees of freedom. This can be illustrated with reference to a one-dimensional linear element of Figure 5.5 as follows.

$$\phi(x) = a_0 + a_1 x \tag{5.50}$$

The generalized coordinates a_0 and a_1 may be found by evaluating this expression at each node and inverting the resulting set of matrix equations. Therefore we have

$$\phi_1 = a_0 + a_1 x_1 \tag{5.51}$$

$$\phi_2 = a_0 + a_1 x_2 \tag{5.52}$$

In matrix notation, equations (5.51) and (5.52) may be expressed as

$$\{\phi\} = (C)\{a\} \tag{5.53}$$

where

$$\{\phi\} = \left\{ \begin{array}{c} \phi_1 \\ \phi_2 \end{array} \right\}, \quad (C) = \left(\begin{array}{cc} 1 & x_1 \\ 1 & x_2 \end{array} \right), \quad \{a\} = \left\{ \begin{array}{c} a_0 \\ a_1 \end{array} \right\} \tag{5.54}$$

It can be seen that the vector of generalized coordinates can be obtained by solving equation (5.53) so that

$$\{a\} = (C)^{-1}\{\phi\} \tag{5.55}$$

If we now express the function in equation (5.50) as a product of a row vector and a column vector, we can write

$$\phi = (D)\{a\} \tag{5.56}$$

where

$$(D) = [1, x] \tag{5.57}$$

Substituting equation (5.55) into (5.56), we obtain

$$\phi = (D)(C)^{-1}\{\phi\} \tag{5.58}$$

Shape Functions

We can rewrite equation (5.58) as

$$\phi = (\xi)\{\phi\} \tag{5.59}$$

where

$$(\xi) = (D)(C)^{-1} \tag{5.60}$$

Although this procedure has been illustrated with a one-dimensional first-order element, it is generally applicable to all straight-sided elements.

It must be pointed out that the original interpolation function of equation (5.50) reduced to matrix form as in equation (5.56) is distinctly different from equation (5.59). The distinction is that $[D][a]$ is an interpolation function that applies to the whole element or region and expresses the behavior of the solution function or field variable in terms of generalized coordinates, whereas the interpolation function (5.60) refers to individual nodes and $[\xi][\phi]$ refers to the expansion of ϕ in terms of the nodal degrees of freedom.

The inverse of the matrix (C) is easily obtained by standard procedure as

$$(C)^{-1} = \frac{1}{x_2 - x_1}\begin{pmatrix} x_2 & -1 \\ -x_1 & 1 \end{pmatrix}^T = \frac{1}{x_2 - x_1}\begin{pmatrix} x_2 & -x_1 \\ -1 & 1 \end{pmatrix} \tag{5.61}$$

and $(D)(C)^{-1}$ will become

$$(\xi) = (D)(C)^{-1} = \frac{[1, x]}{x_2 - x_1}\begin{pmatrix} x_2 & -x_1 \\ -1 & 1 \end{pmatrix} \tag{5.62}$$

or

$$(\xi) = \left(\frac{x_2 - x}{x_2 - x_1}, \frac{-x_1 + x}{x_2 - x_1}\right) \tag{5.63}$$

Substituting equation (5.63) into the expression for ϕ in equation (5.59) and expanding the terms, we have

$$\phi = \frac{x_2 - x}{x_2 - x_1}\phi_1 + \frac{x - x_1}{x_2 - x_1}\phi_2 = \xi_1\phi_1 + \xi_2\phi_2 \tag{5.64}$$

Figure 5.6 Variation of Shape Functions and ϕ Along x

where

$$\begin{aligned} \xi_1 &= \frac{x_2 - x}{x_2 - x_1} \\ \xi_2 &= \frac{x - x_1}{x_2 - x_1} \end{aligned} \quad (5.65)$$

We may note that ξ_1 and ξ_2 are related to the geometry and describe the behavior of the field variable in terms of its nodal values or degrees of freedom in the one-dimensional element. It can also be seen from (5.64) that at $x = x_1$, the value of the field variable

$$\phi = \phi_1 \quad (5.66)$$

and at $x = x_2$, the field variable

$$\phi = \phi_2 \quad (5.67)$$

Figure 5.6 shows the variation of ξ_1 and ξ_2 as a function of the element coordinate x. Each of the graphs describes the shape of the interpolation function and the value of ϕ within the element. For example, from Figure 5.6a it can be seen that the value of ϕ at node 1 is ϕ_1 and its value at node 2 equals zero.

Similarly, from Figure 5.6b, the value of the field variable at node 2 equals ϕ_2, while it is zero at node 1. The superposition of these two distributions yields the variation of ϕ over the whole element.

We should also examine ξ_1 and ξ_2 more closely. First, they are nondimensional and have a value of unity at their respective nodes and a zero value at the others. Thus, ξ_1

at $x = x_1$ has a value of unity and at $x = x_2$ has a zero value. The sum of the shape functions at any point within the element will always equal unity.

We may also rewrite the expressions in equations (5.64) and (5.65) as follows. Supposing x is chosen to be the origin of the element and has a value of zero and $(x_2 - x_1) = L = x_2$, then the shape functions can be written as

$$\xi_1 = \frac{(L-x)}{L} = \left(1 - \frac{x}{L}\right) \quad (5.68)$$

and

$$\xi_2 = \frac{(x-0)}{L} = \frac{x}{L} \quad (5.69)$$

Therefore,

$$\phi = \left(1 - \frac{x}{L}\right)\phi_1 + \frac{x}{L}\phi_2 = (1 - \xi)\phi_1 + \xi\phi_2 \quad (5.70)$$

where ξ is defined as the natural coordinate $\frac{x}{L}$. If we choose the origin midway between the end points of the element, instead of at x_1, we will have

$$\xi_1 = \frac{\xi - 1}{2} \quad \text{and} \quad \xi_2 = \frac{\xi + 1}{2} \quad (5.71)$$

where ξ ranges between -1 and $+1$.

5.4 LAGRANGIAN INTERPOLATION

This method offers an easy way to construct shape functions directly and avoids the monotonous task of writing and solving equations defined in terms of generalized coordinates. This is achieved by determining the coefficients or shape functions of a polynomial series passing through a specified number of points using the Lagrange interpolation technique. The formula for this interpolation scheme as described in reference [57] is as follows.

$$\xi_i = \frac{\prod_{j=0,\ j=i}^{n}(x - x_j)}{\prod_{j=0,\ j=i}^{n}(x_i - x_j)} \quad (5.72)$$

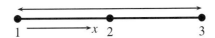

Figure 5.7 One-Dimensional Element with Quadratic Interpolation

where the symbol Π denotes the product of binomials $(x - x_j)$, $(x_i - x_j)$ over the entire range of j.

Expanding (5.72) for a two-noded element of Figure 5.5,

$$\xi_1 = \frac{x - x_2}{x_1 - x_2} \text{ and } \xi_2 = \frac{x - x_1}{x_2 - x_1} \quad (5.73)$$

which are the same as the expressions obtained in equation (5.65). Extending the formula of equation (5.72) for the one-dimensional second-order three-noded element shown in Figure 5.7, we have

$$\begin{aligned}
\xi_1 &= \frac{(x - x_2)(x - x_3)}{(x_1 - x_2)(x_1 - x_3)} \\
\xi_2 &= \frac{(x - x_1)(x - x_3)}{(x_2 - x_1)(x_2 - x_3)} \\
\xi_3 &= \frac{(x - x_1)(x - x_2)}{(x_3 - x_1)(x_3 - x_2)}
\end{aligned} \quad (5.74)$$

If we substitute $x_1 = -a$, $x_2 = 0$, and $x_3 = a$ in the expressions for the shape functions of equation (5.74),

$$\xi_1 = \frac{x(x - a)}{2a^2}, \quad \xi_2 = \frac{-(x + a)(x - a)}{a^2}, \quad \xi_3 = \frac{x(x + a)}{2a^2} \quad (5.75)$$

Shape Functions

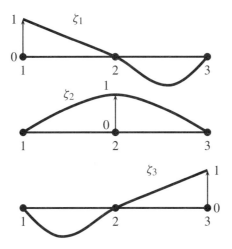

Figure 5.8 Variation of Shape Functions for the Three-Noded One-Dimensional Quadratic Element as a Function of ξ

Defining a nondimensional parameter $\xi = \frac{(x-x_0)}{a}$, where x_0 is the coordinate of the midpoint of the element and ξ varies between -1 and $+1$, the shape functions may be obtained as

$$\xi_1 = -\frac{\xi}{2}(1-\xi), \quad \xi_2 = 1-\xi^2, \quad \xi_3 = \frac{\xi}{2}(1+\xi) \tag{5.76}$$

The variation of the individual shape functions over the interval -1 to $+1$ is shown in Figure 5.8.

Superposition of these distributions, suitably multiplied by the respective nodal values of ϕ, yields the potential distribution over the element.

Referring to equation (5.72) we note that the $\xi(x)$ is a product of n linear factors, and therefore the polynomial is of degree n. We also note that when $x = x_i$, the numerator and denominator of (x) are identical and the polynomial function has a unit

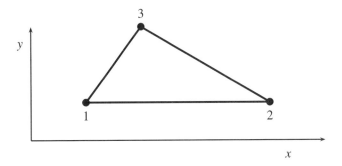

Figure 5.9 First-Order Two-Dimensional Triangular Element

value. But when $x = x_j$ and $i \neq j$, the polynomial has a zero value. Therefore these polynomials, generally termed Lagrangian interpolation polynomials or Lagrangian interpolation coefficients, are interpolatory. Because they also guarantee continuity of ϕ at the connected nodes, they are suitable for ensuring C^0 continuity.

5.5 TWO-DIMENSIONAL ELEMENTS

One-dimensional finite element formulations are of limited value and are not of much use for solving electric and magnetic field problems. In fact, the development of the finite element method was to devise practical tools for the solution of two- and three-dimensional boundary value problems. In this section, we shall extend the method of developing shape functions for two- and three-dimensional problems based on the concepts described for the one-dimensional case.

In a first-order two-dimensional triangular element as shown in Figure 5.9, the potential function may be expressed as

$$\phi = a_1 + a_2 x + a_3 y \tag{5.77}$$

Shape Functions

or in matrix form

$$\phi = [1, x, y]\{a\} \tag{5.78}$$

Equation (5.78) may also be expressed in terms of shape functions and nodal values of the field variable by the summation convention as

$$\phi = \sum_{i=1}^{3} \xi_i \phi_i \tag{5.79}$$

If we write equation (5.78) for each of the nodal values of potential, we have

$$\begin{pmatrix} \phi_1 \\ \phi_2 \\ \phi_3 \end{pmatrix} = \begin{pmatrix} 1 & x_1 & y_1 \\ 1 & x_2 & y_2 \\ 1 & x_3 & y_3 \end{pmatrix} \begin{pmatrix} a_1 \\ a_2 \\ a_3 \end{pmatrix} \tag{5.80}$$

or in matrix notation

$$[\phi_i] = [D]\{a\} \tag{5.81}$$

from which the generalized coordinates are obtained as

$$[a] = [D]^{-1}\{\phi_i\} \tag{5.82}$$

Substituting the values of a from equation (5.82) into equation (5.78)

$$\phi = [1, x, y](D)^{-1}(\phi_i) \tag{5.83}$$

Using the standard matrix inversion procedure described in reference [58],

$$(D)^{-1} = \frac{1}{|D|} \begin{pmatrix} (x_2 y_3 - x_3 y_2) & -(y_3 - y_2) & (x_3 - x_2) \\ -(x_1 y_3 - x_3 y_1) & (y_3 - y_1) & -(x_3 - x_1) \\ (x_1 y_2 - x_2 y_1) & -(y_2 - y_1) & (x_2 - x_1) \end{pmatrix}^T \quad (5.84)$$

where the determinant

$$|D| = (x_2 y_3 - x_3 y_2) + x_1(y_2 - y_3) + y_1(x_3 - x_2) \quad (5.85)$$

Also, this determinant equals twice the area of the triangle, as shown in Appendix B. Rewriting equation (5.84) in terms of new variables p, q, r and the triangle area, where

$$\begin{aligned} p_1 &= (x_2 y_3 - x_3 y_2) \\ q_1 &= (y_2 - y_3) \\ r_1 &= (x_3 - x_2) \end{aligned} \quad (5.86)$$

and so on for (p_2, q_2, r_2) and (p_3, q_3, r_3) in cyclic form

$$(D)^{-1} = \frac{1}{2\Delta} \begin{pmatrix} p_1 & p_2 & p_3 \\ q_1 & q_2 & q_3 \\ r_1 & r_2 & r_3 \end{pmatrix} \quad (5.87)$$

Substituting equation (5.87) into equation (5.83)

$$\phi = \frac{[1, x, y]}{2\Delta} \begin{pmatrix} p_1 & p_2 & p_3 \\ q_1 & q_2 & q_3 \\ r_1 & r_2 & r_3 \end{pmatrix} \begin{pmatrix} \phi_1 \\ \phi_2 \\ \phi_3 \end{pmatrix} \quad (5.88)$$

Shape Functions

Expanding the matrix equation (5.88) and rearranging terms,

$$\phi = \frac{(p_1 + q_1 x + r_1 y)\phi_1}{2\Delta} + \frac{(p_2 + q_2 x + r_2 y)\phi_2}{2\Delta} + \frac{(p_3 + q_3 x + r_3 y)\phi_3}{2\Delta} \quad (5.89)$$

Comparing equations (5.89) and (5.79), the shape functions for this linear triangular element may be expressed as

$$\zeta_i = \frac{p_i + q_i x + r_i y}{2\Delta}, \quad i = 1, 2, 3 \quad (5.90)$$

Expressions ζ_1, ζ_2, and ζ_3 are interpolatory functions, and it can be shown by substituting the appropriate values at $i = 1$ into equation (5.90) that

$$\zeta_1 = 1 \text{ at node 1 and 0 at nodes 2 and 3} \quad (5.91)$$

The preceding relations hold similarly for ζ_2 and ζ_3. This can be generalized by the expression

$$\zeta_j(x_i, y_i) = \delta_{ij} \quad (5.92)$$

where $\delta_{ij} = 1$ for $i = j$ and $\delta_{ij} = 0$ for $i \neq j$.

Because the ζ's are interpolatory and satisfy the conditions of equation (5.92), it can be shown that the sum of the shape functions must equal unity. This can also be observed from Figure 5.10, where the point P represents the intersection of the loci of ζ_1, ζ_2, and ζ_3.

At point P,

$$\zeta_1 = \frac{1}{2}, \quad \zeta_2 = \frac{1}{4}, \quad \zeta_3 = \frac{1}{4} \quad (5.93)$$

so that $\sum \zeta_i = 1$.

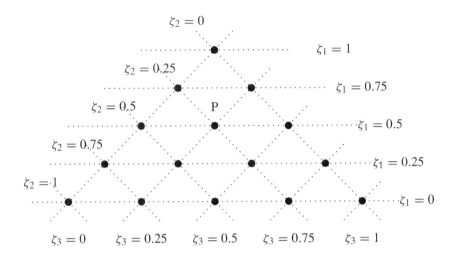

Figure 5.10 Loci of Shape Functions in a First-Order Planar Triangle

Equation (5.93) holds similarly for any other position in the triangle. The solution function for the preceding two-dimensional first-order interpolation in a triangle is shown in geometrical form in Figure 5.11.

5.5.1 Natural Coordinates in Two Dimensions

The development of natural coordinates follows the same procedure used for one-dimensional elements described in equations (5.68), (5.69) and (5.71). As in that case, the goal is to choose normalized coordinates to describe the location of any point P within the triangular element or on its edges. The Cartesian coordinate of a point in a first-order triangular element can be expressed as a linear combination of the natural coordinates and the coordinates of the triangle vertices, so that

$$x = \zeta_1 x_1 + \zeta_2 x_2 + \zeta_3 x_3 \tag{5.94}$$

$$y = \zeta_1 y_1 + \zeta_2 y_2 + \zeta_3 y_3 \tag{5.95}$$

The additional condition required is that the sum of the weighting functions in equations (5.94) and (5.95) that are the natural coordinates must sum to unity.

Shape Functions

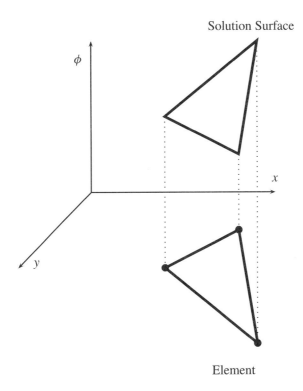

Figure 5.11 Geometrical Representation of the Solutions Function in a First-Order Planar Triangle

Thus,

$$\zeta_1 + \zeta_2 + \zeta_3 = 1 \tag{5.96}$$

It is evident from equation (5.96), that only two of the natural coordinates can be independent, just as in the original coordinate system, where there are only two independent coordinates for a two-dimensional system.

Transforming equations (5.94) through (5.96) to matrix form yields

$$\begin{pmatrix} x \\ y \\ 1 \end{pmatrix} = \begin{pmatrix} x_1 & x_2 & x_3 \\ y_1 & y_2 & y_3 \\ 1 & 1 & 1 \end{pmatrix} \begin{pmatrix} \zeta_1 \\ \zeta_2 \\ \zeta_3 \end{pmatrix} \quad (5.97)$$

Inverting the 3 × 3 matrix in equation (5.97), using the procedure described, we obtain the values of ζ_1, ζ_2, and ζ_3 as

$$\begin{aligned} \zeta_1 &= \frac{1}{2\Delta}(a_1 + b_1 x + c_1 y) \\ \zeta_2 &= \frac{1}{2\Delta}(a_2 + b_2 x + c_2 y) \\ \zeta_3 &= \frac{1}{2\Delta}(a_3 + b_3 x + c_3 y) \end{aligned} \quad (5.98)$$

where the determinant of the matrix equation and a_1, b_1, c_1, are defined as follows.

$$\det = \begin{pmatrix} 1 & x_1 & y_1 \\ 1 & x_2 & y_2 \\ 1 & x_3 & y_3 \end{pmatrix} = 2\Delta \quad (5.99)$$

$a = (x_2 y_3 - x_3 y_2)$, $b = (y_2 - y_3)$, $c = (x_3 - x_2)$, and so on in a cyclic fashion.

The natural coordinates ζ_1, ζ_2, and ζ_3 are the same as the shape functions for the first-order triangular element in two dimensions. We may also consider the natural coordinates as the ratios of areas subtended by a point vertex within the triangle. From Figure 5.12, it can be shown after some algebraic manipulation that

$$\begin{aligned} \frac{\text{area of } \triangle PBC}{\text{area of } \triangle ABC} &= \frac{a_1 + b_1 x + c_1 y}{2\Delta} \\ \frac{\text{area of } \triangle PAC}{\text{area of } \triangle ABC} &= \frac{a_2 + b_2 x + c_2 y}{2\Delta} \\ \frac{\text{area of } \triangle PBA}{\text{area of } \triangle ABC} &= \frac{a_3 + b_3 x + c_3 y}{2\Delta} \end{aligned} \quad (5.100)$$

Shape Functions

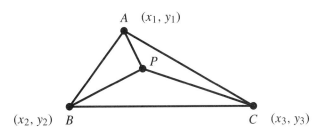

Figure 5.12 Illustration of the Area Coordinate Concept

The loci of these ratios are the same as those of the first-order triangular shape functions shown in Figure 5.10. Also, ζ_1 has a value of unity at A and zero at B and C. Similar relations hold for ζ_2 and ζ_3 also.

In the solution of boundary value problems by the finite element method, derivatives of the field variables with respect to the global coordinates x and y are required, and integrations over the element area also need to be performed. The procedure for these is as follows. Because the field variables can be expressed as a function of the area coordinates ζ_1, ζ_2, and ζ_3, by the chain rule of differentiation,

$$\frac{\partial \phi}{\partial x} = \frac{\partial \phi}{\partial \zeta_1}\frac{\partial \zeta_1}{\partial x} + \frac{\partial \phi}{\partial \zeta_2}\frac{\partial \zeta_2}{\partial x} + \frac{\partial \phi}{\partial \zeta_3}\frac{\partial \zeta_3}{\partial x}$$

$$\frac{\partial \phi}{\partial y} = \frac{\partial \phi}{\partial \zeta_1}\frac{\partial \zeta_1}{\partial y} + \frac{\partial \phi}{\partial \zeta_2}\frac{\partial \zeta_2}{\partial y} + \frac{\partial \phi}{\partial \zeta_3}\frac{\partial \zeta_3}{\partial y} \quad (5.101)$$

where

$$\frac{\partial \zeta_i}{\partial x} = b_i$$

$$\frac{\partial \zeta_i}{\partial y} = c_i \quad \text{for } i = 1, 2, 3 \quad (5.102)$$

Element matrices resulting from functional minimization in finite element analysis require integration of quantities in terms of area coordinates. These integrations are easily performed over the element area using area coordinates with the help of the following integration formula:

$$I = \int_{area} \xi_1^\alpha \xi_2^\beta \xi_3^\gamma \, dx \, dy = \frac{\alpha!\beta!\gamma!}{(\alpha + \beta + \gamma + 2)!} 2\Delta \qquad (5.103)$$

Appendix D gives the values of equation (5.103) for various integers α, β, and γ.

5.6 HIGH-ORDER TRIANGULAR INTERPOLATION FUNCTIONS

The method of deriving high-order interpolation functions for triangular elements was developed by Irons [59] and independently by Silvester [60]. Irons demonstrated that if the interpolation functions for one order of the element are known, the next high-order interpolation could be established by a recurrence relationship. This means that one can start with the interpolation functions for the linear triangle and then construct interpolation functions for all high-order elements. This method was described for a second-order interpolation in reference [61] as follows.

The method due to Irons comprises an algorithm that generates high-order interpolation polynomials over triangular regions by obtaining several different first-order polynomials in the same triangle. In Figure 5.13a, two first-order polynomials $\phi_1(x, y)$ and $\phi_2(x, y)$ are defined. Both functions have a value of unity at the vertex P. Function $\phi(x, y)$ intersects the triangle PQR along the line QR and has zero value along this line. $\phi_2(x, y)$ intersects the same triangle PQR along the line ST. The product of the two linear functions

$$\phi(x, y) = \phi_1(x, y)\phi_2(x, y) \qquad (5.104)$$

at the vertex P has zero value along the line QR and ST. By a similar procedure, a quadratic interpolation polynomial that has a value of unity at point S and zero along triangle edges QR and PR can be constructed as the product of two first-order polynomials $\phi_3(x, y)$ and $\phi_4(x, y)$ having values of unity and zero along QR and PR, as

Shape Functions

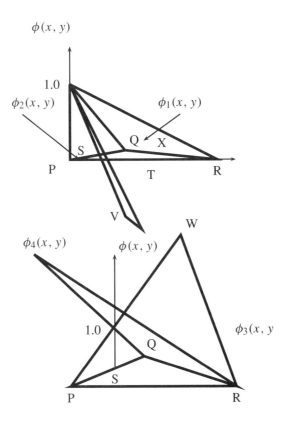

Figure 5.13 Construction of Second-Order Interpolation Functions for a Triangular Element. Vertex-Node Interpolation (top) Side-Node Interpolation (bottom)

shown in Figure 5.13b. Having thus developed a procedure for generating interpolation polynomials at points P and S, by geometric reasoning, a recurrence relationship can be developed to generate quadratic interpolation polynomials corresponding to each of the remaining triangle nodes Q, X, R, and T. Although the foregoing method is general and conceptually appealing, it is cumbersome and requires judicious scaling and normalizing in each application. For these reasons, a more straightforward procedure described by Silvester [62] is generally adopted. This method is as follows.

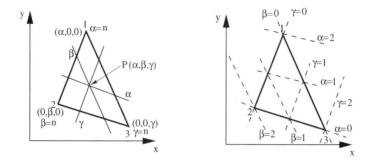

Figure 5.14 (a) Triple-Subscripted Node Identification at Vertices and Inside the Triangle, (b) Loci of α, β, γ for a Six-Noded Triangular Element

5.6.1 Silvester's Method

Using an orderly procedure for designating nodes of high-order triangles, Silvester described a triple-index numbering scheme. This is used to define interpolation functions for any order of a triangular finite element.

Referring to Figure 5.14, Silvester gave a three-digit numbering for each node, where α, β, and γ are integers satisfying the relation $\alpha + \beta + \gamma = n$, the order of the interpolation function for the element. These integers are also the constant coordinate lines in the area coordinate system as shown in Figure 5.14.

We shall use the same triple-subscript designation for the interpolation functions so that $\zeta_{\alpha\beta\gamma}(\xi_1, \xi_2, \xi_3)$ will denote the interpolation or shape function for node $P(\alpha, \beta, \gamma)$, in terms of the area coordinates ξ_1, ξ_2, ξ_3.

Silvester describes the interpolation function for an nth order triangular element by the simple formula

$$\zeta_{\alpha\beta\gamma}(\zeta_1, \zeta_2, \zeta_3) = \zeta_\alpha(\zeta_1)\zeta_\beta(\zeta_2)\zeta_\gamma(\zeta_3) \tag{5.105}$$

where

$$\zeta_\alpha(\zeta_1) = \prod_{i=1}^{\alpha} \frac{n\zeta_1 - i + 1}{i}, \quad \alpha \geq 1 \tag{5.106}$$

Shape Functions

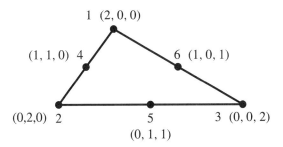

Figure 5.15 Triple Node Numbering for a Second-Order Triangular Element

$$\zeta_\alpha(\zeta_1) = 1, \quad \alpha = 0 \tag{5.107}$$

For $\zeta_\beta(\zeta_2)$ and $\zeta_\gamma(\zeta_3)$, the preceding formula holds with the appropriate changes for the respective area coordinates. The symbol Π signifies, as before, the product of all the terms. Equations (5.106) and (5.107) provide the means for constructing interpolation functions for all higher order elements of this series. For example, if we wish to develop interpolation functions for a six-noded second-order triangular element, $n = 2$ with the three-digit notation for each of the nodes, we need to evaluate ζ_{200}, ζ_{020}, ζ_{003} for the three vertices and ζ_{101}, ζ_{110}, and ζ_{011} for the three midside nodes, respectively (see Figure 5.15).

From equation (5.106), for node 1 ($\alpha = 2$, $\beta = 0$, $\gamma = 0$)

$$\zeta_\alpha = \zeta_2 = \prod_{i=1}^{2} \frac{2\xi_1 - i + 1}{i} \tag{5.108}$$

or

$$\zeta_\alpha = \frac{(2\xi_1 - 1 + 1)}{1}\frac{(2\xi_1 - 2 + 1)}{2} = \xi_1(2\xi_1 - 1) \tag{5.109}$$

$$\zeta_\beta = \zeta_0 = 1 \tag{5.110}$$

$$\zeta_\gamma = \zeta_0 = 1 \tag{5.111}$$

Therefore

$$\zeta_{200} = \zeta_2(\xi_1)\zeta_0(\xi_2)\zeta_0(\xi_3) = \xi_1(2\xi_1 - 1) \tag{5.112}$$

Following the foregoing procedure, it is easy to show the following relations for the shape functions of the remaining nodes. Referring to Figure 5.15,

$$\begin{aligned}\zeta_{020} &= \xi_2(2\xi_2 - 1)\\ \zeta_{002} &= \xi_3(2\xi_3 - 1)\\ \zeta_{101} &= 4\xi_3\xi_1\\ \zeta_{110} &= 4\xi_1\xi_2\\ \zeta_{011} &= 4\xi_2\xi_3\end{aligned} \tag{5.113}$$

It will be appropriate to furnish the following shape functions for the first few orders of shape functions for triangular elements to facilitate their use in practical engineering applications.

Linear Element

$$\zeta_i = \xi_i \quad i = 1, 2, 3 \tag{5.114}$$

Shape Functions

Quadratic Element

$$\zeta_i = \xi_i(2\xi_i - 1) \quad i = 1, 2, 3 \quad \text{corner nodes}$$
$$\zeta_i = 4\xi_i\xi_j \quad i = 1, 2, 3; \quad j = 2, 3, 1 \quad \text{midside nodes} \tag{5.115}$$

Cubic Element

$$\zeta_i = \frac{\xi_i(3\xi_i - 1)(3\xi_i - 2)}{2} \quad \text{corner nodes}$$

$$\zeta_i = \frac{9\xi_i\xi_j(3\xi_i - 1)}{2} \quad \text{1st set midside nodes}$$

$$\zeta_i = \frac{9\xi_i\xi_j(3\xi_j - 1)}{2} \quad \text{2nd set midside nodes}$$

$$\zeta_i = 27\xi_i\xi_j\xi_m \quad \text{internal node} \tag{5.116}$$

5.7 RECTANGULAR ELEMENTS

For planar rectangular elements with sides parallel to the global axes, we can develop interpolation functions based on the Lagrange interpolation method for first-order elements. We shall form products of the functions that hold for the individual one-dimensional coordinate directions. This procedure may be illustrated with reference to the four-noded planar rectangular element of Figure 5.16. Expressing the interpolation in terms of natural coordinates ξ and η and Lagrange interpolation functions L_1 and L_2,

$$\psi(\xi, \eta) = \zeta_1(\xi, \eta)\psi_1 + \zeta_2(\xi, \eta)\psi_2 + \zeta_3(\xi, \eta)\psi_3 + \zeta_4(\xi, \eta)\psi_4 \tag{5.117}$$

where
$$\zeta_i(\xi, \eta) = L_i(\xi)L_i(\eta), \quad i = 1, 2, 3, 4 \tag{5.118}$$

For this four noded-element, Lagrange interpolation yields the following functions in the ξ and η directions.

$$L_1(\xi) = \frac{\xi - \xi_2}{\xi_1 - \xi_2}, \quad L_2(\xi) = \frac{\xi - \xi_1}{\xi_2 - \xi_1} \tag{5.119}$$

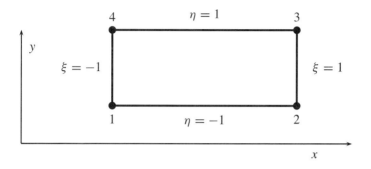

Figure 5.16 A Four-Noded First-Order Rectangular Element

$$L_1(\eta) = \frac{\eta - \eta_4}{\eta_1 - \eta_4}, \quad L_2(\eta) = \frac{\eta - \eta_3}{\eta_2 - \eta_3} \tag{5.120}$$

Thus

$$\zeta_1(\xi, \eta) = L_1(\xi)L_1(\eta) \tag{5.121}$$

or

$$\zeta_1 = \frac{(\xi - \xi_2)(\eta - \eta_4)}{(\xi_1 - \xi_2)(\eta_1 - \eta_4)} \tag{5.122}$$

We also know that for the rectangle shown in Figure 5.16,

$$\xi_1 = -1, \quad \eta_1 = -1$$

Shape Functions

$$\xi_2 = 1, \quad \eta_2 = -1$$
$$\xi_3 = 1, \quad \eta_3 = 1$$
$$\xi_4 = -1, \quad \eta_4 = 1 \tag{5.123}$$

Therefore,

$$\zeta_1 = \frac{(\xi - 1)(\eta - 1)}{(-1 - 1)(-1 - 1)} = \frac{(\xi - 1)(\eta - 1)}{4} \tag{5.124}$$

Similarly, we can evaluate ζ_2, ζ_3, and ζ_4, so that

$$\zeta_2 = -\frac{(\xi + 1)(\eta - 1)}{4}$$
$$\zeta_3 = \frac{(\xi + 1)(\eta + 1)}{4}$$
$$\zeta_4 = -\frac{(\xi - 1)(\eta + 1)}{4} \tag{5.125}$$

Interpolation or shape functions formed by using the procedure just described satisfy the requirements of unit value at the nodes for which they are defined and zero values at other nodes. For example, at node 1, where $\xi = -1$ and $\eta = -1$,

$$L_1(\xi_1) = 1 \tag{5.126}$$

$$\zeta_1(\xi_1, \eta_1) = \frac{(-1 - 1)(-1 - 1)}{4} = 1 \tag{5.127}$$

and at node 2, $\xi_2 = 1$, $\eta_2 = -1$, $L_1(\xi_2) = 0$.

$$\zeta_1(\xi_2, \eta_2) = \frac{(1 - 1)(-1 - 1)}{4} = 0 \tag{5.128}$$

We can similarly show that at nodes 3 and 4

$$L_1(\xi_3) = L_1(\eta_4) = 0 \text{ and } \zeta_1 = 0 \tag{5.129}$$

Along edge 1–2, the interpolations L_1 and ζ_1 vary linearly between the values of 1 and 0, and this satisfies C^0 continuity along that edge. Similar comments hold for the other first-order Lagrange polynomials and shape functions for the other nodes. The Lagrange interpolation function approximations, products of which yield the shape functions, are called bilinear functions for obvious reasons.

Just as an illustration, we may consider a second-order planar rectangular element having four vertex nodes, four midside nodes and one interior node. The interpolation function can be expressed for this case as a biquadratic expansion as

$$\psi(\xi, \eta) = \sum_{j=1}^{N} \zeta_j(\xi_i, \eta_i)\psi_i, \quad -1 < \xi_i < 1, \quad -1 < \eta_i < 1 \tag{5.130}$$

where N is the total number of nodes. Also, as for the first-order case,

$$\zeta_j(\xi_i, \eta_i) = L_j(\xi_i)L_j(\eta_i) \tag{5.131}$$

where the L_j's are one-dimensional second-order Lagrange interpolation functions. We shall illustrate the expansion of equation (5.131) with reference to Figure 5.17 for one corner node and one midside node.

Let us consider node 1 in the figure. The nondimensional parameters or local coordinates ξ, η at node 1 have the values $\xi_1 = -1, \eta_1 = -1$, respectively. The first-order Lagrange interpolation functions for the x and y directions along lines (1,2,3) and (1,4,7), following the procedure described but using ξ, η in place of x, y, are

$$L_1(\xi) = \frac{(\xi - \xi_2)(\xi - \xi_3)}{(\xi_1 - \xi_2)(\xi_1 - \xi_3)} = \frac{(\xi - 0)(\xi - 1)}{(-1 - 0)(-1 - 1)} = \frac{\xi(\xi - 1)}{2} \tag{5.132}$$

Shape Functions

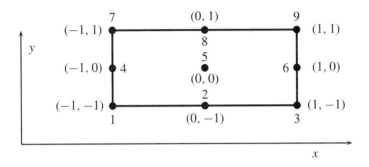

Figure 5.17 A Second-Order Rectangular Element

$$L_1(\eta) = \frac{(\eta - \eta_4)(\eta - \eta_7)}{(\eta_1 - \eta_4)(\eta_1 - \eta_7)} = \frac{(\eta - 0)(\eta - 1)}{(-1 - 0)(-1 - 1)} = \frac{\eta(\eta - 1)}{2} \quad (5.133)$$

Therefore

$$\zeta_1(\xi, \eta) = L_1(\xi)L_1(\eta) = \frac{\xi\eta(\xi - 1)(\eta - 1)}{4} \quad (5.134)$$

Similarly, for node 2, the following relations hold

$$L_2(\xi) = \frac{(\xi - \xi_1)(\xi - \xi_3)}{(\xi_2 - \xi_1)(\xi_2 - \xi_3)} = \frac{(\xi - (-1))(\xi - 1)}{(0 - (-1))(0 - 1)} = 1 - \xi^2 \quad (5.135)$$

$$L_2(\eta) = \frac{(\eta - \eta_5)(\eta - \eta_8)}{(\eta_2 - \eta_5)(\eta_2 - \eta_8)} = \frac{(\eta - 0)(\eta - 1)}{(-1 - 0)(-1 - 1)} = \frac{\eta(\eta - 1)}{2} \quad (5.136)$$

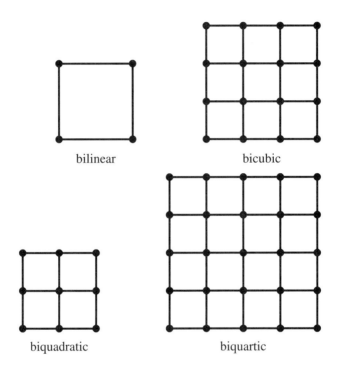

Figure 5.18 Sample of Planar Rectangular Lagrange Elements

Hence

$$\zeta_2(\xi, \eta) = L_2(\xi)L_2(\eta) = \frac{(1-\xi^2)\eta(\eta-1)}{2} \qquad (5.137)$$

Other high-order shape functions for rectangular elements can be formed in precisely the same manner and the procedure is fully described in reference [63]. Figure 5.18 shows a sample of Lagrange elements that may find application.

Shape Functions

We may summarize that irrespective of the order of the expansion, the interpolation function and the potential solution in two dimensions can be expressed in terms of Lagrange interpolation functions as follows.

$$\zeta_j(\xi, \eta) = L_j(\xi) L_j(\eta) \tag{5.138}$$

$$\psi(\xi, \eta) = \sum_{j=1}^{N} L_j(\xi) L_j(\eta) \psi_j \tag{5.139}$$

Although the interpolation functions in equations (5.138) and (5.139) are described in terms of local coordinates, it is a simple matter to recover their form in the global coordinate system by substituting

$$\xi = \frac{x - x_0}{a}, \quad \eta = \frac{y - y_0}{b} \tag{5.140}$$

Also, the order of the polynomials given by equations (5.138) and (5.139) is such that it ensures C^0 continuity along element edges or sides.

Lagrange elements are limited in their application because high-order elements proliferate nodes, a good number of which are interior nodes. Node condensation has been effectively used to eliminate the interior nodes, but it requires extra algebraic manipulation. Another important property of Lagrange elements is that the interpolation functions are not complete polynomials and they possess geometric isotropy only when equal numbers of nodes are used in the x and y directions.

A more useful family of rectangular elements known as *serendipity* elements devised by Ergatoudis et al. [64] is shown in Figure 5.19. These elements do not contain interior nodes and have nodes only along the edges of the rectangle.

Because the number of nodes on each side or edge constitute the required degrees of freedom for a complete polynomial of a given order along the edge, C^0 continuity is satisfied on interelement boundaries. Originally the interpolation functions for serendipity elements were devised by inspection. However, they can be derived rigorously by Lagrange interpolation suitably modified. We shall illustrate this procedure for a quadratic element as follows.

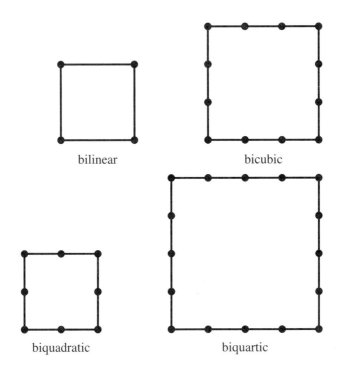

Figure 5.19 Sample Rectangular Serendipity Elements

5.8 DERIVATION OF SHAPE FUNCTIONS FOR SERENDIPITY ELEMENTS

For a planar rectangular element with four nodes, the serendipity element is the same as the Lagrangian element and, therefore, the procedure will not be repeated here. However, if we consider a second-order rectangular element, the procedure is as follows.

As we have already stated, the serendipity element has only nodes on the sides or edges as shown in Figure 5.20.

Shape Functions

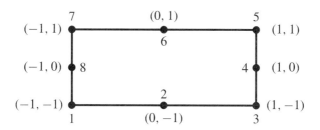

Figure 5.20 An Eight Noded Second-Order Rectangular Element and Local Coordinates of the Nodes

We first derive the shape functions for the four corner nodes 1, 3, 5, and 7, as for a four-noded rectangular element. Thus, as before,

$$\zeta_k = \frac{(1+\xi_p\xi)(1+\eta_p\eta)}{4} \tag{5.141}$$

where $\xi_p = 1$ for nodes 3 and 5, $\xi_p = -1$ for nodes 1 and 7, $\eta_p = 1$ for nodes 5 and 7, and $\eta_p = -1$ for nodes 1 and 3. Next we consider the shape functions for the midside nodes. Let us, for example, derive the shape functions for nodes 2 and 8. The shape function for node 2, by Lagrangian interpolation, can be obtained as the product of linear interpolation along y and quadratic interpolation along x. Therefore,

$$\zeta_2 = L_2(x)L_2(y) \tag{5.142}$$

where

$$L_2(y) = \frac{\eta - \eta_6}{\eta_2 - \eta_6} \tag{5.143}$$

Note that we have used the local coordinates instead of the global coordinates x, y in equation (5.143). The values of the local coordinates as shown in Figure 5.20 are $\eta_2 = -1$, $\eta_6 = 1$. Therefore,

$$L_2(y) = -\frac{(\eta - 1)}{2} \tag{5.144}$$

Similarly,

$$L_2(x) = \frac{(\xi - \xi_1)(\xi - \xi_3)}{(\xi_2 - \xi_1)(\xi_2 - \xi_3)} = \frac{(\xi + 1)(\xi - 1)}{(0 + 1)(0 - 1)} = -(\xi^2 - 1) \tag{5.145}$$

Thus,

$$\zeta_2 = \frac{(\eta - 1)(\xi^2 - 1)}{2} \tag{5.146}$$

We also observe that the linear shape functions at nodes 1 and 3 from equation (5.141) do not vanish at node 2. This is because the local coordinates of node 2 ($\xi = 0$, $\eta = -1$), when substituted in the shape functions for ζ_1 and ζ_3, will be

$$\zeta_1 = \frac{(1 - \xi)(1 - \eta)}{4} = \frac{1}{2} \tag{5.147}$$

$$\zeta_3 = \frac{(1 + \xi)(1 - \eta)}{4} = \frac{1}{2} \tag{5.148}$$

However, the shape functions at nodes 5 and 7 do vanish at node 2. We must, therefore, correct the shape functions ζ_1 and ζ_3 so that they vanish at node 2. This is accomplished by subtracting one half of the value of the shape function at node 2 from the linear shape functions at nodes 1 and 3.

Shape Functions

Thus,

$$\zeta_1 = \frac{(1-\xi)(1-\eta)}{4} - \frac{(\xi^2-1)(\eta-1)}{4} \tag{5.149}$$

which after some algebraic manipulation reduces to the form

$$\zeta_1 = -\frac{\xi(\xi-1)(\eta-1)}{4} \tag{5.150}$$

Similarly,

$$\zeta_3 = -\frac{\xi(\xi+1)(\eta-1)}{4} \tag{5.151}$$

However, this is not the whole story. We must consider the shape function for the midside node 8, which is obtained from the product of the linear Lagrange interpolation along x and quadratic interpolation along y. Therefore,

$$\zeta_8 = \frac{(\xi-\xi_4)(\eta-\eta_1)(\eta-\eta_7)}{(\xi_8-\xi_4)(\eta_8-\eta_1)(\eta_8-\eta_7)} = \frac{(\xi-1)(\eta^2-1)}{2} \tag{5.152}$$

We also note that the partially modified shape function at node 1 does not vanish at node 8, because at node 8 ($\xi = -1$, $\eta = 0$), therefore

$$\zeta_1(new) = -\frac{\xi(\xi-1)(\eta-1)}{4} - \frac{(\xi-1)(\eta^2-1)}{4} \tag{5.153}$$

which after some algebra yields

$$\zeta_1 = \frac{(\eta-1)(\xi-1)(-\xi-\eta-1)}{4} \tag{5.154}$$

Similarly, we can obtain the correct shape function for node 3 as

$$\zeta_3 = \frac{(1+\xi)(1-\eta)}{4} - \frac{1}{2}\zeta_2 - \frac{1}{2}\zeta_4 \qquad (5.155)$$

or

$$\zeta_3 = \frac{(1+\xi)(1-\eta)}{4} - \frac{(\eta-1)(\xi^2-1)}{4} - \frac{(\xi-\xi_8)(\eta-\eta_3)(\eta-\eta_5)}{2(\xi_4-\xi_8)(\eta_4-\eta_3)(\eta_4-\eta_5)} \qquad (5.156)$$

Substituting the appropriate values of $\xi_4, \xi_8, \eta_3, \eta_4, \eta_5$, one obtains

$$\zeta_3 = \frac{(1+\xi)(1-\eta)}{4} - \frac{(\xi+1)(\xi-1)(\eta-1)}{4} - \frac{(\xi-(-1))(\eta-(-1))(\eta-1)}{2(1-(-1))(0-(-1))(0-1)} \qquad (5.157)$$

which after some manipulation reduces to the form

$$\zeta_3 = -\frac{(\xi+1)(\eta-1)(\xi-\eta-1)}{4} \qquad (5.158)$$

Following this procedure, the shape functions for the remaining two corner nodes are obtained as

$$\zeta_5 = \frac{(\xi+1)(\eta+1)(\xi+\eta-1)}{4} \text{ and } \zeta_7 = \frac{(1-\xi)(1+\eta)(-\xi+\eta-1)}{4} \qquad (5.159)$$

Shape Functions

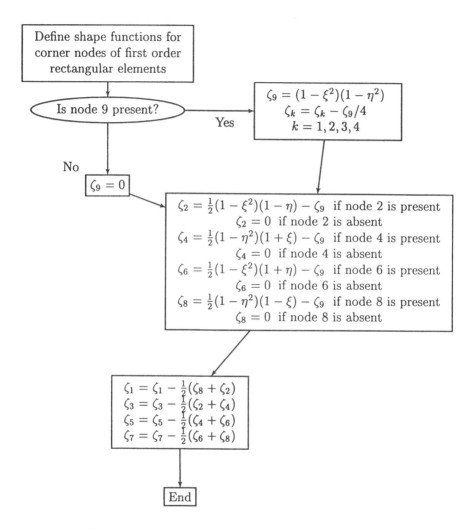

Figure 5.21 Shape Functions for Four- Through Nine-Noded Elements

The method of deriving the shape functions for the second-order rectangular element including the eight-noded serendipity element from first-order element shape functions has been formalized in reference [65] in the form of the flowchart of Figure 5.21. The procedure described enables all combinations of elements from four to nine nodes. The generalization of high-order Lagrangian and serendipity quadrilateral elements are shown in Figures 5.22 and 5.23.

5.9 THREE-DIMENSIONAL FINITE ELEMENTS

In general, the methods described for two-dimensional elements can be extended to three-dimensional finite elements without much difficulty. We shall illustrate the procedure with respect to tetrahedral elements, triangular prisms, and hexahedral elements. Other special types of elements are better handled by isoparametric representation which will be discussed in a later section.

It is well known that a line, a triangle and a tetrahedron constitute a family of simplexes in one-, two-, and three-dimensional coordinate space, respectively. As a result, the number of vertices yield complete polynomials of order one. For the triangle and the tetrahedron, the number of nodes generated by the Pascal arraying procedure for a given order of interpolation will suffice to yield complete polynomials.

5.9.1 Tetrahedral Elements

Let us consider the simple four-noded tetrahedral element shown in Figure 5.24.

The field variable can be expressed by the linear interpolation function

$$\phi = a_0 + a_1 x + a_2 y + a_3 z \tag{5.160}$$

As in the case of the triangle, we can generate the matrix relation for the nodal values of ϕ as

$$\begin{pmatrix} \phi_1 \\ \phi_2 \\ \phi_3 \\ \phi_4 \end{pmatrix} = \begin{pmatrix} 1 & x_1 & y_1 & z_1 \\ 1 & x_2 & y_2 & z_2 \\ 1 & x_3 & y_3 & z_3 \\ 1 & x_4 & y_4 & z_4 \end{pmatrix} \begin{pmatrix} a_0 \\ a_1 \\ a_2 \\ a_3 \end{pmatrix} \tag{5.161}$$

or in matrix notation

$$(\phi_i) = (D)(a) \tag{5.162}$$

Shape Functions 241

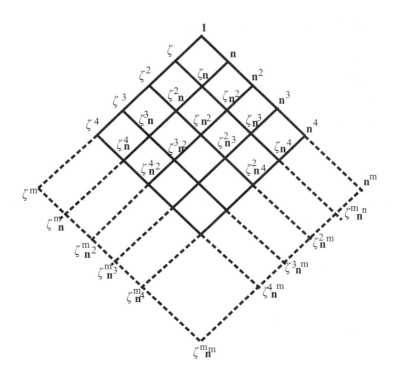

Figure 5.22 Lagrangian Quadrilateral Elements

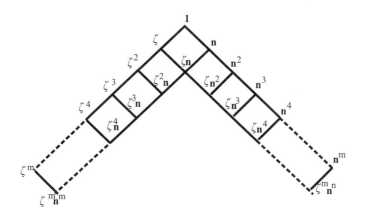

Figure 5.23 Serendipity Quadrilateral Elements

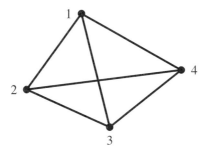

Figure 5.24 A Four-Noded Tetrahedral Element

For the triangle, the shape functions are derived as follows:

$$\phi = [1, x, y, z](D)^{-1}(\phi_i) = (\zeta_i)(\phi_i) \qquad (5.163)$$

yielding

$$(\zeta_i) = [1, x, y, z](D)^{-1} \qquad (5.164)$$

For high-order elements, the interpolation functions can be developed using the arraying system illustrated in reference [49]. For the sake of completeness, a linear, a quadratic, and a cubic element of the tetrahedral family are shown in Figure 5.25.

It is evident that the interpolation function on any face of the tetrahedron will be the same as that for the corresponding triangular element of the same order.

The procedure described earlier for deriving shape functions for tetrahedral elements is laborious, because for each element, the inverse of the matrix $[D]$ must be numerically determined. For high-order elements, computer implementation of the inverses will be prohibitive. A simpler and more elegant procedure was proposed by Silvester [60] using

Shape Functions

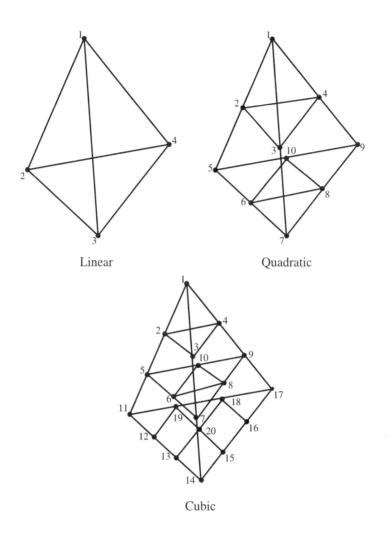

Figure 5.25 First-, Second-, and Third-Order Tetrahedral Elements

a four-digit node numbering scheme for the three-dimensional elements as a natural extension of the three-digit representation for triangular elements. The interpolation function is then expressed as

$$\zeta_{\alpha\beta\gamma\theta}(\xi_1, \xi_2, \xi_3 \xi_4) = \zeta_\alpha(\xi_1)\zeta_\beta(\xi_2)\zeta_\gamma(\xi_3)\zeta_\theta(\xi_4) \quad (5.165)$$

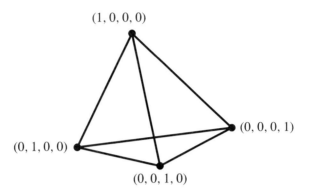

Figure 5.26 A First-Order Tetrahedral Element with Four-Digit Node Numbering

where

$$\zeta_\alpha(\xi_1) = \prod_{i=1}^{\alpha} \frac{\eta\xi_1 - i + 1}{i} \quad \text{for } \alpha \geq 1$$
$$= 1 \quad \text{for } \alpha = 0 \tag{5.166}$$

and n is the order of the interpolation polynomial.

As an example to illustrate this, we shall derive the shape function for node 1 of the first-order tetrahedral element shown in Figure 5.26 with four-digit node numbering.

$$\zeta_1(\xi_1) = \frac{\eta\xi_1 - 1 + 1}{1} \times 1 \times 1 \times 1 = \xi_1 \tag{5.167}$$

Because $\eta = 1$

$$\zeta_0(\xi_2) = 1 \tag{5.168}$$

Shape Functions

Similarly,

$$\zeta_0(\xi_3) = 1, \quad \zeta_0(\xi_4) = 1 \tag{5.169}$$

Therefore

$$\zeta_{1,0,0,0} = \zeta_1(\xi_1)\zeta_0(\xi_2)\zeta_0(\xi_3)\zeta_0(\xi_4) = \xi_1 \tag{5.170}$$

Following the preceding procedure,

$$\zeta_{0,1,0,0} = \xi_2, \quad \zeta_{0,0,1,0} = \xi_3, \quad \text{and} \quad \zeta_{0,0,0,1} = \xi_4 \tag{5.171}$$

Thus the field variable can be expressed as

$$\phi = \xi_1\phi_1 + \xi_2\phi_2 + \xi_3\phi_3 + \xi_4\phi_4 \tag{5.172}$$

We can derive shape functions of any order following this procedure and using the Pascal arraying system discussed before. The following are expressions for shape functions for tetrahedral elements of orders one to three.

Linear four-noded element:

$$\zeta_i = \xi_i, \quad i = 1, 2, 3, 4 \tag{5.173}$$

Quadratic 10 nodes:

$$\begin{aligned}\zeta_i &= (2\xi_i - 1)\xi_i, \quad \text{corner nodes} \\ \zeta_k &= 4\xi_i\xi_j, \quad \text{midside nodes}\end{aligned} \tag{5.174}$$

Cubic 20 nodes:

$$\zeta_i = \frac{1}{2}(3\xi_i - 1)(3\xi_i - 2)\xi_i, \quad \text{corner nodes}$$

$$\zeta_k = \frac{9}{2}\xi_i\xi_j(3\xi_i - 1), \quad \text{midside nodes}$$

$$\zeta_\ell = 27\xi_i\xi_j\xi_m, \quad \text{midface nodes} \tag{5.175}$$

5.9.2 Natural or Volume Coordinates in Three Dimensions

Recalling the procedure for the three-noded triangle, we can define natural or volume coordinates for the linear tetrahedral elements with four nodes. These will be the ratios of volumes inside the tetrahedron.

The volume coordinates of the point P inside the tetrahedron as shown in Figure 5.27 are defined as follows:

$$\xi_1 = \frac{\text{VOL P234}}{\text{VOL 1234}}$$

$$\xi_2 = \frac{\text{VOL P341}}{\text{VOL 1234}}$$

$$\xi_3 = \frac{\text{VOL P124}}{\text{VOL 1234}}$$

$$\xi_4 = \frac{\text{VOL P321}}{\text{VOL 1234}} \tag{5.176}$$

We can obtain the values of the volume coordinates $\xi_1, \xi_2, \xi_3, \xi_4$ by expanding the coordinates of P as

$$x = \xi_1 x_1 + \xi_2 x_2 + \xi_3 x_3 + \xi_4 x_4 \tag{5.177}$$

$$y = \xi_1 y_1 + \xi_2 y_2 + \xi_3 y_3 + \xi_4 y_4 \tag{5.178}$$

$$z = \xi_1 z_1 + \xi_2 z_2 + \xi_3 z_3 + \xi_4 z_4 \tag{5.179}$$

$$\xi_1 + \xi_2 + \xi_3 + \xi_4 = 1 \tag{5.180}$$

Shape Functions

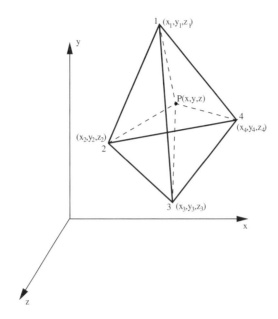

Figure 5.27 Volume Coordinates of a Linear Tetrahedral Element

Note: The auxiliary equation (5.180) is derived from the fact that the sum of the partial volumes within the element equals the total volume of the tetrahedron. The ξ's can be obtained by rewriting equations (5.177) through (5.180) in matrix form, inverting the coefficient matrix thus obtained, and performing the remaining matrix operations as in the case of the triangular element. After some algebra, it can be shown that

$$\xi_i = \frac{a_i + b_i x + c_i y + d_i z}{6V}, \quad i = 1, 2, 3, 4 \tag{5.181}$$

where the determinant of the coefficient matrix is given by

$$\det = 6V = \begin{vmatrix} 1 & x_1 & y_1 & z_1 \\ 1 & x_2 & y_2 & z_2 \\ 1 & x_3 & y_3 & z_3 \\ 1 & x_4 & y_4 & z_4 \end{vmatrix} \tag{5.182}$$

and

$$a_1 = \begin{vmatrix} x_2 & y_2 & z_2 \\ x_3 & y_3 & z_3 \\ x_4 & y_4 & z_4 \end{vmatrix}, \quad b_1 = -\begin{vmatrix} 1 & y_2 & z_2 \\ 1 & y_3 & z_3 \\ 1 & y_4 & z_4 \end{vmatrix},$$

$$c_1 = \begin{vmatrix} 1 & x_2 & z_2 \\ 1 & x_3 & z_3 \\ 1 & x_4 & z_4 \end{vmatrix}, \quad d_1 = -\begin{vmatrix} 1 & x_2 & y_2 \\ 1 & x_3 & y_3 \\ 1 & x_4 & y_4 \end{vmatrix} \quad (5.183)$$

The other geometric coefficients are derived by permuting the subscripts of a, b, c, and d cyclically and assigning the appropriate sign +1 or -1 corresponding to whether the sum of the row and column of the cofactor is even or odd, respectively.

5.9.3 Derivatives and Integration Procedure

In the course of implementing the finite element method, one needs to evaluate the derivatives of the field variable with respect to the global coordinates and perform integrations over the element volume.

The derivatives with respect to x, y, and z are easily obtained by the chain rule of differentiation, so that

$$\frac{\partial \phi}{\partial x, y, \text{ or } z} = \sum_{j=1}^{4} \frac{\partial \phi}{\partial \xi_j} \frac{\partial \xi_j}{\partial x, y, \text{ or } z} = \sum_{j=1}^{4} \frac{\partial \phi}{\partial \xi_j} \frac{b_i, c_i, \text{ or } d_i}{6V} \quad (5.184)$$

Integrations of expressions containing the shape functions and correspondingly the volume coordinates over the element are performed using the following integration formula.

$$I = \frac{1}{V} \int_{vol} \xi_1^\alpha \xi_2^\beta \xi_3^\gamma \xi_4^\theta \, dV = \frac{\alpha! \beta! \gamma! \theta!}{(\alpha + \beta + \gamma + \theta + 3)!} 6V \quad (5.185)$$

Tabulated values of equation (5.185) for different combinations of α, β, γ, and θ are included in Appendix D.

Shape Functions

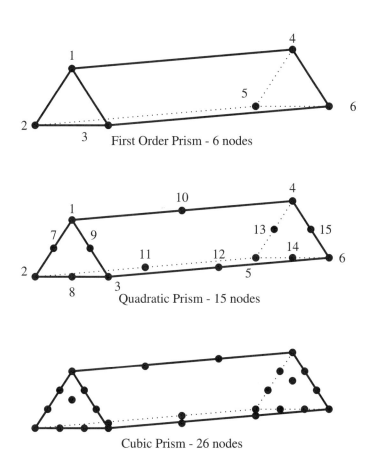

Figure 5.28 Prism Elements

5.9.4 Triangular Prisms

The procedure for obtaining shape functions for triangular prisms from triangular elements is described fully in reference [66]. The method can be illustrated with respect to the first-order triangular prism element shown in Figure 5.28.

The shape functions for the triangular prism are obtained as the product of the shape functions of the triangular faces (cross section) and the shape function for the line element corresponding to the z dimension. This procedure is applicable to both Lagrange and serendipity types of elements. The main advantage of using triangular prisms either by themselves or in combination with hexahedral elements for discretizing the geometry is that they fit irregular boundaries better and more easily than hexahedral elements alone. It must be noted from Figure 5.28 that the sides along the z dimension are rectangles.

For the preceding right triangular first-order prism, the shape functions are obtained as

$$\zeta_i = \xi_i \frac{1 \pm \lambda}{2}, \quad i = 1, 2, 3 \tag{5.186}$$

where ζ_i is the shape function of the triangular prism at node i, ξ_i is the shape function of the triangular face at node i and λ_i is the half length of the prism along the z dimension.

Shape functions for high-order elements can also be obtained by following the same procedure. For purposes of illustration, the shape functions for linear and quadratic prism elements from reference [49] are presented next.

Linear element (6 nodes):

$$\zeta_i = \xi_i \frac{1 \pm \lambda}{2} \tag{5.187}$$

Quadratic element (15 nodes):

$$\zeta_i = \frac{1}{2}\xi_i(2\xi_i - 1)(1 + \lambda) - \frac{1}{2}(1 - \lambda^2)\xi_i \tag{5.188}$$

For the midsides of the triangles,

$$\zeta_k = 2\xi_i \xi_j (1 + \lambda) \tag{5.189}$$

For the midsides of the rectangles,

$$\zeta_k = \xi_i (1 - \lambda^2) \tag{5.190}$$

Shape Functions

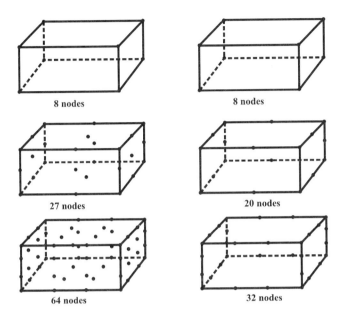

Figure 5.29 Lagrange- and Serendipity-Type Hexahedral Elements

5.9.5 Hexahedral Elements

The method of deriving shape functions for Lagrange- or serendipity-type two-dimensional rectangular elements described in earlier sections can be easily extended to hexahedral elements in three dimensions. Figure 5.29 illustrates the two types of elements for first- through third-order interpolation.

The shape functions for these elements are derived in exactly the same manner as described for the corresponding rectangular elements. For example, the element shape functions for the linear eight-noded Lagrangian element are obtained by taking the product of the corresponding linear interpolation functions in the x, y, and z directions, respectively. Thus, for node 1 for the linear Lagrange element shown in Figure 5.30,

$$\zeta_1 = \zeta_{x1}\zeta_{y1}\zeta_{z1} \tag{5.191}$$

where

$$\zeta_{x1} = \frac{x - x_2}{x_1 - x_2}$$

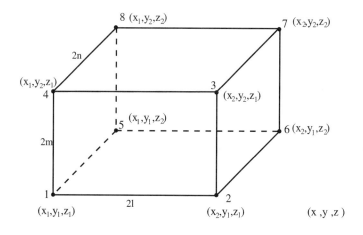

Figure 5.30 Linear Lagrange Element with Node Numbering and Coordinates

$$\zeta_{y1} = \frac{y - y_2}{y_1 - y_2} \qquad (5.192)$$

$$\zeta_{z1} = \frac{z - z_2}{z_1 - z_2}$$

Substituting $x_1 = -l$ and $x_2 = +l$, $y_1 = -m$ and $y_2 = +m$, $z_1 = -n$ and $z_2 = +n$, and introducing nondimensional parameters or natural coordinates $\xi = \frac{x}{l}$, $\eta = \frac{y}{m}$, and $\lambda = \frac{z}{n}$, the shape function for node 1 is obtained as

$$\zeta_1 = \frac{(1-\xi)(1-\eta)(1-\lambda)}{8} \qquad (5.193)$$

Similarly, the shape functions for the other nodes can also be obtained. For high-order elements, Lagrange-type elements involve a very large number of nodes and therefore are used only in special cases. The more commonly used ones are the Serendipity elements. The procedure for obtaining these follow the method described earlier for two-dimensional elements and will not be repeated here. For the reader's convenience, the shape functions for the serendipity-type hexahedral elements for first through third order are given next:

Shape Functions

Linear element:

$$\zeta_i = \frac{1}{8}(1 + \xi_i\xi)(1 + \eta_i\eta)(1 + \lambda_i\lambda), \quad \xi_i = \pm 1, \ \eta_i = \pm 1, \ \lambda_i = \pm 1 \quad (5.194)$$

Quadratic elements: for the corner nodes

$$\zeta_i = \frac{1}{8}(1 + \xi_i\xi)(1 + \eta_i\eta)(1 + \lambda_i\lambda)(\xi_i\xi + \eta_i\eta + \lambda_i\lambda - 2) \quad (5.195)$$

and for the midside nodes

$$\zeta_i = \frac{1}{4}(1 - \xi^2)(1 + \eta_i\eta)(1 + \lambda_i\lambda), \quad \eta_i = \pm 1, \ \lambda_i = \pm 1, \ \xi_i = 0 \quad (5.196)$$

Cubic elements: for the corner nodes

$$\zeta_i = \frac{1}{64}(1+\xi_i\xi)(1+\eta_i\eta)(1+\lambda_i\lambda)[9(\xi^2+\eta^2+\lambda^2)-19], \quad \xi_i = \pm 1, \ \eta_i = \pm 1, \ \lambda_i = \pm 1 \quad (5.197)$$

and for the midside nodes

$$\zeta_i = \frac{9}{64}(1-\xi^2)(1+9\xi_i\xi)(1+\eta_i\eta)(1+\lambda_i\lambda), \quad \xi_i = \pm\frac{1}{3}, \ \eta_i = \pm 1, \ \lambda_1 = \pm 1 \quad (5.198)$$

Irregular hexahedral element shape functions have been derived for certain specified geometric shapes by Armor *et al.* [67], and by assembling groups of tetrahedral elements as described by Clough and Felippa [68]. The general irregular curvilinear tetrahedral and hexahedral elements and prisms are best modeled by isoparametric elements, which will be discussed in the following section.

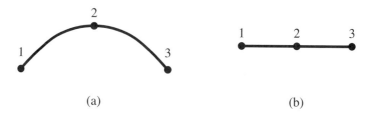

Figure 5.31 Curved Element: (a) Original Element; (b) Parent Element

5.9.6 Isoparametric Elements

Regular quadrilateral and triangular elements in two dimensions and tetrahedral and hexahedral elements and prisms in three-dimensional problems gene-rally suffice to model the geometry including the boundaries, which are often irregular. Although several researchers have developed their own set of irregular two- and three-dimensional elements for specific applications, these were not always found satisfactory for modeling curved geometries and boundaries. The isoparametric element, therefore, has become a much desired tool in describing complex geometrical shapes. Further, because these elements permit modeling of curved boundaries effectively, it would be possible to use a smaller number of large elements and yet achieve an accurate boundary representation. For three-dimensional problems, where the number of unknowns and the resulting matrix size may far exceed the memory of most powerful computers of today, isoparametric elements will prove advantageous by providing the means to reduce the number of elements required for discretizing the field region or domain. The development of isoparametric elements is attributed to Taig [69], Irons [59], Ergatoudis [64] , Zienkiewicz [70] and others.

The term *isoparametric element* refers to the representation of the geometry of the element and the interpolatory or shape functions therein in terms of the same (iso) parameters. The basic notion of isoparametric representation consists of mapping or transforming simple regular geometrical shapes in a local coordinate system into distorted, irregular, or curved shapes in the global coordinate system. The converse is also true; i.e., for example, a curved geometry can be transformed to a regular geometry by a particular manner of description of the interpolation functions. Let us clarify these ideas with reference to the curved line element shown in Figure 5.31.

Shape Functions

We shall assume that the interpolation function is quadratic. Therefore, the number of degrees of freedom is three, denoted by the nodes 1, 2, and 3. If we define a nondimensional parameter

$$\xi = \frac{x}{\ell} \tag{5.199}$$

where 2ℓ is the arc length 1–3, then it is evident that at $x = -l$, the value of $\xi = -1$, and at $x = l$, its value is $+1$. The associated shape functions for nodes 1, 2, and 3, as given previously are

$$\begin{aligned}\zeta_1 &= -\frac{\xi(1-\xi)}{2} \\ \zeta_2 &= 1-\xi^2 \\ \zeta_3 &= \frac{\xi(1+\xi)}{2}\end{aligned} \tag{5.200}$$

In isoparametric mapping, we use the same interpolation or shape functions for the representation of the x coordinate as we would for the function. Therefore,

$$x = \sum \zeta_i x_i \tag{5.201}$$

$$\phi = \sum \zeta_i \phi_i \tag{5.202}$$

Then from equations (5.200) and (5.201), we can obtain the value of x at any point on the arc as

$$x = \frac{\xi(\xi-1)}{2} x_1 + (1-\xi^2) x_2 + \frac{\xi(\xi+1)}{2} x_3 \tag{5.203}$$

If, for example, we wish to evaluate x at $\xi = -\frac{1}{2}$, then

$$x = \left(-\frac{1}{2}\right)\left(-\frac{3}{2}\right)(-l)/2 - \left(\frac{1}{4} - 1\right)(0) + \left(-\frac{1}{2}\right)\left(-\frac{1}{2} + 1\right)\frac{l}{2} = -\frac{l}{2} \quad (5.204)$$

which is midway between nodes 1 and 2, as $\xi = -\frac{1}{2}$ is midway between nodes 1 and 2 in the parent element. Thus we have mapped the point $\xi = -\frac{1}{2}$ from the parent element to its global coordinates in the curved element exactly. This is a one-to-one mapping of the point being considered. Similarly, we can find other values of x in the curved element corresponding to different values of ξ in the parent element.

If the shape functions which define the coordinates, and which can be used to transform them from one system to another, also define the potential function, it is called isoparametric mapping. In reference [49], it is shown that isoparametric elements satisfy C^0 continuity (i.e., at the element interior at least one first-order derivative exists, potential continuity is established at element interfaces, and the completeness criterion $\sum \zeta_i = 1$ is satisfied).

We can also extend these ideas to two- and three-dimensional interpolation. Let us now consider the treatment of the simple irregular four-noded first-order quadrilateral element shown in Figure 5.32b. The parent element is shown in Figure 5.32a, with the appropriate values of the nondimensional or natural coordinates.

The shape functions of the parent element are identical to those of a regular rectangular element and are reproduced here for convenience.

$$\zeta_i = \frac{(1 + \xi_i \xi)(1 + \eta_i \eta)}{4}$$

where $\xi_i = 1$ for nodes 2 and 3, $\xi_i = -1$ for nodes 1 and 4, $\eta_i = 1$ for nodes 3 and 4, and $\eta_i = -1$ for nodes 1 and 2.

As before, for the curved line element, we shall illustrate the isoparametric mapping of the x and y coordinates from the ξ, η plane to the x, y plane.

$$x = \sum_{i=1}^{4} \zeta_i x_i \quad (5.205)$$

Shape Functions

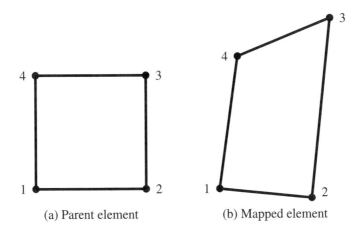

Figure 5.32 Isoparametric Mapping of a Four-Noded Rectangular Element into a Corresponding Linear Quadrilateral Element

$$y = \sum_{i=1}^{4} \zeta_i y_i \qquad (5.206)$$

Suppose we wish to find the values of x, y corresponding to, say, $\xi = 0$ and $\eta = 1$, which is midway between nodes 3 and 4 in the parent element; the values of the shape functions corresponding to the values of ξ, η will be

$$\zeta_1 = \frac{1}{4}(1-0)(1-1) = 0 \qquad (5.207)$$

$$\zeta_2 = \frac{1}{4}(1+0)(1-1) = 0 \qquad (5.208)$$

$$\zeta_3 = \frac{1}{4}(1+0)(1+1) = \frac{1}{2} \qquad (5.209)$$

$$\zeta_4 = \frac{1}{4}(1-0)(1+1) = \frac{1}{2} \qquad (5.210)$$

Hence

$$x = \frac{x_3 + x_4}{2} \qquad (5.211)$$

$$y = \frac{y_3 + y_4}{2} \qquad (5.212)$$

which are the values of a point midway between nodes 3 and 4 in the irregular quadrilateral element. We can show by the same reasoning that every point in the parent element identically maps into the corresponding point in the irregular quadrilateral. The mapping defined by equations (5.205) and (5.206) used linear shape functions, which resulted in the irregular quadrilateral element. The functional representation of the field variables and the functional representation of the irregular boundaries are expressed by interpolation functions of the same order. These ideas also hold for curved elements as illustrated in reference [49] and, for easy understanding, shown in Figure 5.33.

Elements defined by the same interpolation functions for both the functional variable and the global coordinates are called isoparametric elements. If the geometry is described by a lower order polynomial than the one used for the field variable, then the element is called subparametric. If the geometry is described by a higher order interpolation function than that for the functional values, the element is called superparametric. Of the three types of elements of irregular shapes, isoparametric elements are the most commonly used in engineering applications. We shall, therefore, confine further discussion to isoparametric elements.

Equations (5.205) and (5.206) which express the transformation between local and global coordinates, assume that each point in one system is uniquely mapped into a corresponding point in the other. However, if the transformation is not unique, then undesirable distortions in the global system may occur, which would not be acceptable. One easy way to check that the transformation is unique is to ensure that the angle subtended by an arc of a curved element is less than 180°. Further, in the construction of irregular and curved isoparametric elements, it is necessary to preserve continuity

Shape Functions

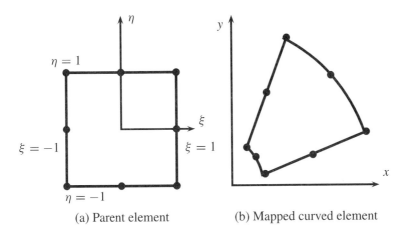

(a) Parent element   (b) Mapped curved element

Figure 5.33 Curved Quadrilateral Element Mapped from a Second-Order Rectangular Element

conditions in the global coordinate system. Some guidelines described in reference [49] are as follows.

- Two irregular or curved elements adjacently placed will preserve continuity if the shape functions of the parent elements from which they are generated satisfy interelement continuity.
- If the interpolations functions expressed in local coordinates preserve continuity of the field variables in the parent element, the field variables in the global coordinate system in the irregular or curved elements will also be continuous.
- If the completeness criterion is satisfied in the parent element, it will also be satisfied in the irregular or curved element.

Formulation of the shape functions is only the first step in implementing the finite element method. The latter involves the evaluation of derivatives and integrals of products of the function and its derivatives or the derivatives alone, which is discussed in the next chapter. It will suffice here for our discussion to consider integrals of functions of the form

$$I_1 = \int_\Gamma F\left(\phi, \frac{\partial \phi}{\partial x}\right) dx \quad \text{for 1D elements}$$

$$I_2 = \int_S F\left(\phi, \frac{\partial \phi}{\partial x}, \frac{\partial \phi}{\partial y}\right) dx\, dy \quad \text{for 2D elements}$$

$$I_3 = \int_V F\left(\phi, \frac{\partial \phi}{\partial x}, \frac{\partial \phi}{\partial y}, \frac{\partial \phi}{\partial z}\right) dx\, dy\, dz \quad \text{for 3D elements} \tag{5.213}$$

where Γ, S, and V are, respectively, the line, area and volume as applicable.

In order to obtain the derivatives of ϕ with respect to x, y and/or z, we need a transformation by which they can be determined from the derivatives with respect to the local coordinates. This is accomplished as follows

$$\frac{\partial \phi}{\partial x} = \sum_{i=1}^n \frac{\partial \zeta_i}{\partial x} \phi_i \;:\; \frac{\partial \phi}{\partial y} = \sum_{i=1}^n \frac{\partial \zeta_i}{\partial y} \phi_i;\; \frac{\partial \phi}{\partial z} = \sum_{i=1}^n \frac{\partial \zeta_i}{\partial z} \phi_i$$

By chain rule of differentiation, we can write the derivatives of the shape functions in equation (5.213) with respect to the local coordinates as

$$\frac{\partial \zeta}{\partial \xi} = \frac{\partial \zeta_i}{\partial x} \frac{dx}{d\xi} + \frac{\partial \zeta_i}{\partial y} \frac{dy}{d\xi} \quad \text{for 2D cases} \tag{5.214}$$

or

$$\frac{\partial \zeta}{\partial \xi} = \frac{\partial \zeta_i}{\partial x} \frac{dx}{d\xi} + \frac{\partial \zeta_i}{\partial y} \frac{dy}{d\xi} + \frac{\partial \zeta_i}{\partial z} \frac{dz}{d\xi} \quad \text{for 3D cases}$$

The one-dimensional case is a subset of the preceding case and is obtained by dropping all terms except those containing x or derivatives thereof. Similarly,

$$\frac{\partial \zeta_i}{\partial \eta} = \frac{\partial \zeta_i}{\partial x} \frac{dx}{d\eta} + \frac{\partial \zeta_i}{\partial y} \frac{dy}{d\eta} \quad \text{for 2D cases} \tag{5.215}$$

or

$$\frac{\partial \zeta_i}{\partial \eta} = \frac{\partial \zeta_i}{\partial x} \frac{dx}{d\eta} + \frac{\partial \zeta_i}{\partial y} \frac{dy}{d\eta} + \frac{\partial \zeta_i}{\partial z} \frac{dz}{d\eta} \quad \text{for 3D cases} \tag{5.216}$$

$$\frac{\partial \zeta_i}{\partial \lambda} = \frac{\partial \zeta_i}{\partial x}\frac{dx}{d\lambda} + \frac{\partial \zeta_i}{\partial y}\frac{dy}{d\lambda} + \frac{\partial \zeta_i}{\partial z}\frac{dz}{d\lambda} \quad \text{for 3D cases} \tag{5.217}$$

In matrix form, equations (5.214)–(5.217) can be expressed as

$$\begin{pmatrix} \frac{\partial \zeta_i}{\partial \xi} \\ \frac{\partial \zeta_i}{\partial \eta} \end{pmatrix} = \begin{pmatrix} \frac{\partial x}{\partial \xi} & \frac{\partial y}{\partial \xi} \\ \frac{\partial x}{\partial \eta} & \frac{\partial y}{\partial \eta} \end{pmatrix} \begin{pmatrix} \frac{\partial \zeta_i}{\partial x} \\ \frac{\partial \zeta_i}{\partial y} \end{pmatrix} \tag{5.218}$$

for 2D problems

or

$$\begin{pmatrix} \frac{\partial \zeta_i}{\partial x} \\ \frac{\partial \zeta_i}{\partial y} \\ \frac{\partial \zeta_i}{\partial z} \end{pmatrix} = \begin{pmatrix} \frac{\partial x}{\partial \xi} & \frac{\partial y}{\partial \xi} & \frac{\partial z}{\partial \xi} \\ \frac{\partial x}{\partial \eta} & \frac{\partial y}{\partial \eta} & \frac{\partial z}{\partial \eta} \\ \frac{\partial x}{\partial \lambda} & \frac{\partial y}{\partial \lambda} & \frac{\partial z}{\partial \lambda} \end{pmatrix} \begin{pmatrix} \frac{\partial \zeta_i}{\partial x} \\ \frac{\partial \zeta_i}{\partial y} \\ \frac{\partial \zeta_i}{\partial z} \end{pmatrix} \tag{5.219}$$

for 3D problems

The square matrices in equations (5.218) and (5.219) are termed the Jacobian matrices or simply Jacobians. The derivatives of the shape functions are then obtained by inverting the Jacobians and multiplying the result by the derivatives of the shape functions with respect to the local coordinates. This process can be formalized as

$$\begin{pmatrix} \frac{\partial \zeta_i}{\partial x} \\ \frac{\partial \zeta_i}{\partial y} \end{pmatrix} = (J)^{-1} \begin{pmatrix} \frac{\partial \zeta_i}{\partial \xi} \\ \frac{\partial \zeta_i}{\partial \eta} \end{pmatrix} \tag{5.220}$$

for 2D problems and

$$\begin{pmatrix} \frac{\partial \zeta_i}{\partial x} \\ \frac{\partial \zeta_i}{\partial y} \\ \frac{\partial \zeta_i}{\partial z} \end{pmatrix} = (J)^{-1} \begin{pmatrix} \frac{\partial \zeta_i}{\partial \xi} \\ \frac{\partial \zeta_i}{\partial \eta} \\ \frac{\partial \zeta_i}{\partial \lambda} \end{pmatrix} \tag{5.221}$$

for 3D problems.

Similarly, we can show as explained in reference [49] that

$$da = dx\,dy = |J|\,d\xi\,d\eta \qquad (5.222)$$

for 2D problems or

$$dV = dx\,dy\,dz = |J|\,d\xi\,d\eta\,d\lambda \qquad (5.223)$$

for 3D problems.

The matrix operations described in equations (5.218) and (5.219) presuppose the existence of the inverse for each of the elements of the discretized field region. The uniqueness of the isoparametric transformation and the acceptable level of element discretization are assured only if the inverse of the Jacobian matrix exists. This can be readily tested, as suggested in reference [71] by checking the sign of the Jacobian matrix. If the sign does not change in the solution region, we can be assured of acceptable transformation and mapping.

Using the preceding transformation, the integrals in equation (5.213) can be written in the form

$$\int_{-1}^{1}\int_{-1}^{1} F'(\xi,\eta)\,d\xi\,d\eta \qquad (5.224)$$

for 2D problems and

$$\int_{-1}^{1}\int_{-1}^{1}\int_{-1}^{1} F'(\xi,\eta,\lambda)\,d\xi\,d\eta\,d\lambda \qquad (5.225)$$

for 3D problems.

It must be noted that F' in the integral is not a simple function that can be evaluated analytically so that the integration is performed in closed form easily. Numerical integration is required in most cases and one seeks to use the method that yields the least error. Further, the numerical integrations have to be accurate enough to ensure convergence and avoid singularities. The reader is referred to Irons [59] and Zienkiewicz [71], who provide useful guidelines for practical engineering applications. As a practical aid to the intended user of this method, we suggest the use of Gaussian integration or quadrature, where the points of integration known as *Gauss points* are the zeros of the Legendre polynomial. As a simple example, a one-dimensional integration is illustrated as follows. Let us suppose we wish to evaluate the following integral.

$$I = \int_0^\ell \frac{\partial \phi}{\partial x_i} \cdot \frac{\partial \phi}{\partial x_j} \, dx \tag{5.226}$$

for $i = j$.

Assuming a linear first-order element, we can write the shape functions and the field variable and its derivative as follows:

$$\zeta_1 = 1 - \xi \tag{5.227}$$

$$\zeta_2 = \xi, \quad \text{where } \xi = 0 \text{ or } 1 \tag{5.228}$$

$$\phi = \zeta_1 x_1 + \zeta_2 x_2 \tag{5.229}$$

Similarly, x and its derivative can be written as

$$x = \zeta_1 x_1 + \zeta_2 x_2 \tag{5.230}$$

$$(J) = \left(\frac{\partial x}{\partial \xi}\right) = -x_1 + x_2 = \ell \tag{5.231}$$

because $x_1 = 0$ and $x_2 = \ell$. Therefore

$$dx = |J| \, d\xi = \ell d\xi \tag{5.232}$$

Then
$$\left(\frac{\partial x}{\partial \xi}\right)^{-1} = \frac{1}{\ell} \tag{5.233}$$

and
$$\frac{\partial \phi}{\partial x} = \frac{1}{\ell} \frac{\partial \phi}{\partial \xi} \tag{5.234}$$

From equation (5.229), by differentiation with respect to ξ,

$$\frac{\partial \phi}{\partial \xi} = -\phi_1 + \phi_2 \tag{5.235}$$

Therefore,
$$\frac{\partial \phi}{\partial x} = \frac{\phi_2 - \phi_1}{\ell} \tag{5.236}$$

Substituting for $\frac{\partial \phi}{\partial x} i$ and $\frac{\partial \phi}{\partial x} j$ from (5.236) in the integral of equation (5.226)

$$I = \int_{-1}^{1} \left(\frac{\phi_2 - \phi_1}{\ell}\right)^2 \ell \, d\xi = K \int_{-1}^{1} d\xi \tag{5.237}$$

where $K = \frac{(\phi_2 - \phi_1)^2}{\ell}$.

Although the preceding integral can be exactly performed, it will be evaluated by Gaussian quadrature as follows:

$$I = K \int_{-1}^{1} d\xi = \sum F(i) W(i) \tag{5.238}$$

where i is the Gauss point, $F(i)$ the function to be integrated, and $W(i)$ the corresponding weight.

Because $F(i) = K$, a constant, the integral of equation (5.238) reduces to the form

$$I = K \sum W(i) \tag{5.239}$$

From Gauss–Legendre tables of Gauss points and weights of Appendix D, $W(i)$ for a two-Gauss-point quadrature will be 1 for each Gauss point. Therefore,

$$I = 2K \tag{5.240}$$

the exact solution.

It must be noted that the choice of two Gauss points was predicated by the choice of the interpolation functions. Because it is a first-order interpolation, only two nodes or end points are required, the function can be evaluated only at these two nodes, and so forth. The integration procedure for two- and three-dimensional problems is an extension of the 1D case for each of the coordinate directions. This is described briefly in the following section.

5.9.7 Two-Dimensional Quadrilateral Element

Let us suppose we wish to evaluate the integral

$$I = \int \left(\frac{\partial \phi}{\partial x} i \cdot \frac{\partial \phi}{\partial x} j + \frac{\partial \phi}{\partial y} i \cdot \frac{\partial \phi}{\partial y} j \right) dx \, dy \tag{5.241}$$

The derivatives of the shape functions $\frac{\partial \phi}{\partial x}$ and $\frac{\partial \phi}{\partial y}$ for the irregular quadrilateral element cannot be evaluated directly. Therefore, we must map the element into its parent element as shown in Figure 5.34. The derivatives in (5.241) can be obtained as follows:

$$\frac{\partial \phi}{\partial x} = \sum_{i=1}^{4} \frac{\partial \zeta_i}{\partial x} \phi_i \tag{5.242}$$

$$\frac{\partial \phi}{\partial y} = \sum_{i=1}^{4} \frac{\partial \zeta_i}{\partial y} \phi_i \tag{5.243}$$

The derivatives with respect to the local coordinates are then obtained using the chain rule of differentiation, so that

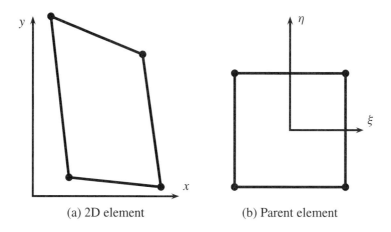

Figure 5.34 A Two-Dimensional Irregular Quadrilateral Element

$$\begin{aligned}\frac{\partial \zeta_i}{\partial \xi} &= \frac{\partial \zeta_i}{\partial x} \cdot \frac{\partial x}{\partial \xi} + \frac{\partial \zeta_i}{\partial y} \cdot \frac{\partial y}{\partial \xi} \\ \frac{\partial \zeta_i}{\partial \eta} &= \frac{\partial \zeta_i}{\partial x} \cdot \frac{\partial x}{\partial \eta} + \frac{\partial \zeta_i}{\partial y} \cdot \frac{\partial y}{\partial \eta} \end{aligned} \quad (5.244)$$

Equation (5.244) has been formalized by the Jacobian transformation and in matrix form will be

$$\begin{pmatrix} \frac{\partial \zeta_i}{\partial \xi} \\ \frac{\partial \zeta_i}{\partial \eta} \end{pmatrix} = (J) \begin{pmatrix} \frac{\partial \zeta_i}{\partial x} \\ \frac{\partial \zeta_i}{\partial y} \end{pmatrix} \quad (5.245)$$

where

$$(J) = \begin{pmatrix} \frac{\partial x}{\partial \xi} & \frac{\partial y}{\partial \xi} \\ \frac{\partial x}{\partial \eta} & \frac{\partial y}{\partial \eta} \end{pmatrix} \quad (5.246)$$

Shape Functions

By inverse transformation of equation (5.246), we obtain

$$\begin{pmatrix} \frac{\partial \zeta_i}{\partial x} \\ \frac{\partial \zeta_i}{\partial y} \end{pmatrix} = (J)^{-1} \begin{pmatrix} \frac{\partial \zeta_i}{\partial \xi} \\ \frac{\partial \zeta_i}{\partial \eta} \end{pmatrix} \quad (5.247)$$

and from equation (5.222)

$$dx\, dy = |J|\, d\xi\, d\eta \quad (5.248)$$

Rewriting the integral (5.241) in terms of the shape functions, we have

$$\begin{aligned} I &= \iint \left(\left(\frac{\partial \phi}{\partial x}\right)_i \left(\frac{\partial \phi}{\partial x}\right)_j + \left(\frac{\partial \phi}{\partial y}\right)_i \left(\frac{\partial \phi}{\partial y}\right)_j \right) dx\, dy \\ &= \iint \left(\sum\sum \left(\frac{\partial \zeta_i}{\partial x} \frac{\partial \zeta_j}{\partial x} + \frac{\partial \zeta_i}{\partial y} \frac{\partial \zeta_j}{\partial y} \right) \phi_j \right) dx\, dy \end{aligned} \quad (5.249)$$

Now, from equation (5.247), the derivatives of the shape functions with respect to the global coordinates can be obtained from the derivatives of the shape functions of the parent element in local coordinates. Therefore, the integral is transformed as

$$I = \int_{-1}^{1} \int_{-1}^{1} f(\xi, \eta) |J|\, d\xi\, d\eta \quad (5.250)$$

Note that the limits of integration -1 to 1 pertain to the limits of integration in the parent element. Using Gaussian quadrature, the integral of equation (5.250) is transformed as the summation as in the one-dimensional case as

$$I = \sum\sum F(\xi, \eta) W_i W_j \quad (5.251)$$

5.9.8 Three-Dimensional Curved Finite Elements

The associated functional for solving the scalar Laplace equation, for example, will be of the form

$$I = \iiint_V \left(\left(\frac{\partial \phi}{\partial x}\right)_i \left(\frac{\partial \phi}{\partial x}\right)_j + \left(\frac{\partial \phi}{\partial y}\right)_i \left(\frac{\partial \phi}{\partial y}\right)_j + \left(\frac{\partial \phi}{\partial z}\right)_i \left(\frac{\partial \phi}{\partial z}\right)_j \right) dx\, dy\, dz \tag{5.252}$$

The respective derivatives in the x, y, and z directions can be expressed in terms of the derivatives of the shape functions for a 20-noded brick element as follows:

$$\frac{\partial \phi}{\partial x} = \sum_{i=1}^{20} \frac{\partial \zeta_i}{\partial x} \phi_i$$

$$\frac{\partial \phi}{\partial y} = \sum_{i=1}^{20} \frac{\partial \zeta_i}{\partial y} \phi_i$$

$$\frac{\partial \phi}{\partial z} = \sum_{i=1}^{20} \frac{\partial \zeta_i}{\partial z} \phi_i \tag{5.253}$$

The derivatives with respect to the local coordinates (ξ, η, λ) are obtained, as before, by the chain rule of differentiation so that

$$\frac{\partial \zeta_i}{\partial \xi} = \frac{\partial \zeta_i}{\partial x} \cdot \frac{\partial x}{\partial \xi} + \frac{\partial \zeta_i}{\partial y} \cdot \frac{\partial y}{\partial \xi} + \frac{\partial \zeta_i}{\partial z} \frac{\partial z}{\partial \xi}$$

$$\frac{\partial \zeta_i}{\partial \eta} = \frac{\partial \zeta_i}{\partial x} \cdot \frac{\partial x}{\partial \eta} + \frac{\partial \zeta_i}{\partial y} \cdot \frac{\partial y}{\partial \eta} + \frac{\partial \zeta_i}{\partial z} \frac{\partial z}{\partial \eta}$$

$$\frac{\partial \zeta_i}{\partial \lambda} = \frac{\partial \zeta_i}{\partial x} \cdot \frac{\partial x}{\partial \lambda} + \frac{\partial \zeta_i}{\partial y} \cdot \frac{\partial y}{\partial \eta} + \frac{\partial \zeta_i}{\partial z} \frac{\partial z}{\partial \lambda} \tag{5.254}$$

Shape Functions

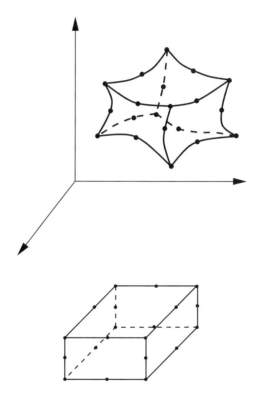

Figure 5.35 Three-Dimensional Second-Order Curved Finite Element (top) Parent Element (bottom)

In matrix form, this set of equations can be written as

$$\begin{pmatrix} \frac{\partial \zeta_i}{\partial \xi} \\ \frac{\partial \zeta_i}{\partial \eta} \\ \frac{\partial \zeta_i}{\partial \lambda} \end{pmatrix} = \begin{pmatrix} \frac{\partial x}{\partial \xi} & \frac{\partial y}{\partial \xi} & \frac{\partial z}{\partial \xi} \\ \frac{\partial x}{\partial \eta} & \frac{\partial y}{\partial \eta} & \frac{\partial z}{\partial \eta} \\ \frac{\partial x}{\partial \lambda} & \frac{\partial y}{\partial \lambda} & \frac{\partial z}{\partial \lambda} \end{pmatrix} \begin{pmatrix} \frac{\partial \zeta_i}{\partial x} \\ \frac{\partial \zeta_i}{\partial y} \\ \frac{\partial \zeta_i}{\partial z} \end{pmatrix} \qquad (5.255)$$

which is identical to (5.219). The square matrix in equation (5.255) is the Jacobian matrix, and the relationship between the derivatives with respect to the global and local coordinate systems may be expressed in the form

$$\begin{pmatrix} \frac{\partial \zeta_i}{\partial x} \\ \frac{\partial \zeta_i}{\partial y} \\ \frac{\partial \zeta_i}{\partial z} \end{pmatrix} = (J)^{-1} \begin{pmatrix} \frac{\partial \zeta_i}{\partial \xi} \\ \frac{\partial \zeta_i}{\partial \eta} \\ \frac{\partial \zeta_i}{\partial \lambda} \end{pmatrix} \qquad (5.256)$$

Also, the element volume

$$dV = dx\, dy\, dz = |J|\, d\xi\, d\eta\, d\lambda \qquad (5.257)$$

Using the global derivatives of the shape functions, the functional is transformed as

$$I = \int_{-1}^{1}\int_{-1}^{1}\int_{-1}^{1} f(\xi, \eta, \lambda)|J|\, d\xi\, d\eta\, d\lambda \qquad (5.258)$$

where the limits of integration are those of the parent element. Using Gaussian quadrature for each of the coordinate directions, this integral is obtained as

$$I = \sum\sum\sum F(\xi_i, \eta_i, \lambda_i) W_i W_j W_k \qquad (5.259)$$

where $i, j, k = 1, ..., n$.

The Gauss points and weights are obtained from the tables in Appendix D.

5.9.9 Triangular Isoparametric Elements

The method for triangular elements is much the same as for the curved 2D or 3D elements. There are, however, differences in the procedure for forming the Jacobian, and these will be illustrated by 2D and 3D examples.

Let us consider a six-noded second-order triangular element, which may be planar or curvilinear. The derivatives of the potential function may be expressed in terms of the derivatives of the shape functions and nodal values of potential as follows.

$$\frac{\partial \phi}{\partial x} = \sum \frac{\partial \zeta}{\partial x} \phi_i$$
$$\frac{\partial \phi}{\partial y} = \sum \frac{\partial \zeta}{\partial y} \phi_i \qquad (5.260)$$

Because the area or local coordinates of a two-dimensional triangular element satisfy the relationship

$$\xi i_1 + \xi i_2 + \xi i_3 = 1 \qquad (5.261)$$

we need only use ξi_1 and ξi_2 as the variables to express the shape functions and their derivatives. The Jacobian is then obtained as

$$\begin{pmatrix} \frac{\partial x}{\partial \xi i_1} & \frac{\partial y}{\partial \xi i_1} \\ \frac{\partial x}{\partial \xi i_2} & \frac{\partial y}{\partial \xi i_2} \end{pmatrix} \qquad (5.262)$$

As before, we can expand the coordinates x and y and their derivatives with respect to the local coordinates in terms of the shape functions and vertex values as

$$x = \sum \zeta_i x_i$$
$$y = \sum \zeta_i y_i \qquad (5.263)$$

$$\frac{\partial x}{\partial \xi_1} = \sum \frac{\partial \zeta_i}{\partial \xi_1} x_i$$

$$\frac{\partial y}{\partial \xi_1} = \sum \frac{\partial \zeta_i}{\partial \xi_1} y_i \quad (5.264)$$

$$\frac{\partial x}{\partial \xi_2} = \sum \frac{\partial \zeta_i}{\partial \xi_2} x_i$$

$$\frac{\partial y}{\partial \xi_2} = \sum \frac{\partial \zeta_i}{\partial \xi_2} y_i \quad (5.265)$$

The rest of the procedure for triangular elements is the same as for the rectangular element.

In the case of the 3D element, for example, for a first-order tetrahedron, the local coordinates satisfy the relationship

$$\xi_1 + \xi_2 + \xi_3 + \xi_4 = 1 \quad (5.266)$$

and, therefore, we can derive the Jacobian in terms of the three variables ξ_1, ξ_2, and ξ_3. The rest of the procedure is an extension of the method for 2D triangular elements and will not be repeated here. The integrations for 2D and 3D triangular elements are performed using Gaussian quadrature in exactly the same manner as for the corresponding rectangular and hexahedral elements, respectively.

5.9.10 Special Finite Element Shape Functions

In this section, we shall discuss singular elements and C^1 continuous elements. Other examples of special shape functions pertain to the use of potentials and their derivatives.

A $1/r$ singularity is encountered in modeling applications in some electrical engineering applications. Examples of these are axisymmetric representation of single phase transformers; and axiperiodic representation of the end region of electrical generators. Let us consider the axisymmetric case and the Poisson equation for magnetostatic

Shape Functions

modeling of a transformer. Expressing the magnetic induction in terms of the curl of a vector potential function, we have

$$\nu \left(\frac{\partial^2 A_\theta}{\partial r^2} + \frac{1}{r} \frac{\partial A_\theta}{\partial r} - \frac{A_\theta}{r^2} + \frac{\partial^2 A_\theta}{\partial z^2} \right) = -J_\theta \tag{5.267}$$

The associated functional to be minimized for obtaining the potential solution for the transformer problem will be

$$\mathcal{F} = \int \nu \left(\left(\left(\frac{\partial A_\theta}{\partial r} \right)^2 + 2 \frac{A_\theta}{r} \frac{\partial A_\theta}{\partial r} + \frac{A_\theta^2}{r^2} + \left(\frac{\partial A_\theta}{\partial z} \right)^2 \right) - 2 J_\theta A_\theta \right) r \, dr \, d\theta \, dz \tag{5.268}$$

or

$$\mathcal{F} = \int \nu \left(\left(r \left(\frac{\partial A_\theta}{\partial r} \right)^2 + 2 A_\theta \frac{\partial A_\theta}{\partial r} + \frac{A_\theta^2}{r} + r \left(\frac{\partial A_\theta}{\partial z} \right)^2 \right) - 2 r J_\theta A_\theta \right) dr \, d\theta \, dz \tag{5.269}$$

It is evident that the presence of the $1/r$ term in the differential equation and its consequent appearance in the functional will cause singularities of the potential function at the origin. Several methods of solving the singularity problem are suggested in references [48]. One of the methods proposed is to replace the radius term in global coordinates by the centroidal value of r in each triangular or rectangular finite element. This procedure yields satisfactory results for problems in which coarse modeling of the field region is considered adequate to achieve an acceptable degree of accuracy. However, for applications in which a high degree of accuracy is required, this procedure is not recommended. A second method proposed is to transform the field variable and the source function in terms of auxiliary variables such that

$$A_\theta = r\phi, \quad J_\theta = r\psi \tag{5.270}$$

or

$$A_\theta = \sqrt{r}\phi, \quad J_\theta = \sqrt{r}\psi \qquad (5.271)$$

Substituting for A_θ and J_θ from (5.270) and (5.271) in the expression for the functional in equation (5.269), one obtains

$$\mathcal{F} = \int_V \left(\nu \left(4r\phi^2 + 4r^2\phi \frac{\partial \phi}{\partial r} + r^3 \left(\frac{\partial \phi}{\partial r} \right)^2 + r^3 \left(\frac{\partial \phi}{\partial z} \right)^2 \right) - 2r^3\psi\phi \right) dr\, d\theta\, d\phi \qquad (5.272)$$

or

$$\mathcal{F} = \int_V \left(\nu \left(\frac{9}{4}\phi^2 + 3r\phi \frac{\partial \phi}{\partial r} + r^2 \left(\frac{\partial \phi}{\partial r} \right)^2 + r^2 \left(\frac{\partial \phi}{\partial z} \right)^2 \right) - 2r^2\psi\phi \right) dr\, d\theta\, dz \qquad (5.273)$$

The integrations are either performed as in reference [48] or by isoparametric mapping and numerical integration using Gaussian quadrature. These methods are useful in eliminating singularities in the potential function and yield highly accurate results. However, the singularity reappears in the expression for the flux density, specifically in the z component, because

$$B_z = \frac{A}{r} + \frac{\partial A}{\partial r} \qquad (5.274)$$

One easy way to avoid the singularity at a point where $r = 0$ is to use l'Hôpital's rule and transform the expression for B_z as

$$B_z = 2\frac{\partial A}{\partial r} \qquad (5.275)$$

Corner singularities are encountered in waveguide and antenna problems, in nondestructive analysis by magnetostatic or eddy current methods, and in fracture mechanics. The procedure most commonly adopted to overcome such problems is to subdivide the field region in the vicinity of the corner with a very large number of small finite elements. A more useful method suggested by structural engineers [65] is to use special finite elements employing midside nodes that are placed off center.

5.9.11 Mixed Elements and C^1 Continuous Elements

In some applications of the finite element method to electrical engineering problems, there is a need to represent the potential function and its derivative at the same node. An example of this is the hybrid finite element and boundary element formulation of the field problem. The development of the method is discussed in a later chapter, and we shall confine ourselves here to the discussion of the special finite element and associated shape functions for this application.

Sometimes derivative continuity at interfaces is required, such as when the continuity of the magnetic field has to be rigorously imposed. Although other methods based on the use of edge elements have been developed and widely used in recent years, it may be in order to discuss shape functions that provide derivative continuity at interfaces, so-called C^1 continuous elements. This is fully described by Gallagher [53].

Mixed Elements

We have already discussed the development of shape functions for a third-order C^0 cubic triangular element. These were readily obtained by using Silvester's formula [62] and the unknown potential function is associated with each node. An alternative method of developing shape functions with cubic variation of the potential function over a triangular element is to define the potential function and its derivatives at each vertex and the potential function alone at an internal node. This is illustrated in Figure 5.36.

In this application, 10 degrees of freedom are available for the cubic polynomial. The potential function may be expressed as

$$\phi = a_1 + a_2 x + a_3 y + a_4 xy + a_5 x^2 + a_6 y^2 + a_7 x^2 y + a_8 xy^2 + a_9 x^3 + a_{10} y^3 \quad (5.276)$$

Substituting the value of the vertex coordinates and the internal node, as explained in earlier sections, one obtains in matrix form

$$\begin{pmatrix} \phi_1 \\ \frac{\partial \phi_1}{\partial x} \\ \frac{\partial \phi_1}{\partial y} \\ \phi_2 \\ \frac{\partial \phi_2}{\partial x} \\ \frac{\partial \phi_2}{\partial y} \\ \phi_3 \\ \frac{\partial \phi_3}{\partial x} \\ \frac{\partial \phi_3}{\partial y} \\ \phi_4 \end{pmatrix} = \begin{pmatrix} 1 & x_1 & y_1 & x_1 y_1 & x_1^2 & y_1^2 & y_1 x_1^2 & x_1 y_1^2 & x_1^3 & y_1^3 \\ 0 & 1 & 0 & y_1 & 2x_1 & 0 & 2x_1 y_1 & y_1^2 & 3x_1^2 & 0 \\ 0 & 0 & 1 & x_1 & 0 & 2y_1 & x_1^2 & 2x_1 y_1 & 0 & 3y_1^2 \\ & & & \cdots & & & & & & \\ 1 & x_4 & y_4 & x_4 y_4 & x_4^2 & y_4^2 & y_4 x_4^2 & x_4 y_4^2 & x_4^3 & y_4^3 \end{pmatrix} \times \begin{pmatrix} a_1 \\ a_2 \\ a_3 \\ a_4 \\ a_5 \\ a_6 \\ a_7 \\ a_8 \\ a_9 \\ a_{10} \end{pmatrix} \quad (5.277)$$

Expressing the vector of ϕ and its x and y derivatives in equation (5.277), we may write in compact form

$$(\phi) = (P)(a) \qquad (5.278)$$

Shape Functions

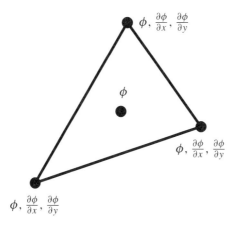

Figure 5.36 Illustration of a Mixed Triangular Element

By inverting this matrix equation and substituting the values of (a) in equation (5.276), one obtains

$$(\phi) = (1,\ x,\ y,\ xy,\ x^2,\ y^2,\ x^2y,\ xy^2,\ ,x^3,\ y^3)(P)^{-1}(a) \tag{5.279}$$

from which the shape functions may be expressed as

$$(\zeta) = (1,\ x,\ y,\ xy,\ x^2,\ y^2,\ x^2y,\ xy^2,\ ,x^3,\ y^3)(P)^{-1} \tag{5.280}$$

In reference [49], the vector of shape functions was derived as follows:

$$(\zeta) = \begin{pmatrix} \xi_1^2(3-\xi_1) - 7\xi_1\xi_2\xi_3 \\ \xi_1^2(c_m\xi_2^2 - c_j\xi_3) + (c_j - c_m)\xi_1\xi_2\xi_3 \\ \xi_1^2(-b_m\xi_2 + b_j\xi_3) - (b_j + b_m)\xi_1\xi_2\xi_3 \\ \dots\dots\dots\dots\dots \\ \dots\dots\dots\dots\dots \\ 27\xi_1\xi_2\xi_3 \end{pmatrix} \tag{5.281}$$

where $b_i = y_j - y_m$, $c_i = x_m - x_j$ in cyclic modulo 3.

5.10 ORTHOGONAL BASIS FUNCTIONS

Orthogonal basis functions such as trigonometric functions, Bessel functions, and Legendre functions can be effectively used in finite element analysis. For the present, the most effective application of these functions appears to be in hybrid finite element formulations, for example, interior finite element analysis coupled to exterior boundary integral or analytical methods. The nonlinear behavior of the iron parts in the discretized region may be modeled by the finite element method while the unbounded region is represented by the boundary integral or closed form expressions.

In this section, we shall expand on the idea of using orthogonal trigonometric functions as basis functions, described in reference [71], in place of polynomial basis functions. Optionally, the method of least squares may be employed to determine the unknown coefficients. A more useful and elegant approach would be to use line elements on the boundary of the finite element region. The resulting coefficient matrix for the exterior region will be symmetric and sparse.

5.10.1 Application of the Method to Laplace's Equation

In general, the finite element region in low-frequency electromagnetic field problems may be subdivided into an interior region comprising sources, iron parts, permanent magnets, dielectrics, and other inhomogeneous media and an exterior region consisting of free space. The practical application of this method to unbounded field problems will be discussed in a later chapter. We shall confine ourselves for the present to solving Laplace's equation in free space. For a two-dimensional exterior region bounding the interior finite element discretized region by an arc of a circle may be represented by a macroelement with global support on the boundary nodes. This is illustrated in Figure 5.37.

The governing differential equation and the functional for the field in the exterior region are given by

$$\nabla \times \nu_0 \nabla \times A = 0 \tag{5.282}$$

$$\mathcal{F} = \int_V \frac{\nu_0}{2} (\nabla \times A)^2 \, dV \tag{5.283}$$

Shape Functions

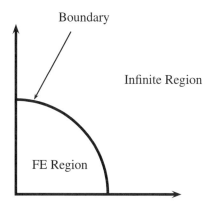

Figure 5.37 FE Region and Exterior

The potential solution in the exterior region may be defined by

$$A = \sum_{n=1}^{\infty}(C_n \cos n\theta + D_n \sin n\theta)\left(\frac{R}{r}\right)^n \qquad (5.284)$$

Expanding the functional of equation (5.283) in polar form yields

$$\mathcal{F} = \int \frac{\nu_0}{2}\left[r^{-2}\left(\frac{\partial A}{\partial \theta}\right)^2 + \left(\frac{\partial A}{\partial r}\right)^2\right] r\, dr\, d\theta\, dz \qquad (5.285)$$

Minimization of equation (5.285) with respect to each of the potentials at the nodes after substituting for A from equation (5.284), yields the matrices to the field problem in the exterior region. These are coupled with the matrices for the interior region to obtain the general solution.

5.10.2 Trigonometric Interpolation

An alternative approach to polynomial interpolation is the use of trigonometric basis functions. Use of these functions in structural mechanics is described by Zienkiewicz [66]. These functions are well suited for macroelements, which has been described in an earlier section. We shall discuss their application to interior finite elements with local support. Suggested applications of these elements are modeling the end regions of rotating electrical machines, some eddy current problems, and the solution of the wave equation for high-frequency analysis.

We shall limit our discussion to two-dimensional scalar or vector problems and illustrate the method with reference to a two-dimensional four-noded rectangular element.

We may recall that the shape functions for a four-noded planar rectangular element in terms of polynomial basis functions are

$$\zeta = \frac{(1+\xi\xi_i)(1+\eta\eta_i)}{4} \tag{5.286}$$

where $\xi = +1$ for nodes 1 and 4, $\xi = -1$ for nodes 2 and 3, $\eta = +1$ for nodes 1 and 2, and $\eta = -1$ for nodes 3 and 4.

If we now replace ξ_i by $\sin \xi_i$ and η_i by $\sin \eta_i$, where $\xi_i = \frac{\pi}{2}x/a$ and $\eta_i = \frac{\pi}{2}y/b$, we obtain the trigonometric shape functions as

$$\zeta = \frac{(1+\xi \sin \xi_i)(1+\eta \sin \eta_i)}{4} \tag{5.287}$$

It is evident that these shape functions satisfy the requirements that $\zeta_i = 1$ at node i and $\zeta_i = 0$ at any other node.

We can also verify that $\sum \zeta_i = 1$ at any point within the element or on its boundary. For example, let us consider the point $x = a/2$, $y = b$.

Then $\xi_i = \pi/4$, $\eta_i = \pi/2$, and

$$\zeta_1 = \frac{(1+\sin \pi/4)(1+\sin \pi/2)}{4} = 0.8535$$

$$\zeta_2 = \frac{(1-\sin \pi/4)(1+\sin \pi/2)}{4} = 0.1465$$

$$\zeta_3 = \frac{(1-\sin \pi/4)(1-\sin \pi/2)}{4} = 0.0$$

Shape Functions

$$\zeta_4 = \frac{(1 + \sin \pi/4)(1 - \sin \pi/2)}{4} = 0.0$$

Therefore,

$$\zeta_1 + \zeta_2 + \zeta_3 + \zeta_4 = 1$$

We can also verify that $\sum \zeta_i = 1$ holds at an interior point. Let us consider an interior point $x = a/3; \ y = -2b/3$.

Therefore,

$$\xi_i = \pi/6; \ \eta_i = -\pi/3$$

Then

$$\zeta_1 = \frac{(1 + \sin \pi/6)(1 + \sin \pi/3)}{4} = 0.050240473$$

$$\zeta_2 = \frac{(1 - \sin \pi/6)(1 + \sin \pi/3)}{4} = 0.016246824$$

$$\zeta_3 = \frac{(1 - \sin \pi/6)(1 + \sin \pi/3)}{4} = 0.233253175$$

$$\zeta_4 = (1 + \sin \pi/6)(1 + \sin \pi/3) = 0.699259526$$

So $\zeta_1 + \zeta_2 + \zeta_3 + \zeta_4 = 1$.

Similarly, we can verify that the sum of the interpolation polynomials at any arbitrary point in the interior or at the boundary will always equal unity. Therefore, we can conclude that the basis functions are interpolatory and continuity is satisfied. We cannot, however, assign an order to the interpolation as we would do with regular interpolation polynomials.

6

THE FINITE ELEMENT METHOD

6.1 INTRODUCTION

The finite element method is a numerical technique and is a special case of Ritz's method discussed before. In this method, the partial differential equations modeling the field problem are transformed into energy-related functionals. Approximate solutions are then sought to the field problem that extremize (or minimize) these functionals. By this procedure, a detailed modeling of the geometry of the field region is possible, and the results obtained are found to be accurate. Unlike those in Ritz's procedure, the boundary conditions need not be associated with the trial functions, and the former can be implemented separately. The trial functions themselves require only potential continuity at interfaces in a majority of cases, thereby facilitating matrix assembly.

Advantages of the finite element method are that

- It is versatile.
- It offers flexibility for modeling complex geometry.
- It yields stable and accurate solutions.
- Natural boundary conditions are implicit in the functional formulation.
- It can handle nonlinearity and eddy currents well.

The steps involved in the finite element method are as follows:

- Define the boundary value problem by partial differential equations.

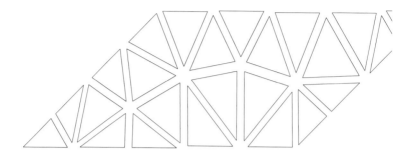

Figure 6.1 Subdivision of the Field Region into Triangular Finite Elements

- Obtain a variational formulation for the partial differential equations in terms of energy-related functionals or weighted residual expressions.
- Subdivide the field region into subregions (finite elements; see Figure 6.1).
- Choose a trial solution in terms of nodal values of the elements and interpolatory functions.
- Minimize the functional with respect to each of the nodal potentials and solve the resulting set of algebraic equations or obtain the equations directly by the Galerkin procedure and solve.
- Convert the potential solution to useful design parameters.

We shall now illustrate this method by application to one- and two-dimensional examples.

6.1.1 One-Dimensional Electrostatic Problem

As an example of a one-dimensional electrostatic problem, we shall consider the parallel plate capacitor shown in Figure 6.2.

The governing equations for the electrostatic field problem are

$$\nabla \cdot D = \rho \tag{6.1}$$

$$D = \epsilon E \tag{6.2}$$

The Finite Element Method

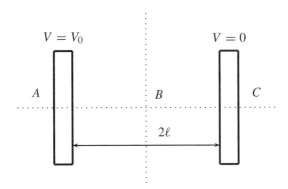

Figure 6.2 Parallel Plate Capacitor

$$E = -\nabla V \tag{6.3}$$

where D is the electrostatic flux density, ρ is volumetric charge density, ϵ is the permittivity, E is the electric field and V is the potential.

Assuming that the applied charge density is zero and substituting equations (6.2) and (6.3) in equation (6.1), we obtain the scalar Laplace equation for the electrostatic field as

$$\nabla \cdot \nabla V = 0 \tag{6.4}$$

We are required to solve equation (6.4) subject to the boundary conditions

$$\begin{aligned} V &= V_0 \text{ at } A \text{ and} \\ V &= 0 \text{ at } C \end{aligned} \tag{6.5}$$

The functional for this problem is obtained from equation (4.71) by the procedure described in earlier sections, and this expression is of the form

$$\mathcal{F} = \int \frac{\epsilon}{2}(\nabla V)^2 dx \tag{6.6}$$

Element Discretization

Let us assume that the solution to the potential V can be described by the polynomial

$$V = a_1 + a_2 x + a_3 y \tag{6.7}$$

Referring to Figure 6.2, we can write the following:

$$\begin{aligned} V_A &= a_1 + 0 + 0 \\ V_B &= a_1 + a_2 \ell + a_3 \ell^2 \\ V_C &= a_1 + 2a_2 \ell + 4a_3 \ell^2 \end{aligned} \tag{6.8}$$

Solving the set of simultaneous equations (6.8) for a_1, a_2, and a_3 we obtain

$$V = V_A \left(1 - \frac{x}{\ell}\right)\left(1 - \frac{x}{2\ell}\right) + V_B \left(\frac{x}{\ell}\right)\left(2 - \frac{x}{\ell}\right) - V_C \left(\frac{x}{2\ell}\right)\left(1 - \frac{x}{\ell}\right) \tag{6.9}$$

If we designate the terms in parentheses of equation (6.9) as shape functions ξ_A, ξ_B, and ξ_C, we may write

$$\begin{aligned} \xi_A &= \left(1 - \frac{x}{\ell}\right)\left(1 - \frac{x}{2\ell}\right) \\ \xi_B &= \left(\frac{x}{\ell}\right)\left(2 - \frac{x}{\ell}\right) \\ \xi_C &= -\left(\frac{x}{2\ell}\right)\left(1 - \frac{x}{\ell}\right) \end{aligned} \tag{6.10}$$

Then
$$V = \xi_A V_A + \xi_B V_B + \xi_C V_C \qquad (6.11)$$

and
$$\frac{\partial V}{\partial x} = \frac{\partial \xi_A}{\partial x} V_A + \frac{\partial \xi_B}{\partial x} V_B + \frac{\partial \xi_c}{\partial x} V_C \qquad (6.12)$$

or
$$\frac{\partial V}{\partial x} = \left(\frac{-3}{2\ell} + \frac{x}{\ell^2}\right) V_A + \left(\frac{2}{\ell} - \frac{2x}{\ell^2}\right) V_B + \left(-\frac{1}{2\ell} + \frac{x}{\ell^2}\right) V_C \qquad (6.13)$$

Because
$$\mathcal{F} = \int \frac{\epsilon}{2} (\nabla V)^2 \, dx \qquad (6.14)$$

functional minimization yields
$$\frac{\partial \mathcal{F}}{\partial V_A} = \epsilon \int \frac{\partial V}{\partial x} \frac{\partial}{\partial V_A} \frac{\partial V}{\partial x} \, dx = 0 \qquad (6.15)$$

Now
$$\frac{\partial}{\partial V_A}\left(\frac{\partial V}{\partial x}\right) = \frac{-3}{2\ell} + \frac{x}{\ell^2} \qquad (6.16)$$

Therefore,
$$\frac{\partial \mathcal{F}}{\partial V_A} = \epsilon \int_0^{2\ell} \left[\left(\frac{-3}{2\ell} + \frac{x}{\ell^2}\right)^2 V_A \right.$$
$$\left. + \left(\frac{-3}{2\ell} + \frac{x}{\ell^2}\right)\left(\frac{2}{\ell} - \frac{2x}{\ell^2}\right) V_B + \left(\frac{-3}{2\ell} + \frac{x}{\ell^2}\right)\left(\frac{-1}{2\ell} + \frac{x}{\ell^2}\right) V_C \right] dx \qquad (6.17)$$

or
$$\frac{\partial \mathcal{F}}{\partial V_A} = \frac{\epsilon}{6\ell}(7V_A - 8V_B + V_C) = \frac{\epsilon}{6\ell}(7, \; -8, \; 1)\begin{pmatrix} V_A \\ V_B \\ V_C \end{pmatrix} \qquad (6.18)$$

Similarly

$$\frac{\partial \mathcal{F}}{\partial V_B} = \frac{\epsilon}{6\ell}(-8, \; 16, \; -8)\begin{pmatrix} V_A \\ V_B \\ V_C \end{pmatrix} \quad (6.19)$$

$$\frac{\partial \mathcal{F}}{\partial V_C} = \frac{\epsilon}{6\ell}(1, \; -8, \; 7)\begin{pmatrix} V_A \\ V_B \\ V_C \end{pmatrix} \quad (6.20)$$

Combining equations (6.18) through (6.20)

$$\begin{pmatrix} \frac{\partial \mathcal{F}}{\partial V_A} \\ \frac{\partial \mathcal{F}}{\partial V_B} \\ \frac{\partial \mathcal{F}}{\partial V_C} \end{pmatrix} = \frac{\epsilon}{6\ell}\begin{pmatrix} 7 & -8 & 1 \\ -8 & 16 & -8 \\ 1 & -8 & 7 \end{pmatrix}\begin{pmatrix} V_A \\ V_B \\ V_C \end{pmatrix} = 0 \quad (6.21)$$

Boundary conditions are

$$V_A = V_0 \quad \text{and} \quad V_C = 0 \quad (6.22)$$

Substituting (6.22) in (6.21) and expanding the resulting matrix equation and transposing terms, we have

$$\frac{\epsilon}{6\ell}\begin{pmatrix} 0 & -8 & 0 \\ 0 & 16 & 0 \\ 0 & 0 & 0 \end{pmatrix}\begin{pmatrix} V_A \\ V_B \\ V_C \end{pmatrix} = \frac{\epsilon}{6\ell}\begin{pmatrix} -7V_0 \\ 8V_0 \\ -V_0 \end{pmatrix} \quad (6.23)$$

But the first and third rows of equation (6.23) corresponding to the potentials V_A and V_C should be eliminated, yielding

$$V_A = V_0, \quad V_B = \frac{V_0}{2}, \quad V_C = 0 \quad (6.24)$$

The Finite Element Method

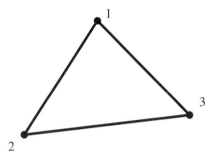

Figure 6.3 Triangular Finite Element

A more elegant method of obtaining the result of equation (6.24) is achieved by setting the diagonal terms corresponding to V_A and V_C in matrix equation (6.21) to unity and the corresponding off-diagonal terms to zero. The rows corresponding to V_A and V_C in the right-hand vector are replaced by V_0 and 0, respectively. Also, the right-hand term corresponding to V_B is set as $\frac{\epsilon}{6\ell} 8 V_0$ as shown in equation (6.23). Therefore,

$$\frac{\epsilon}{6\ell} \begin{pmatrix} 1 & 0 & 0 \\ 0 & 16 & 0 \\ 0 & 0 & 1 \end{pmatrix} \begin{pmatrix} V_A \\ V_B \\ V_C \end{pmatrix} = \frac{\epsilon}{6\ell} \begin{pmatrix} V_0 \\ 8 V_0 \\ 0 \end{pmatrix} \qquad (6.25)$$

Solving (6.25), we obtain $V_A = V_0$, $V_B = \frac{V_0}{2}$, $V_C = 0$.

6.1.2 Two-Dimensional Finite Element Analysis

Let us consider the solution of the two-dimensional Poisson equation in the field region as shown in Figure 6.1. Let us for the present restrict ourselves to considering a single element as shown in Figure 6.3, in which a linear variation of the potential is defined in terms of the vertex values.

We can write the variation of the potential in the triangular region as

$$A = \alpha_1 + \alpha_2 x + \alpha_3 y \tag{6.26}$$

Substituting for x, y, and A the values x_1, y_1, A_1; x_2, y_2, A_2; x_3, y_3, and A_3 in succession, we obtain the following set of three equations:

$$\begin{aligned} A_1 &= \alpha_1 + \alpha_2 x_1 + \alpha_3 y_1 \\ A_2 &= \alpha_1 + \alpha_2 x_2 + \alpha_3 y_2 \\ A_3 &= \alpha_1 + \alpha_2 x_3 + \alpha_3 y_3 \end{aligned} \tag{6.27}$$

Writing the set of equations (6.27) in matrix form

$$\begin{pmatrix} 1 & x_1 & y_1 \\ 1 & x_2 & y_2 \\ 1 & x_3 & y_3 \end{pmatrix} \begin{pmatrix} \alpha_1 \\ \alpha_2 \\ \alpha_3 \end{pmatrix} = \begin{pmatrix} A_1 \\ A_2 \\ A_3 \end{pmatrix} \tag{6.28}$$

Inverting the matrix equation (6.28), we have

$$\begin{pmatrix} \alpha_1 \\ \alpha_2 \\ \alpha_3 \end{pmatrix} = \begin{pmatrix} 1 & x_1 & y_1 \\ 1 & x_2 & y_2 \\ 1 & x_3 & y_3 \end{pmatrix}^{-1} \begin{pmatrix} A_1 \\ A_2 \\ A_3 \end{pmatrix} \tag{6.29}$$

or

$$[\alpha] = [S]^{-1}[A_i] \tag{6.30}$$

But from equation (6.26)

$$A = \alpha_1 + \alpha_2 x + \alpha_3 y = (1, x, y) \begin{pmatrix} \alpha_1 \\ \alpha_2 \\ \alpha_3 \end{pmatrix} = (1, x, y)(\alpha) \tag{6.31}$$

The Finite Element Method

substituting for (α) from (6.30) in (6.31)

$$A = (1, x, y)(S)^{-1}(A_i) \tag{6.32}$$

The inverse matrix $(S)^{-1}$ can be evaluated by rules of matrix algebra as

$$(S)^{-1} = \frac{1}{\det} \begin{pmatrix} x_2 y_3 - x_3 y_2 & x_3 y_1 - x_1 y_3 & x_1 y_2 - x_2 y_1 \\ y_2 - y_3 & y_3 - y_1 & y_1 - y_2 \\ x_3 - x_2 & x_1 - x_3 & x_2 - x_1 \end{pmatrix} \tag{6.33}$$

or

$$(S)^{-1} = \frac{1}{\det} \begin{pmatrix} a_1 & a_2 & a_3 \\ b_1 & b_2 & b_3 \\ c_1 & c_2 & c_3 \end{pmatrix} \tag{6.34}$$

where

$$\det = \begin{vmatrix} 1 & x_1 & y_1 \\ 1 & x_2 & y_2 \\ 1 & x_3 & y_3 \end{vmatrix} = 2 \times \text{triangle area } \Delta$$

Substituting for $[S]^{-1}$ from equation (6.34) into equation (6.30)

$$\begin{aligned} \alpha_1 &= \frac{a_1 A_1 + a_2 A_2 + a_3 A_3}{2\Delta} \\ \alpha_2 &= \frac{b_1 A_1 + b_2 A_2 + b_3 A_3}{2\Delta} \\ \alpha_2 &= \frac{c_1 A_1 + c_2 A_2 + c_3 A_3}{2\Delta} \end{aligned} \tag{6.35}$$

From equations (6.31) and (6.35)

$$(A) = [1, x, y] \begin{pmatrix} \alpha_1 \\ \alpha_2 \\ \alpha_3 \end{pmatrix} \tag{6.36}$$

or

$$A = \frac{a_1 + b_1 x + c_1 y}{2\Delta} A_1 + \frac{a_2 + b_2 x + c_2 y}{2\Delta} A_2 + \frac{a_3 + b_3 x + c_3 y}{2\Delta} A_3 \quad (6.37)$$

It must be noted that the coefficients of A_1, A_2, and A_3, the shape functions, all depend on the geometry of the triangular element and are interpolatory functions. These are denoted by ζ_1, ζ_2, and ζ_3, respectively. The vector potential A can be expressed succinctly using the summation convention, so that

$$A = \sum_{i=1}^{3} \zeta_i A_i \quad (6.38)$$

Also, for this first-order triangular element, ζ_1, ζ_2, and ζ_3 are the same as the area coordinates ξ_1, ξ_2, and ξ_3.

The two-dimensional Poisson equation for the magnetostatic problem and the corresponding functional are given as follows.

$$\nu \nabla \cdot \nabla A = -J \quad (6.39)$$

$$\mathcal{F} = \iint \left(\left(\frac{\nu}{2}\right)(\nabla A)^2 - J \cdot A \right) dx\, dy \quad (6.40)$$

Substituting $(\nabla A)^2 = \left(\frac{\partial A}{\partial x}\right)^2 + \left(\frac{\partial A}{\partial y}\right)^2$

$$\mathcal{F} = \iint \left(\frac{\nu}{2} \left[\left(\frac{\partial A}{\partial x}\right)^2 + \left(\frac{\partial A}{\partial y}\right)^2 \right] - J \cdot A \right) dx\, dy \quad (6.41)$$

Minimizing (6.41) with respect to each vertex value of A,

The Finite Element Method

$$\frac{\partial \mathcal{F}}{\partial A_k} = \iint \left(\nu \left(\frac{\partial A}{\partial x}\right) \frac{\partial}{\partial A_k}\left(\frac{\partial A}{\partial x}\right) + \left(\frac{\partial A}{\partial y}\right) \frac{\partial}{\partial A_k}\left(\frac{\partial A}{\partial y}\right) \right.$$
$$\left. - J \cdot \frac{\partial A}{\partial A_k} \right) dx\, dy = 0 \qquad (6.42)$$

Now from (6.38), we obtain by differentiation

$$\frac{\partial A}{\partial x} = \sum_{i=1}^{3} \frac{\partial \zeta_i}{\partial x} A_i$$

$$\frac{\partial A}{\partial y} = \sum_{i=1}^{3} \frac{\partial \zeta_i}{\partial y} A_i \qquad (6.43)$$

Substituting for A_k successively A_1, A_2, A_3 in equation (6.42) and using (6.43), we obtain in matrix form

$$\begin{pmatrix} \frac{\partial \mathcal{F}}{\partial A_1} \\ \frac{\partial \mathcal{F}}{\partial A_2} \\ \frac{\partial \mathcal{F}}{\partial A_2} \end{pmatrix} =$$

$$\int \nu \begin{pmatrix} (\frac{\partial \zeta_1}{\partial x})^2 & (\frac{\partial \zeta_1}{\partial x})(\frac{\partial \zeta_2}{\partial x}) & (\frac{\partial \zeta_1}{\partial x})(\frac{\partial \zeta_3}{\partial x}) \\ & (\frac{\partial \zeta_2}{\partial x})^2 & (\frac{\partial \zeta_2}{\partial x})(\frac{\partial \zeta_3}{\partial x}) \\ \text{symmetric} & & (\frac{\partial \zeta_3}{\partial x})^2 \end{pmatrix} \begin{pmatrix} A_1 \\ A_2 \\ A_3 \end{pmatrix} dx\, dy$$

$$+ \int \nu \begin{pmatrix} (\frac{\partial \zeta_1}{\partial y})^2 & (\frac{\partial \zeta_1}{\partial y})(\frac{\partial \zeta_2}{\partial y}) & (\frac{\partial \zeta_1}{\partial y})(\frac{\partial \zeta_3}{\partial y}) \\ & (\frac{\partial \zeta_2}{\partial y})^2 & (\frac{\partial \zeta_2}{\partial y})(\frac{\partial \zeta_3}{\partial y}) \\ \text{symmetric} & & (\frac{\partial \zeta_3}{\partial y})^2 \end{pmatrix} \begin{pmatrix} A_1 \\ A_2 \\ A_3 \end{pmatrix} dx\, dy$$

$$- \iint J \begin{pmatrix} \zeta_1 \\ \zeta_2 \\ \zeta_3 \end{pmatrix} dx\, dy \qquad (6.44)$$

Because $\zeta_1 = \dfrac{a_1 + b_1 x + c_1 y}{2\Delta}$, $\zeta_2 = \dfrac{a_2 + b_2 x + c_2 y}{2\Delta}$, and $\zeta_3 = \dfrac{a_3 + b_3 x + c_3 y}{2\Delta}$, we obtain

$$\dfrac{\partial \zeta_1}{\partial x} = \dfrac{b_1}{2\Delta}, \quad \dfrac{\partial \zeta_2}{\partial x} = \dfrac{b_2}{2\Delta}, \quad \dfrac{\partial \zeta_3}{\partial x} = \dfrac{b_3}{2\Delta}$$

$$\dfrac{\partial \zeta_1}{\partial y} = \dfrac{c_1}{2\Delta}, \quad \dfrac{\partial \zeta_2}{\partial y} = \dfrac{c_2}{2\Delta}, \quad \dfrac{\partial \zeta_3}{\partial y} = \dfrac{c_3}{2\Delta} \qquad (6.45)$$

Substituting for the derivatives of ζ_1, ζ_2, and ζ_3 with respect to x and y from (6.45) into (6.44)

$$\begin{pmatrix} \dfrac{\partial \mathcal{F}}{\partial A_1} \\ \dfrac{\partial \mathcal{F}}{\partial A_2} \\ \dfrac{\partial \mathcal{F}}{\partial A_3} \end{pmatrix}$$

$$= \iint \dfrac{v}{4\Delta^2} \begin{pmatrix} b_1^2 + c_1^2 & b_1 b_2 + c_1 c_2 & b_1 b_3 + c_1 c_3 \\ & b_2^2 + c_2^2 & b_2 b_3 + c_2 c_3 \\ \text{symmetric} & & b_3^2 + c_3^2 \end{pmatrix} \begin{pmatrix} A_1 \\ A_2 \\ A_3 \end{pmatrix} dx\, dy$$

$$- \int J \begin{pmatrix} \zeta_1 \\ \zeta_2 \\ \zeta_3 \end{pmatrix} dx\, dy \qquad (6.46)$$

Because b_1, b_2, b_3, c_1, c_2, and c_3 are all functions of the coordinates of the vertices of the triangle and $\iint dx\, dy = \Delta$, the first integral in (6.46) yields

$$\dfrac{v}{4\Delta} \begin{pmatrix} s_{11} & s_{22} & s_{33} \\ s_{21} & s_{22} & s_{33} \\ s_{31} & s_{32} & s_{33} \end{pmatrix} \begin{pmatrix} A_1 \\ A_2 \\ A_3 \end{pmatrix} \qquad (6.47)$$

where $s_{11} = (b_1^2 + c_1^2)$, $s_{12} = (b_1 b_2 + c_1 c_2)$, and so on. Now it remains for us to evaluate the second integral in (6.46). We have shown that the area coordinates vary between the limits of 0 and 1. Also, because the area coordinates sum to unity, only two of these coordinates are independent. Therefore, assuming the limits of ζ_1 are 0 and 1, the limits of ζ_2 will be 0 and $1 - \zeta_1$. Further, we have

$$dx\, dy = 2\Delta d\zeta_1\, d\zeta_2 \qquad (6.48)$$

Hence the second integral in (6.46) becomes

$$2\Delta \int_0^1 \int_0^{1-\zeta_1} J \begin{pmatrix} \zeta_1 \\ \zeta_2 \\ \zeta_3 \end{pmatrix} d\xi_1\, d\xi_2 \qquad (6.49)$$

As has already been stated, for this first-order triangular finite element ζ_1, ζ_2, and ζ_3 are equal to ξ_1, ξ_2, and ξ_3, respectively. Therefore, the integral of (6.49) is evaluated as follows.

$$2\Delta \int_0^1 \int_0^{1-\xi_1} J \begin{pmatrix} \xi_1 \\ \xi_2 \\ \xi_3 \end{pmatrix} d\xi_1\, d\xi_2$$

$$= 2\Delta J \int_0^1 \begin{pmatrix} \xi_1(1-\xi_1) \\ \frac{(1-\xi_1)^2}{2} \\ (1-\xi_1) - \xi_1(1-\xi_1) - \frac{(1-\xi_1)^2}{2} \end{pmatrix} d\xi_1 = \frac{J\Delta}{3} \begin{pmatrix} 1 \\ 1 \\ 1 \end{pmatrix} \qquad (6.50)$$

6.2 FUNCTIONAL MINIMIZATION AND GLOBAL ASSEMBLY

So far we have discussed the energy formulation of the partial differential equation and the minimization of the functional with respect to a single finite element. We shall extend the analysis to a number of finite elements spanning the discretized space. Because the energy that the functional represents is a scalar quantity, the global energy in the field region can be considered to be the sum of the energies in the individual finite elements. Let us, for simplicity, consider two triangles as spanning the discretized field region, as shown in Figure 6.4.

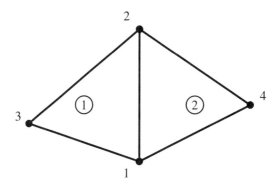

Figure 6.4 Two Finite Element Assembly

The total energy in the system, therefore, is given as

$$W = W_1 + W_2 \tag{6.51}$$

where W is the global energy, and W_1 and W_2 are the energies in elements 1 and 2, respectively.

Assuming a potential ϕ defined at the nodes of the two elements, the minimization of the global energy yields

$$\frac{\partial W}{\partial \phi_k} = \begin{pmatrix} \frac{\partial W_1}{\partial \phi_1} \\ \frac{\partial W_1}{\partial \phi_2} \\ \frac{\partial W_1}{\partial \phi_3} \\ 0 \end{pmatrix} + \begin{pmatrix} \frac{\partial W_2}{\partial \phi_1} \\ \frac{\partial W_2}{\partial \phi_2} \\ 0 \\ \frac{\partial W_2}{\partial \phi_4} \end{pmatrix} \quad \text{for } k = 1, 2, 3, 4 \tag{6.52}$$

or

$$\frac{\partial W}{\partial \phi_k}\bigg|_{k=1,2,3,4} = \frac{\partial W_1}{\partial \phi_\ell}\bigg|_{\ell=1,2,3} + \frac{\partial W_2}{\partial \phi_m}\bigg|_{m=1,2,4} \tag{6.53}$$

The Finite Element Method

We can, therefore, generalize that

$$\frac{\partial W}{\partial \phi_k} = \sum_{\ell=1}^{N} \frac{\partial W_\ell}{\partial \phi_\ell} \tag{6.54}$$

that is, the minimization of the total energy in the field region equals the sum of the energy minima in the respective discretized subregions or elements.

Global Assembly

Let us consider the solution, ϕ, of Poisson's equation

$$\nabla \cdot \nabla \phi = -K \tag{6.55}$$

The solution to the preceding partial differential equation also minimizes the functional

$$\mathcal{F} = \int_V (\phi^T \nabla \cdot \nabla \phi - 2\phi^T K) \, dV \tag{6.56}$$

In discretized space, the functional in each element can be expressed in matrix form as

$$\mathcal{F} = (\phi)^T (S)(\phi) - 2(\phi)^T (K) \tag{6.57}$$

Let us now consider two elements as before as in Figure 6.4. The total functional will then equal the sum of the functionals in the two respective elements, so that

$$\mathcal{F}_{\text{total}} = \mathcal{F}_1 + \mathcal{F}_2 \tag{6.58}$$

We may also express the total potential vector (ϕ) as the sum of the disjoint potential vectors of each of the elements by the use of a connection matrix, such that

$$(\phi)_{e1} + (\phi)_{e2} = (C)(\phi) \tag{6.59}$$

where the subscripts $e1$ and $e2$ refer to the element contributions. We may illustrate equation (6.59) as follows.

$$\begin{pmatrix} \phi_1^{(1)} \\ \phi_2^{(1)} \\ \phi_3^{(1)} \\ \phi_1^{(2)} \\ \phi_2^{(2)} \\ \phi_4^{(2)} \end{pmatrix} = \begin{pmatrix} 1 & 0 & 0 & 0 \\ 0 & 1 & 0 & 0 \\ 0 & 0 & 1 & 0 \\ 1 & 0 & 0 & 0 \\ 0 & 1 & 0 & 0 \\ 0 & 0 & 0 & 1 \end{pmatrix} \begin{pmatrix} \phi_1 \\ \phi_2 \\ \phi_3 \\ \phi_4 \end{pmatrix} \tag{6.60}$$

Substituting for (ϕ) from equation (6.59) into equation (6.57) for the global functional

$$\mathcal{F} = (\phi)^T (C)^T (S)(C)(\phi) - 2(\phi)^T (C)^T (K) \tag{6.61}$$

Let us now denote the modified coefficient matrix and forcing function as

$$(S') = (C)^T (S)(C) \text{ and } (K') = (C)^T (K) \tag{6.62}$$

where (S') is the total coefficient matrix, (S) is the assembly of disjoint matrices corresponding to the respective elements, (K') is the total forcing function, and (K) is the vector of disjoint forcing functions for the respective elements.

The Finite Element Method

Equation (6.61) applied to the two-element assembly may be expressed as follows:

$$(S) = \begin{pmatrix} s_{11}^{(1)} & s_{12}^{(1)} & s_{13}^{(1)} & 0 & 0 & 0 \\ s_{21}^{(1)} & s_{22}^{(1)} & s_{23}^{(1)} & 0 & 0 & 0 \\ s_{31}^{(1)} & s_{32}^{(1)} & s_{33}^{(1)} & 0 & 0 & 0 \\ 0 & 0 & 0 & s_{11}^{(2)} & s_{12}^{(2)} & s_{14}^{(2)} \\ 0 & 0 & 0 & s_{21}^{(2)} & s_{22}^{(2)} & s_{24}^{(2)} \\ 0 & 0 & 0 & s_{41}^{(2)} & s_{42}^{(2)} & s_{44}^{(2)} \end{pmatrix} \quad (6.63)$$

$$(C)^T = \begin{pmatrix} 1 & 0 & 0 & 1 & 0 & 0 \\ 0 & 1 & 0 & 0 & 1 & 0 \\ 0 & 0 & 1 & 0 & 0 & 0 \\ 0 & 0 & 0 & 0 & 0 & 1 \end{pmatrix} \quad (6.64)$$

Therefore

$$(S') = \begin{pmatrix} 1 & 0 & 0 & 1 & 0 & 0 \\ 0 & 1 & 0 & 0 & 1 & 0 \\ 0 & 0 & 1 & 0 & 0 & 0 \\ 0 & 0 & 0 & 0 & 0 & 1 \end{pmatrix} \begin{pmatrix} s_{11}^{(1)} & s_{12}^{(1)} & s_{13}^{(1)} & 0 & 0 & 0 \\ s_{21}^{(1)} & s_{22}^{(1)} & s_{23}^{(1)} & 0 & 0 & 0 \\ s_{31}^{(1)} & s_{32}^{(1)} & s_{33}^{(1)} & 0 & 0 & 0 \\ 0 & 0 & 0 & s_{11}^{(2)} & s_{12}^{(2)} & s_{14}^{(2)} \\ 0 & 0 & 0 & s_{21}^{(2)} & s_{22}^{(2)} & s_{24}^{(2)} \\ 0 & 0 & 0 & s_{41}^{(2)} & s_{42}^{(2)} & s_{44}^{(2)} \end{pmatrix}$$

$$\times \begin{pmatrix} 1 & 0 & 0 & 0 \\ 0 & 1 & 0 & 0 \\ 0 & 0 & 1 & 0 \\ 1 & 0 & 0 & 0 \\ 0 & 1 & 0 & 0 \\ 0 & 0 & 0 & 1 \end{pmatrix} \quad (6.65)$$

or

$$(S') = \begin{pmatrix} s_{11}^{(1)} + s_{11}^{(2)} & s_{12}^{(1)} + s_{12}^{(2)} & s_{13}^{(1)} & s_{14}^{(2)} \\ s_{21}^{(1)} + s_{21}^{(2)} & s_{22}^{(1)} + s_{22}^{(2)} & s_{23}^{(1)} & s_{24}^{(2)} \\ s_{31}^{(1)} & s_{32}^{(1)} & s_{33}^{(1)} & 0 \\ s_{41}^{(2)} & s_{42}^{(2)} & 0 & s_{44}^{(2)} \end{pmatrix} \quad (6.66)$$

Similarly, for the forcing function

$$(K') = \begin{pmatrix} 1 & 0 & 0 & 1 & 0 & 0 \\ 0 & 1 & 0 & 0 & 1 & 0 \\ 0 & 0 & 1 & 0 & 0 & 0 \\ 0 & 0 & 0 & 0 & 0 & 1 \end{pmatrix} \begin{pmatrix} K_1^{(1)} \\ K_2^{(1)} \\ K_3^{(1)} \\ K_1^{(2)} \\ K_2^{(2)} \\ K_4^{(2)} \end{pmatrix} = \begin{pmatrix} K_1^{(1)} + K_1^{(2)} \\ K_2^{(1)} + K_2^{(2)} \\ K_3^{(1)} \\ K_4^{(2)} \end{pmatrix} \quad (6.67)$$

The connection matrix just described clearly illustrates the assembly of the element matrices into the global coefficient matrix. The resulting global matrix implies continuity of the potential at the interface of two adjoining elements (already established by node numbering), the so-called C^0 continuity. Further, the connection matrix, although it is seldom evaluated explicitly in finite element computation, serves as a useful tool to study various constraints on the potential function such as the constant potential condition at certain interfaces, periodic or anti-periodic boundary conditions, and others.

6.2.1 Implementation of the Solution for the Global System

We have discussed at some length the procedure for solving the global system of equations for the connected problem by a numerical example. In this section, we shall make an attempt to formalize the procedure so that it is valid generally. This section closely follows the work done by Silvester [48].

Referring to the previous numerical example, we solved a vector potential problem with a source current in a two-dimensional field region subject to specified zero potentials

The Finite Element Method

at the boundary. We shall now examine the case of the solution of Laplace's equation with nonzero specified potentials on the boundary. As before, we minimize the energy-related functional in the entire field region with respect to each of the nodal potentials. Without the boundary conditions, this procedure will yield zero everywhere, emphasizing the result of unconstrained minimization. However, in well-posed boundary value problems, there are always known potentials that are specified at some points. Therefore a portion of the solution vector can be assumed to be known or specified. For convenience, if we number the nodes in the discretized field region in such a manner that the free potentials (i.e., unknown and yet to be determined potentials) come first and the prescribed or known potentials come last, then we may write the energy minimization procedure as follows.

$$\frac{\partial \mathcal{F}}{\partial \phi_k} = \frac{\partial}{\partial (\phi_f)_k} (\phi_f, \phi_p) \begin{pmatrix} S_{ff} & S_{fp} \\ S_{pf} & S_{pp} \end{pmatrix} \begin{pmatrix} \phi_f \\ \phi_p \end{pmatrix} = 0 \qquad (6.68)$$

where ϕ_f and ϕ_p refer to free and prescribed potentials, and the corresponding submatrices S_f and S_p refer to those that correspond to free potentials and those that correspond to prescribed potentials, respectively. The differentiation in equation (6.68) is carried out only over terms containing free potentials because there can be no variation in the prescribed or known potentials. This gives

$$S_{ff}\phi_f + S_{fp}\phi_p = 0 \qquad (6.69)$$

or

$$\phi_f = -S_{ff}^{-1} S_{fp} \phi_p \qquad (6.70)$$

The matrix corresponding to free potentials, is a square symmetric matrix that is nonsingular. The potential vector is then obtained as

$$\phi = \begin{pmatrix} -S_{ff}^{-1} S_{fp} \phi_p \\ \phi_p \end{pmatrix} \qquad (6.71)$$

It must be noted that the solution to the problem is not limited to the nodes corresponding to free potentials but exists everywhere in the domain which can be obtained by interpolation.

6.3 SOLUTION TO THE NONLINEAR MAGNETOSTATIC PROBLEM WITH FIRST-ORDER TRIANGULAR FINITE ELEMENTS

Because the permeability of the iron parts may be nonlinear and field dependent, one cannot obtain the solution to the field problem by solving the matrix equation for the discretized field region only once. Further, because the solution to the problem is piecewise continuous, we cannot *a priori* determine the permeability in the different subregions or finite elements spanning the iron parts. It is therefore necessary to resort to some iterative method. The solution of nonlinear equations is fully discussed in a later chapter, and in this section we shall discuss only two methods, the latter of which is almost universally used. These methods are the chord method and the Newton–Raphson method.

6.3.1 The Chord Method

In this method the initial solution is obtained by assuming a value or values of permeability in the iron region. For simplicity, a constant value of permeability for the entire iron region of one type of material is adequate. From the initial solution, one can determine the flux density in iron region and obtain the respective permeabilities or reluctivities as the case may be. These material parameters are then updated to correspond to the ones given by the material characteristic curves ($B - H$ curves). In the chord method, the permeability or reluctivity is often updated in the nonlinear regions using multipliers known as relaxation factors. Therefore, we may write

$$\nu_{new} = \nu_{old} + \beta(\nu_{new} - \nu_{old}) \tag{6.72}$$

As is evident, this procedure reduces the error, which is the difference between the reluctivity obtained in this step and its value in the previous step of the iteration procedure, linearly. Although this procedure is generally stable and assured to yield convergence of the iterative scheme (convergence means homing in on the solution, in this case reluctivity, to the desired value of acceptable accuracy), it is time consuming and computationally uneconomical.

6.3.2 The Newton–Raphson Method

As a result of the slowness of the chord method or its related variants, engineers and researchers have tended to look for fast and less expensive methods. One such is the so-called Newton method or more popularly the Newton–Raphson method.

The N-R method is based on derivatives of the function or quantity to be updated, and the procedure yields rapid convergence, with the error in a given step decreasing as the square of the error in the previous step. It is, therefore, classified as a second-order method. The application of this procedure consists of finding a change in the potential as a result of a change in the reluctivity, which again varies as a function of the potential, the chicken or the egg situation. This can be illustrated by the following Poisson equation defined in terms of a single component (z-directed) vector potential, and the solution is sought in a two-dimensional region that includes current sources k.

$$\nu(A)\nabla \cdot \nabla A = -k \tag{6.73}$$

After discretization of the field region into finite elements with an unknown vector potential defined in each element, and after minimization of the applicable energy-related functional, equation (6.73) is transformed into a matrix equation as

$$\nu(A)[S][A] - [f] = 0 \tag{6.74}$$

where (S) is the coefficient matrix, $\nu(A)$ are the field-dependent nonlinear reluctivities, and (f) is the forcing function.

Differentiating equation (6.74) throughout by each of the nodal potentials A_k, one obtains

$$(J) = \nu(A)[S] + \frac{\partial \nu}{\partial A_k}[S][A] \tag{6.75}$$

where (J) is called the Jacobian matrix of derivatives.

Now the derivative $\frac{\partial \nu}{\partial A_k}$ can be further expanded by the chain rule of differentiation as follows.

$$\frac{\partial \nu(A)}{\partial A_k} = \frac{\partial \nu(A)}{\partial |B|^2} \cdot \frac{\partial |B|^2}{\partial A_k} \tag{6.76}$$

Also, for the two-dimensional region, the flux density B in each element is defined in terms of the x, y derivatives of the vector potential A as

$$|B|^2 = \left(\frac{\partial A}{\partial x}\right)^2 + \left(\frac{\partial A}{\partial y}\right)^2 \qquad (6.77)$$

Substituting the right-hand side of equation (6.77) for B in the derivative term of B with respect to A_k in equation (6.76), and performing the necessary algebraic operations, one obtains

$$\frac{\partial \nu}{\partial A_k} = 2\frac{\partial \nu}{\partial |B|^2}\left(\left(\frac{\partial A}{\partial x}\right)\frac{\partial}{\partial A_k}\left(\frac{\partial A}{\partial x}\right) + \left(\frac{\partial A}{\partial y}\right)\frac{\partial}{\partial A_k}\left(\frac{\partial A}{\partial y}\right)\right)$$

$$= 2\frac{\partial \nu}{\partial |B|^2}((S)(A))^T \qquad (6.78)$$

Finally, substituting the result from equation (6.78) into (6.75), the Jacobian is obtained as

$$(J) = \nu(A)(S) + 2\frac{\partial \nu(A)}{\partial |B|^2}((S)(A))((S)(A))^T \qquad (6.79)$$

The N-R algorithm can then be written as

$$(J)(\delta A) = -(R) \qquad (6.80)$$

where (R) is the residual vector given in terms of the vector potential at the start of this iteration as

$$\nu(A)(S)(A) - (f) \qquad (6.81)$$

Solving the set of algebraic equations resulting from (6.80) in each iterative step of the N-R scheme, one obtains the change in the vector potential A. The new vector potential in the $(k+1)$st iterative step is found in terms of its value in the kth step as follows.

$$(A)^{k+1} = (A)^k + (\delta A)^k \qquad (6.82)$$

From the vector potential obtained in the $(k+1)$st iteration, the flux density B is calculated and the new value of the derivative of the reluctivity with respect to B^2 is found from the material characteristic curve. The finite element solution is once again obtained as described in equations (6.74) through (6.81) and the new value of

The Finite Element Method

the vector potential is computed. When the magnitude of the residual vector given by equation (6.81) is found to be less than an acceptable limit, the iterative process is discontinued, and the solution vector is saved and used for postprocessing. The detailed algebraic steps of forming the Jacobian and residual vector for each N-R step are given later in the chapter.

Numerical Example

We shall illustrate the N-R algorithm by solving the following set of simultaneous equations.

$$g(x, y) = x^2 + y^2 - 4 = 0 \tag{6.83}$$

$$h(x, y) = -x^2 + y^2 - 1 = 0 \tag{6.84}$$

The Jacobian matrix is obtained from the partial derivatives of the functions g and h of equations (6.83) and (6.84) with respect to x and y. Therefore

$$(J) = \begin{pmatrix} 2x & 2y \\ -2x & 2y \end{pmatrix} \tag{6.85}$$

The residual vector will be

$$(R) = \begin{pmatrix} g(x, y) \\ h(x, y) \end{pmatrix} \tag{6.86}$$

The N-R algorithm as stated in equation (6.80) yields the matrix equation

$$(J)(\delta A) = -(R) \tag{6.87}$$

where the vector (A) is the vector of the variables x and y. The new values of x and y are then obtained for the $(k+1)$st iteration by adding the increment in A to the vector A in the kth iteration. Formally, this can be expressed, as before, as follows.

$$(A)^{k+1} = (A)^k + (\delta A)^k \tag{6.88}$$

We shall assume values for x and y as $(1, 1)$ and perform the sequence of steps as described until the residual vector becomes sufficiently small when the iteration scheme can be terminated. This is illustrated in Table 6.1.

Table 6.1 Newton–Raphson Iterations for Example

It.	x_0	y_0	g	h	$x = x_0 + \delta x$	$y = y_0 + \delta y$
1	1	1	-2	-1	1.25	1.75
2	1.25	1.75	0.625	0.5	1.225	1.589286
3	1.225	1.589286	2.645419E-2	2.520408E-2	1.224745	1.581160
4	1.224745	1.581160	6.61405E-5	6.592215E-5	1.224745	1.581139
5	1.224745	1.581139	1.698096E-7	-4.854105E-8	1.224745	1.581139

6.4 APPLICATION OF THE NEWTON–RAPHSON METHOD TO A FIRST-ORDER ELEMENT

The first application of the Newton-Raphson method to the nonlinear magnetic finite element problem was by Silvester and Chari [72]. With fairly minor changes this remains the standard formulation used today. We consider here the two dimensional nonlinear Poisson's equation for the magnetic vector potential. We have found that the stiffness matrix for a two-dimensional first-order triangle is

$$\frac{\nu}{4\Delta} \begin{pmatrix} s_{ii} & s_{ij} & s_{ik} \\ s_{ji} & s_{jj} & s_{jk} \\ s_{ki} & s_{kj} & s_{kk} \end{pmatrix} \begin{pmatrix} A_i \\ A_j \\ A_k \end{pmatrix} = \frac{\Delta}{3} \begin{pmatrix} J_e \\ J_e \\ J_e \end{pmatrix} \qquad (6.89)$$

Let F represent the first equation, G the second equation, and H the third equation. Then,

$$\begin{aligned} F &= \frac{\nu}{4\Delta}[s_{ii}\, s_{ij}\, s_{ik}] \begin{pmatrix} A_i \\ A_j \\ A_k \end{pmatrix} - \frac{J_e \Delta}{3} \\ G &= \frac{\nu}{4\Delta}[s_{ji}\, s_{jj}\, s_{jk}] \begin{pmatrix} A_i \\ A_j \\ A_k \end{pmatrix} - \frac{J_e \Delta}{3} \\ H &= \frac{\nu}{4\Delta}[s_{ki}\, s_{kj}\, s_{kk}] \begin{pmatrix} A_i \\ A_j \\ A_k \end{pmatrix} - \frac{J_e \Delta}{3} \end{aligned} \qquad (6.90)$$

The Finite Element Method

To find the derivatives necessary for the Newton–Raphson method we differentiate these equations with respect to the nodal vector potentials. For F this gives

$$\frac{\partial F}{\partial A_i} = \frac{\nu}{4\Delta} s_{ii} + \frac{1}{4\Delta}[s_{ii}A_i + s_{ij}A_j + s_{ik}A_k]\frac{\partial \nu}{\partial B^2}\frac{\partial B^2}{\partial A_i} \quad (6.91)$$

$$\frac{\partial F}{\partial A_j} = \frac{\nu}{4\Delta} s_{ij} + \frac{1}{4\Delta}[s_{ii}A_i + s_{ij}A_j + s_{ik}A_k]\frac{\partial \nu}{\partial B^2}\frac{\partial B^2}{\partial A_j} \quad (6.92)$$

$$\frac{\partial F}{\partial A_k} = \frac{\nu}{4\Delta} s_{ik} + \frac{1}{4\Delta}[s_{ii}A_i + s_{ij}A_j + s_{ik}A_k]\frac{\partial \nu}{\partial B^2}\frac{\partial B^2}{\partial A_k} \quad (6.93)$$

Here we have used the chain rule to replace $\frac{\partial \nu}{\partial A}$ by $\frac{\partial \nu}{\partial B^2}\frac{\partial B^2}{\partial A}$.

The Newton–Raphson equation is

$$\frac{\partial F}{\partial A_i}\Delta A_i + \frac{\partial F}{\partial A_j}\Delta A_j + \frac{\partial F}{\partial A_k}\Delta A_k = -F(A_i, A_j, A_k) \quad (6.94)$$

Substituting equations (6.91) – (6.93) into (6.94) and collecting terms

$$\frac{\nu}{4\Delta}[s_{ii}\Delta A_i + s_{ij}\Delta A_j + s_{ik}\Delta A_k]$$
$$+\frac{1}{4\Delta}[s_{ii}A_i + s_{ij}A_j + s_{ik}A_k][\frac{\partial \nu}{\partial B^2}][\frac{\partial B^2}{\partial A_i}\Delta A_i + \frac{\partial B^2}{\partial A_j}\Delta A_j + \frac{\partial B^2}{\partial A_k}\Delta A_k] \quad (6.95)$$
$$= \frac{-\nu}{4\Delta}[s_{ii}A_i + s_{ij}A_j + s_{ik}A_k] + \frac{J_e\Delta}{3}$$

In matrix notation this becomes

$$\frac{\nu}{4\Delta}(s_{ii}, s_{ij}, s_{ik})\begin{pmatrix}\Delta A_i \\ \Delta A_j \\ \Delta A_k\end{pmatrix}$$
$$+\frac{1}{4\Delta}\frac{\partial \nu}{\partial B^2}\left[\left(\sum_{n=i}^{k} s_{in}A_n\right)\frac{\partial B^2}{\partial A_i}, \left(\sum_{n=i}^{k} s_{in}A_n\right)\frac{\partial B^2}{\partial A_j}, \left(\sum_{n=i}^{k} s_{in}A_n\right)\frac{\partial B^2}{\partial A_k}\right]\begin{pmatrix}\Delta A_i \\ \Delta A_j \\ \Delta A_k\end{pmatrix}$$
$$= \frac{-\nu}{4\Delta}(s_{ii}, s_{ij}, s_{ik})\begin{pmatrix}A_i \\ A_j \\ A_k\end{pmatrix} + \frac{J_e\Delta}{3}$$

(6.96)

We now do the same for equations G and H to get the element equation

$$\frac{\nu}{4\Delta} \begin{pmatrix} s_{ii} & s_{ij} & s_{ik} \\ s_{ji} & s_{jj} & s_{jk} \\ s_{ki} & s_{kj} & s_{kk} \end{pmatrix} \begin{pmatrix} \Delta A_i \\ \Delta A_j \\ \Delta A_k \end{pmatrix}$$

$$+ \frac{1}{4\Delta} \frac{\partial \nu}{\partial B^2} \begin{pmatrix} \sum_{n=i}^{k} s_{in} A_n & \sum_{n=i}^{k} s_{in} A_n & \sum_{n=i}^{k} s_{in} A_n \\ \sum_{n=i}^{k} s_{jn} A_n & \sum_{n=i}^{k} s_{jn} A_n & \sum_{n=i}^{k} s_{jn} A_n \\ \sum_{n=i}^{k} s_{kn} A_n & \sum_{n=i}^{k} s_{kn} A_n & \sum_{n=i}^{k} s_{kn} A_n \end{pmatrix}$$

$$\times \begin{pmatrix} \frac{\partial B^2}{\partial A_i} & & \\ & \frac{\partial B^2}{\partial A_j} & \\ & & \frac{\partial B^2}{\partial A_k} \end{pmatrix} \begin{pmatrix} \Delta A_i \\ \Delta A_j \\ \Delta A_k \end{pmatrix} \quad (6.97)$$

$$= \frac{-\nu}{4\Delta} \begin{pmatrix} s_{ii} & s_{ij} & s_{ik} \\ s_{ji} & s_{jj} & s_{jk} \\ s_{ki} & s_{kj} & s_{kk} \end{pmatrix} \begin{pmatrix} A_i \\ A_j \\ A_k \end{pmatrix} + \frac{J_e \Delta}{3} \begin{pmatrix} 1 \\ 1 \\ 1 \end{pmatrix}$$

The vector potentials here are taken from the previous iteration. We find $\frac{\partial \nu}{\partial B^2}$ from the saturation curve representation (for example, a cubic spline). To evaluate $\frac{\partial B^2}{\partial A_i}$ we proceed as follows:

$$B^2 = \left(\frac{\partial A}{\partial x}\right)^2 + \left(\frac{\partial A}{\partial y}\right)^2 \quad (6.98)$$

For first-order elements

$$A = \frac{a_i + b_i x + c_i y}{2\Delta} A_i + \frac{a_j + b_j x + c_j y}{2\Delta} A_j + \frac{a_k + b_k x + c_k y}{2\Delta} A_k \quad (6.99)$$

Then

$$\begin{aligned}\frac{\partial A}{\partial x} &= \frac{A_i b_i + A_j b_j + A_k b_k}{2\Delta} \\ \frac{\partial A}{\partial y} &= \frac{A_i c_i + A_j c_j + A_k c_k}{2\Delta}\end{aligned} \tag{6.100}$$

The square of the flux density is

$$B^2 = \frac{(A_i b_i + A_j b_j + A_k b_k)^2 + (A_i c_i + A_j c_j + A_k c_k)^2}{4\Delta^2} \tag{6.101}$$

Therefore

$$\frac{\partial B^2}{\partial A_i} = \frac{2b_i(A_i b_i + A_j b_j + A_k b_k) + 2c_i(A_i c_i + A_j c_j + A_k c_k)}{4\Delta^2} \tag{6.102}$$

We can summarize the process as follows:

1. Assume a value for ν and A for each element and node.
2. Evaluate the matrices in equation (6.97) using these values and the geometric coefficients.
3. Assemble the matrix in the normal way.
4. Apply boundary conditions and solve for the ΔA vector.
5. Find the new A by adding ΔA to the previous value of A.
6. Apply a stopping or convergence test, such as that the relative change in ΔA is smaller than ϵ.
7. If the test fails, recompute the matrices in equation (6.97) and repeat the process from step 2.

Note that for elements with constant permeability, the second term in equation (6.97) is zero because of the zero derivative. In this case the equation reduces to that obtained for the linear case.

6.5 DISCRETIZATION OF TIME BY THE FINITE ELEMENT METHOD

We have seen in Chapter 3 that time can be discretized by finite differences. It is also possible to discretize time by the finite element method. This method was developed by Zienkiewicz and others [71]. In this section the method is summarized. Let us consider the partial differential equation for the two dimensional linear homogeneous eddy current problem.

$$\nu \nabla^2 A - \sigma \frac{\partial A}{\partial t} = -J_s \quad (6.103)$$

Treating time as an independent variable, we can write

$$\alpha = \sum \zeta_i \alpha_i \quad (6.104)$$

where α_i are the nodal values of the potentials, A, at time t. We shall consider $\zeta_i(t)$ as first-order shape functions. Then for $0 \leq \xi \leq 1$ and $\xi = \frac{t}{\Delta t}$,

$$\begin{aligned} \zeta_i &= 1 - \xi : & \frac{\partial \zeta_i}{\partial t} &= -\frac{\partial \xi}{\partial t} = -\frac{1}{\Delta t} \\ \zeta_{i+1} &= \xi; & \frac{\partial \zeta_{i+1}}{\partial t} &= \frac{\partial \xi}{\partial t} = \frac{1}{\Delta t} \end{aligned} \quad (6.105)$$

Figure 6.5 illustrates the variation of the shape functions over the interval Δt. Substituting equation (6.104) into equation (6.103), we obtain

$$\nu \nabla^2 \sum \zeta_i \alpha_i - \sigma \frac{\partial}{\partial t} \sum \zeta_i \alpha_i = -J_s \quad (6.106)$$

We may now write the weighted residual form of equation (6.106) as

$$\int \left(W_j \nu \nabla^2 \sum \zeta_i \alpha_i - W_j \sigma \frac{\partial A}{\partial t} + W_j J_s \right) dx \, dy = 0 \quad (6.107)$$

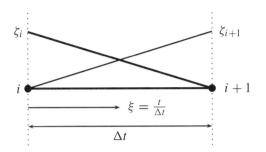

Figure 6.5 Variation of Time-Dependent Shape Functions

The first term in the integrand of equation (6.107) can be separated by integration by parts such that

$$\nabla \cdot W_j \nabla A = W_j \nabla^2 A + \nabla W_j \cdot \nabla A \tag{6.108}$$

where $A = \sum \zeta_i \alpha_i$. Therefore equation (6.107) becomes

$$\int \left(\nu \nabla W_j \cdot \nabla A + W_j \sigma \frac{\partial A}{\partial t} - W_j J_s \right) dx\, dy - \oint W_j \frac{\partial A}{\partial n} d\ell = 0 \tag{6.109}$$

or

$$\int \left(\nu \nabla W_j \cdot \nabla \sum \zeta_i \alpha_i + W_j \sigma \frac{\partial}{\partial t} \sum \zeta_i \alpha_i - W_j J_s \right) dx\, dy = 0 \tag{6.110}$$

Equation (6.109) may be formally written as

$$PA + Q \frac{\partial A}{\partial t} + R_s = 0 \tag{6.111}$$

Substituting equation (6.104) into equation (6.111), we get

$$P \sum \zeta_i \alpha_i + Q \frac{\partial}{\partial t} \sum \zeta_i \alpha_i + R_s = 0 \qquad (6.112)$$

or

$$P \sum \zeta_i \alpha_i + Q \sum \frac{\partial \zeta_i}{\partial t} \alpha_i + R_s = 0 \qquad (6.113)$$

Over the interval Δt

$$P(\zeta_i \alpha_i + \zeta_{i+1} \alpha_{i+1}) + Q \left(\frac{\partial \zeta_i}{\partial t} \alpha_i + \frac{\partial \zeta_{i+1}}{\partial t} \alpha_{i+1} \right) + R_s = 0 \qquad (6.114)$$

Substituting (6.105) into (6.114)

$$P((1 - \xi)\alpha_i + \xi \alpha_{i+1}) + Q \left(\frac{-\alpha_i}{\Delta t} + \frac{\alpha_{i+1}}{\Delta t} \right) + R_s = 0 \qquad (6.115)$$

The weighted residual form of equation (6.115) will be

$$P \int_0^1 W_j ((1 - \xi)\alpha_i + \xi \alpha_{i+1}) \, d\xi + Q \int_0^1 W_j \left(\frac{-\alpha_i}{\Delta t} + \frac{\alpha_{i+1}}{\Delta t} \right) d\xi$$

$$+ \int_0^1 W_j R_s \, d\xi = 0 \qquad (6.116)$$

or

$$P \int_0^1 W_j (1 - \xi)\alpha_i \, d\xi - Q \int_0^1 \frac{W_j \alpha_i}{\Delta t} \, d\xi + \int_0^1 W_j \xi \alpha_{i+1} \, d\xi$$

$$+ Q \int_0^1 \frac{W_j \alpha_{i+1}}{\Delta t} \, d\xi + \int_0^1 W_j R_s \, d\xi = 0 \qquad (6.117)$$

It must be noted that P and Q are independent of time.

6.6 AXISYMMETRIC FORMULATION FOR THE EDDY CURRENT PROBLEM USING VECTOR POTENTIAL

For the axisymmetric case of the vector potential, both A and J have only θ components.

$$\begin{aligned} A &= A_\theta \\ J &= J_\theta \end{aligned} \tag{6.118}$$

The functional becomes

$$\mathcal{F} = \frac{1}{2} \iint_R \{\nu[(\nabla \times A_\theta) \cdot (\nabla \times A_\theta)] - 2A_\theta \cdot J_\theta + j\omega\sigma A_\theta^2\} dR$$

$$+ \oint_C A_\theta \left(\frac{\partial A_\theta}{\partial n} + \frac{A_\theta}{r} \cdot \frac{\partial r}{\partial n} \right) dc \tag{6.119}$$

In cylindrical coordinates

$$\mathcal{F} = 2\pi \iint_R \left(\frac{\nu r}{2} \left[\left(\frac{\partial A_\theta}{\partial r} \right)^2 + \left(\frac{\partial A_\theta}{\partial z} \right)^2 \right] + \nu A_\theta \frac{\partial A_\theta}{\partial r} + \frac{\nu A_\theta^2}{2r} - J_\theta r A_\theta \right) dr\, dz$$

$$+ 2\pi \iint_R \frac{j\omega\sigma A_\theta^2 r}{2} dr\, dz + \oint_C A_\theta \left(\frac{\partial A_\theta}{\partial n} + \frac{A_\theta}{r} \cdot \frac{\partial r}{\partial n} \right) dc \tag{6.120}$$

For the most common types of problems we can eliminate the line integral. It is our aim now to minimize $\mathcal{F}$ over the region of interest. To this end we discretize the region by dividing it into triangular elements as in Figure 6.6.

The first-order shape functions are

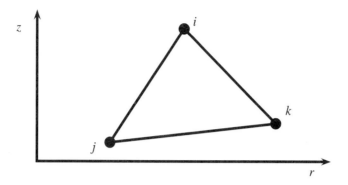

Figure 6.6 Triangular Finite Element

$$\begin{aligned} \zeta_i &= (a_i + b_i z + c_i r)/2\Delta \\ \zeta_j &= (a_j + b_j z + c_j r)/2\Delta \\ \zeta_k &= (a_k + b_k z + c_k r)/2\Delta \end{aligned} \qquad (6.121)$$

where

$$\begin{aligned} a_i &= \begin{vmatrix} z_j & z_k \\ r_j & r_k \end{vmatrix} \\ b_i &= r_j - r_k \\ c_i &= z_k - z_j \end{aligned} \qquad (6.122)$$

As before, all indices are cyclic modulo 3.

The Finite Element Method

Therefore inside each triangle

$$A = \sum_{i,j,k} A_i \zeta_i = \sum_{i,j,k} \frac{(a_i + b_i z + c_i r) A_i}{2\Delta} \tag{6.123}$$

We now minimize (6.120) by taking the first variation and setting it equal to zero.

$$\begin{aligned}\frac{\delta F}{\delta A_i} &= 2\pi \iint_R \left\{ vr \left[\left(\frac{\partial A}{\partial r}\right) \frac{\partial}{\partial A_i} \left(\frac{\partial A}{\partial r}\right) + \left(\frac{\partial A}{\partial z}\right) \frac{\partial}{\partial A_i} \left(\frac{\partial A}{\partial z}\right) \right] \right. \\ &+ v \left(\frac{\partial A}{\partial A_i}\right) \left(\frac{\partial A}{\partial r}\right) + vA \frac{\partial}{\partial A_i} \left(\frac{\partial A}{\partial r}\right) + \frac{vA}{r} \left(\frac{\partial A}{\partial A_i}\right) + j\omega\sigma r A \left(\frac{\partial A}{\partial A_i}\right) \\ &- \left. Jr \left(\frac{\partial A}{\partial A_i}\right) \right\} dr\, dz = 0 \end{aligned} \tag{6.124}$$

Substituting (6.123) into (6.124)

$$\begin{aligned}\sum_{i=1}^{n} 2\pi \iint_R \{ &vr \left[\left(\frac{\partial \zeta_i}{\partial r}\right) \left(\frac{\partial \zeta_j}{\partial r}\right) + \left(\frac{\partial \zeta_i}{\partial z}\right) \left(\frac{\partial \zeta_j}{\partial z}\right) \right] [A_i] \\ &+ v \left[\zeta_i \left(\frac{\partial \zeta_i}{\partial r}\right) + (\zeta_i)^T \left(\frac{\partial \zeta_i}{\partial r}\right)^T \right] [A_i] \\ &+ v \left[\frac{\zeta_i \zeta_j}{r} \right] [A_i] + j\omega\sigma r \left[\zeta_i \zeta_j \right] [A_i] \\ &- Jr\zeta_i \} dR = 0 \end{aligned} \tag{6.125}$$

Equation (6.125) can be written in matrix form as

$$v[S][A] + v[D][A] + v[E][A] + j\omega\sigma[T][A] = [T][J] \tag{6.126}$$

The matrices $[S]$, $[D]$, $[E]$, and $[T]$ have elements

$$\begin{aligned}
S_{ij} &= \iint r\left[\left(\frac{\partial \zeta_i}{\partial r}\right)\left(\frac{\partial \zeta_j}{\partial r}\right) + \left(\frac{\partial \zeta_i}{\partial z}\right)\left(\frac{\partial \zeta_j}{\partial z}\right)\right] dr\, dz \\
D_{ij} &= \iint \left[\zeta_i\left(\frac{\partial \zeta_j}{\partial r}\right) + \zeta_j\left(\frac{\partial \zeta_i}{\partial r}\right)\right] dr\, dz \\
E_{ij} &= \iint \frac{1}{r}\zeta_i \zeta_j\, dr\, dz \\
T_{ij} &= \iint r\zeta_i \zeta_j\, dr\, dz
\end{aligned} \qquad (6.127)$$

Notice that one term here goes to infinity at the $r = 0$ axis. To evaluate these integrals, Konrad and Silvester [73] suggested a change of variables that has proven useful. Let

$$\begin{aligned}
A &= \sqrt{r}\phi \\
J &= \sqrt{r}\psi
\end{aligned} \qquad (6.128)$$

Substituting these in the matrix equation (6.124) we get

$$\begin{aligned}
S_{ij} &= \iint r^2\left[\left(\frac{\partial \zeta_i}{\partial r}\right)\left(\frac{\partial \zeta_j}{\partial r}\right) + \left(\frac{\partial \zeta_i}{\partial z}\right)\left(\frac{\partial \zeta_j}{\partial z}\right)\right] dr\, dz \\
D_{ij} &= \iint \frac{3}{2}r\left[\zeta_i\left(\frac{\partial \zeta_i}{\partial r}\right) + \zeta_i\left(\frac{\partial \zeta_i}{\partial r}\right)\right] dr\, dz \\
E_{ij} &= \iint \frac{9}{4}\zeta_i \zeta_j\, dr\, dz \\
T_{ij} &= \iint r^2 \zeta_i \zeta_j\, dr\, dz
\end{aligned} \qquad (6.129)$$

Note that the singularity in equation (6.127) has disappeared in (6.129). The matrices in equation (6.129) are evaluated as follows for first-order elements.

$$\zeta_m = \frac{a_m + b_m z + c_m r}{2\Delta}, \quad m = i, j, k$$

The Finite Element Method

$$\frac{\partial \zeta_m}{\partial r} = \frac{c_m}{2\Delta}, \quad m = i, j, k$$
$$\frac{\partial \zeta_m}{\partial z} = \frac{b_m}{2\Delta}, \quad m = i.j, k. \tag{6.130}$$

r can be written as a linear combination of its vertex values

$$r = \sum_{m=i,j,k} \zeta_m r_m \tag{6.131}$$

Upon substitution, this gives

$$S_{ij} = 2\pi \iint (\zeta_i r_i + \zeta_j r_j + \zeta_k r_k)^2 \left[\frac{(b_i b_j + c_i c_j)}{4\Delta^2}\right] dr\, dz \tag{6.132}$$

We now use

$$dr\, dz = 2\Delta d\zeta_i d\zeta_j \tag{6.133}$$

and

$$\zeta_i + \zeta_j + \zeta_k = 1 \tag{6.134}$$

where the limits of ζ_i are 0 to 1, and the limits of ζ_j are 0 to $(1 - \zeta_i)$. Substituting into equation (6.132).

$$\begin{aligned}S_{ij} = \tfrac{2\pi\nu}{2\Delta} \int_0^1 \int_0^{1-\zeta_i} &\{\zeta_i^2 (r_i - r_k)^2 + \zeta_j^2 (r_j - r_k)^2 + r_k^2 \\ &+ 2\zeta_i \zeta_j (r_i - r_k)(r_j - r_k) + 2r_k \zeta_j (r_j - r_k) \\ &+ 2r_k \zeta_i (r_i - r_k)\}[b_i b_j + c_i c_j]\, d\zeta_1 d\zeta_j\end{aligned} \tag{6.135}$$

The integrals may now be evaluated. The first integral becomes

$$\frac{2\pi v}{2\Delta} \int_0^1 \int_0^{1-\zeta_i} \zeta_i^2 (r_i - r_k)^2 \, d\zeta_i \, d\zeta_j = \frac{\pi v}{12\Delta}(r_i - r_k)^2 \qquad (6.136)$$

The other terms are similarly evaluated with the result that

$$[S] = \frac{\pi v}{12}(r_i^2 + r_j^2 + r_k^2 + r_i r_j + r_j r_k + r_k r_i)$$

$$\times \begin{bmatrix} b_i^2 + c_i^2 & b_i b_j + c_i c_j & b_i b_k + c_i c_k \\ & b_j^2 + c_j^2 & b_j b_k + c_j c_k \\ \text{symmetric} & & b_k^2 + c_k^2 \end{bmatrix} \qquad (6.137)$$

To evaluate the elements of $[D]$ consider

$$D_{ij} = 4\pi\Delta \iint \frac{3}{2}(r_i\zeta_i + r_j\zeta_j + r_k\zeta_k)\left[\zeta_i\frac{\partial \zeta_j}{\partial r} + \zeta_j\frac{\partial \zeta_i}{\partial r}\right] d\zeta_i \, d\zeta_j \qquad (6.138)$$

Multiplying out the terms

$$\begin{aligned} D_{ij} &= 6\pi\Delta \iint \left[r_i\zeta_i^2 \left(\tfrac{\partial \zeta_j}{\partial r}\right) + r_j\zeta_i\zeta_j \left(\tfrac{\partial \zeta_j}{\partial r}\right) + r_k\zeta_i\zeta_k \left(\tfrac{\partial \zeta_j}{\partial r}\right)\right] \partial \zeta_i \, \partial \zeta_j \\ &\quad + \left[r_i\zeta_i\zeta_j\left(\tfrac{\partial \zeta_i}{\partial r}\right) + r_j\zeta_j^2\left(\tfrac{\partial \zeta_i}{\partial r}\right) + r_k\zeta_j\zeta_k\left(\tfrac{\partial \zeta_i}{\partial r}\right)\right] d\zeta_i \, d\zeta_j \end{aligned} \qquad (6.139)$$

Integrating ζ_i from 0 to 1 and ζ_j from 0 to $(1 - \zeta_i)$ gives

$$D_{ij} = \frac{\pi}{8}\{c_i r_j + c_j r_i - c_k(r_i + r_j + r_k)\} \qquad (6.140)$$

The Finite Element Method

Therefore D becomes

$$D = \frac{v\pi}{8} \begin{bmatrix} 2c_i(2r_i + r_j + r_k) & c_ir_j + c_jr_i & -c_j(r_i + r_j + r_k) \\ & -c_k(r_i + r_j + r_k) & +c_ir_k + c_kr_i \\ & 2c_j(r_i + 2r_j + r_k) & -c_i(r_i + r_j + r_k) \\ & +c_jr_k + c_kr_j & \\ \text{symmetric} & & 2c_k(r_i + r_j + 2r_k) \end{bmatrix}$$
(6.141)

An element of the T matrix is found as

$$\begin{aligned} T_{ij} &= j2\pi\omega\sigma \iint r^2 \phi \left(\frac{\partial \phi}{\partial \phi_k}\right) \\ &= j2\pi\omega\sigma \iint 2\Delta(\zeta_i r_i + \zeta_j r_j + \zeta_k r_k)^2 \sum \zeta_i \zeta_j \phi_i \, d\zeta_i \, d\zeta_j \end{aligned}$$
(6.142)

Integrating over the same limits, the first row of T is

$$\begin{aligned} j\omega T_{ii}, T_{ij}, T_{ik} = \tfrac{j2\pi\omega\sigma\Delta}{180}[&(12r_i^2 + 6r_ir_j + 6r_ir_k + 2r_j^2 + 2r_jr_k + 2r_k^2, \\ &(3r_i^2 + 4r_ir_j + 2r_ir_k + 3r_j^2 + r_k^2), \\ &3r_i^2 + 2r_ir_j + 4r_ir_k + r_j^2 + 2r_jr_k + 3r_k^2)] \end{aligned}$$
(6.143)

The second row of the matrix (the jj and jk terms only because the matrix is symmetric) is

$$\begin{aligned} j\omega T_{jj} &= \tfrac{j2\pi\omega\sigma\Delta}{180}(2r_i^2 + 6r_ir_j + 2r_ir_k + 12r_j^2 + 6r_jr_k + 2r_k^2) \\ j\omega T_{kj} &= \tfrac{j2\pi\omega\sigma\Delta}{180}(r_i^2 + 2r_ir_j + 2r_ir_k + 3r_j^2 + 4r_jr_k + 3r_k^2) \end{aligned}$$
(6.144)

Finally, the T_{kk} term is

$$j\omega\sigma T_{kk} = \frac{j2\pi\omega\sigma\Delta}{180}(2r_i^2 + 2r_ir_j + 6r_ir_k + 2r_j^2 + 6r_jr_k + 12r_k^2) \qquad (6.145)$$

Evaluating the E matrix

$$E_{ij} = \frac{9\pi\nu}{2}\iint \phi\frac{\partial\phi}{\partial\phi_k} = \frac{9\pi\nu}{2}\int_0^1\int_0^{1-\zeta_i} 2\Delta(\zeta_i^2, \zeta_i\zeta_j, \zeta_i\zeta_k)[\phi]\,d\zeta_i\,d\zeta_j \qquad (6.146)$$

Integrating each term in the preceding equation gives

$$\frac{9\pi\nu}{2}\iint \phi\left(\frac{\partial\phi}{\partial\phi_k}\right) = \frac{3\pi\nu\Delta}{8}[2, 1, 1,][\phi] \qquad (6.147)$$

Integrating for i, j, and k,

$$[E] = \frac{3\pi\nu\Delta}{8}\begin{bmatrix} 2 & 1 & 1 \\ 1 & 2 & 1 \\ 1 & 1 & 2 \end{bmatrix}\begin{pmatrix} \phi_i \\ \phi_j \\ \phi_k \end{pmatrix} \qquad (6.148)$$

6.7 FINITE DIFFERENCE AND FIRST-ORDER FINITE ELEMENTS

It is interesting to note that the five-point finite difference formula described in Chapter 3 gives practically the same global system of equations as a first-order finite element formulation given that the nodes are equally spaced. As an example, consider the four-element problem illustrated in Figure 6.7. We will solve the two-dimensional Poisson problem described by

$$k\nabla^2\phi = f \qquad (6.149)$$

The Finite Element Method

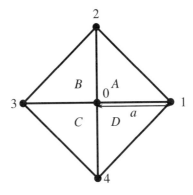

Figure 6.7 Four First-Order Triangular Elements

As shown previously, for first-order elements we can find the stiffness matrix from the nodal coordinates. For element A we have nodes $(0, 1, 2)$ and the matrix elements become

$$s_{00} = \frac{k}{4\Delta}((y_2 - y_1)(y_2 - y_1) + (x_2 - x_1)(x_2 - x_1)) = k\frac{2a^2}{2a^2} = k \quad (6.150)$$

In equation (6.150) we note that the result is independent of the size of the element. Only the shape is important. This is why in all subsequent terms the dimension a will disappear. It is also true that elements $A, B, C,$ and D are identical. They are only rotated by $90°$ and therefore have the same stiffness matrices. Continuing,

$$s_{01} = \frac{k}{4\Delta}((a)(-a) + (a)(0)) = -\frac{k}{2}$$

$$s_{02} = \frac{k}{4\Delta}((-a)(0) + (-a)(a)) = -\frac{k}{2}$$

$$s_{11} = \frac{k}{4\Delta}((-a)(-a) + (0)(0)) = \frac{k}{2}$$

$$etc. \quad (6.151)$$

The stiffness matrix for element A is then

$$s_A = k \begin{pmatrix} 1 & -1/2 & -1/2 \\ -1/2 & 1/2 & 0 \\ -1/2 & 0 & 1/2 \end{pmatrix} \quad (6.152)$$

We now take advantage of the fact that all of the elements are congruent and the stiffness matrices of, for example, $A(012)$ and $B(023)$ are equal to obtain the equation for the potential at node 0.

$$4\phi_0 - \phi_1 - \phi_2 - \phi_3 - \phi_4 = 0 \quad (6.153)$$

This is exactly the equation we would obtain from the finite difference expansion. If we are dealing with Poisson's equation and there is a source density term f, then the finite element equation would have $4 \times \frac{1}{3} \times \frac{a^2}{2} \times f = \frac{2}{3} a^2 f$. The 4 is from the four elements; the 1/3 times the area of the element comes from integrating the first-order shape function. In the finite difference equation we obtain the area of the cell times the source function or $a^2 f$. Over the entire problem we obtain the same total source but the distributions are somewhat different in the two methods.

6.8 GALERKIN FINITE ELEMENTS

We will now see that the Galerkin method will yield the same finite element equations as the variational method in cases in which the variational principle is known.

As an example, we begin with the time harmonic form of the diffusion equation,

$$\frac{1}{\mu}\frac{d^2 A}{dx^2} + \frac{1}{\mu}\frac{d^2 A}{dy^2} = -J_o + j\omega\sigma A \quad (6.154)$$

Substituting an approximation $\hat{A}$ for A gives a residual R.

$$R = \frac{1}{\mu}\frac{\partial^2 \hat{A}}{\partial x^2} + \frac{1}{\mu}\frac{d^2 \hat{A}}{dy^2} - j\omega\sigma\hat{A} + J_o \quad (6.155)$$

The Finite Element Method

Multiplying by a weighting function and setting the integral to zero

$$\int_\Omega RW \, dx \, dy = 0 \tag{6.156}$$

where Ω is the region of interest.

Substituting

$$\int W \frac{1}{\mu}\left(\frac{d^2\hat{A}}{dx^2} + \frac{d^2\hat{A}}{dy^2}\right) dx \, dy - j\omega\sigma \int W\hat{A} \, dx \, dy = \int WJ_o \, dx \, dy \tag{6.157}$$

Integrating the first term by parts,

$$\int_D W\left(\frac{1}{\mu}\frac{d^2\hat{A}}{dx^2} + \frac{1}{\mu}\frac{d^2\hat{A}}{dy^2}\right) dx \, dy =$$
$$-\int_R \frac{1}{\mu}\left(\frac{dW}{dx}\frac{d\hat{A}}{dx} + \frac{dW}{dy}\frac{d\hat{A}}{dy}\right) dx \, dy + \oint_C \frac{1}{\mu} W \frac{d\hat{A}}{dn} dc \tag{6.158}$$

Substituting into (6.157), we now break the surface integrals into summations over the triangles (elements)

$$\sum_E \left\{\frac{1}{\mu^e}\int_{R_e}\frac{dW^e}{dx}\frac{dA^e}{dx} + \frac{dW^e}{dy}\frac{dA^e}{dy}\right\} dx \, dy + j\omega\sigma^e \int_{R_e} W^e A^e dx \, dy$$
$$- \tfrac{1}{\mu^e}\tfrac{dA}{dn}\int W^e dc = J_o \int W^e dx \, dy \tag{6.159}$$

The line integral needs to be evaluated only over elements that have a side in common with the boundaries of the problem. Normally this integral is simply set to zero, which gives the so-called *natural* boundary condition. However, this integral is often used in problems in which the finite element regions are coupled to other regions such as that of the hybrid method in which the finite elements are coupled to boundary elements (see [74]) or the use of *air gap* elements described in [75]. We consider the interior elements only (i.e., without the line integral) and assume

$$A = a + bx + cy \tag{6.160}$$

We have previously shown that we can write

$$A^e = [S_i^e, S_j^e, S_k^e] \begin{bmatrix} A_i^e \\ A_j^e \\ A_k^e \end{bmatrix} \tag{6.161}$$

where

$$\begin{aligned} S_i^e &= (a_i^e + b_i^e x + c_i^e y)/2\Delta \\ S_j^e &= (a_j^e + b_j^e x + c_j^E y)/2\Delta \\ S_k^e &= (a_k^e + b_k^e x + c_k^e y)/2\Delta \end{aligned} \tag{6.162}$$

In the Galerkin method

$$W^e = \begin{bmatrix} S_i^e \\ S_j^e \\ S_k^e \end{bmatrix} \tag{6.163}$$

$$\begin{aligned} \frac{dA}{dx} &= \frac{1}{2\Delta}[b_i^e, b_j^e, b_k^e] \begin{bmatrix} A_i \\ A_j \\ A_k \end{bmatrix} \\ \frac{dA}{dy} &= \frac{1}{2\Delta}[c_i^e, c_j^e, c_k^e] \begin{bmatrix} A_i \\ A_j \\ A_k \end{bmatrix} \end{aligned} \tag{6.164}$$

The Finite Element Method

$$\frac{\partial W^e}{\partial x} = \frac{1}{2\Delta} \begin{bmatrix} b_i^e \\ b_j^e \\ b_k^e \end{bmatrix}$$

$$\frac{\partial W^e}{\partial y} = \frac{1}{2\Delta} \begin{bmatrix} b_i^e \\ b_j^e \\ b_k^e \end{bmatrix} \qquad (6.165)$$

Using these, the first integral on the left of (6.159) is

$$\frac{1}{\mu^e}\left(\frac{\partial W^e}{\partial x}\frac{\partial \hat{A}}{\partial x} + \frac{\partial W}{\partial y}\frac{\partial \hat{A}}{\partial y}\right)\int dx\,dy, \text{ but } \int dx\,dy = \Delta \qquad (6.166)$$

Substituting

$$\frac{1}{\mu}\int\left(\frac{\partial W^e}{\partial x}\frac{\partial \hat{A}^e}{\partial x} + \frac{\partial W^e}{\partial y}\frac{\partial \hat{A}^e}{\partial y}\right) dx\,dy$$

$$= \frac{1}{4\mu\Delta}\begin{bmatrix} b_i^2 + c_i^2 & b_i b_j + c_i c_j & b_i b_k + c_i c_k \\ b_i b_j + c_i c_j & b_j^2 + c_j^2 & b_j b_k + c_j c_k \\ b_i b_k + c_i c_k & b_j b_k + c_j c_k & b_k^2 + c_k^2 \end{bmatrix}\begin{bmatrix} \hat{A}_i \\ \hat{A}_j \\ \hat{A}_k \end{bmatrix} \qquad (6.167)$$

The second integral becomes

$$j\omega\sigma^e \int W^e\,dx\,dy = j\omega\sigma^e \int \begin{bmatrix} S_i^e \\ S_j^e \\ S_k^e \end{bmatrix}[S_i^e\,S_j^e\,S_k^e]\begin{bmatrix} \hat{A}_i \\ \hat{A}_j \\ \hat{A}_k \end{bmatrix} dx\,dy$$

$$= \frac{j\omega\sigma^e\Delta}{12}\begin{bmatrix} 2 & 1 & 1 \\ 1 & 2 & 1 \\ 1 & 1 & 2 \end{bmatrix}\begin{bmatrix} \hat{A}_i \\ \hat{A}_j \\ \hat{A}_k \end{bmatrix} \qquad (6.168)$$

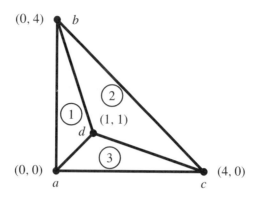

Figure 6.8 Three-Element Problem

The forcing function becomes

$$J_o^e \int W^e \, dx \, dy = \frac{J_o \Delta}{3} \begin{bmatrix} 1 \\ 1 \\ 1 \end{bmatrix} \qquad (6.169)$$

We see that these are the same results as we obtained for the variational method.

6.9 THREE-ELEMENT MAGNETOSTATIC PROBLEM

To illustrate the application of the finite element method, we will solve a simple three-element problem in a number of different ways [76]. Consider the three-element linear magnetostatic problem illustrated in Figure 6.8. We shall solve for the magnetic vector potentials at the nodes. We have a constant current density of 1000 amperes/square inch in triangle 1. The reluctivity is constant at 80,000 inch/henry. The boundary conditions are A_a and A_b are equal to zero. We will find the potentials at nodes c and d and the element flux densities.

The Finite Element Method

Table 6.2 Definitions and Coordinates for Three-Element Example

Elem.	x_i	y_i	x_j	y_j	x_k	y_k	b_i	b_j	b_k	c_i	c_j	c_k	Δ
1(bad)	0	4	0	0	1	1	−1	−3	4	1	−1	0	2
2(bdc)	0	4	1	1	4	0	1	−4	3	3	−4	1	4
3(acd)	0	0	4	0	1	1	−1	1	0	−3	−1	4	2

Recall that we have for the first-order element

$$\nu \begin{pmatrix} s_{ii} & s_{ij} & s_{ik} \\ s_{ji} & s_{jj} & s_{jk} \\ s_{ki} & s_{kj} & s_{kk} \end{pmatrix} \begin{pmatrix} A_i \\ A_j \\ A_k \end{pmatrix} = \frac{\Delta}{3} \begin{pmatrix} J_i \\ J_j \\ J_k \end{pmatrix} \qquad (6.170)$$

The i, j, k definitions are given in the table.

This gives the following individual element matrices.

For element 1 (*bad*)

$$(S)_1 = \nu_1 \begin{pmatrix} 0.25 & 0.25 & -0.5 \\ 0.25 & 1.25 & -1.5 \\ -0.5 & -1.5 & 2 \end{pmatrix} \qquad (6.171)$$

For element 2 (*bdc*)

$$(S)_2 = \nu_2 \begin{pmatrix} 0.625 & -1 & 0.375 \\ -1 & 2 & -1 \\ 0.375 & -1 & 0.625 \end{pmatrix} \qquad (6.172)$$

and for element 3 (*acd*)

$$(S)_3 = \nu_3 \begin{pmatrix} 1.25 & 0.25 & -1.5 \\ 0.25 & 0.25 & -0.5 \\ -1.5 & -0.5 & 2 \end{pmatrix} \qquad (6.173)$$

Assembling the matrix into the global system

$$S =$$

$$\begin{pmatrix} 0.25v_1 + 0.625v_2 & 0.25v_1 & 0.375v_2 & -0.5v_1 - v_2 \\ 0.25v_1 & 1.25v_1 + 1.25v_3 & .25v_3 & -1.5v_1 - 1.5v_3 \\ 0.375v_2 & 0.25v_3 & 0.625v_2 + 0.25v_3 & -v_2 - 0.5v_3 \\ -0.5v_1 - v_2 & -1.5v_1 - 1.5v_3 & -v_2 - 0.5v_3 & 2v_1 + 2v_2 + 2v_3 \end{pmatrix}$$

(6.174)

Here $v_1 = v_2 = v_3 = v_0$ so

$$(S) = v_0 \begin{pmatrix} 0.875 & 0.25 & 0.375 & -1.5 \\ 0.25 & 2.5 & .25 & -3 \\ 0.375 & 0.25 & 0.875 & -1.5 \\ -1.5 & -3 & -1.5 & 6 \end{pmatrix}$$

(6.175)

The right-hand-side vectors are

$$(J)_1 = \begin{pmatrix} 2000/3 \\ 2000/3 \\ 2000/3 \end{pmatrix}, \quad (J)_2 = \begin{pmatrix} 0 \\ 0 \\ 0 \end{pmatrix}, \quad (J)_3 = \begin{pmatrix} 0 \\ 0 \\ 0 \end{pmatrix}$$

(6.176)

Therefore the global set of equations is

$$v_0 \begin{pmatrix} 0.875 & 0.25 & 0.375 & -1.5 \\ 0.25 & 2.5 & .25 & -3 \\ 0.375 & 0.25 & 0.875 & -1.5 \\ -1.5 & -3 & -1.5 & 6 \end{pmatrix} \begin{pmatrix} A_b \\ A_a \\ A_c \\ A_d \end{pmatrix} = \begin{pmatrix} 2000/3 \\ 2000/3 \\ 0 \\ 2000/3 \end{pmatrix}$$

(6.177)

The Finite Element Method

Applying the boundary conditions $A_a = 0.0$ and $A_b = 0.0$,

$$v_0 \begin{pmatrix} 0.0 & 0.0 & 0.0 & 0.0 \\ 0.0 & 0.0 & 0.0 & 0.0 \\ 0.0 & 0.0 & 0.875 & -1.5 \\ 0 & 0 & -1.5 & 6 \end{pmatrix} \begin{pmatrix} A_b \\ A_a \\ A_c \\ A_d \end{pmatrix} = \begin{pmatrix} 0 \\ 0 \\ 0 \\ 2000/3 \end{pmatrix} \quad (6.178)$$

Solving, we get $A_c = 4.1667 \times 10^{-3}$ webers/inch and $A_d = 2.4306 \times 10^{-3}$ webers/inch.

Solving for the flux densities,

$$\begin{aligned} B_x &= \frac{\partial A}{\partial y} = \frac{A_i c_i + A_j c_j + A_k c_k}{2\Delta} \\ B_y &= -\frac{\partial A}{\partial x} = -\frac{A_i b_i + A_j b_j + A_k b_k}{2\Delta} \\ B &= \sqrt{B_x^2 + B_y^2} \end{aligned} \quad (6.179)$$

For element 1

$$\begin{aligned} B_x &= \frac{0+0+0}{2 \times 2} = 0 \\ B_y &= -\frac{0 + 0 + 2.4306 \times 10^{-3} \times 4}{2 \times 2} = 2.4306 \times 10^{-3} \\ B &= 2.4306 \times 10^{-3} \end{aligned} \quad (6.180)$$

For element 2

$$\begin{aligned} B_x &= \frac{0 + 2.4306 \times 10^{-3}(-4) + 4.1667 \times 10^{-3}(1)}{8} = -0.694 \times 10^{-3} \\ B_y &= -\frac{0 + 2.4306 \times 10^{-3}(-4) + 4.1667 \times 10^{-3}(3)}{8} = 0.347 \times 10^{-3} \\ B &= 0.775916 \times 10^{-3} \end{aligned} \quad (6.181)$$

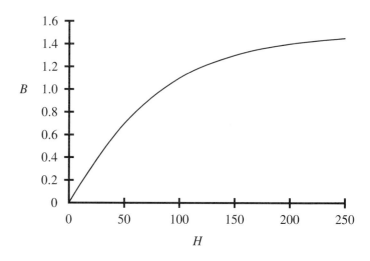

Figure 6.9 B-H Curve for Material in Element 2

For element 3

$$B_x = \frac{0 + 4.1667 \times 10^{-3}(-1) + 2.4306 \times 10^{-3}(4)}{4} = 1.389 \times 10^{-3}$$

$$B_y = -\frac{0 + 4.1667 \times 10^{-3}(1) + 0}{4} = 1.042 \times 10^{-3}$$

$$B = 1.7364 \times 10^{-3} \tag{6.182}$$

6.9.1 Nonlinear Example

We will now rework the first-order problem with element 2 composed of a nonlinear material with B–H characteristic shown in Figure 6.9.

First Iteration

For the first iteration let us set all of the potentials to zero so that $A_a = A_b = A_c = A_d = 0.0$. All initial values of B, the flux density, are then zero and $\frac{\partial v}{\partial B^2} = 0.0$. In this case, the local Jacobian matrices are the same as the local element stiffness matrices. We have $v_1 = v_3 = 80000$ in./H and $v_2 = 17846$ in./H from the saturation curve.

The Finite Element Method

The local Jacobians (including the $\frac{1}{4\Delta}$ terms) are therefore

$$S_{1bad} = 80,000 \begin{pmatrix} 0.25 & 0.25 & -0.5 \\ 0.25 & 1.25 & -1.5 \\ -0.5 & -1.5 & 2.0 \end{pmatrix} \tag{6.183}$$

$$S_{2bdc} = 17,846 \begin{pmatrix} 0.625 & -1.0 & 0.375 \\ -1.0 & 2.0 & -1.0 \\ 0.375 & -1.0 & 0.625 \end{pmatrix} \tag{6.184}$$

$$S_{3acd} = 80,000 \begin{pmatrix} 1.25 & 0.25 & -1.5 \\ -0.25 & 0.25 & -0.5 \\ -1.5 & -0.5 & 2.0 \end{pmatrix} \tag{6.185}$$

We now combine the local Jacobian matrices to form the global Jacobian ($bacd$).

$$S_{bacd} = \begin{pmatrix} 31.1543 & 20 & 6.6926 & -57.847 \\ 20 & 200 & 20 & -240 \\ 6.6926 & 20 & 31.1543 & -57.847 \\ -57.847 & -240 & -57.847 & 355.6936 \end{pmatrix} \times 10^3 \tag{6.186}$$

The local current density vectors for the three elements are

$$J_1 = \begin{pmatrix} 2000/3 \\ 2000/3 \\ 0 \\ 2000/3 \end{pmatrix} \tag{6.187}$$

$$J_2 = J_3 = \begin{pmatrix} 0 \\ 0 \\ 0 \\ 0 \end{pmatrix} \tag{6.188}$$

which gives a global current input vector of

$$J = \begin{pmatrix} 2000/3 \\ 2000/3 \\ 0 \\ 2000/3 \end{pmatrix} \tag{6.189}$$

The system of equations to solve then becomes

$$\begin{pmatrix} 31.1543 & 20 & 6.6926 & -57.847 \\ 6.6926 & 200 & 20 & -240 \\ 6.6926 & 20 & 31.1543 & -57.847 \\ -57.847 & -240 & -57.847 & 355.6939 \end{pmatrix} \times 10^3 \begin{pmatrix} \Delta A_b \\ \Delta A_a \\ \Delta A_c \\ \Delta A_d \end{pmatrix} = \begin{pmatrix} 2000/3 \\ 2000/3 \\ 0 \\ 2000/3 \end{pmatrix}$$

$$\tag{6.190}$$

The Dirichlet boundary condition $A_a = A_b = 0.0$ force $\Delta_a = \Delta_b = 0.0$ and the global equations (6.190) then become

$$\begin{pmatrix} 31.1543 & -57.847 \\ -57.847 & 355.6939 \end{pmatrix} \times 10^3 \begin{pmatrix} \Delta A_c \\ \Delta A_d \end{pmatrix} = \begin{pmatrix} 0 \\ 2000/3 \end{pmatrix} \tag{6.191}$$

Solving for the two remaining unknowns, we get

$$\begin{pmatrix} \Delta A_c \\ \Delta A_d \end{pmatrix} = \begin{pmatrix} 4.98566 \times 10^{-3} \\ 2.6851 \times 10^{-3} \end{pmatrix} \tag{6.192}$$

Second Iteration

We now find the values of vector potential after the first iteration as $A^{k+1} = A^k + \Delta A^k$ so

$$A_a = A_b = 0.0$$

$$A_c = 0.0 + 4.98566 \times 10^{-3} = 4.98566 \times 10^{-3} \qquad (6.193)$$
$$A_d = 0.0 + 2.6851 \times 10^{-3} = 2.6851 \times 10^{-3}$$

We now compute $|B|^2 = \vec{B} \cdot \vec{B}$ in each element. Because this is a two-dimensional problem

$$B^2 = B_y^2 + B_x^2 = \left(\frac{A_i b_i + A_j b_j + A_k b_k}{2\Delta}\right)^2 + \left(\frac{A_i c_i + A_j c_j + A_k c_k}{2\Delta}\right)^2 \qquad (6.194)$$

By multiplying out this expression we can find that

$$B^2 = \frac{1}{4\Delta}(A_i \ A_j \ A_k)\begin{pmatrix} b_i^2 + c_i^2 & b_i b_j + c_i c_j & b_i b_k + c_i c_k \\ b_i b_j + c_i c_j & b_j^2 + c_j^2 & b_j b_k + c_j c_k \\ b_i b_k + c_i c_k & b_j b_k + c_j c_k & b_k^2 + c_k^2 \end{pmatrix}\begin{pmatrix} A_i \\ A_j \\ A_k \end{pmatrix}$$
$$(6.195)$$

This matrix is, of course, the geometric coefficients of the local stiffness matrix. Therefore, in element 1 we have

$$B^2 = \frac{1}{\Delta}(0 \ 0 \ 2.6851 \times 10^{-3})$$
$$\times \begin{pmatrix} 0.25 & 0.25 & -0.5 \\ 0.25 & 1.25 & -1.5 \\ -0.5 & -1.5 & 2.0 \end{pmatrix}\begin{pmatrix} 0.0 \\ 0.0 \\ 2.6851 \times 10^{-3} \end{pmatrix} = 7.20974 \ \frac{Wb^2}{in^4} \qquad (6.196)$$

For element 2

$$B^2 = \frac{1}{\Delta}(0, \ 2.6851 \times 10^{-3}, \ 4.985 \times 10^{-3}) \qquad (6.197)$$

$$\times \begin{pmatrix} 0.625 & -1.0 & 0.375 \\ -1.0 & 2.0 & -1.0 \\ 0.375 & -1.0 & 0.625 \end{pmatrix} \begin{pmatrix} 0.0 \\ 2.6851 \times 10^{-3} \\ 4.985 \times 10^{-3} \end{pmatrix}$$

$$= 0.795258 \times 10^{-9} \frac{Wb^2}{in^4}$$

Similarly, for element 3 we find that $B^2 = 3.62335 \frac{Wb^2}{in^4}$.

For the nonlinear element we now need ν and $\frac{\partial \nu}{\partial B^2}$. From equation (6.197) we have $B^2 = 1.91061$ T^2 or $B = 1.38$ T. From the saturation curve we get $\nu = 691$ m/H or 27.1969 in./H and $\frac{\partial \nu}{\partial B^2} = 499.8$ m/H/T^2 or $47.2745 \times 10^9 \frac{in.^5}{HT^2}$

For element 2 the $\sum_i$ terms are

$$\Sigma_b = A_b b_b + A_d b_d + A_c b_c$$
$$= 0 + 2.685 \times 10^{-3} \times (-4) + 4.985 \times 10^{-3} \times (3) = 4.217 \times 10^{-3} \quad (6.198)$$

$$\Sigma_c = A_b c_b + A_d c_d + A_c c_c$$
$$= 0 + 2.685 \times 10^{-3} \times (-4) + 4.985 \times 10^{-3} \times (1) = -5.755 \times 10^{-3} \quad (6.199)$$

The derivatives are

$$\frac{\partial B^2}{\partial A_i} = (2 b_i \Sigma_b + 2 c_i \Sigma_c)/(4\Delta^2)$$
$$= (2 \times 1 \times 4.217 \times 10^{-3} + 2 \times 3 \times (-5.755 \times 10^{-3}))/(4 \times 16)$$
$$= -407.74 \times 10^{-6} \quad (6.200)$$

$$\frac{\partial B^2}{\partial A_j} = (2 b_j \Sigma_b + 2 c_j \Sigma_c)/(4\Delta^2)$$
$$= (2 \times (-4) \times 4.217 \times 10^{-3} + 2 \times (-4) \times (-5.755 \times 10^{-3}))/(4 \times 16)$$
$$= 192.27 \times 10^{-6} \quad (6.201)$$

The Finite Element Method

$$\frac{\partial B^2}{\partial A_k} = (2b_k \Sigma_b + 2c_k \Sigma_c)/(4\Delta^2)$$
$$= (2 \times 3 \times 4.217 \times 10^{-3} + 2 \times 1 \times (-5.755 \times 10^{-3}))/(4 \times 16)$$
$$= 215.47 \times 10^{-6} \quad (6.202)$$

We now form the sums involving the first computed values of vector potential

$$\frac{1}{4\Delta}\sum_{n=1}^{k} s_{in} A_n = \frac{1}{4\Delta}(s_{aa} A_a + s_{ad} A_d + s_{ac} A_c)$$
$$= 0 + (-1)(2.685 \times 10^{-3} + 0.375 \times 4.98566 \times 10^{-3} = -815 \times 10^{-6} \quad (6.203)$$

$$\Sigma_j = \frac{1}{4\Delta}\sum_{n=1}^{k} s_{jn} A_n = \frac{1}{4\Delta}(s_{da} A_b + s_{ad} A_d + s_{dc} A_c)$$
$$= 0 + 2.0 \times 2.685 \times 10^{-3} - 1 \times 4.98566 \times 10^{-3}) = 384.53 \times 10^{-6} \quad (6.204)$$

$$\Sigma_k = \frac{1}{4\Delta}\sum_{n=1}^{k} s_{kn} A_n = \frac{1}{4\Delta}(s_{cb} A_b + s_{cd} A_d + s_{cc} A_c)$$
$$= 0 - 1.0 \times 2.685 \times 10^{-3} + 0.625 \times 4.98566 \times 10^{-3} = 430.94 \times 10^{-6} \quad (6.205)$$

The nonlinear part of the Jacobian for element 2 then becomes

$$\frac{\partial \nu}{\partial B^2}\begin{pmatrix} \frac{\partial B^2}{\partial A_i}\Sigma_i & \frac{\partial B^2}{\partial A_j}\Sigma_i & \frac{\partial B^2}{\partial A_k}\Sigma_i \\ \frac{\partial B^2}{\partial A_i}\Sigma_j & \frac{\partial B^2}{\partial A_j}\Sigma_j & \frac{\partial B^2}{\partial A_k}\Sigma_j \\ \frac{\partial B^2}{\partial A_i}\Sigma_k & \frac{\partial B^2}{\partial A_j}\Sigma_k & \frac{\partial B^2}{\partial A_k}\Sigma_k \end{pmatrix} \quad (6.206)$$

This matrix becomes

$$\begin{pmatrix} 15.72 & -7.41 & -8.31 \\ -7.41 & 3.495 & 3.971 \\ -8.307 & 3.917 & 4.39 \end{pmatrix} \times 10^3 \qquad (6.207)$$

We now add this to the linear part of the Jacobian to get the final nonlinear element Jacobian

$$27.1969 \times 10^3 \begin{pmatrix} 0.625 & -1.0 & 0.375 \\ -1.0 & 2.0 & -1.0 \\ 0.375 & -1.0 & 0.625 \end{pmatrix}$$

$$+ \begin{pmatrix} 15.72 & -7.41 & -8.307 \\ -7.41 & 3.495 & 3.971 \\ -8.307 & 3.917 & 4.39 \end{pmatrix} \times 10^3$$

$$= \begin{pmatrix} 32.718 & -34.61 & 1.892 \\ -34.61 & 57.89 & -23.28 \\ 1.892 & -23.28 & 21.39 \end{pmatrix} \times 10^3 \qquad (6.208)$$

The updated global Jacobian now becomes

$$\begin{pmatrix} 52.781 & 20.0 & 1.892 & -74.61 \\ 20.0 & 200.0 & 20.0 & -240.0 \\ 1.892 & 20.0 & 41.39 & -63.28 \\ -74.61 & -240.0 & -63.28 & 377.89 \end{pmatrix} \qquad (6.209)$$

The component of the right-hand side depending on the potentials (excluding the current sources) is given by

$$\frac{-\nu}{4\Delta}(S)\begin{pmatrix} A_i \\ A_j \\ A_k \end{pmatrix} \qquad (6.210)$$

The Finite Element Method

For element 1 this is

$$-80000 \begin{pmatrix} 0.25 & 0.25 & -0.5 \\ 0.25 & 1.25 & -1.5 \\ -0.5 & -1.5 & 2.0 \end{pmatrix} \begin{pmatrix} 0.0 \\ 0.0 \\ 2.6851 \times 10^{-3} \end{pmatrix} = \begin{pmatrix} 107.4 \\ 322.2 \\ -429.6 \end{pmatrix} \quad (6.211)$$

For element 2 we have

$$-27196.9 \begin{pmatrix} 0.625 & -1.0 & 0.375 \\ -1.0 & 2.0 & -1.0 \\ 0.375 & -1.0 & 0.625 \end{pmatrix} \begin{pmatrix} 0.0 \\ 2.6851 \times 10^{-3} \\ 4.9857 \times 10^{-3} \end{pmatrix} = \begin{pmatrix} 22.18 \\ -10.46 \\ -11.72 \end{pmatrix}$$

$$(6.212)$$

And for element 3

$$-80000 \begin{pmatrix} 1.25 & 0.25 & -1.5 \\ 0.25 & 0.25 & -0.5 \\ -1.5 & -0.5 & 2.0 \end{pmatrix} \begin{pmatrix} 0.0 \\ 4.9857 \times 10^{-3} \\ 2.6851 \times 10^{-3} \end{pmatrix} = \begin{pmatrix} 222.5 \\ 7.691 \\ -230.19 \end{pmatrix}$$

$$(6.213)$$

The total right-hand side (including the input current) is

$$\begin{pmatrix} 107.4 + 22.18 + 2000/3 \\ 322.2 + 222.5 + 2000/3 \\ -11.72 + 7.691 + 0 \\ -10.46 - 230.19 - 429.6 + 2000/3 \end{pmatrix} = \begin{pmatrix} 796.247 \\ 1211.4 \\ -4.03 \\ -3.58 \end{pmatrix} \quad (6.214)$$

The Newton–Raphson equation is then

$$10^3 \begin{pmatrix} 52.781 & 20.0 & 1.892 & -74.61 \\ 20.0 & 200.0 & 20.0 & -240.0 \\ 1.892 & 20.0 & 41.39 & -63.28 \\ -74.61 & -240.0 & -63.28 & 377.89 \end{pmatrix} \begin{pmatrix} A_b \\ A_a \\ A_c \\ A_d \end{pmatrix} = \begin{pmatrix} 796.247 \\ 1211.4 \\ -4.03 \\ -3.58 \end{pmatrix} \quad (6.215)$$

After applying the boundary conditions, we solve the 2 × 2 system

$$10^3 \begin{pmatrix} 41.39 & -63.28 \\ -63.28 & 377.89 \end{pmatrix} \begin{pmatrix} \Delta A_c \\ \Delta A_d \end{pmatrix} = \begin{pmatrix} -4.03 \\ -3.58 \end{pmatrix} \quad (6.216)$$

from which $\Delta A_c = -150.34 \times 10^{-6}$ and $\Delta A_d = -34.65 \times 10^{-6}$. The vector potentials are therefore

$A_a = A_b = 0.0$, $A_c = 4.98566 \times 10^{-3} - 150.34 \times 10^{-6} = 4.8353 \times 10^{-3}$, $A_d = 2.6851 \times 10^{-3} - 34.65 \times 10^{-6} = 2.65045 \times 10^{-3}$

This procedure is then repeated until the convergence criterion is satisfied.

6.10 PERMANENT MAGNETS

For 2D magnetostatic analysis the nonlinear Poisson equation for the vector potential is

$$\nabla \times \nu(\nabla \times A) = J + \nabla \times (\nu \mu_0 M) \quad (6.217)$$

The Galerkin form is then

$$\iint_\Omega W \cdot (\nabla \times \nu(\nabla \times A)) \, dx \, dy - \iint_\Omega W \cdot J \, dx \, dy \quad (6.218)$$
$$- \iint_\Omega W \cdot (\nabla \times (\nu \mu_0 M)) \, dx \, dy = 0$$

or

$$\iint_\Omega \nabla \times (\nu \nabla \times A - \nu\mu_0 M) \cdot W \, dx \, dy - \iint_\Omega J \cdot W dx \, dy = 0 \qquad (6.219)$$

As before, we choose the weighting function W to be the same as the element shape function, N. We now integrate by parts. Using the vector identity

$$(\nabla \times F) \cdot G = \nabla \cdot (F \times G) + F \cdot (\nabla \times G) \qquad (6.220)$$

and with

$$\begin{aligned} F &= \nu \nabla \times A - \nu\mu_0 M \\ G &= N \end{aligned} \qquad (6.221)$$

we can write the first term as

$$\iint_\Omega (\nabla \times (\nu \nabla \times A) - \nu\mu_0 M) \cdot N \, dx \, dy = \qquad (6.222)$$
$$\iint_\Omega (\nu \nabla \times A - \nu\mu_0 M) \cdot (\nabla \times N) \, dx \, dy$$
$$+ \iint_\Omega \nabla \cdot ((\nu \nabla \times A - \nu\mu_0 M) \times N) \, dx \, dy$$

The last term can be written as a line integral using the divergence theorem.

$$\iint_\Omega \nabla \cdot (\nu \nabla \times A - \nu\mu_0 M) \times N \, dS = \oint_C \{(\nu \nabla \times A - \nu\mu_0 M) \times N\} \cdot \hat{n} \, dC \qquad (6.223)$$

Using identities

$$F \times G = -G \times F \qquad (6.224)$$

and

$$(F \times G) \cdot T = F \cdot (G \times T) \qquad (6.225)$$

the line integral becomes

$$\oint_C \{(\nu \nabla \times A - \nu\mu_0 M) \times N\} \cdot \hat{n} \, dC = \oint_C N \cdot \{(\nu \nabla \times A - \nu\mu_0 M) \times \hat{n}\} \, dC \qquad (6.226)$$

If we ignore this line integral, the integral must be zero for all choices of N so that quantity in brackets must be zero. Since this quantity is the tangential component of H we have imposed a homogeneous Neumann boundary condition. We are left with

$$\iint_\Omega \nu(\nabla \times A) \cdot (\nabla \times N) \, dx \, dy = \iint_\Omega \nu\mu_0 M \cdot (\nabla \times N) \, dx \, dy + \iint_\Omega N \cdot J \, dx \, dy$$

(6.227)

For the two-dimensional Cartesian case, A and J have only z components and M has only x and y components. Substituting the curl of the weighting function in the first term on the right hand side of equation (6.227) we obtain

$$\iint_\Omega \nu \left(\frac{\partial A}{\partial x} \frac{\partial N}{\partial x} + \frac{\partial A}{\partial y} \frac{\partial N}{\partial y} \right) dx \, dy = \qquad (6.228)$$
$$\iint_\Omega \left(\nu\mu_0 \left(M_x \frac{\partial N}{\partial y} - M_y \frac{\partial N}{\partial x} \right) + J \cdot N \right) dx \, dy$$

We have seen the left-hand side and the second term on the right-hand side of this equation earlier. The term representing the permanent magnet is

$$\iint_\Omega \nu\mu_0 \left(M_x \frac{\partial N}{\partial y} - M_y \frac{\partial N}{\partial x} \right) dx \, dy \qquad (6.229)$$

For first-order triangles

$$\frac{\partial N_i}{\partial x} = \frac{b_i}{2\Delta} \text{ and }$$
$$\frac{\partial N_i}{\partial y} = \frac{c_i}{2\Delta} \qquad (6.230)$$

Substituting into equation (6.229), the integral becomes

$$\iint_\Omega \frac{\nu\mu_0}{2\Delta} \left(M_x \begin{pmatrix} c_i \\ c_j \\ c_k \end{pmatrix} - M_y \begin{pmatrix} b_i \\ b_j \\ b_k \end{pmatrix} \right) dx \, dy \qquad (6.231)$$

The integrand is a constant so the double integral becomes

$$\iint_\Omega dx\, dy = \Delta \tag{6.232}$$

The final result is then

$$\frac{\nu\mu_0}{2}\left(M_x \begin{pmatrix} c_i \\ c_j \\ c_k \end{pmatrix} - M_y \begin{pmatrix} b_i \\ b_j \\ b_k \end{pmatrix}\right) \tag{6.233}$$

6.10.1 Three-Element Example With Permanent-Magnet Excitation

For this illustration we shall use the three element example of the previous section. In place of a source current in element 1, permanent magnet excitation with only a y directed component is used.

The element matrices will be identical to those in the previous section. The following fully assembled matrix with $\nu_1 = \nu_2 = \nu_3 = \nu_0$ will also be the same.

$$(S) = \nu_0 \begin{pmatrix} 0.875 & 0.25 & 0.375 & -1.5 \\ 0.25 & 2.5 & 0.25 & -3.0 \\ 0.375 & 0.25 & 0.875 & -1.5 \\ -1.5 & -3.0 & -1.5 & 6.0 \end{pmatrix} \begin{matrix} b \\ a \\ c \\ d \end{matrix} \tag{6.234}$$

The forcing function, however, will be given by

$$(f) = \frac{-M_y}{2\Delta} \begin{pmatrix} b_i \\ b_j \\ b_m \end{pmatrix} \tag{6.235}$$

Using the previous values for b_i, b_j, b_k, and Δ, for a value of magnetization, $M_y = -100,000$ A-t/m for the permanent magnet in element 1, the global forcing function will be

$$(f) = \frac{100000}{2 \times 2} \begin{pmatrix} -1 \\ -3 \\ 0 \\ 4 \end{pmatrix} \qquad (6.236)$$

Because the potentials at b and a are zero, the finite element equation reduces to

$$\nu_0 \begin{pmatrix} 0.875 & -1.5 \\ -1.5 & 6.0 \end{pmatrix} \begin{pmatrix} A_c \\ A_d \end{pmatrix} = 25000 \begin{pmatrix} 0 \\ 4 \end{pmatrix} \qquad (6.237)$$

Because $\nu_0 = \frac{1}{4\pi \times 10^{-7}} \approx 800000$, the solution of the equation yields

$$A_c = 0.125, \quad A_d = 0.05208 \qquad (6.238)$$

Thus the solution vector will be (W/m)

$$A = \begin{pmatrix} 0 \\ 0 \\ 0.125 \\ 0.05208 \end{pmatrix} \qquad (6.239)$$

The flux densities can now be found by the procedures described before.

6.11 NUMERICAL EXAMPLE OF MATRIX FORMATION FOR ISOPARAMETRIC ELEMENTS

We have seen that the isoparametric approach is a more general method of evaluating the integrals to compute the element matrix coefficients.[1] We will now use the three-element example problem and use the isoparametric approach to evaluate the matrices

---

[1] It is simpler for high-order elements where numerical integration is required.

The Finite Element Method 343

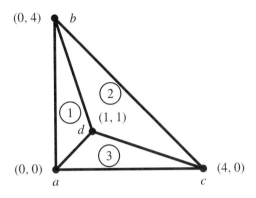

Figure 6.10 Three-Element Isoparametric Problem

[76]. Because these elements are first-order triangles the isoparametric approach is really not required and closed form integration, as in the previous example, is actually simpler. The example, however, will illustrate all the steps required in the approach. The only difference for high-order elements is the number of Gauss points necessary to evaluate the integrals accurately.

The geometry is repeated in Figure 6.10. The nodes (i, j, k) for each element are defined in the following.

We recall that the values of the element matrices are found by evaluating the equation

$$S_{ij} = \iint_\Omega v \left(\frac{\partial N_i}{\partial x} \frac{\partial N_j}{\partial x} + \frac{\partial N_i}{\partial y} \frac{\partial N_j}{\partial y} \right) dx\, dy \qquad (6.240)$$

where N_i and N_j are the shape functions. In the isoparametric formulation we wish to integrate over the local coordinates, (ξ, η) of the master element. This gives

$$S_{ij} = \iint_\Omega v \left(\frac{\partial N_i}{\partial x} \frac{\partial N_j}{\partial x} + \frac{\partial N_i}{\partial y} \frac{\partial N_j}{\partial y} \right) |J|\, d\xi\, d\eta \qquad (6.241)$$

We have seen that the partial derivatives are related by

$$\begin{pmatrix} \frac{\partial N_i}{\partial x} \\ \frac{\partial N_j}{\partial y} \end{pmatrix} = (J)^{-1} \begin{pmatrix} \frac{\partial N_i}{\partial \xi} \\ \frac{\partial N_j}{\partial \eta} \end{pmatrix} \qquad (6.242)$$

where

$$|J| = \left(\sum_{p=1}^{m} \frac{\partial N_p}{\partial \xi} x_p\right)\left(\sum_{p=1}^{m} \frac{\partial N_p}{\partial \eta} y_p\right) - \left(\sum_{p=1}^{m} \frac{\partial N_p}{\partial \eta} x_p\right)\left(\sum_{p=1}^{m} \frac{\partial N_p}{\partial \xi} y_p\right) \qquad (6.243)$$

and

$$(J)^{-1} = \frac{1}{|J|}\begin{pmatrix} \left(\sum_{p=1}^{m} \frac{\partial N_p}{\partial \eta} y_p\right) & -\left(\sum_{p=1}^{m} \frac{\partial N_p}{\partial \xi} y_p\right) \\ -\left(\sum_{p=1}^{m} \frac{\partial N_p}{\partial \eta} x_p\right) & \left(\sum_{p=1}^{m} \frac{\partial N_p}{\partial \xi} x_p\right) \end{pmatrix} \qquad (6.244)$$

and m is the number of nodes in the element.

The shape functions for the triangle are most easily expressed in local triangle coordinates. The relationships between these coordinates and the general coordinates (ξ, η) are

$$\begin{aligned} \xi &= L_1 \\ \eta &= L_2 \\ 1 - \eta - \xi &= L_3 \end{aligned} \qquad (6.245)$$

From the chain rule we can now evaluate the partial derivatives of equation (6.242).

$$\frac{\partial N_1}{\partial \xi} = \frac{\partial N_1}{\partial L_1}\frac{\partial L_1}{\partial \xi} + \frac{\partial N_1}{\partial L_2}\frac{\partial L_2}{\partial \xi} + \frac{\partial N_1}{\partial L_3}\frac{\partial L_3}{\partial \xi}$$

The Finite Element Method

$$
\begin{aligned}
\frac{\partial N_1}{\partial \eta} &= \frac{\partial N_1}{\partial L_1} - \frac{\partial N_1}{\partial L_3} \\
&= \frac{\partial N_1}{\partial L_1}\frac{\partial L_1}{\partial \eta} + \frac{\partial N_1}{\partial L_2}\frac{\partial L_2}{\partial \eta} + \frac{\partial N_1}{\partial L_3}\frac{\partial L_3}{\partial \eta} \\
&= \frac{\partial N_1}{\partial L_2} - \frac{\partial N_1}{\partial L_3}
\end{aligned}
\tag{6.246}
$$

We must also change the limits of integration in equation (6.241) so that it becomes

$$
S_{ij} = \int_0^1 \int_0^{1-L_1} \nu \left(\frac{\partial N_i}{\partial x}\frac{\partial N_j}{\partial x} + \frac{\partial N_i}{\partial y}\frac{\partial N_j}{\partial y} \right) |J|\, dL_2\, dL_1 \tag{6.247}
$$

where we use the fact that $d\xi\, d\eta = dL_1\, dL_2$.

As we have seen, for first-order triangles the shape function is just the local coordinate so

$$
\begin{aligned}
N_i &= L_1 \\
N_j &= L_2 \tag{6.248} \\
N_k &= L_3 \tag{6.249}
\end{aligned}
$$

The partial derivatives of the shape functions with respect to the general coordinates are now for node i

$$
\begin{aligned}
\frac{\partial N_i}{\partial \xi} &= \frac{\partial N_i}{\partial L_1} - \frac{\partial N_i}{\partial L_3} = 1 - 0 = 1 \\
\frac{\partial N_i}{\partial \eta} &= \frac{\partial N_i}{\partial L_2} - \frac{\partial N_i}{\partial L_3} = 0 - 0 = 0
\end{aligned}
\tag{6.250}
$$

Similarly, for node j

$$
\frac{\partial N_j}{\partial \xi} = \frac{\partial N_j}{\partial L_1} - \frac{\partial N_j}{\partial L_3} = 0
$$

$$\frac{\partial N_j}{\partial \eta} = \frac{\partial N_j}{\partial L_2} - \frac{\partial N_j}{\partial L_3} = 1 \tag{6.251}$$

and for node k

$$\frac{\partial N_k}{\partial \xi} = \frac{\partial N_k}{\partial L_1} - \frac{\partial N_k}{\partial L_3} = -1$$
$$\frac{\partial N_k}{\partial \eta} = \frac{\partial N_k}{\partial L_2} - \frac{\partial N_k}{\partial L_3} = -1 \tag{6.252}$$

The summations of the partial derivatives now become

- Element 1:

Node	i	j	k
	b	a	d
Coord.	$x_i = 0.0$	$x_j = 0.0$	$x_k = 1.0$
	$y_i = 4.0$	$y_j = 0.0$	$y_k = 1.0$

$$\sum_{n=i}^{k} \frac{\partial N_n}{\partial \xi} x_n = \frac{\partial N_i}{\partial \xi} x_i + \frac{\partial N_j}{\partial \xi} x_j + \frac{\partial N_k}{\partial \xi} x_k$$
$$= (1)(0) + (0)(0) + (-1)(1) = -1.0$$

$$\sum_{n=i}^{k} \frac{\partial N_n}{\partial \eta} y_n = \frac{\partial N_i}{\partial \eta} y_i + \frac{\partial N_j}{\partial \eta} y_j + \frac{\partial N_k}{\partial \eta} y_k$$
$$= (0)(4) + (1)(0) + (-1)(1) = -1$$

$$\sum_{n=i}^{k} \frac{\partial N_n}{\partial \eta} x_n = \frac{\partial N_i}{\partial \eta} x_i + \frac{\partial N_j}{\partial \eta} x_j + \frac{\partial N_k}{\partial \eta} x_k$$
$$= (0)(0) + (1)(0) + (-1)(1) = -1$$

$$\sum_{n=i}^{k} \frac{\partial N_n}{\partial \xi} y_n = \frac{\partial N_i}{\partial \xi} y_i + \frac{\partial N_j}{\partial \xi} y_j + \frac{\partial N_k}{\partial \xi} y_k$$
$$= (1)(4) + (0)(0) + (-1)(1) = 3$$
$$|J| = (-1)(-1) - (-1)(3) = 4$$

The Finite Element Method

$$(J)^{-1} = \frac{1}{4} \begin{pmatrix} -1 & -3 \\ 1 & -1 \end{pmatrix} \qquad (6.253)$$

- Element 2:

Node	i	j	k
	b	d	c
Coord.	$x_i = 0.0$	$x_j = 1.0$	$x_k = 4.0$
	$y_i = 4.0$	$y_j = 1.0$	$y_k = 0.0$

$$\sum_{n=i}^{k} \frac{\partial N_n}{\partial \xi} x_n = \frac{\partial N_i}{\partial \xi} x_i + \frac{\partial N_j}{\partial \xi} x_j + \frac{\partial N_k}{\partial \xi} x_k$$
$$= (1)(0) + (0)(1) + (-1)(4) = -4.0$$

$$\sum_{n=i}^{k} \frac{\partial N_n}{\partial \eta} y_n = \frac{\partial N_i}{\partial \eta} y_i + \frac{\partial N_j}{\partial \eta} y_j + \frac{\partial N_k}{\partial \eta} y_k$$
$$= (0)(4) + (1)(1) + (-1)(0) = 1$$

$$\sum_{n=i}^{k} \frac{\partial N_n}{\partial \eta} x_n = \frac{\partial N_i}{\partial \eta} x_i + \frac{\partial N_j}{\partial \eta} x_j + \frac{\partial N_k}{\partial \eta} x_k$$
$$= (0)(0) + (1)(1) + (-1)(4) = -3$$

$$\sum_{n=i}^{k} \frac{\partial N_n}{\partial \xi} y_n = \frac{\partial N_i}{\partial \xi} y_i + \frac{\partial N_j}{\partial \xi} y_j + \frac{\partial N_k}{\partial \xi} y_k$$
$$= (1)(4) + (0)(1) + (-1)(0) = 4$$

$$|J| = (-4)(1) - (-3)(4) = 8$$

$$(J)^{-1} = \frac{1}{8} \begin{pmatrix} 1 & -4 \\ 3 & -4 \end{pmatrix} \qquad (6.254)$$

- Element 3:

Node	i	j	k
	a	c	d
Coord.	$x_i = 0.0$	$x_j = 4.0$	$x_k = 1.0$
	$y_i = 0.0$	$y_j = 0.0$	$y_k = 1.0$

$$\sum_{n=i}^{k} \frac{\partial N_n}{\partial \xi} x_n = \frac{\partial N_i}{\partial \xi} x_i + \frac{\partial N_j}{\partial \xi} x_j + \frac{\partial N_k}{\partial \xi} x_k$$
$$= (1)(0) + (0)(4) + (-1)(1) = -1.0$$
$$\sum_{n=i}^{k} \frac{\partial N_n}{\partial \eta} y_n = \frac{\partial N_i}{\partial \eta} y_i + \frac{\partial N_j}{\partial \eta} y_j + \frac{\partial N_k}{\partial \eta} y_k$$
$$= (0)(0) + (1)(0) + (-1)(1) = -1$$
$$\sum_{n=i}^{k} \frac{\partial N_n}{\partial \eta} x_n = \frac{\partial N_i}{\partial \eta} x_i + \frac{\partial N_j}{\partial \eta} x_j + \frac{\partial N_k}{\partial \eta} x_k$$
$$= (0)(0) + (1)(4) + (-1)(1) = 3$$
$$\sum_{n=i}^{k} \frac{\partial N_n}{\partial \xi} y_n = \frac{\partial N_i}{\partial \xi} y_i + \frac{\partial N_j}{\partial \xi} y_j + \frac{\partial N_k}{\partial \xi} y_k$$
$$= (1)(0) + (0)(0) + (-1)(1) = -1$$
$$|J| = (-1)(-1) - (3)(-1) = 4$$
$$(J)^{-1} = \frac{1}{4} \begin{pmatrix} -1 & 1 \\ -3 & -1 \end{pmatrix} \tag{6.255}$$

The derivatives with respect to the global coordinates (x, y) can now be written as

- Element 1:

$$\begin{pmatrix} \frac{\partial N_i}{\partial x} \\ \frac{\partial N_i}{\partial y} \end{pmatrix} = \frac{1}{4} \begin{pmatrix} -1 & -3 \\ 1 & -1 \end{pmatrix} \begin{pmatrix} 1 \\ 0 \end{pmatrix} = \begin{pmatrix} -\frac{1}{4} \\ \frac{1}{4} \end{pmatrix}$$

$$\begin{pmatrix} \frac{\partial N_j}{\partial x} \\ \frac{\partial N_j}{\partial y} \end{pmatrix} = \frac{1}{4} \begin{pmatrix} -1 & -3 \\ 1 & -1 \end{pmatrix} \begin{pmatrix} 0 \\ 1 \end{pmatrix} = \begin{pmatrix} -\frac{3}{4} \\ -\frac{1}{4} \end{pmatrix}$$

$$\begin{pmatrix} \frac{\partial N_k}{\partial x} \\ \frac{\partial N_k}{\partial y} \end{pmatrix} = \frac{1}{4} \begin{pmatrix} -1 & -3 \\ 1 & -1 \end{pmatrix} \begin{pmatrix} -1 \\ -1 \end{pmatrix} = \begin{pmatrix} 1 \\ 0 \end{pmatrix}$$

- Element 2:

$$\begin{pmatrix} \frac{\partial N_i}{\partial x} \\ \frac{\partial N_i}{\partial y} \end{pmatrix} = \frac{1}{8}\begin{pmatrix} 1 & -4 \\ 3 & -4 \end{pmatrix}\begin{pmatrix} 1 \\ 0 \end{pmatrix} = \begin{pmatrix} \frac{1}{8} \\ \frac{3}{8} \end{pmatrix}$$

$$\begin{pmatrix} \frac{\partial N_j}{\partial x} \\ \frac{\partial N_j}{\partial y} \end{pmatrix} = \frac{1}{8}\begin{pmatrix} 1 & -4 \\ 3 & -4 \end{pmatrix}\begin{pmatrix} 0 \\ 1 \end{pmatrix} = \begin{pmatrix} -\frac{1}{2} \\ -\frac{1}{2} \end{pmatrix}$$

$$\begin{pmatrix} \frac{\partial N_k}{\partial x} \\ \frac{\partial N_k}{\partial y} \end{pmatrix} = \frac{1}{8}\begin{pmatrix} 1 & -4 \\ 3 & -4 \end{pmatrix}\begin{pmatrix} -1 \\ -1 \end{pmatrix} = \begin{pmatrix} \frac{3}{8} \\ \frac{1}{8} \end{pmatrix}$$

- Element 3:

$$\begin{pmatrix} \frac{\partial N_i}{\partial x} \\ \frac{\partial N_i}{\partial y} \end{pmatrix} = \frac{1}{4}\begin{pmatrix} -1 & 1 \\ -3 & -1 \end{pmatrix}\begin{pmatrix} 1 \\ 0 \end{pmatrix} = \begin{pmatrix} -\frac{1}{4} \\ -\frac{3}{4} \end{pmatrix}$$

$$\begin{pmatrix} \frac{\partial N_j}{\partial x} \\ \frac{\partial N_j}{\partial y} \end{pmatrix} = \frac{1}{4}\begin{pmatrix} -1 & 1 \\ -3 & -1 \end{pmatrix}\begin{pmatrix} 0 \\ 1 \end{pmatrix} = \begin{pmatrix} \frac{1}{4} \\ -\frac{1}{4} \end{pmatrix}$$

$$\begin{pmatrix} \frac{\partial N_k}{\partial x} \\ \frac{\partial N_k}{\partial y} \end{pmatrix} = \frac{1}{4}\begin{pmatrix} -1 & 1 \\ -3 & -1 \end{pmatrix}\begin{pmatrix} -1 \\ -1 \end{pmatrix} = \begin{pmatrix} 0 \\ 1 \end{pmatrix}$$

Having evaluated the partial derivatives, we now integrate to find the stiffness matrix

$$S_{ij} = \int_0^1 \int_0^{1-L_1} v\left[\left(\frac{\partial N_i}{\partial x}\frac{\partial N_j}{\partial x}\right) + \left(\frac{\partial N_i}{\partial y}\frac{\partial N_j}{\partial y}\right)\right]|J|dL_2 dL_1 \qquad (6.256)$$

For first-order elements the partial derivatives in the integrand are constants and we can evaluate equation (6.256) in closed form. For illustration we will use a one-point Gaussian quadrature formula. In this case an integral of the form[2]

$$\iint f\, dA \qquad (6.257)$$

---

[2] Recall that $dA = |J|dL_2 dL_1$.

is approximated by the summation

$$A \sum_{i=1}^{N} w_i f(L_{1i}, L_{2i}, L_{3i}) \qquad (6.258)$$

The points (L_{1i}, L_{2i}, L_{3i}) at which the function are evaluated are the Gauss points and the w_i values are weighting factors. Because the integrand is a constant we can pick one Gauss point in, for example, the center of the element and with the weighting factor equal to 1 the integration is exact. The elements of the local stiffness matrix are then evaluated as

- Element 1, $\Delta = 2.0$

$$S_{ii} = f = \nu \left(\left(\frac{\partial N_i}{\partial x} \right)^2 + \left(\frac{\partial N_i}{\partial y} \right)^2 \right) = \nu \left(\frac{1}{16} + \frac{1}{16} \right)$$

$$S_{ii} = 2 \sum_{i=1}^{1} \frac{\nu}{8} = 0.25\nu$$

$$S_{ij} = f = \nu \left(\frac{\partial N_i}{\partial x} \frac{\partial N_j}{\partial x} + \frac{\partial N_i}{\partial y} \frac{\partial N_j}{\partial y} \right) = \nu \left(\frac{3}{16} - \frac{1}{16} \right)$$

$$S_{ij} = 2 \sum_{i=1}^{1} \frac{\nu}{8} = 0.25\nu$$

$$S_{ik} = f = \nu \left(\frac{\partial N_i}{\partial x} \frac{\partial N_k}{\partial x} + \frac{\partial N_i}{\partial y} \frac{\partial N_k}{\partial y} \right) = \nu \left(\frac{-1}{4} - 0 \right)$$

$$S_{ik} = 2 \sum_{i=1}^{1} \frac{-\nu}{4} = -0.5\nu$$

The Finite Element Method

$$S_{jj} = f = v\left(\left(\frac{\partial N_j}{\partial x}\right)^2 + \left(\frac{\partial N_j}{\partial y}\right)^2\right) = v\left(\frac{9}{16} + \frac{1}{16}\right)$$

$$S_{jj} = 2\sum_{i=1}^{1} \frac{5v}{8} = 1.25v$$

$$S_{jk} = f = v\left(\frac{\partial N_j}{\partial x}\frac{\partial N_k}{\partial x} + \frac{\partial N_j}{\partial y}\frac{\partial N_k}{\partial y}\right) = v\left(\frac{-3}{4} + 0\right)$$

$$S_{ik} = 2\sum_{i=1}^{1} \frac{-3v}{4} = -1.5v$$

$$S_{kk} = f = v\left(\left(\frac{\partial N_k}{\partial x}\right)^2 + \left(\frac{\partial N_k}{\partial y}\right)^2\right) = v \quad (6.259)$$

$$S_{kk} = 2\sum_{i=1}^{1} v = 2.0v$$

The local stiffness matrix for element 1 is therefore

$$(S)_1 = v_1 \begin{pmatrix} 0.25 & 0.25 & -0.5 \\ 0.25 & 1.25 & -1.5 \\ -0.5 & -1.5 & 2.0 \end{pmatrix} \quad (6.260)$$

- Element 2, $\Delta = 4.0$

The reader may verify that by the same procedure

$$(S)_2 = v_2 \begin{pmatrix} 0.625 & -1.0 & 0.375 \\ 0.375 & 2.0 & -1.0 \\ 0.375 & -1.0 & 0.625 \end{pmatrix} \quad (6.261)$$

- Element 3, $\Delta = 2.0$

Similarly,

$$(S)_3 = \nu_3 \begin{pmatrix} 1.25 & 0.25 & -1.5 \\ 0.25 & 0.25 & -0.5 \\ -1.5 & -0.5 & 2.0 \end{pmatrix} \quad (6.262)$$

These results are the same as those we found using closed form integration.

We now consider the source term. We have a source (uniform current density) J_s in element 1. The contribution to the finite element system is

$$J_s \frac{\Delta}{3} \begin{pmatrix} 1 \\ 1 \\ 1 \end{pmatrix} \quad (6.263)$$

In isoparametric form this becomes

$$\iint J_s \begin{pmatrix} L_1 \\ L_2 \\ L_3 \end{pmatrix} |J| d\xi d\eta \quad (6.264)$$

Because $|J| d\xi d\eta = dS$ and using one-point Gauss integration and the fact that the area (S_1) of element 1 is 2.0,

$$\begin{aligned} J_s \left(S_1 \sum_{i=1}^{1} L_{1i} \right) &= J_s S_1 \left(\frac{1}{3} \right) = \frac{2}{3} J_s \\ J_s \left(S_1 \sum_{i=1}^{1} L_{2i} \right) &= J_s S_1 \left(\frac{1}{3} \right) = \frac{2}{3} J_s \\ J_s \left(S_1 \sum_{i=1}^{1} L_{3i} \right) &= J_s S_1 \left(\frac{1}{3} \right) = \frac{2}{3} J_s \end{aligned} \quad (6.265)$$

We note that this result is also the same as the one we found by closed form integration.

The Finite Element Method   353

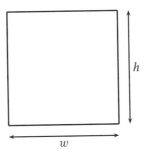

Figure 6.11 2D First-Order Rectangular Edge Element

6.12 EDGE ELEMENTS

The shape functions we have seen so far have been associated with nodal values for the degrees of freedom. In this way we have formulated the problem to solve for the value of the potential or field at the nodes of the element. There is an alternative approach when dealing with vector functions in which we solve for the vector component along the element edges. These so-called *edge elements* have grown in popularity in recent years in the analysis of high-frequency problems and for eddy current analysis. The approach is consistent with the one we have previously established, and in this section we outline the procedure for finding the element matrices for first-order rectangular and triangular edge elements.

6.12.1 First-Order Rectangular Elements

For simplicity let us consider the rectangular element of Figure 6.11. The coordinates of the centroid are (x_c, y_c), the height of the rectangle is h, and the width is w. We will define vector quantities along the four edges of the rectangle. Let these be $A_1..A_4$. We then write the vector A at any point in the element as a linear combination of the values at the edges. Therefore

$$A_x = \frac{1}{h}\left[y_c + \frac{h}{2} - y\right]A_1 + \frac{1}{h}\left[y - y_c + \frac{h}{2}\right]A_2 \qquad (6.266)$$

and

$$A_y = \frac{1}{w}\left[x_c + \frac{w}{2} - x\right]A_3 + \frac{1}{w}\left[x - x_c + \frac{w}{2}\right]A_4 \qquad (6.267)$$

Note that, for example, on edge 1 the value of the vector is equal to A_1. The coefficients of the edge vectors are the shape functions. As in the case of the nodal shape functions, these coefficients are equal to unity along their respective edges and vanish on all other edges. The vector variable in the element is then expressed as

$$A_e = \sum_{i=1}^{4} \xi_i A_i \qquad (6.268)$$

The four vector shape functions are

$$\begin{aligned}
\xi_1 &= \frac{1}{h}\left(y_c + \frac{h}{2} - y\right)\hat{a}_x \\
\xi_2 &= \frac{1}{h}\left(y - y_c + \frac{h}{2}\right)\hat{a}_x \\
\xi_3 &= \frac{1}{w}\left(x_c + \frac{w}{2} - x\right)\hat{a}_y \\
\xi_4 &= \frac{1}{w}\left(x - x_c + \frac{w}{2}\right)\hat{a}_y
\end{aligned} \qquad (6.269)$$

The integrals resulting from the Galerkin formulation for the vector fields involve integrating the "curl–curl" operator. The element matrices are therefore of the form

$$S_{ij} = \int_\Omega \nabla \times \xi_i \cdot \nabla \times \xi_j \, d\Omega \qquad (6.270)$$

In the case of a first-order rectangle the curl of the shape function is a constant. For example, look at the curl of ξ_1.

$$\nabla \times \xi_1 = \begin{vmatrix} \hat{a}_x & \hat{a}_y & \hat{a}_z \\ \frac{\partial}{\partial x} & \frac{\partial}{\partial y} & \frac{\partial}{\partial y} \\ \xi_x & 0 & 0 \end{vmatrix} = \frac{1}{h}\hat{a}_z \quad (6.271)$$

The integration is therefore quite simple. For example, let us evaluate S_{11}. The magnitude of the curl is $\frac{1}{h}$ so the integrand is $\frac{1}{h^2}d\Omega$. The integral of $d\Omega$ is just wh, so we have $S_{11} = \frac{w}{h}$. Similarly, the 4×4 matrix is

$$S = \begin{pmatrix} \frac{w}{h} & -\frac{w}{h} & -1 & -1 \\ -\frac{w}{h} & \frac{w}{h} & 1 & -1 \\ -1 & 1 & \frac{h}{w} & -\frac{h}{w} \\ 1 & -1 & -\frac{h}{w} & \frac{h}{w} \end{pmatrix} \quad (6.272)$$

We should also note here that the divergence of the shape functions is zero so that the gauge condition is implied in the finite element formulation. For example, consider $\nabla \cdot \xi_1$. Because there is only an $\hat{a}_x$ component and this is a function only of y, the divergence vanishes. It is also interesting that in this case the tangential continuity of the vector is satisfied exactly.

The second integral that interests us is the T matrix of the form

$$T_{ij} = \int_\Omega \xi_i \cdot \xi_j d\Omega \quad (6.273)$$

Again the integration here is straightforward. Consider the first row of T.

$$T_{11} = \int_\Omega \xi_1 \xi_1 d\Omega = \int_\Omega \frac{1}{h^2}\left(y_c + \frac{h}{2} - y\right)^2 dx\, dy = \frac{hw}{3} \quad (6.274)$$

The other terms can be found in the same way. Note that because the shape function are vectors, the dot product results in zero for the (1,3) and (1,4) terms and any terms involving the product of orthogonal vectors.

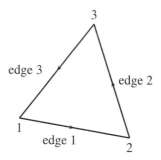

Figure 6.12 2D First-Order Triangular Edge Element

The entire matrix is then

$$T = hw \begin{pmatrix} 2 & 1 & 0 & 0 \\ 1 & 2 & 0 & 0 \\ 0 & 0 & 2 & 1 \\ 0 & 0 & 1 & 2 \end{pmatrix} \tag{6.275}$$

6.12.2 Triangular Edge Elements

Jin [77] gives a method of finding the shape function for the triangular element in terms of the area coordinates of the triangle, which were introduced in Chapter 5.

Consider the triangular element of Figure 6.12, with area coordinates L_1, L_2, and L_3.

The area coordinate L_1 has magnitude 1 at node 1 and is zero at node 2. Similarly, L_2 is 1 at node 2 and zero at node 1. This is illustrated in Figure 6.13. If edge 1 (connecting vertex 1 and 2) is of length ℓ_{12}, then the gradient of L_1 along the edge is simply $\frac{-1}{\ell_{12}}$. By a similar argument, the gradient of L_2 along the edge is $\frac{1}{\ell_{12}}$. We also note that the sum of L_1 and L_2 at any point along the edge is equal to 1. Thus if we construct a vector function of the form

$$w_{12} = L_1 \nabla L_2 - L_2 \nabla L_1 \tag{6.276}$$

The Finite Element Method

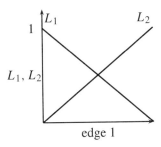

Figure 6.13 Area Coordinates L_1 and L_2 over Edge 1

we find that this function has a constant component along edge 1 equal to $\frac{1}{\ell_{12}}$. Also, because the area coordinates are zero on the opposite edges, the function w_{12} has zero component along edges 2 and 3. Let us examine further the properties of w_{12}. Taking the divergence, we see

$$\begin{aligned} \nabla \cdot w_{12} &= \nabla \cdot (L_1 \nabla L_2) - \nabla \cdot (L_2 \nabla L_1) \\ &= L_1 \nabla^2 L_2 + \nabla L_1 \cdot \nabla L_2 - \nabla^2 L_1 - \nabla L_2 \cdot \nabla L_1 = 0 \end{aligned} \quad (6.277)$$

Now taking the curl, (and using the identity $\nabla \times (a\hat{F}) = (a\nabla \times \hat{F} - \hat{F} \times \nabla a)$, we have

$$\begin{aligned} \nabla \times w_{12} &= \nabla \times (L_1 \nabla L_2) - \nabla \times L_2 \nabla L_1 \\ &= L_1 \nabla \times \nabla L_2 - \nabla L_2 \times \nabla L_1 - L_2 \nabla \times \nabla L_1 + \nabla L_1 \times \nabla L_2 \\ &= 2 \nabla L_1 \times \nabla L_2 \end{aligned} \quad (6.278)$$

Here we have also used the fact that the curl of the gradient of any scalar function is zero.

We can now take for a shape function of edge 1, ℓ_{12} times w_{12}. This is to scale the result so that the function has a value of 1 along its associated edge. In this way the three shape functions for the triangle are

$$\begin{aligned} \xi_1 &= (L_1 \nabla L_2 - L_2 \nabla L_1)\ell_{12} \\ \xi_2 &= (L_2 \nabla L_3 - L_3 \nabla L_2)\ell_{23} \\ \xi_3 &= (L_3 \nabla L_1 - L_1 \nabla L_3)\ell_{31} \end{aligned} \quad (6.279)$$

7

INTEGRAL EQUATIONS

7.1 INTRODUCTION

We have discussed the finite difference and finite element methods. These methods are sometimes called *differential methods*. The methods rely on approximating the differential equation over small regions. Referring to the finite element method, the shape functions described in Chapter 5 are not exact solutions to the operator equation. Also, the element shape functions apply only over the individual element (i.e., they are not defined outside the element). For this reason we say that the basis functions provide *local support*. An alternative formulation of electromagnetics problems is in terms of integral equations in which the field or potential is approximated by a series of functions which are usually exact solutions of the differential equations and have *global support* (their actions are over the entire problem domain). We shall see, however, that while the solutions to the differential equations are exact, the boundary conditions may be only approximately satisfied. This is in contrast to the FEM, in which the Dirichlet boundary conditions are satisfied exactly but the operator equation is satisfied only approximately.

Integral equations can be classified into a number of subcategories. We will discuss the method of moments, the charge simulation method, and the boundary element method.

7.2 BASIC INTEGRAL EQUATIONS

The integral equations that result from Laplace's or Poisson's equation can be represented in the form

$$a(x)\phi(x) + f(x) = \int_a^b K(x,\xi)\phi(\xi)\,d\xi \qquad (7.1)$$

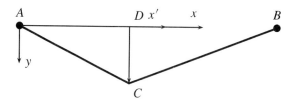

Figure 7.1 String with Point Load

In this equation the function ϕ is the unknown. The equation is a linear integral equation because ϕ appears in a linear form (i.e., we do not have terms like ϕ^2). If $a = 0$ then we have a Fredholm integral equation of the first kind. In these equations the unknown appears only in the integral term. If $a \neq 0$ then we have a Fredholm integral equation of the second kind in which the unknown appears both in the integral and outside it. We shall see that we can formulate electromagnetics problems as equations of the first or second kind.

To see how we obtain these equations it is useful to begin with a problem from mechanics [78]. Consider the string with a point force as shown in Figure 7.1. The string has a certain tension, T. We assume that the displacement is small enough not to affect the tension. A point load P is placed at $x = x'$. The distance $CD = \Delta$ and the length of the string $L = AB \gg \Delta$. Balancing the vertical forces we get

$$P = T\frac{\Delta}{x'} + T\frac{\Delta}{L - x'} \tag{7.2}$$

Solving for Δ we get

$$\Delta = \frac{P(L - x')x'}{TL} \tag{7.3}$$

Integral Equations

Now let the displacement

$$y(x) = P \cdot G(x, x') \tag{7.4}$$

where

$$G(x) = \frac{x(L - x')}{TL} \qquad 0 \le x \le x'$$
$$G(x) = \frac{(L - x)x'}{TL} \qquad x' \le x \le L \tag{7.5}$$

We can check from equation (7.5) that $G(x, x') = G(x', x)$. The function G is the solution if the input is a point unit load. In the theory of integral equations, functions such as this, that is, the solution of the operator equation to a point source, are often found under the integral sign.

Let us now say that the force is continuously distributed and represented by a force density function $\rho(x')$. A small section, between x' and $x' + \delta x'$, of the string will have a force equal to $\rho(x') \, \delta x'$. Then the position of the string $y(x)$ will be

$$y(x) = \int_0^L G(x, x')\rho(x') \, dx' \tag{7.6}$$

If we pose the problem so that the shape of the string is known ($y(x)$ is given), then we need to find the force distribution $\rho(x)$. This is a Fredholm integral equation of the first kind. The function $G(x, x')$ is called the kernel of the integral equation.

Now let us assume that the force is sinusoidally time varying and that we are at steady state. Because we have a linear problem, for a force density $\rho(x') \sin \omega t$ we have a response $y(x, t) = y(x) \sin \omega t$. Let the mass density be $\sigma(x')$. The inertial force is then $-\sigma(x') \, dx' \frac{\partial^2 y}{\partial t^2} = \sigma(x')y(x')\omega^2 \sin \omega t \, dx'$. We now add the applied force $\rho(x') \sin \omega t \, dx'$. This gives an integral equation

$$y(x) \sin \omega t = \int_0^L G(x, x')[\rho(x') \sin \omega t + \omega^2 \sigma(x')y(x') \sin \omega t] \, dx' \tag{7.7}$$

Cancelling out the sin ωt terms in equation (7.7) and defining

$$f(x) = \int_0^L G(x, x')\rho(x')\, dx', \quad K(x, x') = G(x, x')\sigma(x'), \quad \text{and } \lambda = \omega^2$$

we have

$$y(x) = \lambda \int_0^L K(x, x') y(x')\, dx' + f(x) \tag{7.8}$$

If the force distribution $\rho(x')$ and therefore $f(x)$ is given, then we have a Fredholm equation of the second kind for determining $y(x)$.

7.3 METHOD OF MOMENTS

A very popular numerical method based on integral equations is called the method of moments (MOM). For an extensive and excellent discussion of the MOM we refer the reader to the classic book by Harrington [79]. To introduce the method of moments and relate it to the weighted residual methods that we have discussed and to the boundary element method, we will begin with a definition. We define an inner product as an operation between two functions u and v that results in a scalar quantity. This is represented as $\langle u, v \rangle$. The inner product has the following properties:

1. $\langle u, v \rangle = \langle v, u \rangle$.

2. For a third function w we have

$$\langle \alpha u + \beta v, w \rangle = \alpha \langle u, w \rangle + \beta \langle v, w \rangle$$

3. $\langle u^*, u \rangle \geq 0$ for $u \neq 0$.

4. $\langle u^*, u \rangle = 0$ for $u = 0$.

Integral Equations

A possible definition of the inner product that satisfies these conditions is

$$\langle u, v \rangle = \int_0^1 u(x)v(x)\, dx \qquad (7.9)$$

We make use of the concept of an inner product in the method of moments as follows. First consider the operator equation

$$\mathcal{L}\phi = f \qquad (7.10)$$

We expand ϕ as a series of orthogonal functions

$$\phi = \sum_{i=1}^{N} \alpha_i \phi_i \qquad (7.11)$$

where α_i are constants. Substituting into (7.10) gives

$$\sum_{i=1}^{N} \alpha_i \mathcal{L}(\phi_i) = f \qquad (7.12)$$

We now define a set of weighting functions w_j and take the inner product of each of these with equation (7.11).

$$\sum_{i=1}^{N} \alpha_i \langle w_j, \mathcal{L}(\phi_i) \rangle = \langle w_j, f \rangle \qquad (7.13)$$

As $j = 1, 2, 3...$ we obtain a set of simultaneous equations for the α_is.

$$(S)\{\alpha\} = \{F\} \qquad (7.14)$$

The elements of (S) are

$$s_{ij} = \langle w_i, \mathcal{L}\phi_j \rangle \tag{7.15}$$

and the elements of F are

$$F_i = \langle w_i, f \rangle \tag{7.16}$$

One-Dimensional Poisson Equation

To illustrate the steps involved and the level of approximation let us consider the one-dimensional Poisson equation on the interval $[0, 1]$

$$\nabla^2 \phi = \frac{\partial^2 \phi}{\partial x^2} = -x \tag{7.17}$$

with boundary conditions $\phi(0) = \phi(1) = 0$.

This problem is easily solved by direct integration. The exact solution is

$$\phi = \frac{1}{6}(x - x^3) \tag{7.18}$$

For this type of problem Harrington [79] suggests a set of trial functions of the form

$$\phi_i = x - x^{i+1} \tag{7.19}$$

Note that these functions satisfy the boundary conditions. Our solution is then of the form

$$\phi = \sum_{i=1}^{N} \alpha_i (x - x^{i+1}) \tag{7.20}$$

Integral Equations

If we now choose the weighting function to be the same as the trial function, we have, using the inner product of equation (7.9), the Galerkin method. In fact, Harrington shows that the MOM can be interpreted as an error minimization process and is equivalent to the variational and weighted residual methods.

The integrals are evaluated as follows:

$$\langle w_i, \mathcal{L}\phi_j \rangle = j(j+1) \int_0^1 (x - x^{i+1}) x^{j-1} \, dx = \frac{ij}{i+j+1} \qquad (7.21)$$

and for the right-hand side

$$\langle w_i, f \rangle = \int_0^1 x(x - x^{i+1}) \, dx = \frac{1}{3} - \frac{1}{i+3} \qquad (7.22)$$

To see how the method of moments is applied, we begin with a single polynomial expression. This corresponds to the case where $N = 1$. The use of equations (7.21) and (7.22) gives a single equation for α_1.

$$\left(\frac{1}{3}\right)\{\alpha_1\} = \left\{\frac{1}{12}\right\} \qquad (7.23)$$

The solution is $\alpha_1 = \frac{1}{4}$, which gives for the solution

$$\phi = \frac{1}{4}(x - x^2) \qquad (7.24)$$

This is plotted along with the exact solution in Figure 7.2 The error minimization interpretation is that we have found the *best possible* coefficient for our choice of test function.

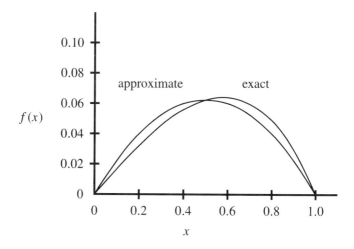

Figure 7.2 Exact and Approximate $N = 1$ Solutions

If we now repeat the solution for $N = 2$, we obtain a 2×2 set of equations for the two unknown α's.

The set of equations is

$$\begin{pmatrix} \frac{1}{3} & \frac{1}{2} \\ \frac{1}{2} & \frac{4}{5} \end{pmatrix} \left\{ \begin{array}{c} \alpha_1 \\ \alpha_2 \end{array} \right\} = \left\{ \begin{array}{c} \frac{1}{12} \\ \frac{2}{15} \end{array} \right\} \tag{7.25}$$

The solution is $\alpha_1 = 0$ and $\alpha_2 = 0.167$, which is the exact solution. Choosing $N = 3$ or higher will also give the exact solution because it corresponds to one of our shape functions.

7.3.1 Charged Wire Example

Consider the cylindrical wire of Figure 7.3. We assume that $\ell \gg a$. The wire is an equipotential, say V_0, and the potential and surface charge density are related by

$$V_0 = \frac{1}{4\pi \epsilon_0} \int \frac{\rho_s(a, \phi', z')}{r} ds' \tag{7.26}$$

Integral Equations

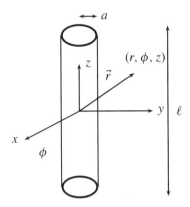

Figure 7.3 Charged Wire

As in the charge simulation method, we solve for the charge distribution on the surface and then use this to find the field everywhere. Because of the axial symmetry of the problem we expect the charge distribution to be a function only of z. We now integrate out the ϕ dependence.

$$V_0 = \frac{1}{4\pi\epsilon_0} \int_{-\frac{\ell}{2}}^{\frac{\ell}{2}} \rho_s(z') \int_0^{2\pi} \frac{a \, d\phi' \, dz'}{r} \tag{7.27}$$

where $r = \sqrt{2a^2 - 2a^2 \cos\phi' + (z-z')^2} = \sqrt{4a^2 \sin^2\frac{\phi'}{2} + (z-z')^2}$.

Performing the integration over ϕ' we get

$$V_0 = \frac{a}{2\epsilon_0} \int_{-\frac{\ell}{2}}^{\frac{\ell}{2}} \frac{\rho_s(z')}{\sqrt{a^2 + (z-z')^2}} dz' \tag{7.28}$$

We now expand the unknown charge distribution into a summation of orthogonal trial functions

$$\rho(z') = \sum_{n=1}^{N} \alpha_n \chi_n(z') \tag{7.29}$$

The values of α are to be determined. We now have

$$V_0 = \frac{a}{2\epsilon_0} \sum_{n=1}^{N} \alpha_n \int_{-\frac{\ell}{2}}^{\frac{\ell}{2}} \frac{\xi_n(z')\,dz'}{\sqrt{a^2 + (z-z')^2}} \qquad (7.30)$$

We now have a set of N simultaneous equations for the unknown α's. In equation 7.30 the denominator can get very small when $z = z'$. Therefore it may be of some advantage to rearrange the integral as follows [80]:

$$\begin{aligned}
I_n &= \int_{-\frac{\ell}{2}}^{\frac{\ell}{2}} \frac{\xi_n(z')\,dz'}{\sqrt{a^2+(z-z')^2}} = \int_{-\frac{\ell}{2}}^{\frac{\ell}{2}} \frac{\xi_n(z') - \xi_n(z) + \xi_n(z)\,dz'}{\sqrt{a^2+(z-z')^2}} \\
&= \xi_n(z) \int_{-\frac{\ell}{2}}^{\frac{\ell}{2}} \frac{dz'}{\sqrt{a^2+(z-z')^2}} + \int_{-\frac{\ell}{2}}^{\frac{\ell}{2}} \frac{\xi_n(z') - \xi_n(z)}{\sqrt{a^2+(z-z')^2}} dz' \\
&= \xi_n(z) \ln \frac{(z+\frac{\ell}{2})\sqrt{a^2(z+\frac{\ell}{2})^2)}}{(z-\frac{\ell}{2}) + \sqrt{a^2+(z-\frac{\ell}{2})^2}} \\
&\quad + \int_{-\frac{\ell}{2}}^{\frac{\ell}{2}} \frac{\xi_n(z') - \xi_n(z)}{\sqrt{a^2+(z-z')^2}}\,dz' \qquad (7.31)
\end{aligned}$$

The integral is now better behaved. We obtain the N simultaneous equations by multiplying the integral equation by a weighting function w_m (which will be a function of z) and integrating from $-\frac{\ell}{2}$ to $\frac{\ell}{2}$. These integrals are called the moments, and the order of the method depends on the order of the weighting function. One possibility is the pulse function

$$w_m = U(z - z_m) = 1, \quad \text{for } z = z_m, \quad 0 \text{ otherwise} \qquad (7.32)$$

This is the collocation method and we are essentially evaluating the integral at N points. Using equal intervals we obtain

Integral Equations

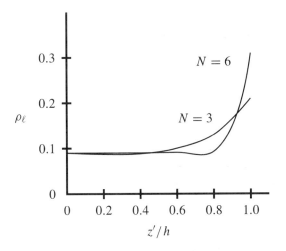

Figure 7.4 Charge Density versus Position

$$\frac{2\epsilon_0 V_0}{a} = \sum_{n=1}^{N} \alpha_n (\xi_n(z) \ln \frac{(z+\frac{\ell}{2})a^2(z+\frac{\ell}{2})^2)^{\frac{1}{2}}}{(z-\frac{\ell}{2}) + \sqrt{a^2 + (z-\frac{\ell}{2})^2}}$$
$$+ \int_{-\frac{\ell}{2}}^{\frac{\ell}{2}} \frac{\xi_n(z') - \xi_n(z)}{\sqrt{a^2 + (z-z')^2}} dz') \quad (7.33)$$

We still must choose the function $\xi_n(z')$. Because we know that the charge density will peak at the ends of the wire and that the function is symmetric around $z = 0$, we can choose

$$\xi_n(z') = \left(\frac{z'}{\ell/2}\right)^{2n} \quad (7.34)$$

Results for this choice are shown in Figure 7.4.

A popular choice that results in relatively simple integration is to let the charge density be rectangular pulses as shown in Figure 7.5. In this case

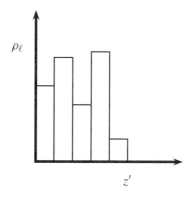

Figure 7.5 Charge Density along the Wire with Pulse Functions

$$\rho_s(z) = \sum_{n=1}^{N} a_n \left\{ u\left[z' - \frac{(n-1)h}{N}\right] - u\left[z' - \frac{nh}{N}\right] \right\} \tag{7.35}$$

where u is the unit step function. Note that the charge density in this case is discontinuous. Generally speaking, we need more subdivisions to obtain a good result. Figure 7.6 shows the results for $N = 30$ rectangular pulses.

7.4 THE CHARGE SIMULATION METHOD

The charge simulation method is reminiscent of the method of images and is based on the uniqueness theorem. Consider the example in Figure 7.7. In the figure on the left we have a charge and a semi-infinite conductor. Because the conductor surface is an equipotential we can replace the conductor by a negative image charge as in the right-hand figure. Because the problem at the right, the field of two charges, is described by the same differential equation and the boundary conditions are the same, the solutions in the upper region are identical (see the uniqueness theorem in Chapter 1). This concept of replacing a conducting (equipotential) surface by equivalent charges is the basis of the charge simulation method.

Integral Equations

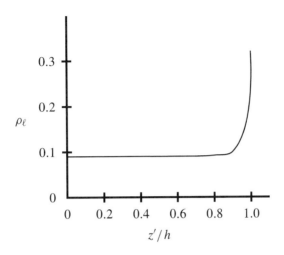

Figure 7.6 Charge Density versus Position for 30 Equally Spaced Pulses

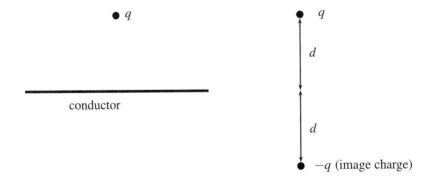

Figure 7.7 Replacing a Conducting Surface with an Image Charge

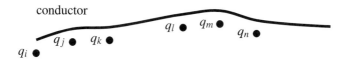

Figure 7.8 Conductor Surface with Equivalent Charges

The uniqueness theorem states that if we can find a solution that satisfies Laplace's equation and the boundary condition $V = V_0$ on Γ, this is the only solution. In the charge simulation method we seek equivalent (fictitious) charges near the surface of the conductor as illustrated in Figure 7.8.

The potential due to the ensemble of charges clearly satisfies Laplace's equation. If we now choose the charges so that the boundary condition is satisfied as well, we can remove the conductor and solve for the field of the charges by Coulomb's law. This solution is valid outside the conductor. Inside the conductor the field is zero and the potential is a constant, V_0. We note also that this is an open boundary problem. We require that $V = 0$ at infinity. This condition is automatically satisfied by the collection of equivalent charges.

The potential at point j due to a charge q at source point i is

$$V_j = \frac{q_i}{4\pi \epsilon_0 r_{ij}} \tag{7.36}$$

where r_{ij} is the distance from point i to point j. If we now pick a field point (x', y', z') on the boundary of the conductor and locate N fictitious charges inside the conductor and (generally) close (but not too close) to the boundary, then we have for the potential at this field point

$$V(x_j, y_j, z_j) = \frac{1}{4\pi \epsilon_0} \sum_{i=1}^{N} \frac{q_i}{r_{ij}} \tag{7.37}$$

Integral Equations

If we now pick N points on the boundary, we obtain a set of N simultaneous algebraic equations in terms of the unknown test charges, q_i.

$$\begin{pmatrix} V_0 \\ \vdots \\ V_0 \end{pmatrix} = \frac{1}{4\pi\epsilon_0} \begin{pmatrix} \frac{1}{r_{11}} & \frac{1}{r_{12}} & \cdots \\ \frac{1}{r_{21}} & \frac{1}{r_{22}} & \cdots \\ \vdots & \vdots & \vdots \\ \frac{1}{r_{N1}} & \cdots & \frac{1}{r_{NN}} \end{pmatrix} \begin{pmatrix} q_1 \\ \vdots \\ q_N \end{pmatrix} \quad (7.38)$$

The coefficient matrix in equation (7.38) is full and symmetric. Full matrices are typical of integral equation methods and are due to the global support of the fundamental solution. The potential at all points is affected by each charge and this is what leads to the full matrix. Once we solve equation (7.38) for the unknown q_i's, the potential at any field point (η, χ) can be found by adding the contributions of the individual charges. The potential will be

$$V(\eta, \chi) = \frac{1}{4\pi\epsilon_0} \sum_{i=1}^{N} \frac{q_i}{r_i} \quad (7.39)$$

where $r_i = \sqrt{(x_i - \eta)^2 + (y_i - \chi)^2}$. We also see that the field at (η, χ) can be found by direct differentiation of the fundamental solution. So

$$\vec{E}(\eta, \chi) = \frac{1}{4\pi\epsilon_0} \sum_{i=1}^{N} \frac{q_i \hat{r}_i}{|r_i|^2} \quad (7.40)$$

The charge simulation method (and integral equations in general) yields very smooth values for the field, unlike the finite element method, in which we find the field by differentiating the (low-order polynomial) shape functions.

We see that the CSM gives an exact solution to the operator equation. The boundary conditions are only approximately satisfied on the conductor, being met only at the N nodes. Between these points the potential will generally not be V_0. As we pick more field points and more source points, we obtain a better approximation to the boundary condition and therefore to the original problem.

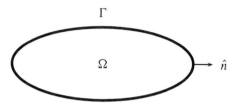

Figure 7.9 2D Region and Surface

The question remains, where should one locate the fictitious charges? For electrostatic problems the charges are on the surface and are continuously distributed. This will be the model used in the boundary element method. In the charge simulation method the location of the charges depends on which results are required. If we need the far field, we can use fewer charges and locate them farther from the boundary. If we need the field near the surface of the conductor, we need more charges and we need them closer to the surface. If we get too close, however, the evaluation of the singular matrix elements ($\frac{1}{r}$) will become less accurate. A compromise is necessary and, as with any of these methods, the results must be carefully checked.

7.5 BOUNDARY ELEMENT EQUATIONS FOR POISSON'S EQUATION IN TWO DIMENSIONS

In this section we will develop the boundary element relations for Poisson's equation in two dimensions. Consider the region and boundary shown in Figure 7.9. We begin with

$$\nabla^2 \phi = f(x, y) \tag{7.41}$$

Integral Equations

Consider now the function $G(r)$, which satisfies

$$\nabla^2 G = \delta(r) \tag{7.42}$$

where G is the Green's function and $\delta(r)$ is defined by

$$\delta(r) = \begin{cases} 0 & \text{for } r \neq 0 \\ \infty & \text{for } r = 0 \end{cases} \tag{7.43}$$

and has the property that

$$\int_{-\infty}^{\infty} \delta(r)\,dr = 1 \tag{7.44}$$

where $r = \sqrt{(\xi - x)^2 + (\eta - y)^2}$, and (ξ, η) and (x, y) are the source and field points, respectively.

Integrating the difference of equation (7.41) times G and equation (7.42) multiplied by ϕ over the region of interest, Ω, we obtain

$$\iint_\Omega (G\nabla^2\phi - \phi\nabla^2 G)\,d\Omega = \iint_\Omega fG\,d\Omega \tag{7.45}$$

The function G defined by equation (7.42) is the potential due to an infinite line source located at $r = 0$. G can be found as the solution of

$$\frac{1}{r}\frac{\partial}{\partial r}r\frac{\partial G}{\partial r} = \delta(r) \tag{7.46}$$

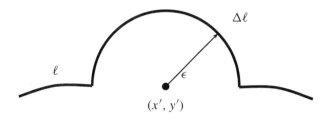

Figure 7.10 Integrating around the Singularity

The solution of this equation is

$$G = -\frac{1}{2\pi} \ln(r) \tag{7.47}$$

Applying Green's theorem to the left-hand side of equation (7.45) we obtain

$$\int_\ell \left(G \frac{\partial \phi}{\partial n} - \phi \frac{\partial G}{\partial n} \right) d\ell = \iint_\Omega fG \, d\Omega \tag{7.48}$$

The kernel of the line integral in equation (7.48) contains a singularity when the source and field points coincide. The integral, however, is finite. To evaluate the line integral we divide it into two parts, one singularity free and the other containing the singularity as shown in Figure 7.10. We then evaluate

$$\int_\ell \left(G \frac{\partial \phi}{\partial n} - \phi \frac{\partial G}{\partial n} \right) d\ell + \lim_{\Delta\ell \to 0} \int_{\Delta\ell} \left(G \frac{\partial \phi}{\partial n} - \phi \frac{\partial G}{\partial n} \right) d\Delta\ell = \iint_\Omega fG \, d\Omega \tag{7.49}$$

Integral Equations

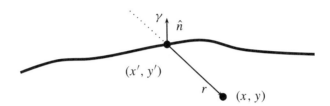

Figure 7.11 Definition of Boundary Terms

The contour $\Delta\ell$, shown in Figure 7.10, is a semicircle of radius ϵ with center on the singularity. On the contour $\Delta\ell$ we have $\frac{\partial}{\partial n} = -\frac{\partial}{\partial r}$ and $d\Delta\ell = \epsilon d\theta$. Using these, the integral over $\Delta\ell$ becomes

$$\lim_{\Delta\ell \to 0} \int_{\Delta\ell} \left(G \frac{\partial \phi}{\partial n} - \phi \frac{\partial G}{\partial n} \right) d\Delta\ell = \frac{1}{2\pi} \lim_{\epsilon \to 0} \int_0^{2\pi} \left(\frac{\partial \phi}{\partial r} \epsilon \ln \epsilon - \phi \right) d\theta = -\phi \tag{7.50}$$

Referring to Figure 7.11, we see that

$$\frac{\partial r}{\partial n} = \cos \gamma \tag{7.51}$$

and

$$\frac{\partial G}{\partial n} = \frac{\partial G}{\partial r} \frac{\partial r}{\partial n} = -\frac{\cos \gamma}{2\pi r} \tag{7.52}$$

Substituting into equation (7.49)

$$\phi(x, y) = \frac{1}{2\pi} \int_\ell \left(\phi \frac{\cos \gamma}{r} - \frac{\partial \phi}{\partial n} \ln r \right) d\ell + \frac{1}{2\pi} \iint_\Omega f \ln r \, d\Omega \qquad (7.53)$$

Equation (7.53) is an expression for the potential at any position in the region in terms of the potential and its normal derivative at the boundaries plus a contribution due to the sources in the region. If the sources are known, as is frequently the case, the surface integral in (7.53) can be carried out numerically or if possible analytically. In an important class of problems, the surface integral can be transformed into a boundary integral and the potential can be expressed entirely in terms of values at the boundary.

Let us assume that the forcing function is harmonic, that is, is a solution of Laplace's equation.

$$\nabla^2 f = 0 \qquad (7.54)$$

An important example would be a constant source. We now define a function, g, that satisfies

$$\nabla^2 g = G \qquad (7.55)$$

We now consider the last term of equation (7.53), which is of the form

$$\iint_\Omega f G \, d\Omega \qquad (7.56)$$

Equation (7.56) is equal to the integral of (7.55) times f minus the integral of (7.54) (which is zero) times g. Thus

$$\iint_\Omega f G \, d\Omega = \iint_\Omega (f \nabla^2 g - g \nabla^2 f) \, d\Omega \qquad (7.57)$$

Integral Equations

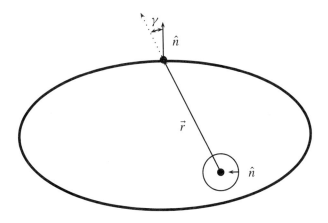

Figure 7.12 Definitions for Integration at Internal Point

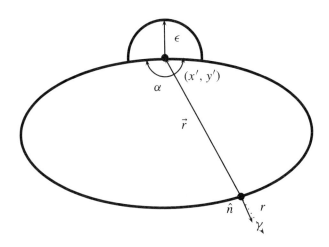

Figure 7.13 Integration for a Field Point Located on the Boundary

We use Green's theorem again to obtain

$$\iint_\Omega fG \, d\Omega = \int_\ell \left(f\frac{\partial g}{\partial n} - g\frac{\partial f}{\partial n} \right) d\ell + \lim_{\Delta\ell \to 0} \int_{\Delta\ell} \left(f\frac{\partial g}{\partial n} - g\frac{\partial f}{\partial n} \right) d\ell \qquad (7.58)$$

As before we integrate around the singularity at $r = 0$ by distorting the boundary into a circular arc. We then take the limit as the radius of the circle goes to zero.

We evaluate g in equation (7.55) by solving

$$\frac{1}{r}\frac{\partial}{\partial r}r\frac{\partial g}{\partial r} = -\frac{1}{2\pi}\ln r \qquad (7.59)$$

The solution is

$$g = \frac{r^2}{8\pi}(1 - \ln r) \qquad (7.60)$$

The last integral in equation (7.58) now becomes

$$\frac{1}{8\pi}\lim_{\epsilon \to 0}\int_0^{2\pi}\left(f\epsilon^2(2\ln\epsilon - 1) + \frac{\partial f}{\partial r}\epsilon^3(1 - \ln\epsilon)\right) d\theta = 0 \qquad (7.61)$$

We see that

$$\frac{\partial g}{\partial n} = \frac{\partial g}{\partial r}\frac{\partial r}{\partial n} = \frac{r}{8\pi}(1 - 2\ln r)\cos\gamma \qquad (7.62)$$

Substituting equations (7.61) and (7.62) into (7.58) and the result into equation (7.53)

$$\phi(x, y) = \frac{1}{2\pi}\int_\ell \left(\phi\frac{\cos\gamma}{r} - \frac{\partial\phi}{\partial n}\ln r\right) d\ell$$
$$- \frac{1}{8\pi}\int_\ell \left(\frac{\partial f}{\partial n}r^2(1 - \ln r) - fr(1 - 2\ln r)\cos\gamma\right) d\ell \qquad (7.63)$$

Integral Equations 381

The singularity in this last integral is removed by integrating around the singular point in a circular arc and taking the limit as the radius of the arc goes to zero as shown previously. So we have

$$\lim_{\Delta\ell \to 0} \int_{\Delta\ell} \left(\frac{\partial f}{\partial n} r^2 (1 - \ln r) - f \frac{\partial}{\partial r} r^2 (1 - \ln r) \right) d\Delta\ell \quad (7.64)$$

Finally, we obtain an expression for the potential at any point (x', y') on the contour as in Figure 7.13,

$$\alpha \phi(x', y') + \int_\ell \left(\frac{\partial \phi}{\partial n} \ln r - \phi \frac{\cos \gamma}{r} \right) d\ell$$

$$= \frac{1}{4} \int_\ell \left(\frac{\partial f}{\partial n} r^2 (1 - \ln r) - f r (1 - 2 \ln r) \cos \gamma \right) d\ell \quad (7.65)$$

Note that the integral should not be evaluated at the singularity because the (finite) contribution to the integral has already been evaluated. Equation (7.65) contains two unknowns, the potential and its normal derivative on the contour. Once these are known, equation (7.63) can be used to find the potential at any point in the region Ω.

7.6 EXAMPLE OF BEM SOLUTION OF A TWO-DIMENSIONAL POTENTIAL PROBLEM

Let us consider the Laplacian problem of Figure 7.14, where we have a rectangular region with homogeneous material of property k, and a potential $\phi = 100$ (Dirichlet condition) is set on the left boundary and $\phi = 0$ is set on the right side. With Neumann conditions of $\frac{\partial \phi}{\partial n} = 0$ on the top and bottom boundaries we actually have a one-dimensional problem. Laplace's equation can easily be integrated in this case to give

$$\phi(x) = 100 - 50x \quad (7.66)$$

We will, however, treat this as a two-dimensional problem to see the steps involved in the boundary element method. We discretize the boundary of Figure 7.14 into four linear boundary elements as shown in Figure 7.15.

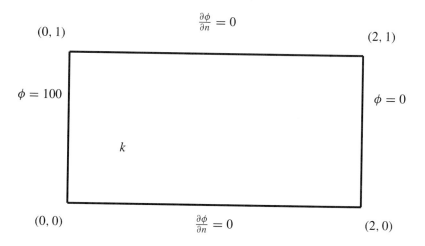

Figure 7.14 Two-Dimensional Homogeneous Example Problem

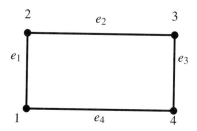

Figure 7.15 Four Linear Boundary Elements

Integral Equations

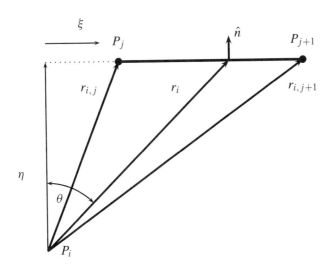

Figure 7.16 Coordinate System and Definitions

For the shape functions of the elements we select linear interpolation functions. Referring to Figure 7.16, we have the field point P_i and the nodes of the element P_j and P_{j+1}. The local coordinate ξ is in the direction from P_j to P_{j+1} and the normal $\hat{n}$ is pointing outward. We then can express ϕ and $\frac{\partial \phi}{\partial n}$ in terms of the nodal values as

$$\phi = \frac{(\phi_{j+1} - \phi_j)\xi + \xi_{j+1}\phi_j - \xi_j\phi_{j+1}}{\xi_{j+1} - \xi_j} \tag{7.67}$$

and

$$\frac{\partial \phi}{\partial n} = \frac{(\frac{\partial \phi}{\partial n}|_{j+1} - \frac{\partial \phi}{\partial n}|_j)\xi + \xi_{j+1}\frac{\partial \phi}{\partial n}|_j - \xi_j\frac{\partial \phi}{\partial n}|_{j+1}}{\xi_{j+1} - \xi_j} \tag{7.68}$$

In forming the global matrix we must evaluate the integral

$$\int_{\xi_j}^{\xi_{j+1}} \left(\frac{\phi}{r} \frac{\partial r_i}{\partial n} - \ln r_i \frac{\partial \phi}{\partial n} \right) d\xi \tag{7.69}$$

Substituting equations (7.67) and (7.68) into (7.69), we obtain for the integral

$$\frac{1}{\xi_{j+1}-\xi_j}\int_{\xi_j}^{\xi_{j+1}}\left(\frac{(\phi_{j+1}-\phi_j)\xi+\xi_{j+1}\phi_j-\xi_j\phi_{j+1}}{r}\frac{\partial r_i}{\partial n}\right.$$
$$-\ln r_i\left(\left(\frac{\partial\phi}{\partial n}|_{j+1}-\frac{\partial\phi}{\partial n}|_j\right)\xi+\xi_{j+1}\frac{\partial\phi}{\partial n}|_j-\xi_j\frac{\partial\phi}{\partial n}|_{j+1}\right)\right)d\xi \quad (7.70)$$

We now break this integral into four parts, each of which can be evaluated in closed form. We note from Figure 7.16 that

$$r_i = \sqrt{\eta_i^2 + \xi_i^2} \quad (7.71)$$

and

$$\frac{\partial r_i}{\partial n} = \frac{\partial r_i}{\partial \eta} = \frac{\eta_i}{r_i} \quad (7.72)$$

Note that this is also $\cos\theta$ in Figure 7.16.

1. The first integral is

$$\frac{1}{\xi_{j+1}-\xi_j}\int_{\xi_j}^{\xi_{j+1}}\frac{1}{r_i}\frac{\partial r_i}{\partial n}\xi\,d\xi = \frac{1}{2}\frac{\eta_i\ln\left(\frac{\eta_i^2+\xi_{j+1}^2}{\eta_i^2+\xi_j^2}\right)}{\xi_{j+1}-\xi_j} \quad (7.73)$$

2. The second integral is

$$\frac{1}{\xi_{j+1}-\xi_j}\int_{\xi_j}^{\xi_{j+1}}\frac{1}{r_i}\frac{\partial r}{\partial \eta}d\xi = \tan^{-1}\frac{\xi_{j+1}}{\eta_i} - \tan^{-1}\frac{\xi_j}{\eta_i} \quad (7.74)$$

Integral Equations

3. The third integral is

$$\frac{1}{\xi_{j+1}-\xi_j}\int_{\xi_j}^{\xi_{j+1}} \ln r_i \xi \, d\xi = \frac{1}{4(\xi_{j+1}-\xi_j)}((\eta_i^2 + \xi_{j+1}^2)(\ln(\eta_i^2 + \xi_{j+1}^2)-1) \\ -(\eta_i^2 + \xi_j^2)(\ln(\eta_i^2 + \xi_j^2)-1)) \quad (7.75)$$

4. And finally the fourth term is

$$\frac{1}{\xi_{j+1}-\xi_j}\int_{\xi_j}^{\xi_{j+1}} \ln r_i \, d\xi = \frac{1}{2(\xi_{j+1}-\xi_j)}((\xi_{j+1})(\ln(\eta_i^2+\xi_{j+1}^2)) \\ -\xi_j \ln(\eta_i^2+\xi_j^2)) - 2(\xi_{j+1}-\xi_j) \\ +2\eta_i\left(\tan^{-1}\left(\frac{\xi_{j+1}}{\eta_i}\right) - \tan^{-1}\left(\frac{\xi_j}{\eta_i}\right)\right) \quad (7.76)$$

We now develop the equation for node 1 in Figure 7.15. At this node

$$\frac{\pi}{2}\phi_1 = \int_\Gamma \left(\phi \frac{1}{r}\frac{\partial r}{\partial n} - \ln r \frac{\partial \phi}{\partial n}\right) d\Gamma \quad (7.77)$$

In the coordinates shown x, y are equivalent to ξ, η. Consider the first term in the integral of equation (7.77) evaluated between nodes 1 and 2.

$$\int \phi \frac{1}{r}\frac{\partial r}{\partial n} d\Gamma = \int_1^2 \phi \frac{1}{r}\frac{\partial r}{\partial n} d\Gamma = \int_1^2 \phi \frac{1}{y}\frac{\partial y}{\partial x} d\Gamma = 0 \quad (7.78)$$

We also have

$$\int_2^3 \frac{\phi}{r}\frac{\partial r}{\partial n} d\Gamma = \int_2^3 \phi \frac{1}{r}\frac{\partial r}{\partial y} dx$$

$$\int_3^4 \frac{\phi}{r}\frac{\partial r}{\partial n} d\Gamma = \int_3^4 \phi \frac{1}{r}\frac{\partial r}{\partial x} dy$$

$$\int_4^1 \frac{\phi}{r}\frac{\partial r}{\partial n} d\Gamma = \int_4^1 \phi \frac{1}{x}\frac{\partial r}{\partial y} dx = 0 \quad (7.79)$$

The first and fourth terms in equation (7.79) are zero because $\cos\theta = 0$. Between nodes 2 and 3 we have

$$\phi = \phi_2\left(\frac{2-x}{2}\right) + \phi_3\left(\frac{x}{2}\right) \tag{7.80}$$

Also noting that $r = \sqrt{x^2+1}$ and $\frac{\partial r}{\partial n} = \frac{\partial r}{\partial y} = \frac{y}{\sqrt{x^2+y^2}}$, the second term becomes

$$\int_0^2 (\phi_2 \frac{2-x}{2} + \phi_3 \frac{x}{2}) \frac{dx}{x^2+1}$$

$$= \phi_2(\tan^{-1} x - \frac{1}{4}\ln(x^2+1))\bigg|_{x=0}^{x=2} + \phi_3\left(\frac{1}{4}\ln(x^2+1)\right)\bigg|_{x=0}^{x=2}$$

$$= 0.70479\phi_2 + 0.40236\phi_3 \tag{7.81}$$

For element 3 from points 3 to 4 we have $\phi = \phi_3 y + \phi_4(1-y)$. To find the geometrical quantities from node 1 we note that $r = \sqrt{4+y^2}$ and $\frac{\partial r}{\partial n} = \frac{\partial r}{\partial x} = \frac{2}{\sqrt{4+y^2}}$. Using these, we find

$$-\int_1^0 (\phi_3 y + \phi_4(1-y))\frac{2}{\sqrt{4+y^2}}dy$$

$$= \phi_3(\ln(4+y^2))\big|_0^1 + \phi_4\left(\tan^{-1}\frac{y}{2} - \ln(4+y^2)\right)\bigg|_0^1$$

$$= 0.22314\phi_3 + 0.24050\phi_4 \tag{7.82}$$

We also break the second term of equation (7.77) into four parts

$$\int \ln r \frac{\partial\phi}{\partial n}d\Gamma$$

$$= \int_0^1 \ln y \frac{\partial\phi}{\partial n}dy + \int_0^2 \ln r \frac{\partial\phi}{\partial n}dx + \int_1^0 \ln r \frac{\partial\phi}{\partial n}(-dy) + \int_0^2 \ln x \frac{\partial\phi}{\partial n}(-dx) \tag{7.83}$$

Integral Equations

Between nodes 1 and 2 we can express the normal derivatives as

$$\frac{\partial \phi}{\partial n} = \frac{\partial \phi}{\partial n}\Big|_1 (1-y) + \frac{\partial \phi}{\partial n}\Big|_2 y \tag{7.84}$$

Along element 1 we have

$$\int_0^1 \ln y \left(\frac{\partial \phi}{\partial n}\Big|_1 (1-y) + \frac{\partial \phi}{\partial n}\Big|_2 \right) dy$$
$$= \frac{\partial \phi}{\partial n}\Big|_1 (y \ln y - y - \frac{y^2}{2}\left(\ln y - \frac{1}{2}\right))\Big|_{y=0}^{y=1} \tag{7.85}$$
$$+ \frac{\partial \phi}{\partial n}\Big|_2 (\frac{y^2}{2} \ln y - \frac{y^2}{4})\Big|_0^1$$
$$= -0.75 \frac{\partial \phi}{\partial n}\Big|_1 - 0.25 \frac{\partial \phi}{\partial n}\Big|_2$$

Along element 3, we can express the normal derivative as

$$\frac{\partial \phi}{\partial n} = \frac{\partial \phi}{\partial n}\Big|_3 y + \frac{\partial \phi}{\partial n}\Big|_4 (1-y) \tag{7.86}$$

where $r = \sqrt{4 + y^2}$.

The integral then becomes

$$\int_0^1 \ln(4+y^2)^{\frac{1}{2}} (\frac{\partial \phi}{\partial n}\Big|_3 y + \frac{\partial \phi}{\partial n}\Big|_4 (1-y))(-dy)$$
$$= \frac{\partial \phi}{\partial n}\Big|_3 (\frac{1}{4}(4+y^2)\ln(4+y^2) - 1 + 2y$$
$$+ 4\tan^{-1}\frac{y}{2} - \frac{1}{4}(4+y^2)(\ln(4y^2) - 1))\Big|_0^1$$
$$= 0.37550 \frac{\partial \phi}{\partial n}\Big|_3 + 0.35651 \frac{\partial \phi}{\partial n}\Big|_4 \tag{7.87}$$

The equation for node 0 is then

$$1.57080\phi_1 - 0.70479\phi_2 - 0.62550\phi_3 - 0.24050\phi_4$$
$$+0.75\frac{\partial \phi}{\partial n}|_1 + 0.25\frac{\partial \phi}{\partial n}|_2 - 0.37550\frac{\partial \phi}{\partial n}|_3 - 0.35651\frac{\partial \phi}{\partial n}|_4 \qquad (7.88)$$

By a similar means the reader may show that for node 2

$$-0.7047\phi_1 + 1.5708\phi_2 - 0.2406\phi_3 - 0.62545\phi_4$$
$$+0.25\frac{\partial \phi}{\partial n}|_1 + 0.75\frac{\partial \phi}{\partial n}|_2 - 0.35650\frac{\partial \phi}{\partial n}|_3 - 0.37550\frac{\partial \phi}{\partial n}|_4 \qquad (7.89)$$

For node 3

$$-0.62545\phi_1 - 0.2406\phi_2 + 1.5708\phi_3 - 0.70475\phi_4$$
$$-0.37550\frac{\partial \phi}{\partial n}|_1 - 0.3565\frac{\partial \phi}{\partial n}|_2 + 0.75\frac{\partial \phi}{\partial n}|_3 + 0.25\frac{\partial \phi}{\partial n}|_4 \qquad (7.90)$$

and for node 4

$$-0.24055\phi_1 - 0.6255\phi_2 - 0.70475\phi_3 + 1.5708\phi_4$$
$$-0.35650\frac{\partial \phi}{\partial n}|_1 - 0.37550\frac{\partial \phi}{\partial n}|_2 + 0.25\frac{\partial \phi}{\partial n}|_3 + 0.75\frac{\partial \phi}{\partial n}|_4 \qquad (7.91)$$

This is a system of four equations and eight unknowns. We must specify either the potential or its normal derivative at each point in order to solve the system[1] of equations.

---

[1] It is possible to specify a linear combination of potential and normal derivative so that one may be substituted out in terms of another.

Integral Equations

In this case we set $\phi_1 = \phi_2 = 100$ and $\phi_3 = \phi_4 = 0$. The four normal derivatives are the unknowns. This gives us the 4×4 system

$$\begin{pmatrix} 0.7500 & 0.2500 & -0.3755 & -0.3565 \\ 0.2500 & 0.7500 & -0.3565 & -0.3755 \\ -0.3755 & -0.3565 & 0.02500 & 0.7500 \\ -0.3565 & -0.3755 & 0.7500 & 0.2500 \end{pmatrix} \begin{pmatrix} \frac{\partial \phi_1}{\partial n} \\ \frac{\partial \phi_2}{\partial n} \\ \frac{\partial \phi_3}{\partial n} \\ \frac{\partial \phi_4}{\partial n} \end{pmatrix} = \begin{pmatrix} -86.606 \\ -86.606 \\ 86.606 \\ 86.606 \end{pmatrix} \quad (7.92)$$

The solution of 7.92 is

$$\begin{pmatrix} \frac{\partial \phi_1}{\partial n} \\ \frac{\partial \phi_2}{\partial n} \\ \frac{\partial \phi_3}{\partial n} \\ \frac{\partial \phi_4}{\partial n} \end{pmatrix} = \begin{pmatrix} 50 \\ 50 \\ -50 \\ -50 \end{pmatrix} \quad (7.93)$$

We know that the solution should vary linearly in the x direction from 100 at element 1 ($x = 0$) to 0 at element 3 ($x = 2$). Therefore the normal derivatives in the x direction at elements 1 and 3 should be 50 or -50 because the separation is 2. Now that the potential and normal derivatives are known at all points on the boundary, the potential at any point can be found from equation (7.63).

7.7 AXISYMMETRIC INTEGRAL EQUATIONS FOR MAGNETIC VECTOR POTENTIAL

This development parallels the two-dimensional case [81]. The main difference is that the kernel of the integral equation, instead of being a logarithm, is a linear combination of elliptic integrals. We begin with the linear equation

$$\nabla^2 A - \left(\frac{1}{r^2} + \alpha^2\right) A = -\mu J \quad (7.94)$$

Introducing a function G and using Green's theorem we obtain

$$\int_S r(G\nabla^2 A - A\nabla^2 G)\, dS = \oint_C r\left(G\frac{\partial A}{\partial n} - A\frac{\partial G}{\partial n}\right) dC \qquad (7.95)$$

If we now require that G is the solution of

$$\nabla^2 G - \left(\frac{1}{r^2} + \alpha^2\right) G = -\delta \qquad (7.96)$$

The solution of (7.96) is

$$G = \sqrt{\frac{r_i}{r_j}}\frac{1}{\pi k}\left(\left(1 - \frac{1}{2}k^2\right) K(m) - E(m)\right) \qquad (7.97)$$

where

$$k^2 = \frac{4 r_i r_j}{\sqrt{(r_i + r_j)^2 + (z_i - z_j)^2}} \qquad (7.98)$$

and

$$m = 1 - k^2 \qquad (7.99)$$

Substituting (7.96) into (7.95) gives

$$r A(r, z) = \oint_C rG\frac{\partial A}{\partial n}\, dC - \oint_C rA\frac{\partial G}{\partial n}\, dC \qquad (7.100)$$

Integral Equations

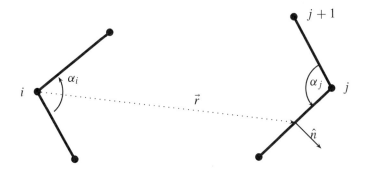

Figure 7.17 Boundary Definitions for Axisymmetric Vector Potential

If we allow the field point to approach the boundary, then equation (7.100) becomes

$$\frac{\alpha}{2\pi} r A(r', z') = \oint_C rG \frac{\partial A}{\partial n} \, dC - \oint_C r A \frac{\partial G}{\partial n} \, dC \qquad (7.101)$$

where the primed coordinates refer to the boundary. The geometry is illustrated in Figure 7.17.

We now require that A and $\frac{\partial A}{\partial n}$ vary linearly on the segment. In matrix form this becomes

$$A = [1\ell][\Gamma]^{-1} \begin{bmatrix} A_i \\ A_j \end{bmatrix} \qquad (7.102)$$

$$\frac{\partial A}{\partial n} = [1\ell][\Gamma]^{-1} \begin{bmatrix} \frac{\partial A_i}{\partial n} \\ \frac{\partial A_j}{\partial n} \end{bmatrix} \qquad (7.103)$$

where ℓ is the length of the boundary element and

$$\Gamma = \begin{pmatrix} 1 & \ell_i \\ 1 & \ell_j \end{pmatrix} \tag{7.104}$$

We now substitute equations (7.102) and (7.103) into equation (7.101), which is integrated numerically. This gives the matrix equation

$$(B \ C) \begin{pmatrix} A \\ \frac{\partial A}{\partial n} = (F) \end{pmatrix} \tag{7.105}$$

where

$$b_{ij} = r_i(1-\alpha)\delta_{ij} - L_j \sum_{k=1}^{5} \ell(k)\frac{\partial G}{\partial n}H(k)r_{jk}$$
$$- \left(L_{j-1} \sum_{k=1}^{5} \ell(6-k)\frac{\partial G}{\partial n}H(k)r_{j-1,k} \right) \tag{7.106}$$

and

$$c_{ij} = L_j \sum_{k=1}^{5} \ell(k) G H(k) r_{jk}$$
$$+ L_{j-1} \sum_{k=1}^{5} \ell(6-k) G H(k) r_{j-1,k} \tag{7.107}$$

We define ℓ and H in Table 7.1.

Integral Equations

Table 7.1 Definition of ℓ and H

k	$\ell(k)$	$H(k)$
1	0.0469101	0.1184635
2	0.2307655	0.2393145
3	0.5000000	0.2844445
4	0.7692345	0.2393145
5	0.953090	0.1184635

7.8 TWO-DIMENSIONAL EDDY CURRENTS WITH $T - \Omega$

In two-dimensional Cartesian coordinates we assume that the eddy currents are in the (x, y) plane and the field is applied in the z direction. In this case T, defined by [82]

$$\nabla \times T = J \tag{7.108}$$

has only one component (z). Using $\nabla \cdot T = 0$, the governing differential equation is

$$\frac{1}{\sigma}\nabla^2 T - j\omega\mu T = j\omega\mu H_0 \tag{7.109}$$

Here H_0 represents the applied field in the z direction.

For homogeneous media we can write equation (7.109) as

$$\nabla^2 T - \alpha^2 T = \alpha^2 H_0 \tag{7.110}$$

where

$$\alpha^2 = j\omega\mu\sigma \tag{7.111}$$

Consider the Green's function that satisfies the equation

$$\nabla^2 G - \alpha^2 G = \delta(\xi - x, \eta - y) \qquad (7.112)$$

where (x, y) are the field points and (ξ, η) are the source points. Multiplying equation (7.112) by T and equation (7.110) by G, subtracting, and integrating over the region R, we obtain

$$\int_R T\delta(\xi - x, \eta - y)\, d\xi\, d\eta = \alpha^2 \int_R H_0 G\, d\xi\, d\eta + \int_R (T\nabla^2 G - G\nabla^2 T)\, d\xi\, d\eta \qquad (7.113)$$

Using Green's theorem on the last term of equation (7.113) gives

$$T(x, y) = \alpha^2 \int_R H_0 G\, d\xi\, d\eta + \int_C \left(T\frac{\partial G}{\partial n} - G\frac{\partial T}{\partial n} \right) dC \qquad (7.114)$$

To evaluate the Green's function we consider equation (7.112), which becomes

$$\frac{\partial^2 G}{\partial r^2} + \frac{1}{r}\frac{\partial G}{\partial r} - \alpha^2 G = \delta(\xi - x, \eta - y) \qquad (7.115)$$

where

$$r = \sqrt{(\xi - x)^2 + (\eta - y)^2} \qquad (7.116)$$

The solution of equation (7.115) is

$$G(\xi, \eta; x, y) = C I_0(\alpha r) + D K_0(\alpha r) \qquad (7.117)$$

where I_0 and K_0 are modified Bessel functions of the first and second kinds of order zero, respectively. From the behavior of these functions at infinity we deduce that $C = 0$. The constant D is found to be $\frac{1}{2\pi}$ by integrating over a small disk centered at $r = 0$.

The term $\frac{\partial G}{\partial n}$ is evaluated as

$$\frac{\partial G}{\partial n} = \frac{\partial G}{\partial r} \frac{\partial r}{\partial n} \tag{7.118}$$

where

$$\frac{\partial G}{\partial r} = -\frac{\alpha}{2\pi} K_1(\alpha r) \tag{7.119}$$

and

$$\frac{\partial r}{\partial n} = -\cos \psi \tag{7.120}$$

where K_1 is the modified Bessel function of the first kind of order 1. The boundary terms are defined in Figure 7.18.

Using equations (7.119) and (7.120) we obtain

$$T(x, y) = \frac{\alpha^2}{2\pi} \int_R H_0 K_0(\alpha r) \, d\xi \, d\eta + \int_C T K_1(\alpha r) \cos \psi \, dC - \frac{1}{2\pi} \int_C \frac{\partial T}{\partial n} K_0(\alpha r) \, dC \tag{7.121}$$

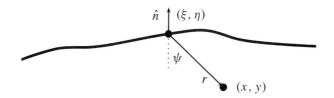

Figure 7.18 Boundary Definitions

As shown before, the surface integral involving the source can be transformed into a line integral if we limit H_0 to the class of functions that satisfy Laplace's equation. Otherwise we must integrate over the region R. If we have

$$\nabla^2 H_0 = 0 \tag{7.122}$$

then we define a function g such that

$$\nabla^2 g = K_0(\alpha r) \tag{7.123}$$

The first integral in equation (7.121) can be written as

$$\int_R H_0 K_0(\alpha r)\, dR = \int_R (H_0 \nabla^2 g - g \nabla^2 H_0)\, dR \tag{7.124}$$

Integral Equations

Applying Green's theorem we get

$$\int_R H_0 K_0(\alpha r)\, dR = \int_C \left(H_0 \frac{\partial g}{\partial n} - g \frac{\partial H_0}{\partial n} \right) dC \qquad (7.125)$$

The solution to equation (7.123) is

$$g = -\frac{1}{\alpha^2} K_0(\alpha r) \qquad (7.126)$$

We also have

$$\frac{\partial g}{\partial n} = -\frac{K_1(\alpha r) \cos \psi}{\alpha} \qquad (7.127)$$

Substituting equations (7.126) and (7.127) into equation (7.125) gives

$$\int_R H_0 K_0(\alpha r)\, dR = -\frac{1}{\alpha} \int_C H_0 K_1(\alpha r)\, dR + \frac{1}{\alpha^2} \int_C \frac{\partial H_0}{\partial n} K_0(\alpha r)\, dC \qquad (7.128)$$

Substituting this into equation (7.121)

$$T(x, y) = -\frac{\alpha}{2\pi} \int_C H_0 K_1(\alpha) \cos \psi\, dC + \frac{1}{2\pi} \int_C \frac{\partial H_0}{\partial n} K_0(\alpha r)\, dC$$
$$+ \frac{\alpha^2}{2\pi} \int_C T K_1(\alpha r) \cos \psi\, dC - \frac{1}{2\pi} \int_C \frac{\partial T}{\partial n} K_0(\alpha r)\, dC \qquad (7.129)$$

Therefore the potential at any point in the region is expressed entirely in terms of values on the boundary. As before, we must remove the singularities that occur on the boundary. Referring to Figure 7.19, we integrate around a singularity along an arc of

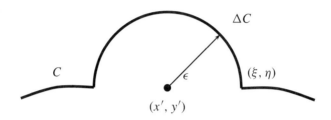

Figure 7.19 Integrating around the Singularity

a circle. We allow (x, y) to approach the point (x', y') on the boundary. Therefore T may be written as

$$T(x, y) = -\frac{\alpha}{2\pi} \int_{C-\Delta C} H_0 K_1(\alpha) \cos \psi \, dC - \frac{\alpha}{2\pi} \int_{\Delta C} H_0 K_1(\alpha) \cos \psi \, dC$$
$$+ \frac{1}{2\pi} \int_{C-\Delta C} \frac{\partial H_0}{\partial n} K_0(\alpha r) \, dC + \frac{1}{2\pi} \int_C \frac{\partial H_0}{\partial n} K_0(\alpha r) \, dC$$
$$+ \frac{\alpha^2}{2\pi} \int_{C-\Delta C} T K_1(\alpha r) \cos \psi \, dC + \frac{\alpha^2}{2\pi} \int_C T K_1(\alpha r) \cos \psi \, dC$$
$$- \frac{1}{2\pi} \int_{C-\Delta C} \frac{\partial T}{\partial n} K_0(\alpha r) \, dC$$
$$- \frac{1}{2\pi} \int_C \frac{\partial T}{\partial n} K_0(\alpha r) \, dC \quad (7.130)$$

All singularities have been isolated in the three integrals around ΔC. Using the asymptotic property of K_0 that

$$K_0(\alpha r) \to -\ln r \quad (7.131)$$

as $r \to 0$ in these integrals we get:

Integral Equations

For the first integral

$$\lim_{\epsilon \to 0} -\frac{\alpha}{2\pi} \int_{\Delta C} H_0 K_1(\alpha r) \cos \psi \, dC = \frac{1}{2\pi} \int_{\Delta C} H_0 \frac{\partial}{\partial n} K_0(\alpha r) \, dC$$

$$= \frac{1}{2\pi} \int_{\Delta C} H_0 \frac{\partial}{\partial n} (-\ln r) \, dC$$

$$= \frac{1}{2\pi} \int_0^{\pi} H_0 \frac{1}{\epsilon} (\epsilon \, d\theta) = \frac{H_0}{2} \quad (7.132)$$

For the second integral

$$\lim_{\epsilon \to 0} \int_{\Delta C} \frac{\partial H_0}{\partial n} K_0(\alpha r) \, dC = \int_{\Delta C} O(\epsilon \ln \epsilon) \, dC = 0 \quad (7.133)$$

For the third integral

$$\lim_{\epsilon \to 0} \frac{\alpha}{2\pi} \int_{\Delta C} T K_1(\alpha r') \cos \psi \, dC = \frac{1}{2\pi} \int_0^{\pi} T \frac{1}{\epsilon} (\epsilon \, d\theta) = \frac{T(x', y')}{2} \quad (7.134)$$

And for the the fourth integral

$$\lim_{\epsilon \to 0} -\frac{1}{2\pi} \int_{\Delta C} \frac{\partial T}{\partial n} K_0(\alpha r) \, dC = \int_{\Delta C} O(\epsilon \ln \epsilon) \, dC = 0 \quad (7.135)$$

We can now express T at the boundary as

$$T(x', y') = -H_0 - \alpha \int_C H_0 K_1(\alpha r) \cos \psi \, dC +$$

$$\frac{\alpha}{\pi} \int_C T K_1(\alpha r) \cos \psi \, dC$$

$$-\frac{1}{\pi} \int_C \frac{\partial T}{\partial n} K_0(\alpha r) \, dC \quad (7.136)$$

We now develop a system of simultaneous equations by assuming that T and $\frac{\partial T}{\partial n}$ are constant on the straight line boundary elements (pulse functions). Writing the integrations in equation (7.136) as summations, we get

$$T_i = -\pi H_0 - \alpha \sum_{\substack{j=1 \\ j \neq i}}^{N} H_0 K_1(\alpha r_{ij}) \cos \psi_{ij} \Delta C_j$$

$$+ \frac{\alpha}{\pi} \sum_{\substack{j=1 \\ j \neq i}}^{N} T_j K_1(\alpha r_{ij}) \cos \psi_{ij} \Delta C_j$$

$$- \frac{1}{\pi} \sum_{\substack{j=1 \\ j \neq i}}^{N} \frac{\partial T_j}{\partial n} K_0(\alpha r_{ij}) \Delta C_j + \frac{\alpha T_i}{\pi} \int_i K_1(\alpha r) \cos \psi \, dC$$

$$- \frac{1}{\pi} \int_i K_0(\alpha r) \, dC \qquad (7.137)$$

The two integrals in equation (7.137) contain singularities. The first integral is zero because the direction cosine is zero. The second integral has been evaluated by Luke [83] and is

$$\frac{1}{\pi} \int_i K_0(\alpha r) \cos \psi \, dC = \frac{2}{\alpha \pi} \left(\sum_{k=0}^{2} d_k \beta^{2k+1} - \ln \beta \sum_{k=0}^{2} c_k \beta^{2k+1} \right) \qquad (7.138)$$

where $\beta = \frac{\alpha \Delta C_i}{4}$ and the coefficients c_k and d_k are defined in Table 7.2.

Thus using equation (7.137) we obtain the set of simultaneous equations

$$(J)\{T\} + (K)\left(\frac{\partial T}{\partial n}\right) = \{F\} \qquad (7.139)$$

At each point on the boundary either T or $\frac{\partial T}{\partial n}$ must be specified in order to make the problem well posed.

Table 7.2 Definition of c and d

k	c_k	d_k
0	2.0	0.8456
1	0.6667	0.5041
2	0.100	0.1123

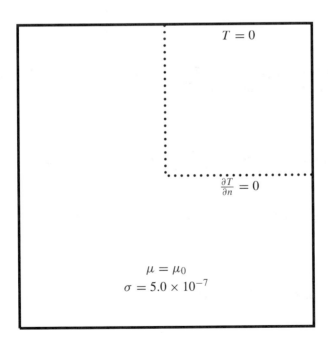

Figure 7.20 Example Problem

Figure 7.21 Real Part of T

Example Problem

As an example of the formulation we consider a square conductor of dimension 10 mm on a side shown in Figure 7.20. A uniform magnetic field H_0 is applied in the z direction. Because of symmetry we represent only one fourth of the geometry. The boundary conditions are $T = 0$ on the conductor surface and $\frac{\partial T}{\partial n}$ on the two lines of symmetry. So we see that either T or $\frac{\partial T}{\partial n}$ is specified at each point on the boundary. In this case the boundary was divided into 72 elements and equation (7.139) was solved. Equation (7.129) was then used to find T in the conductor. The results are shown in Figure 7.21.

7.9 BEM FORMULATION OF THE SCALAR POISSON EQUATION IN THREE DIMENSIONS

We now consider Poisson's equation in three dimensions [84].

$$\nabla^2 \phi = f(x, y, z) \tag{7.140}$$

where

$$\nabla^2 \phi = \frac{\partial^2 \phi}{\partial \xi^2} + \frac{\partial^2 \phi}{\partial \eta^2} + \frac{\partial^2 V}{\partial \zeta^2} = \phi_{\xi\xi} + \phi_{\eta\eta} + \phi_{\zeta\zeta} \tag{7.141}$$

and $\phi_{\xi\xi} = \frac{\partial^2 \phi}{\partial \xi^2}$, $\phi_{\eta\eta} = \frac{\partial^2 \phi}{\partial \eta^2}$ and $\phi_{\zeta\zeta} = \frac{\partial^2 \phi}{\partial \zeta^2}$.

We may multiply this term by a function G and integrate over the volume

$$\int_\tau G \nabla^2 \phi \, d\tau = \int_\tau G(\phi_{\xi\xi} + \phi_{\eta\eta} + \phi_{\zeta\zeta}) \, d\tau \tag{7.142}$$

The first term on the right

$$\int_\tau G \phi_{\xi\xi} \, d\tau = \int_\tau G \phi_{\xi\xi} \, d\xi \, d\eta \, d\zeta$$

can be integrated by parts. Choosing $u = G$ and $dv = \phi_{\xi\xi} \, d\xi$ we have

$$\int_\tau G \phi_{\xi\xi} \, d\xi \, d\eta \, d\zeta = \iint_{\eta\zeta} G \phi_\xi \, d\eta \, d\zeta - \int_\tau G_\xi \phi_\xi \, d\xi \, d\eta \, d\zeta \tag{7.143}$$

where $\phi_\xi = \frac{\partial \phi}{\partial \xi}$. Integrating the second term on the right by parts with $u = G_\xi$ and $dv = \phi_\xi \, d\xi$ gives

$$\int_\tau G \phi_{\xi\xi} \, d\tau = \iint_{\eta\zeta} (G \phi_\xi - G_\xi \phi) \, d\eta \, d\zeta + \int_\tau \phi G_{\xi\xi} \, d\tau \tag{7.144}$$

Similarly,

$$\int_\tau G\phi_{\eta\eta}\,d\tau = \iint_{\xi\zeta}(G\phi_\eta - G_\eta\phi)\,d\xi\,d\zeta + \int_\tau \phi G_{\eta\eta}\,d\tau \qquad (7.145)$$

and

$$\int_\tau G\phi_{\zeta\zeta}\,d\tau = \iint_{\xi\eta}(G\phi_\zeta - G_\zeta\phi)\,d\xi\,d\eta + \int_\tau \phi G_{\zeta\zeta}\,d\tau \qquad (7.146)$$

Using equations (7.143)—(7.146)

$$\begin{aligned}\int_\tau G\nabla^2\phi\,d\tau &= \iint_{\eta\zeta}(G\phi_\xi - G_\xi\phi)\,d\eta\,d\zeta + \iint_{\xi\zeta}(G\phi_\eta - G_\eta\phi)\,d\xi\,d\zeta \\ &\quad + \iint_{\xi\eta}(G\phi_\zeta - G_\zeta\phi)\,d\xi\,d\eta + \int_\tau \phi\nabla^2 G\,d\tau\end{aligned} \qquad (7.147)$$

The first three terms on the right of (7.147) can be written as an integral over the surface S, which encloses the volume τ.

$$\iint_B (G\nabla\phi - \phi\nabla G)\cdot\hat{n}\,d\sigma = \iint_B (G\phi_n - \phi G_n)\,d\sigma \qquad (7.148)$$

where $\hat{n}$ is the unit normal at the surface, $\phi_n = \frac{\partial\phi}{\partial n}$, and $G_n = \frac{\partial G}{\partial n}$. G, Green's function, can be defined as the potential due to a point source (such as a charge) of unit magnitude. Thus

$$\nabla^2 G = \delta(\xi - x, \eta - y, \zeta - z) \qquad (7.149)$$

and

$$\int_\tau \phi\nabla^2 G\,d\tau = \phi(x, y, z) \qquad (7.150)$$

Integral Equations

Using (7.148) and (7.149) in (7.147) gives

$$\int_\tau G\nabla^2\phi \, d\tau = \iint_b (G\phi_n - \phi G_n)d\sigma + \phi(x, y, z) \tag{7.151}$$

The potential due to a point source is

$$G(\xi, \eta, \zeta; x, y, z) = -\frac{1}{4\pi r} \tag{7.152}$$

where

$$r = \sqrt{(\xi - x)^2 + (\eta - y)^2 + (\zeta - z)^2} \tag{7.153}$$

Substituting (7.152) into (7.151) and using $\nabla^2\phi = f$,

$$-\int_\tau \frac{f}{4\pi r} d\tau = \frac{1}{4\pi}\iint_B \left[-\frac{1}{r}\frac{\partial\phi}{\partial n} + \phi\frac{\partial}{\partial n}(1/r)\right] d\sigma + \phi(x, y, z) \tag{7.154}$$

The first term in equation (7.154) is the contribution to the potential of the source distribution in the volume. This distribution is assumed to be known and therefore the integral may be easily evaluated. For simplicity we may drop this term and add it back in later because nothing that will be done in the next argument will have an effect on it. Expression (7.154) relates the potential at any point in the volume to the potential and its derivative at the surface. If we call the potential at the boundary Φ, then we have

$$\phi(x, y, z) = \frac{1}{4\pi}\iint_B \left[\frac{1}{r}\frac{\partial\Phi}{\partial n} - \Phi\frac{\partial}{\partial n}\left(\frac{1}{r}\right)\right] d\sigma \tag{7.155}$$

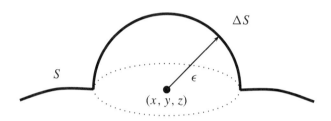

Figure 7.22 Integrating around the Singularity in Three Dimensions

Now let (x, y, z) approach the boundary.

$$\Phi = \frac{1}{4\pi} \iint \left[\frac{1}{r} \frac{\partial \Phi}{\partial n} - \Phi \frac{\partial}{\partial n} \left(\frac{1}{r} \right) \right] d\sigma \qquad (7.156)$$

In evaluating (7.156) care must be taken as (x, y, z) approaches (ξ, η, ζ) that is, as r goes to zero. In order to avoid this difficulty we will isolate the singularity into a separate integral and integrate around a hemisphere of arbitrarily small radius with (x, y, z) at its center as in Figure 7.22. As the radius goes to zero we approach the original problem.

Equation (7.156) now becomes

$$\begin{aligned}
\Phi &= \frac{1}{4\pi} \iint_B \frac{1}{r} \frac{\partial \Phi}{\partial n} d\sigma - \frac{1}{4\pi} \iint_S \Phi \frac{\partial}{\partial n} \frac{1}{r} d\sigma \\
\Phi &= \frac{1}{4\pi} \iint_S \frac{1}{r} \frac{\partial \Phi}{\partial n} d\sigma \\
&\quad - \frac{1}{4\pi} \iint_{S-\Delta S} \Phi \frac{\partial}{\partial n} \left(\frac{1}{r} \right) d\sigma - \frac{1}{4\pi} \iint_{\Delta S} \Phi \frac{\partial}{\partial n} \frac{1}{r} d\sigma \qquad (7.157)
\end{aligned}$$

Integral Equations

We will now take the limit as ΔS approaches zero. Consider now the last term in equation (7.157).

$$\frac{\partial}{\partial n}\frac{1}{r} = -\frac{1}{r^2} = -\frac{1}{a^2} \tag{7.158}$$

and

$$d\sigma = \frac{r^2 d\omega}{2} = \frac{a^2 d\omega}{2} \tag{7.159}$$

Using (7.159)

$$\lim_{\Delta S \to 0} \frac{1}{4\pi} \iint_{\Delta S} \Phi \frac{\partial}{\partial n}\left(\frac{1}{r}\right) d\sigma = \lim_{a \to 0} \frac{1}{4\pi} \int \Phi(Q) \left(\frac{-1}{a^2}\right)\left(\frac{a^2 d\omega}{2}\right)$$

$$= \lim_{a \to 0} \frac{1}{4\pi} \int \Phi(Q) \left(-\frac{d\omega}{2}\right) \tag{7.160}$$

Expanding $\Phi(Q)$ around the center of the hemisphere, P, and ignoring all but the first term gives

$$\lim_{a \to 0} \frac{-1}{4\pi} \int \Phi(Q) \frac{d\omega}{2} = \lim_{a \to 0} -\frac{\Phi(P)}{8\pi} \int_0^{4\pi} d\omega = -\frac{\Phi(P)}{2} \tag{7.161}$$

Substituting (7.161) into (7.157)

$$\frac{\Phi(P)}{2} = \frac{1}{4\pi} \iint \left[\frac{1}{r(P,Q)} \frac{\partial \Phi}{\partial n}\bigg|_Q - \Phi(Q) \frac{\partial}{\partial n}\frac{1}{r(P,Q)}\right] d\sigma \tag{7.162}$$

Boundary conditions must be specified in order to obtain a unique solution. Suppose that on part of the boundary S_1, the potential is known and is ϕ_K, and on the rest of the boundary S_2, the derivative $\frac{\partial \phi_K}{\partial n}$ is known. Then from (7.162)

$$\frac{\Phi(P)}{2} = \frac{1}{4\pi} \iint_{S_1} \left[\frac{1}{r} \frac{\partial \Phi}{\partial n} \bigg|_Q - \phi_K(Q) \frac{\partial}{\partial n}\left(\frac{1}{r}\right) \right] d\sigma$$

$$+ \frac{1}{4\pi} \iint_{S_2} \left[\frac{1}{r} \frac{\partial \phi_K}{\partial n} \bigg|_Q - \Phi(Q) \frac{\partial}{\partial n}\left(\frac{1}{r}\right) \right] d\sigma \quad (7.163)$$

Now break the boundary into patches, ΔS, over which Φ and $\frac{\partial \Phi}{\partial n}$ are constant. The integral equation (7.163) can be approximated by the summation

$$\Phi(P) = \frac{1}{2\pi} \sum_{Q,P}^{S_1} \left\{ \frac{1}{r(P,Q)} \frac{\partial \Phi}{\partial n} \bigg|_Q - \phi_K(Q) \frac{\cos \psi}{r^2(P,Q)} \right\} \Delta S \quad (7.164)$$

$$+ \frac{1}{2\pi} \sum_{Q,P}^{S_2} \left\{ \frac{1}{r(P,Q)} \frac{\partial \phi_K}{\partial n} \bigg|_Q - \phi(Q) \frac{\cos \psi}{r^2(P,Q)} \right\} \Delta S \quad (7.165)$$

where the normal derivative of $1/r$ has been expressed in terms of the direction cosine.

This will give a set of simultaneous equations of the form

$$[A_1] \begin{pmatrix} \frac{\partial \Phi}{\partial n} \\ \frac{\partial \phi_K}{\partial n} \end{pmatrix} + [A_2] \begin{bmatrix} \phi_K \\ \Phi \end{bmatrix} = \begin{bmatrix} 0 \end{bmatrix} \quad (7.166)$$

After solving for $\frac{\partial \Phi}{\partial n}$ and Φ, the potential ϕ can be found anywhere as

$$\phi(P) = \frac{1}{4\pi} \sum_{P,Q}^{S_1} \left\{ \frac{1}{r} \frac{\partial \Phi}{\partial n} - \phi_K \frac{\cos \psi}{r^2} \right\} \Delta S$$

$$+ \frac{1}{4\pi} \sum_{P,Q}^{S_2} \left\{ \frac{1}{r} \frac{\partial \phi_K}{\partial n} - \Phi \frac{\cos \psi}{r^2} \right\} \Delta S \quad (7.167)$$

7.10 GREEN'S FUNCTIONS FOR SOME TYPICAL ELECTROMAGNETICS APPLICATIONS

7.10.1 One-Dimensional Case

For the one-dimensional Laplace equation

$$\frac{\partial^2 \phi}{\partial x^2} = \delta(x) \tag{7.168}$$

we have

$$G(x) = \frac{|x|}{2} \tag{7.169}$$

For the one-dimensional Helmholtz equation

$$\frac{\partial^2 \phi}{\partial x^2} + k^2 \phi = \delta(x) \tag{7.170}$$

we have

$$G(x) = \frac{1}{2k} \sin(kx) \tag{7.171}$$

For the one-dimensional wave equation

$$c^2 \frac{\partial^2 \phi}{\partial x^2} - \frac{\partial^2 \phi}{\partial t^2} = \delta(x, t) \tag{7.172}$$

we have

$$G(x) = \frac{-1}{2c} U(ct - x) \tag{7.173}$$

where U is the unit step function.

For the one-dimensional diffusion equation

$$\frac{\partial^2 \phi}{\partial x^2} - \frac{1}{k} \frac{\partial \phi}{\partial t} = \delta(x, t) \tag{7.174}$$

we have

$$G(x) = \frac{U(t)}{\sqrt{4\pi kt}} \exp\left(\frac{-x^2}{4kt}\right) \tag{7.175}$$

7.10.2 Two-Dimensional Case

For the two-dimensional Laplace equation

$$\frac{\partial^2 \phi}{\partial x^2} + \frac{\partial^2 \phi}{\partial y^2} = \delta(r) \tag{7.176}$$

we have

$$G(x) = \frac{-1}{2\pi} \ln(\frac{1}{r}) \tag{7.177}$$

where $r = \sqrt{x^2 + y^2}$.

For the two-dimensional Helmholtz equation

$$\frac{\partial^2 \phi}{\partial x^2} + \frac{\partial^2 \phi}{\partial y^2} + k^2 \phi = \delta(r) \tag{7.178}$$

we have

$$G(x) = \frac{-1}{4j} H_0^{(2)}(kr) \tag{7.179}$$

where H is a Hankel function.

For the two-dimensional wave equation

$$c^2 \left(\frac{\partial^2 \phi}{\partial x^2} + \frac{\partial^2 \phi}{\partial y^2} \right) - \frac{\partial^2 \phi}{\partial t^2} = \delta(r, t) \tag{7.180}$$

we have

$$G(x) = \frac{-U(ct - r)}{2\pi c(c^2 t^2 - r^2)} \tag{7.181}$$

7.10.3 Three-Dimensional Case

For the three-dimensional Laplace equation

$$\frac{\partial^2 \phi}{\partial x^2} + \frac{\partial^2 \phi}{\partial y^2} + \frac{\partial^2 \phi}{\partial z^2} = \delta(r) \tag{7.182}$$

Integral Equations

we have

$$G(x) = \frac{-1}{4\pi r} \tag{7.183}$$

where $r = \sqrt{x^2 + y^2 + z^2}$.

For the three-dimensional Helmholtz equation

$$\frac{\partial^2 \phi}{\partial x^2} + \frac{\partial^2 \phi}{\partial y^2} + \frac{\partial^2 \phi}{\partial z^2} + k^2 \phi = \delta(r) \tag{7.184}$$

we have

$$G(x) = \frac{-1}{4\pi r} \exp(-jkr) \tag{7.185}$$

For the three-dimensional wave equation

$$c^2 \left(\frac{\partial^2 \phi}{\partial x^2} + \frac{\partial^2 \phi}{\partial y^2} + \frac{\partial^2 \phi}{\partial z^2} \right) - \frac{\partial^2 \phi}{\partial t^2} = \delta(r, t) \tag{7.186}$$

we have

$$G(x) = \frac{-\delta(t - r/c)}{4\pi r} \tag{7.187}$$

8

OPEN BOUNDARY PROBLEMS

8.1 INTRODUCTION

In many engineering applications such as dielectric stress evaluation in insulator strings, surge arrestor voltage distribution, magnetic field computation in accelerator magnets, end region fields in electrical machinery, actuators, contactors, and others, open boundary problems arise. It is uneconomical to solve such problems by bounding the field region at large distances from sources, magnetic and conducting materials, because a large number of finite difference or finite element nodes would be required to model the problem adequately. Also there is no *a priori* knowledge of the boundary of the field region so as not to affect the solution accuracy. Other design criteria that impose limits on the leakage field in order to minimize electromagnetic interference and noise make it imperative that the field distribution is predicted accurately at the design stage.

Several methods are in vogue such as ballooning [85], infinitesimal scaling [86], boundary element method [87], hybrid boundary element and finite element method [88], hybrid harmonic finite element method [57], infinite element method [89], and others. In this chapter, we shall cover and illustrate each method with numerical examples where possible.

8.2 HYBRID HARMONIC FINITE ELEMENT METHOD

This method combines a harmonic solution for the free space region with finite element discretization of the interior region with sources and materials. A macro element is employed for the exterior region that requires only potential continuity at the boundary nodes, with the interior finite elements. In this respect, it differs from other methods

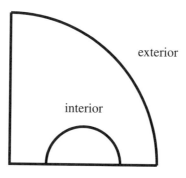

Figure 8.1 Two-Dimensional Field Region

in which either the geometry is ballooned or both potential and normal derivative continuity is enforced at the boundary.

Figure 8.1 shows a two-dimensional field region separated into an interior subregion comprising iron parts and source currents and an exterior homogeneous region. The field distribution may be described in terms of a vector potential or scalar potential or any modifications thereof by appropriate partial differential equations. In the present method a vector potential is used that satisfies the partial differential equations (8.1) and (8.2) for the interior and exterior regions, respectively.

$$\nabla \times \nu \nabla \times A = J \tag{8.1}$$

$$\nabla \times \nu_0 \nabla \times A = 0 \tag{8.2}$$

In the remainder of the discussion, A will be assumed to have a z directed component only.

Open Boundary Problems

Interior Region

Finite element discretization and the minimization of the energy-related functional for the interior region in the traditional manner, as in Chapter 4, yield the integral expression (8.3). These equations are presented here only for the sake of completeness.

$$\frac{\partial \mathcal{F}}{\partial A_k} = \sum_{l=1}^{N} \int_{vol} \nu \left(\frac{\partial A_k}{\partial x} \cdot \frac{\partial A_l}{\partial x} + \frac{\partial A_k}{\partial y} \cdot \frac{\partial A_l}{\partial y} \right) dv \qquad (8.3)$$

Exterior Region

Let us assume that the exterior region is represented by a macro element with global support on the boundary nodes at $r = R$. The potential solution may be defined by

$$A = \sum_{n=1}^{\infty} (C_n \cos n\theta + D_n \sin n\theta)(R/r)^n \qquad (8.4)$$

Expanding equation (8.2) in polar form and substituting in the function for the exterior region yields

$$\mathcal{F} = \int \frac{\nu_0}{2} \left(r^{-2} \left(\frac{\partial A}{\partial \theta} \right)^2 + \left(\frac{\partial A}{\partial r} \right)^2 \right) r \, dr \, \theta \, dz \qquad (8.5)$$

Minimization of (8.5) with respect to each of the potentials at the nodes yields

$$\frac{\partial \mathcal{F}}{\partial A_k} = \int \nu_0 \left(r^{-2} \left(\frac{\partial A}{\partial \theta} \right) \frac{\partial}{\partial A_k} \left(\frac{\partial A}{\partial \theta} \right) + \left(\frac{\partial A}{\partial r} \right) \frac{\partial}{\partial A_k} \left(\frac{\partial A}{\partial r} \right) \right) dv \qquad (8.6)$$

Differentiating (8.4) with respect to r and θ, respectively,

$$\frac{\partial A}{\partial r} = -\sum_{n=1}^{\infty} n(C_n \cos n\theta + D_n \sin n\theta) \left(\frac{1}{r} \right)^{n+1} R^n \qquad (8.7)$$

$$\frac{\partial A}{\partial \theta} = -\sum_{n=1}^{\infty} n(C_n \sin n\theta - D_n \cos n\theta)\left(\frac{R}{r}\right)^n \qquad (8.8)$$

Substituting (8.7) and (8.8) into (8.6) and collecting terms, we obtain

$$\frac{\partial \mathcal{F}}{\partial A_k} = \sum_{n=1}^{\infty} \frac{\partial}{\partial A_k} \int v_0 \frac{n^2(C_n^2 + D_n^2)\, dr\, d\theta\, dz}{2r^{2n+1}} R^{2n} \qquad (8.9)$$

For 1/4 symmetry, only the C_n coefficients with n odd are required. The limits of integration of the values of (r, θ, z) are $R < r < \infty$, $0 < \theta < \pi$, and $0 < z < 1$. Equation (8.9) with C_n's only will be

$$\frac{\partial \mathcal{F}}{\partial A_k} = \sum_{n=1}^{\infty} \frac{\partial}{\partial A_k} \int_R^{\infty}\int_0^{\pi}\int_0^1 v_0 \frac{n^2 C_n^2\, dr\, d\theta\, dz}{2r^{2n+1}} R^{2n} \qquad (8.10)$$

Performing the indicated integrations in (8.10), one obtains

$$\frac{\partial \mathcal{F}}{\partial A_k} = \frac{\partial}{\partial A_k} \frac{v_0 \pi l}{4} \sum_{n=1}^{\infty} n C_n^2 \qquad (8.11)$$

Now it remains only to determine the coefficients C_n. This is done by matching the values of A from the expansion (8.4) at the nodes on the boundary. Typically, on the boundary at $\theta = \theta_1$, we may write

$$A_1 = \sum_{n=1}^{\infty} C_n \cos n\theta_1 \left(\frac{R}{r}\right)^n \qquad (8.12)$$

Open Boundary Problems

Similarly, $A_2, A_3, \ldots, A_n$ can be determined, from which the following matrix equation is obtained.

$$\begin{pmatrix} A_1 \\ A_2 \\ \vdots \\ A_n \end{pmatrix} = \begin{pmatrix} \cos\theta_1 & \cos 3\theta_1 & \cdots & \cos(2n-1)\theta_1 \\ \cos\theta_2 & \cos 3\theta_2 & \cdots & \cos(2n-1)\theta_2 \\ \vdots & \vdots & & \vdots \\ \cos\theta_n & \cos 3\theta_n & \cdots & \cos(2n-1)\theta_n \end{pmatrix} \begin{pmatrix} C_1 \\ C_2 \\ \vdots \\ C_{2n-1} \end{pmatrix} \quad (8.13)$$

It is evident that the coefficients $C_1, C_3, \ldots, C_n$ can then be determined from (8.13) by matrix inversion. Equation (8.12) with the coefficients C obtained from (8.13) yields the coefficient matrix for the exterior region. This can be coupled to the finite element matrix for the interior region at the boundary nodes in the usual manner. The resulting set of equations are solved in the traditional manner. This procedure is a general one and applies to linear, nonlinear, scalar, and vector problems in two dimensions. In case the exterior boundary is not circular, R is taken to be the average radius, and equations (8.12) and (8.13) are modified accordingly.

8.2.1 Applications

Figures 8.2 and 8.3 illustrate the application of the method to conductors in free space, and show the field solution obtained by the finite element method alone and the new hybrid harmonic finite element method. It is seen that the new method facilitates the use of only a few interior finite elements as compared with nearly three times as many elements for the traditional finite element method in order to obtain the corresponding flux distribution.

8.3 INFINITE ELEMENTS

This work is mainly due to Bettes [89] and is adapted in this section for application to electromagnetic field problems. In this method Lagrange interpolation functions are associated with exponential decay terms to represent field variation at infinity. Many such problems occur in electrostatics, electrodynamics, and wave equation modeling where the boundary extends to infinity and, usually but not always, the field decays to zero at the exterior boundary.

418 CHAPTER 8

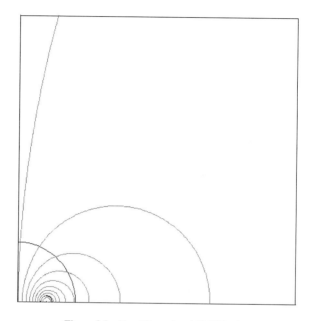

Figure 8.2 Two-Dimensional Field Region

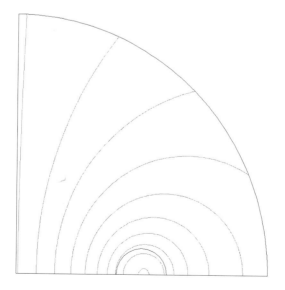

Figure 8.3 Hybrid Harmonic Finite Element Solution

Open Boundary Problems

As stated before, the shape functions are based on Lagrange polynomials and multiplied by exponential decay terms. A set of N points spanning the domain is considered, the first $n - 1$ having finite coordinates and the nth point infinitely distant. The shape functions are defined from 1 to $n - 1$ as follows.

$$\zeta_i = e^{\frac{x_i - x}{\lambda}} \prod_{j=1}^{n-1} \frac{x_j - x}{x_j - x_i} \qquad (8.14)$$

where λ is an arbitrary nonzero distance giving a measure of the severity of the exponential decay. Equation (8.14) satisfies the usual conditions imposed on shape functions, which are

$$\begin{aligned} \zeta_i(x_j) &= \delta_{ij} = 0 \ \text{for} \ i \neq j \\ \zeta_i(x_j) &= 1 \ \text{for} \ i = j \\ \sum_i^n \zeta_i &= 1 \end{aligned} \qquad (8.15)$$

Equation (8.14) holds for any nonzero λ. ζ_n is obtained from the relation

$$\zeta_n = 1 - \sum_{i=1}^{n-1} \zeta_i \qquad (8.16)$$

We shall illustrate the method by a one-dimensional example of a nonlinear resistor whose resistance varies inversely as the applied voltage. The characteristics of the resistor satisfy the differential equation

$$\frac{d^2 R}{dV^2} = \frac{2}{V^3} \qquad (8.17)$$

subject to the boundary conditions

$$R = 1 \ \text{for} \ V = 1, \quad R = 0 \ \text{as} \ V \to \infty \qquad (8.18)$$

The solution to equation (8.17) can be obtained directly by integration, so that

$$\frac{dR}{dV} = -\frac{1}{V^2} + C \tag{8.19}$$

$$R = \frac{1}{V} + CV + D \tag{8.20}$$

Because $R = 0$ as $V \to \infty$ we have $C = 0$. Also, because $R = 1$ at $V = 1$, $D = 0$. Therefore, the solution to (8.17) is

$$R = \frac{1}{V} \tag{8.21}$$

Solution by Infinite Elements

Let us consider an element extending from 1 to infinity and having an intermediate node at $V = 2$ and a single variable R_2 associated with that node. Then we may express the resistance over the domain as

$$R = \zeta_1 + \zeta_2 R_2 \tag{8.22}$$

where R_2 is the value of R at $V = 2$ and

$$\zeta_1 = \frac{V_2 - V}{V_2 - V_1} e^{\frac{V_1 - V}{\lambda}} \tag{8.23}$$

$$\zeta_2 = \frac{V_1 - V}{V_1 - V_2} e^{\frac{V_2 - V}{\lambda}} \tag{8.24}$$

Open Boundary Problems

Because $V_1 = 1$ and $V_2 = 2$ from the stated boundary conditions, equations (8.23) and (8.24) can be written, respectively, as

$$\zeta_1 = (2 - V)e^{\frac{1-V}{\lambda}} \tag{8.25}$$

$$\zeta_2 = (V - 1)e^{\frac{2-V}{\lambda}} \tag{8.26}$$

where λ is an arbitrary nonzero distance.

Expressing equation (8.17) in variational terms by means of a functional, we have

$$\mathcal{F} = \int_1^\infty \left(R\frac{d^2 R}{dV^2} - \frac{4R}{V^3} \right) dV \tag{8.27}$$

The first term in the integral of (8.27) can be transformed by integration by parts so that

$$\int_1^\infty R\frac{d^2 R}{dV^2} = -\int_1^\infty \left(\frac{dR}{dV} \right)^2 dV + R\frac{dR}{dV}\bigg|_1^\infty \tag{8.28}$$

The second term on the right hand-side of equation (8.28) vanishes identically for the applied boundary conditions that at $R = 1$, $V = 1$ and $V = \infty$ at $R = 0$. Therefore, $\frac{dR}{dV} = 0$ at $V = 1$.

Substituting (8.28) into equation (8.27) for the first term in the integrand of (8.27), we have

$$\mathcal{F} = \int_1^\infty \left(\left(\frac{dR}{dV}\right)^2 + \frac{4R}{V^3} \right) dV \tag{8.29}$$

The solution to (8.29) is obtained by minimizing the functional with respect to the unknown value of $R = R_2$, so that

$$\frac{dF}{dR_2} = \int_1^\infty \left[\left(\frac{dR}{dV}\right) \frac{d}{dR_2} \left(\frac{dR}{dV}\right) + \frac{2}{V^3} \left(\frac{dR}{dR_2}\right) \right] dV = 0 \quad (8.30)$$

From equations (8.22) and (8.30), one obtains

$$\frac{dF}{dR_2} =$$
$$\int_1^\infty \left(\left(\frac{d\zeta_1}{dV} + \frac{d\zeta_2}{dV} R_2 \right) \frac{d}{dR_2} \left(\frac{d\zeta_1}{dV} + \frac{d\zeta_2}{dV} R_2 \right) + \frac{2\zeta_2}{V^3} \right) dV = 0 \quad (8.31)$$

which after some algebra reduces to the form

$$\frac{dF}{dR_2} = \int_1^\infty \left(\frac{d\zeta_1}{dV} \frac{d\zeta_2}{dV} + \left(\frac{d\zeta_2}{dV}\right)^2 R_2 + \frac{2\zeta_2}{V^3} \right) dV = 0 \quad (8.32)$$

Differentiating (8.25) and (8.26) with respect to V, we have

$$\frac{d\zeta_1}{dV} = -e^{\frac{1-V}{\lambda}} - \frac{2-V}{\lambda} e^{\frac{1-V}{\lambda}} = -e^{\frac{1-V}{\lambda}} \frac{2-V+\lambda}{\lambda} \quad (8.33)$$

$$\frac{d\zeta_2}{dV} = e^{\frac{2-V}{\lambda}} - \frac{V-1}{\lambda} e^{\frac{2-V}{\lambda}} = e^{\frac{2-V}{\lambda}} \frac{\lambda - V + 1}{\lambda} \quad (8.34)$$

Substituting (8.33) and (8.34) into (8.32), one obtains

$$\frac{dF}{dR_2} = \int_1^\infty -\left(\frac{2+\lambda-V}{\lambda} \frac{1+\lambda-V}{\lambda} e^{\frac{3-2V}{\lambda}} \right.$$
$$\left. + R_2 \frac{(1+\lambda-V)^2}{\lambda^2} e^{\frac{4-2V}{\lambda}} - 2 \frac{1-V}{V^3} e^{\frac{2-V}{\lambda}} \right) dV = 0 \quad (8.35)$$

Open Boundary Problems

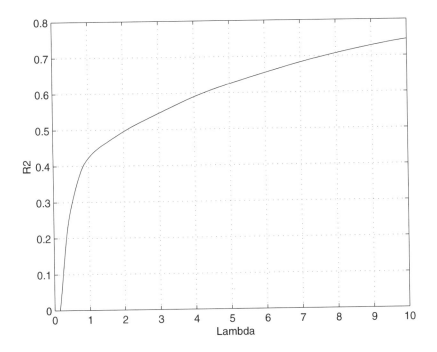

Figure 8.4 Variation of R_2 versus λ

The integrations in (8.35) were performed by Simpson's rule. The solution obtained for R_2 as a function of λ, and the variation of R with V are illustrated in Figures 8.4 and 8.5 respectively.

A Second Example

This example with three elements illustrates the assembly of two one-dimensional standard Lagrange elements and the infinite element, spanning the domain. The boundary conditions are the solution P (at $x = 0$) $= 1$ and P at infinity equals zero. The domain of integration is shown in Figure 8.6.

The differential equation to be solved is of the form

$$\frac{d^2 P}{dx^2} = 0 \qquad (8.36)$$

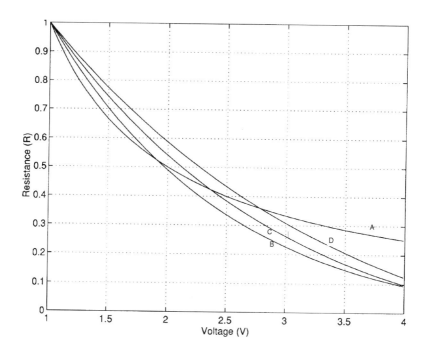

Figure 8.5 Variation of Resistance versus Voltage. A = Analytic, B = $\lambda = 2$, C = $\lambda = 3$, D = $\lambda = 4$

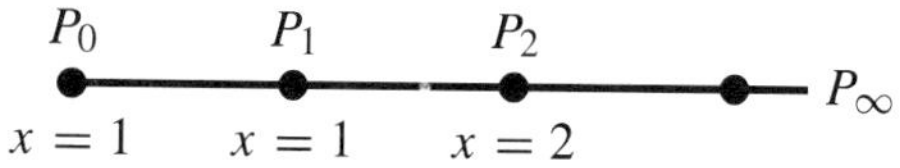

Figure 8.6 Domain of Integration

The shape functions for the locations 1, 2, and 3 are, respectively, as follows.

$$\text{element 1:} \quad \zeta_1 = 1 - x; \quad \zeta_2 = x$$
$$\text{element 2:} \quad \zeta_1 = 2 - x; \quad \zeta_2 = x - 1$$
$$\text{element 3:} \quad \zeta_1 = (3 - x)e^{\frac{2-x}{L}}; \quad \zeta_2 = (x - 2)e^{\frac{3-x}{L}} \quad (8.37)$$

Open Boundary Problems

The element matrices, obtained in the same manner as in the previous example for this application, are as follows:

$$\text{element 1:} \begin{pmatrix} 1 & -1 \\ -1 & 1 \end{pmatrix} \begin{pmatrix} P_0 \\ P_1 \end{pmatrix}$$

$$\text{element 2:} \begin{pmatrix} 1 & -1 \\ -1 & 1 \end{pmatrix} \begin{pmatrix} P_1 \\ P_2 \end{pmatrix}$$

$$\text{element 3:} \begin{pmatrix} 1.25 & -0.75e^{0.5} \\ -0.75e^{0.5} & 0.5e^{1} \end{pmatrix}$$

Assembling the three matrices term by term

$$\begin{pmatrix} 1 & -1 & 0 & 0 \\ -1 & 2 & -1 & 0 \\ 0 & -1 & 2.25 & -1.2365 \\ 0 & 0 & -1.2365 & 1.3591 \end{pmatrix} \begin{pmatrix} P_0 \\ P_1 \\ P_2 \\ P_3 \end{pmatrix} = 0 \qquad (8.38)$$

Substituting $P_0 = 1$ in equation (8.38), transposing the first column to the right-hand side, and eliminating the first row and column, the following matrix equations obtained.

$$\begin{pmatrix} 2 & -1 & 0 \\ -1 & 2.25 & -1.2365 \\ 0 & -1.2365 & 1.3591 \end{pmatrix} \begin{pmatrix} P_1 \\ P_2 \\ P_3 \end{pmatrix} = \begin{pmatrix} 1 \\ 0 \\ 0 \end{pmatrix} \qquad (8.39)$$

Solving (8.39) simultaneously

$$\begin{pmatrix} P_0 \\ P_1 \\ P_2 \\ P_3 \end{pmatrix} = \begin{pmatrix} 1.0 \\ 0.8999 \\ 0.7999 \\ 0.7278 \end{pmatrix} \qquad (8.40)$$

The solution of the $P's$ in elements 1, 2 and 3 are then obtained from equations (8.37) and (8.40) as follows.

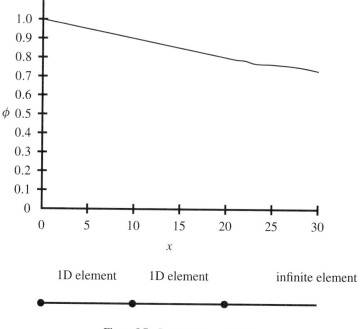

Figure 8.7 Potential versus Position

$$\begin{aligned}
&\text{element 1:} & P &= 1 - 0.1x \\
&\text{element 2:} & P &= 1 - 0.1x \\
&\text{element 3:} & P &= 0.7999(3-x)e^{\frac{2-x}{2}} + 0.7278(x-2)e^{\frac{3-x}{2}}
\end{aligned} \qquad (8.41)$$

This method can easily be extended to two- and three-dimensional open boundary problems. If the shape functions are combined with conventional Lagrange elements in the y direction, rectangular element shape functions are obtained that extend to infinity in the x direction. If Lagrange shape functions are combined with infinite shape functions in the x and y directions, rectangular elements extending to infinity in both x and y directions are obtained. Extension of the method to isoparametric elements is also possible. Combinations of the aforesaid elements can be usefully employed to discretize the field domain, and the integrations can easily be performed in closed form for the Lagrange elements and by Simpson's rule for the infinite elements. In the case of 3D elements, the infinite shape functions may extend in the x, y, and z dimensions or along any two orthogonal directions as required for modeling the domain.

Open Boundary Problems

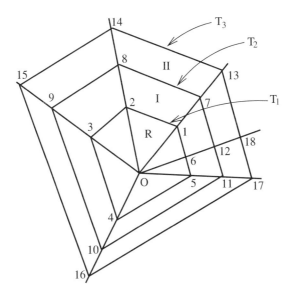

Figure 8.8 Two-Dimensional Interior and Exterior Region

8.4 BALLOONING

The ballooning method has the advantage of simplicity and efficiency and is based on a finite element recursion technique to model the exterior region by a superelement. The method was developed for two-dimensional field problems by Silvester et al. [85], [90]. The method is best illustrated with reference to Figure 8.8.

The interior finite element region consisting of current sources or electric charges and magnetic or dielectric materials is represented by region R. The exterior free space region where Laplace's equation holds is represented by the annuli I and II. The number of nodes on the inner and outer boundaries T_1, T_2, and T_3 are the same and lie along lines radiating from the interior node 0. The inner boundary T_1 separates the interior and exterior regions. The coordinates of the sets of nodes 1–6, 7–12, and 13–18 differ by a fixed ratio called the mapping ratio. Hence the coordinates of vertex 7, for example, will be m times the corresponding values of the coordinates of vertex 1. Similarly, the coordinates of vertex 13 will be m times the corresponding values of vertex 7 and therefore m^2 times the values of the coordinates of vertex 1 and so on.

Formally, we may write

$$(x_n^2, y_n^2) = m(x_n, y_n) \qquad (8.42)$$

on the outer boundary of region I and

$$(x_n^3, y_n^3) = m^2(x_n, y_n) \qquad (8.43)$$

on the outer boundary of region II.

It is evident that because the elements, for example, 1, 2, 7, 8 and 7, 8, 14, 13, are similar and the corresponding vertices differ by the mapping ratio, the two elements will yield the same coefficient matrix. This is true of corresponding elements in the annuli I and II. Further, there is no restriction regarding the discretization of the annuli by first- or higher order triangular or quadrilateral elements.

In each step of the process underlying the ballooning technique, the coefficient matrices of the two annuli are combined, each of which has been derived by combining element coefficient matrices. When successive annuli are added, the effect is one of increasing the mean outer diameter of the annulus at successive steps in the ratio

$$m, m^2, ..., m^{2n} \qquad (8.44)$$

The linear size of the elements and the total region thus constituted increase rapidly. Assuming that the mapping ratio is of the order of $2m$, the successive scale factors will increase rapidly as follows: 2, 4, 16, 256, 65536, 4294967296, 1.84E19, 3.4E38,.... Thus, in a few steps of adding successive annuli, the outer boundary will be at an astronomical distance from T_1.

We shall now describe the process of combining the element matrices for a pair of annuli. We shall denote the potential on T_1 as ϕ_i and on T_2 as ϕ_c. The coefficient matrix of the annular region I is S^I. Similarly, the potential on T_3 will be denoted ϕ_e and the coefficient matrix for the annular region II is S^{II}.

Open Boundary Problems

Because $S^I = S^{II}$, the combination of the element matrices for the two annuli can be written as

$$S^I + S^{II} = \begin{pmatrix} s_{11} & s_{12} \\ s_{21} & s_{22} \end{pmatrix} \begin{pmatrix} \phi_i \\ \phi_c \end{pmatrix} + \begin{pmatrix} s_{11} & s_{12} \\ s_{21} & s_{22} \end{pmatrix} \begin{pmatrix} \phi_c \\ \phi_e \end{pmatrix} = \begin{pmatrix} 0 \\ 0 \end{pmatrix} \quad (8.45)$$

Equation (8.45) may be written after combining the corresponding matrices as

$$\begin{pmatrix} s_{11} & s_{12} & 0 \\ s_{21} & s_{11}+s_{22} & s_{12} \\ 0 & s_{21} & s_{22} \end{pmatrix} \begin{pmatrix} \phi_i \\ \phi_c \\ \phi_e \end{pmatrix} = \begin{pmatrix} 0 \\ 0 \\ 0 \end{pmatrix} \quad (8.46)$$

The potential ϕ_2 is associated with the potentials on the interface T_2 between the two annuli. From equation (8.46) we have, corresponding to the potential ϕ_2, the relationship

$$s_{21}\phi_i + (s_{22}+s_{11})\phi_c + s_{12}\phi_e = 0 \quad (8.47)$$

or

$$\phi_c = -(s_{11}+s_{22})^{-1}s_{21}\phi_i - (s_{11}+s_{22})^{-1}s_{12}\phi_e \quad (8.48)$$

Substituting ϕ_c from equation (8.48) into equation (8.46) we have

$$s_{11}\phi_i + s_{12}\phi_c = s_{11}\phi_i - s_{12}(s_{11}+s_{22})^{-1}s_{21}\phi_i - s_{12}(s_{11}+s_{22})^{-1}s_{12}\phi_e \quad (8.49)$$

Rewriting equation (8.49) in matrix form,

$$(s_{11} - s_{12}(s_{11} + s_{22})^{-1}s_{21}, -s_{12}(s_{11} + s_{22})^{-1}s_{12})\begin{pmatrix} \phi_i \\ \phi_e \end{pmatrix} \quad (8.50)$$

Similarly,

$$s_{21}\phi_c + s_{22}\phi_e = (-s_{21}(s_{11} + s_{22})^{-1}s_{21}, s_{22} - s_{21}(s_{11} + s_{22})^{-1}s_{12})\begin{pmatrix} \phi_i \\ \phi_e \end{pmatrix} \quad (8.51)$$

Combining equations (8.50) and (8.51)

$$\begin{pmatrix} s_{11} - s_{12}(s_{11} + s_{22})^{-1}s_{21}, & -s_{12}(s_{11} + s_{22})^{-1}s_{12} \\ -s_{21}(s_{11} + s_{22})^{-1}s_{21}, & s_{22} - s_{21}(s_{11} + s_{22})^{-1}s_{12} \end{pmatrix} \begin{pmatrix} \phi_i \\ \phi_e \end{pmatrix} \quad (8.52)$$

We note that in the process of assembling the matrix of equation (8.52), we have eliminated the boundary T_2, and the boundary of the combined annuli is now T_3. This process can be carried on recursively until the outer boundary is quite far away. The resulting matrix can then be combined with the coefficient matrix for the interior region. The outer boundary is so far away that we can consider the potentials on it to be zero.

8.4.1 Example

Let us consider an interior finite element region and the exterior annulus as shown in Figure 8.9.

The interior region may be discretized by any number of finite elements but is bounded on T_1 by the nodes 1 and 2. The exterior annulus in our example will be discretized by two elements each. We shall assume that the vertices 1 and 2 have coordinates (0,1) and (1,1) respectively. We shall also assume that the mapping ratio equals 2. Therefore, the coordinates of vertices 3 and 4 will be (0,2) and (2,2), respectively. Similarly, the coordinates of 5 and 6 will be (0,4) and (4,4), respectively.

Open Boundary Problems

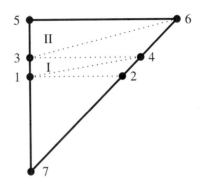

Figure 8.9 Interior and Exterior 2D Regions

Using first-order triangular elements, the assembly of the element matrices for region I yields the matrix equation

$$(s_0)^I (\phi)^I = \begin{pmatrix} 2.0 & -1.5 & -1.0 & 0.5 \\ -1.5 & 2.5 & 0.0 & -1.0 \\ -1.0 & 0.0 & 2.5 & -0.25 \\ 0.5 & -1.0 & -0.25 & 0.75 \end{pmatrix} \begin{pmatrix} \phi_1 \\ \phi_2 \\ \phi_3 \\ \phi_4 \end{pmatrix} \quad (8.53)$$

The corresponding submatrices are

$$S_{11}^I = \begin{pmatrix} 2.0 & -1.5 \\ -1.5 & 2.5 \end{pmatrix} \quad S_{12}^I = \begin{pmatrix} -1.0 & 0.5 \\ 0.0 & -1.0 \end{pmatrix}$$

$$S_{21}^I = \begin{pmatrix} -1.0 & 0.0 \\ 0.5 & -1.0 \end{pmatrix} \quad S_{22}^I = \begin{pmatrix} 2.5 & -0.25 \\ -0.25 & 0.75 \end{pmatrix} \quad (8.54)$$

As stated earlier, the coefficient matrix and the submatrices for region II are identical to those of region I. Therefore, we may write

$$(S_0)^{II} (\phi)^{II} = (S_0)^I (\phi)^I \quad (8.55)$$

Combining equations (8.53) and (8.55)

$$(S0)^{I}(\phi)^{I} + (S0)^{II}(\phi)^{II} = \begin{pmatrix} s_{11}^{I} & s_{12}^{I} & 0 \\ s_{21}^{I} & s_{22}^{I} + s_{11}^{II} & s_{12}^{II} \\ 0 & s_{12}^{II} & s_{22}^{II} \end{pmatrix} \begin{pmatrix} \phi_i \\ \phi_c \\ \phi_e \end{pmatrix} \qquad (8.56)$$

Substituting for the respective submatrices in equation (8.56) from (8.54), we can evaluate the inverse

$$(s_{22}^{I} + s_{11}^{I})^{-1} = \begin{pmatrix} (2.0 + 2.5) & -(1.5 + 0.25) \\ -(1.5 + 0.25) & (2.5 + 0.75) \end{pmatrix}^{-1} = \begin{pmatrix} 0.433 & 0.233 \\ 0.233 & 0.433 \end{pmatrix} \qquad (8.57)$$

From equations (8.52), (8.54), and (8.57), the new matrix for the combined regions I and II, after eliminating the potentials on the interface T_2 and dropping the superscripts I and II, is obtained as

$$(S)_{\text{ext}} = \begin{pmatrix} 1.692 & -1.517 & -0.3167 & 0.1417 \\ -1.517 & 2.067 & -0.2333 & -0.3167 \\ -0.3167 & -0.2333 & 0.8167 & -0.2667 \\ 0.1417 & -0.3167 & -0.2667 & 0.4417 \end{pmatrix} \qquad (8.58)$$

If we add two annuli at a time successively, after 10 such recursive steps, the exterior matrix will reduce to

$$(S)_{\text{ext}} = \begin{pmatrix} 1.633211 & -1.633 & -0.000196 & -0.0000459 \\ -1.633 & 1.634 & -0.000835 & -0.000196 \\ -0.000196 & -0.000835 & 0.3836 & -0.3826 \\ 0.0000459 & -0.000196 & -0.3826 & 0.3828 \end{pmatrix} \qquad (8.59)$$

Open Boundary Problems

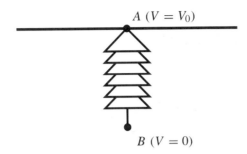

Figure 8.10 Axisymmetric Cross Section of Insulator String

It is evident from equation (8.59) that the coupling or off-diagonal submatrices are small. After 20 recursive steps they become negligibly small, yielding only the diagonal matrices. At this stage, we can set the outer boundary potentials ϕ_{ext} to zero and add only the first submatrix term in equation (8.59) to the matrix for the internal region to solve the field problem.

8.5 INFINITESIMAL SCALING

The advantages of this method that was developed and presented by Hurwitz [86] are that it is efficient, requires less computer storage and execution times and is more accurate than the traditional finite element method, because the basis functions used in the exterior field approximation exactly satisfy Laplace's equation. The order of dimensionality of the problem is reduced by the application of the infinitesimal scaling technique. Therefore, only the contributions to the finite element coefficient matrix due to the boundary nodes need be considered.

We shall consider an electrostatic problem of a power line insulator string of axisymmetric cross section, as shown in Figure 8.10, as an illustration of the method.

We shall assume that the power line is at potential V_0 and the ground wire is at zero potential.

For the electrostatic field distribution at power frequencies excluding displacement currents, the following scalar Laplace equation holds.

$$\epsilon_r \epsilon_0 \nabla^2 V = 0 \qquad (8.60)$$

subject to the boundary conditions that $V = V_0$ at A and $V = 0$ at B. Here ϵ_r is the relative permittivity of the insulating material, ϵ_0 is the permittivity of free space, and V_0 is the voltage on the power line.

The solution to equation (8.60) is obtained by minimizing an energy-related functional

$$\mathcal{F} = \frac{1}{2} \int \epsilon_r \epsilon_0 (\nabla V)^2 dv \qquad (8.61)$$

Expressing the approximation to V in terms of nodal potentials, one obtains the functional (8.61) in matrix form as

$$\mathcal{F} = V^T S V \qquad (8.62)$$

where V^T is the transpose of V.

Adopting the procedure of subdividing the field region into internal and external regions, the contributions to the functional (8.62) may be expressed as

$$\mathcal{F} = V_i^T S V_i + V_e^T C V_e \qquad (8.63)$$

where the subscripts i and e stand for internal and external regions, respectively. Because the minimization of the first term on the right-hand side of (8.63) is carried out in the traditional manner as described in earlier chapters and in [91], it will not be repeated here. We shall describe only the evaluation of the second term in equation (8.63).

Referring to Figure 8.11, the exterior region is represented by scaled surfaces such as R and R_δ, and the potential at R_δ is expressed in terms of the potential at R such that

$$V_{R_\delta} = V_R + \delta \psi \qquad (8.64)$$

Open Boundary Problems

Figure 8.11 Representation of the Exterior Region by Nested Scaled Surfaces

where ψ is the derivative of V along the direction of the scale δ, and

$$\psi = \frac{dV}{d\delta} \tag{8.65}$$

The contribution to the energy-related functional (8.63) due to the exterior region is then obtained as the sum of the contributions due to the region external to R and the annulus between R and R_δ. Therefore,

$$V_e^T C V_e = (V_R + \delta\psi)^T C_\delta (V_R + \delta\psi) +$$
$$\text{energy contribution due to the annulus} \tag{8.66}$$

The contribution due to the annulus may be expressed as being equal to

$$\delta(V_R^T M V + \psi^T N \psi + 2\psi^T Q V_R) \tag{8.67}$$

where M, N, and Q are matrices that are the coefficients of (V_{Ri}, V_{Rj}), (ψ_i, ψ_j), and (ψ_i, V_{Rj}). Substituting (8.67) into (8.66) and eliminating ψ by differentiating with respect to ψ yields a nonlinear differential equation of the form

$$\frac{dC}{ds} = M - (C + Q)^T N^{-1}(C + Q) \tag{8.68}$$

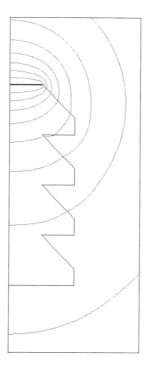

Figure 8.12 Field Plot with Infinitesimal Scaling and Finite Elements for the Insulator String

Equation (8.68) is called the fundamental equation of infinitesimal scaling, where s is the logarithm of a characteristic dimension, the change in s being denoted by δ.

Solution of equation (8.68) yields matrix C, from which the contribution to the functional (8.63) can be evaluated. Minimization of (8.63) in the usual manner results in a matrix algebraic equation representing the discretization of the total field.

The method just described was applied to an insulator string and the solution obtained for the potential distribution is shown in Figure 8.12. The results obtained by this method for the electric field were compared with those obtained by the traditional FE method with the boundary a large distance away from the insulator, with good agreement. Figure 8.12 indicates that the boundary can be truncated close to the insulator string by the use of infinitesimal scaling without loss of accuracy.

8.6 HYBRID FINITE ELEMENT–BOUNDARY ELEMENT METHOD

One very effective method of solving open boundary problems is the hybrid finite element–boundary element method. We have seen that the boundary element method is well suited to open boundary problems because no mesh to infinity is required and the far-field behavior (the potential or field approaches zero at infinity) is built into the Green's functions. Until now we have used only free space Green's functions, but in this section we shall extend the method to include special but practical cases. The hybrid method has the advantages that all nonlinear regions can be represented by finite elements and the space between objects and the exterior region can be represented by integral equations.

8.6.1 Finite Element Equations

When we apply the Galerkin method to the Poisson equation, we have seen that the result is

$$\int_\Omega \nu \left(\frac{\partial N}{\partial x} \frac{\partial A}{\partial x} + \frac{\partial N}{\partial y} \frac{\partial A}{\partial y} \right) d\Omega$$
$$\oint_\Gamma \nu N \frac{\partial A}{\partial n} d\Gamma = \int N J \, d\Omega \tag{8.69}$$

where N is the weighting function.

The line integral term is normally set to zero to impose the homogeneous Neumann boundary condition.

$$A = a + bx + cy \tag{8.70}$$

Writing the equations for the potential at nodes i, j, and k and solving for the potential in terms of the nodal coordinates, we get

$$A = (S_i \; S_j \; S_k) \begin{pmatrix} A_i \\ A_j \\ A_k \end{pmatrix} \tag{8.71}$$

where

$$S_i = \frac{a_i + b_i x + c_i y}{2\Delta}$$

$$S_j = \frac{a_j + b_j x + c_j y}{2\Delta} \qquad (8.72)$$

$$S_k = \frac{a_k + b_k x + c_k y}{2\Delta}$$

and

$$\begin{aligned}
a_i &= x_j y_k - y_j x_k, & b_i &= y_j - y_k, & c_i &= x_k - x_j \\
a_j &= y_i x_k - y_k x_i, & b_j &= y_k - y_i, & c_j &= x_i - x_k \\
a_k &= x_i y_j - y_i x_j, & b_k &= y_i - y_j, & c_k &= x_j - x_i \\
\Delta &= \frac{1}{2}(a_i + b_i x_i + c_i y_i) &&&&
\end{aligned} \qquad (8.73)$$

In the Galerkin method we have

$$N = \begin{pmatrix} S_i \\ S_j \\ S_k \end{pmatrix} \qquad (8.74)$$

The coefficient matrix multiplied by A is equal to

$$\frac{\nu}{4\Delta} \begin{pmatrix} b_i^2 + c_i^2 & b_i b_j + c_i c_j & b_i b_k + c_i c_k \\ b_i b_j + c_i c_j & b_j^2 + c_j^2 & b_j b_k + c_j c_k \\ b_i b_k + c_i c_k & b_j b_k + c_j c_k & b_k^2 + c_k^2 \end{pmatrix} \begin{pmatrix} A_i \\ A_j \\ A_k \end{pmatrix} \qquad (8.75)$$

Open Boundary Problems

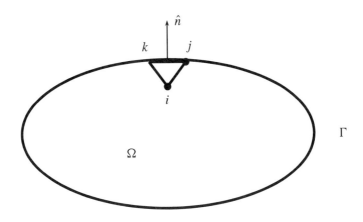

Figure 8.13 Finite Element Region and Boundary

The boundary term is evaluated for elements which have an edge in common with the exterior boundary as in Figure 8.13.

$$-\nu \frac{\partial A}{\partial n} \oint_c N \, dc = -\nu \begin{pmatrix} 0 \\ L/2 \\ L/2 \end{pmatrix} \tag{8.76}$$

where $L = \sqrt{(x_k - x_j)^2 + (y_k - y_j)^2}$.

The right-hand side is

$$J \int_s W \, ds = J \int_s \begin{pmatrix} S_i \\ S_j \\ S_k \end{pmatrix} = \frac{J\Delta}{3} \begin{pmatrix} 1 \\ 1 \\ 1 \end{pmatrix} \tag{8.77}$$

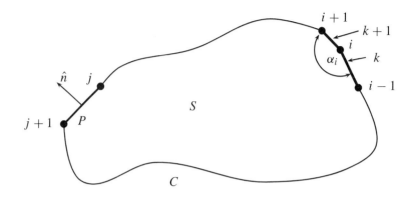

Figure 8.14 Boundary Element Definitions

8.6.2 Boundary Integral Equations

We have seen in Chapter 7 that for Poisson's equation we have

$$\int_S (G\nabla^2 A - A\nabla^2 G)\, ds = -\int_S (A\delta(r) + G\mu J)\, ds \qquad (8.78)$$

where G is the free space Green's function

$$G = \frac{1}{2\pi} \ln r \qquad (8.79)$$

Using equations (8.78) and (8.79), a set of simultaneous equations for the unknown potentials and normal derivatives at the boundary can be found as follows: First, Green's theorem is applied to equation (8.78). This results in an integral equation for the potential in terms of the potential and the normal derivative on the boundary. The potential is then assumed to vary linearly and the normal derivative is taken to be constant on each segment. The resulting set of simultaneous equations is as follows (see Figure 8.14)

Open Boundary Problems

For segment k

$$(\alpha_i/2 \ \ 0)\begin{pmatrix} A_i \\ A_{i-1} \end{pmatrix} + [L_k(\ln L_k - 1)]\frac{\partial A_k}{\partial n} = 0 \qquad (8.80)$$

For segment $k+1$

$$(0 \ \ \alpha_i/2)\begin{pmatrix} A_{i+1} \\ A_i \end{pmatrix} + [L_{k+1}(\ln L_{k+1} - 1)]\frac{\partial A_{k+1}}{\partial n} = 0 \qquad (8.81)$$

For segment P

$$(S_{ij} \ \ S_{i,j+1})\begin{pmatrix} A_j \\ A_{j+1} \end{pmatrix} + q_{ip}\frac{\partial A_p}{\partial n} = f_{ip} \qquad (8.82)$$

The coefficients are evaluated by a three-point Gauss quadrature formula as

$$S_{ij} = -\frac{L_p}{4}\sum_{m=1}^{3} W_m(1 - Z_m)\frac{\cos \beta_m}{r_m} \qquad (8.83)$$

$$S_{i,j+1} = -\frac{L_p}{4}\sum_{m=1}^{3} W_m(1 + Z_m)\frac{\cos \beta_m}{r_m} \qquad (8.84)$$

$$q_{ip} = \frac{L_p}{2}\sum_{m=1}^{3} W_m \ln r_m \qquad (8.85)$$

$$f_{ip} = \frac{\mu J L_p}{8}\sum_{m=1}^{3} W_m r_m (2\ln r_m - 1)\cos \beta_m \qquad (8.86)$$

Table 8.1 Three-Point Gauss Integration

M	Z_m	W_m
1	−0.77459667	0.55555556
2	0.0	0.88888889
3	0.77459667	0.55555556

where

$$r_m = \sqrt{(x - x_i)^2 + (y - y_i)^2} \qquad (8.87)$$

$$\cos \beta_m = \frac{(x - x_i)n_{xp} + (y - y_i)n_{yp}}{r_m} \qquad (8.88)$$

$$x = [(1 - Z_m)x_j + (1 + Z_m)x_{j+1}]/2 \qquad (8.89)$$

$$y = [(1 - Z_m)y_j + (1 + Z_m)y_{j+1}]/2 \qquad (8.90)$$

Z and W are defined in Table 8.1.

8.6.3 Exterior Integral Equation for Node i with Adjacent Source Region

We may have the case where there is a known source region exterior to the finite element region. In this case we can account for the source without increasing the size of the global system matrix. We discretize the region as shown in Figure 8.15 and find the contribution of the source on each boundary element.

Open Boundary Problems

For segment k we have

$$(2\pi - \alpha_i)/2 \; 0)\begin{pmatrix} A_i \\ A_{i-1} \end{pmatrix} + [L_k(1 - \ln L_k)]\frac{\partial A_k}{\partial n} = f_i \qquad (8.91)$$

where

$$f_i = \mu J \sum_{m=1}^{M} \Delta \ln r_m \qquad (8.92)$$

and

$$r_m = \sqrt{(x_i - x_c)^2 + (y_i - y_c)^2} \qquad (8.93)$$

$$\Delta_m = \frac{(x_d - x_b)(y_a - y_b) - (x_b - x_d)(y_b - y_a)}{2} \qquad (8.94)$$

$$\begin{aligned} x_c &= \frac{x_a + x_b + x_d}{3} \\ y_c &= \frac{y_a + y_b + y_c}{3} \end{aligned} \qquad (8.95)$$

For segment $k+1$

$$(0 \; (2\pi - \alpha_i)/2)\begin{pmatrix} A_{i+1} \\ A_i \end{pmatrix} + [L_k(1 - \ln L_{k+1})]\frac{\partial A_{k+1}}{\partial n} \qquad (8.96)$$

For segment P

$$(S_{ij} - S_{i,j+1})\begin{pmatrix} A_j \\ A_{j+1} \end{pmatrix} + (-q_{ip})\frac{\partial A_p}{\partial n} = 0 \qquad (8.97)$$

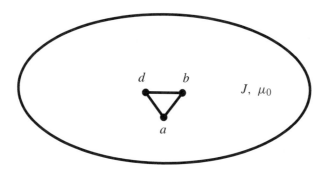

Figure 8.15 Exterior Boundary Element Region with Source

8.6.4 Coupling at the Boundary

The boundary element equations and the finite element equations are coupled by requiring that the potential is continuous so

$$A_{\text{interior}} = A_{\text{exterior}} \tag{8.98}$$

and that the tangential components of H are continuous.

$$\frac{1}{\mu_i}\frac{\partial A}{\partial n}\bigg|_i = \frac{1}{\mu_e}\frac{\partial A}{\partial n}\bigg|_e \tag{8.99}$$

The final set of equations is

$$\begin{pmatrix} \int \nabla N_i \cdot \nabla N_j \, dS & \int N_i N_j \, dS \\ \int N_j \frac{\partial G_i}{\partial n} \, dc & \int N_j G_i \, dc \end{pmatrix} \begin{pmatrix} A \\ \frac{\partial A}{\partial n} \end{pmatrix} = \begin{pmatrix} f \end{pmatrix} \tag{8.100}$$

Open Boundary Problems 445

8.6.5 Examples

To illustrate the hybrid method we refer to the following figures. In Figure 8.16 we see a parallel DC transmission line in free space. In this example we have finite elements in the regions displayed. The exterior of the rectangular region surrounding the conductors is represented by a boundary element layer. The corners of this region are rounded off to avoid the errors that may occur where the normal derivatives are not well defined [92]. The flux lines are plotted only in the conductors for comparison. In Figure 8.17 we have the same problem but now analyzed with finite elements only in the conductors. The rectangular region surrounding the conductors is shown only for comparison with the upper figure. We have boundary element equations on the exterior of each conductor. We see by the crowding of the flux toward the interior that the proximity effect is included by these equations. In Figure 8.18 we again have the twin conductors but only one represented by finite elements. The left-hand conductor is represented by interior and exterior boundary element equations. We have all of the information and effects with a much smaller system.

Figure 8.19 illustrates the same geometry but with AC at 60 hertz. The conductors are made of copper. The right-hand conductor is meshed and is the only finite element region. The left-hand conductor is represented by boundary elements. The interior equations use the modified Bessel functions as the kernel as described in Chapter 7. The exterior integral equations use the free space Green's function. The figure for the real part of the vector potential.

For examples of the hybrid method in nonlinear problems see [88].

8.6.6 Hybrid Method with Half-Space Green's Functions

In the previous section we used free space Green's functions as the kernel of the exterior integral equations. There are a number of practical problems in which modified Green's functions are more appropriate or more convenient. Consider the geometry shown in Figure 8.20.

The Green's function, which is the solution of

$$\nabla^2 G = -\delta(X, Y) \tag{8.101}$$

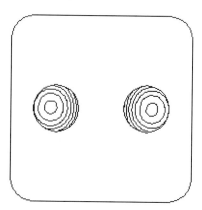

Figure 8.16 DC Transmission Line with Finite Elements in Rectangular Region. Flux Lines Plotted Only in Conductors

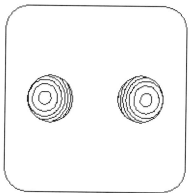

Figure 8.17 DC Transmission Line with Finite Elements in Both Conductors

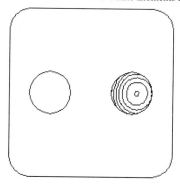

Figure 8.18 DC Transmission Line with Finite Elements in Right Conductor

Open Boundary Problems

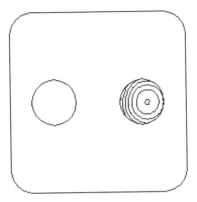

Figure 8.19 Hybrid AC Conductors

μ_1

infinite half space (μ_2)

Figure 8.20 FE Region above Exterior Half-Space

for the half-space problem, becomes

$$G(x, y) = \frac{1}{4\pi}\left[\ln((y-a)^2 + x^2) + \frac{\mu_2 - \mu_1}{\mu_2 + \mu_1}\ln((y+a)^2 + x^2)\right] \quad (8.102)$$

where the source point is located at $(0, a)$. Following the procedures outlined in the previous section, we obtain a set of simultaneous equations of the form

$$\begin{pmatrix} K & B \\ C & D \end{pmatrix}\begin{pmatrix} A \\ \frac{\partial A}{\partial n} \end{pmatrix} = \begin{pmatrix} F \\ 0 \end{pmatrix} \quad (8.103)$$

The K matrix is the regular finite element stiffness matrix. The B matrix for each element is (for first-order elements)

$$\begin{pmatrix} B_{11} & B_{12} \\ B_{21} & B_{22} \end{pmatrix}\begin{pmatrix} \frac{\partial A}{\partial n_i} \\ \frac{\partial A}{\partial n_{i+1}} \end{pmatrix} \quad (8.104)$$

where

$$B_{11} = B_{12} = -\frac{d}{3\mu}$$

$$b_{21} = b_{22} = \frac{B_{11}}{2} \quad (8.105)$$

The C matrix is evaluated as

$$C_{ij} = \frac{\alpha_{ij}}{2\pi}\delta_{ij} - d_j \sum_{i=1}^{5} G(i, j)_k \cdot \zeta_{1,k} \cdot w_{Gk} + d_{j-1}\sum_{k=1}^{5} G(i, j)_k \cdot \zeta_{2k} \cdot w_{Gk} \quad (8.106)$$

where

$$G(i, j)_k = -\frac{1}{2\pi}(\ln r + \beta \ln r') \quad (8.107)$$

and $\beta = \frac{\mu_2 - \mu_1}{\mu_2 + \mu_1}$. Here the w's are the Gauss weights and the ζ's are the shape functions.

Open Boundary Problems

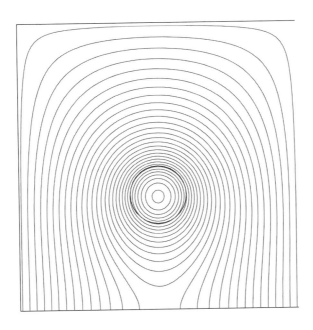

Figure 8.21 Hybrid AC Conductors

Figure 8.21 illustrates the procedure. A conductor above a highly permeable half space is excited with DC current. The rectangular region shown is modeled with finite elements. The boundary element equations on the exterior contain the information concerning the location and permeability of the half space. As we see from the figure, the flux lines are being distorted by the nearby magnetic material.

9

HIGH-FREQUENCY PROBLEMS WITH FINITE ELEMENTS

9.1 INTRODUCTION

This chapter presents a finite element formulation for the wave equation based on nodal based elements and either boundary elements or absorbing boundary conditions for the exterior. Practical applications of the wave equation are frequently open boundary problems. The near field is important because of the often complex structure of the antenna or the scattering object. The far-field pattern is also important in these applications. There is a large class of important problems that are linear, and only these will be treated in this chapter.

Some advantages of the finite element method for wave problems are that

- Complex geometries are easily modeled.
- Boundary conditions are implicit.
- Nonhomogeneous materials are easily accommodated.
- The coefficient matrix is banded, sparse, symmetric, and positive definite.

Some disadvantages of the finite element method are that

- It is difficult to represent open boundaries.
- The unknowns must be solved for throughout the whole domain even if the solution is required only at a few points.

We will use the Galerkin method to obtain the finite element formulation. Recall that this is a particular weighted residual method for which the weighting functions are the same as the shape functions.

9.2 FINITE ELEMENT FORMULATION IN TWO DIMENSIONS

Two different polarizations are possible for two-dimensional electromagnetic field analysis. These polarizations are transverse electric (TE) and transverse magnetic (TM). The vector Helmholtz differential equations for the TE and TM cases, respectively, are derived from Maxwell's equations in free space.

$$\nabla \times \mathbf{E} = -j\omega\mu\mathbf{H} \qquad (9.1)$$

and

$$\nabla \times \mathbf{H} = j\omega\epsilon\mathbf{E} \qquad (9.2)$$

where

- $\mathbf{E}$ is the complex electric field
- $\mathbf{H}$ is the complex magnetic field
- μ is the magnetic permeability
- ϵ is the permittivity
- ω is the angular frequency
- j is the square root of -1

We now use equations (9.1) and (9.2). We obtain an equation for $\mathbf{E}$ by dividing equation (9.1) by $j\omega\mu$, taking the curl, and substituting the result into equation (9.2).

High-Frequency Problems with Finite Elements 453

For **H**, in a similar manner we divide equation (9.2) by $j\omega\epsilon$, take the curl, and substitute the result into equation (9.1). These operations result in the following second-order partial differential equations for either **E** or **H**:

$$\nabla \times \frac{1}{j\omega\mu} \nabla \times \mathbf{E} + j\omega\epsilon \mathbf{E} = 0 \qquad (9.3)$$

$$\nabla \times \frac{1}{j\omega\epsilon} \nabla \times \mathbf{H} + j\omega\mu \mathbf{H} = 0. \qquad (9.4)$$

9.2.1 TM Polarization

Let us assume that equivalent magnetic current density, $\mathbf{J}_m$, is the forcing vector. Then equation (9.1) becomes

$$\nabla \times \mathbf{E} = -\mathbf{J}_m - j\omega\mu \mathbf{H} \qquad (9.5)$$

and equation (9.4) becomes

$$\nabla \times \frac{1}{j\omega\epsilon} \nabla \times \mathbf{H} + j\omega\mu \mathbf{H} = -\mathbf{J}_m \qquad (9.6)$$

By applying the Galerkin method to equation (9.6), the finite element formulation, equation (9.7), is obtained.

$$\int_\Omega \mathbf{W} \cdot \left(\nabla \times \frac{1}{j\omega\epsilon} \nabla \times \mathbf{H} + j\omega\mu \mathbf{H} \right) d\Omega = -\int_\Omega \mathbf{W} \cdot \mathbf{J}_m d\Omega \qquad (9.7)$$

Integrating equation (9.7) by parts results in

$$\oint_\Gamma \mathbf{W} \cdot \left(\mathbf{n} \times \frac{1}{j\omega\epsilon} \nabla \times \mathbf{H} \right) d\Gamma$$
$$+ \int_\Omega \left(\frac{1}{j\omega\epsilon} (\nabla \times \mathbf{W}) \cdot (\nabla \times \mathbf{H}) + j\omega\mu \mathbf{W} \cdot \mathbf{H} \right) d\Omega$$
$$= - \int_\Omega \mathbf{W} \cdot \mathbf{J}_m d\Omega \tag{9.8}$$

where $\mathbf{W}$ in equation (9.8) is a vector weighting function and

$$\mathbf{E}_t = \mathbf{n} \times \frac{1}{j\omega\epsilon} \nabla \times \mathbf{H} \tag{9.9}$$

Ω is the two-dimensional domain and Γ is its boundary.

Rearranging equation (9.8) leads to

$$\int_\Omega \left(\frac{1}{j\omega\epsilon} (\nabla \times \mathbf{W}) \cdot (\nabla \times \mathbf{H}) + j\omega\mu \mathbf{W} \cdot \mathbf{H} \right) d\Omega - \oint_\Gamma \mathbf{W} \cdot \mathbf{E_t} \, dS$$
$$= - \int_\Omega \mathbf{W} \cdot \mathbf{J}_m \, d\Omega \tag{9.10}$$

In two-dimensional analysis, the unknown quantity, $\mathbf{H}$, has only one component, which is perpendicular to the plane of analysis while $\mathbf{E}$ has two components in the plane of analysis.

We use the following notation for the single-component vector:

$$\text{For } \mathbf{H}, \quad H_z \equiv H \tag{9.11}$$
$$\text{For } \mathbf{J}_m, \quad J_{m_z} \equiv J_m \tag{9.12}$$
$$\text{For } \mathbf{W}, \quad W_z \equiv W \tag{9.13}$$

High-Frequency Problems with Finite Elements

Substituting equations (9.11) – (9.13) into equation (9.10) we find

$$\int_\Omega \left(\nabla W \cdot \frac{1}{j\omega\epsilon} \nabla H + j\omega\mu W \cdot H \right) d\Omega - \oint_\Gamma W \cdot E_t d\Gamma = -\int_\Omega W \cdot J_m d\Omega \quad (9.14)$$

Now H, E_t, J_m, and W in equation (9.14) are expanded in terms of the basis function, $N(x, y)$, such that

$$H = \sum_{k=1}^{n} N_k H_k \quad (9.15)$$

$$E_t = \sum_{k=1}^{n} N_k E_{t_k} \quad (9.16)$$

$$J_m = \sum_{k=1}^{n} N_k J_{m_k} \quad (9.17)$$

$$W = \sum_{k=1}^{n} N_k \quad (9.18)$$

where n is the number of unknowns. H_k, E_{t_k}, and J_{m_k} are the nodal values of node k corresponding to H, E_t, and J_m, respectively. The numerical discretization is completed with the enforcement of equation (9.14) for all finite elements. Substitution of equations (9.15), (9.16), (9.17) and (9.18) into equation (9.14) results in the matrix formulation

$$[S]\{H\} - [T]\{E_t\} = \{F\} \quad (9.19)$$

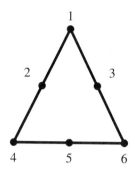

Figure 9.1 A Six-Node Second-Order Triangular Finite Element

where

$$[S] = \sum_{\text{all elements}} \int_{\Omega_e} \left(\frac{1}{j\omega\epsilon} \nabla N^T \cdot \nabla N + j\omega\mu \, N^T \cdot N \right) d\Omega \qquad (9.20)$$

$$[T] = \sum_{\text{all boundary elements}} \int_{\Gamma_e} N^T \cdot N d\Gamma \qquad (9.21)$$

$$\{F\} = -[\sum_{\text{all elements}} \int_{\Omega_e} N^T \cdot N \, d\Omega] \cdot \{J_m\} \qquad (9.22)$$

where N is a row vector, $\{N_1, N_2, ...\}$, and N^T is a transpose vector of N.

For the examples given, the matrices $[S]$ and $\{F\}$ are calculated using the second-order interpolation shape functions (six-noded triangular element) and the matrix $[T]$ is calculated using the ten-point Gauss quadrature method (three-node line element) for the boundary element region.

9.2.2 TE Polarization

The formulation for the case of the TE polarization is similar to that of the TM case. First, we assume the source is an equivalent electric current density, $\mathbf{J}_e$. Equation (9.1) then becomes

$$\nabla \times \mathbf{H} = \mathbf{J}_e + j\omega\epsilon\mathbf{E} \tag{9.23}$$

and equation (9.4) becomes

$$\nabla \times \frac{1}{j\omega\mu}\nabla \times \mathbf{E} - j\omega\epsilon\mathbf{E} = \mathbf{J}_e \tag{9.24}$$

By applying the Galerkin method to equation (9.24), the integral form of equation (9.25) is obtained.

$$\int_\Omega \mathbf{W} \cdot \left(\nabla \times \frac{1}{j\omega\mu}\nabla \times \mathbf{E} + j\omega\epsilon\mathbf{E}\right) d\Omega = -\int_\Omega \mathbf{W} \cdot \mathbf{J}_e d\Omega \tag{9.25}$$

Integrating equation (9.25) by parts results in

$$\oint_\Gamma \mathbf{W} \cdot \left(\mathbf{n} \times \frac{1}{j\omega\mu}\nabla \times \mathbf{E}\right) d\Gamma$$
$$+ \int_\Omega \left(\frac{1}{j\omega\mu}(\nabla \times \mathbf{W}) \cdot (\nabla \times \mathbf{E}) + j\omega\epsilon\mathbf{W} \cdot \mathbf{E}\right) d\Omega$$
$$= -\int_\Omega \mathbf{W} \cdot \mathbf{J}_e \, d\Omega \tag{9.26}$$

where $\mathbf{W}$ in equation (9.26) is a vector weighting function and

$$\mathbf{H}_t = -\mathbf{n} \times \frac{1}{j\omega\mu}\nabla \times \mathbf{E} \tag{9.27}$$

Rearranging equation (9.26) leads to

$$\int_\Omega \left(\frac{1}{j\omega\mu} (\nabla \times \mathbf{W}) \cdot (\nabla \times \mathbf{E}) + j\omega\epsilon \mathbf{W} \cdot \mathbf{E} \right) d\Omega - \oint_\Gamma \mathbf{W} \cdot \mathbf{H}_t d\Gamma$$
$$= -\int_\Omega \mathbf{W} \cdot \mathbf{J}_e \, d\Omega \qquad (9.28)$$

Again, noting that in two dimensions we have a single degree of freedom,

$$\int_\Omega \left(\nabla W \cdot \frac{1}{j\omega\mu} \nabla E + j\omega\epsilon W \cdot E \right) d\Omega - \oint_\Gamma W \cdot H_t d\Gamma = -\int_\Omega W \cdot J_e \, d\Omega \quad (9.29)$$

The variables E, H_t, J_e, and W in equation (9.29) are expanded in terms of the basis function, $N(x, y)$, such that

$$E = \sum_{k=1}^{n} N_k E_k \qquad (9.30)$$

$$H = \sum_{k=1}^{n} N_k H_{t_k} \qquad (9.31)$$

$$J = \sum_{k=1}^{n} N_k J_{e_k} \qquad (9.32)$$

$$W = \sum_{k=1}^{n} N_k \qquad (9.33)$$

The numerical discretization is completed with the enforcement of equation (9.29) for all finite elements. Substitution of equations (9.30), (9.31), (9.32) and (9.33) into equation (9.29) results in the matrix formulation

$$[S]\{E\} - [T]\{H_t\} = \{F\} \tag{9.34}$$

where

$$[S] = \sum_{\text{all elements}} \int_{\Omega_e} \left(\frac{1}{j\omega\mu} \nabla N^T \cdot \nabla N + j\omega\epsilon N^T \cdot N \right) d\Omega \tag{9.35}$$

$$[T] = \sum_{\text{all boundary elements}} \int_{\Gamma_e} N^T \cdot N \, d\Gamma \tag{9.36}$$

$$\{F\} = -[\sum_{\text{all elements}} \int_{\Omega_e} N^T \cdot N \, d\Omega\,] \cdot \{J_e\} \tag{9.37}$$

The matrices $[S]$, $[T]$, and $\{F\}$ are obtained using the same procedures as in the TM case.

9.3 BOUNDARY ELEMENT FORMULATION

As we have seen in Chapter 7, the boundary integral or boundary element method is a technique for solving a partial differential equation using Green's theorem and a free space Green's function for the homogeneous unbounded region. There are various integral-type methods, depending on the formulation and the approximation procedures.

In 1963, Jawson and Symm [93][94] developed a technique for the numerical solution of singular integral equations in two-dimensional potential problems. In 1977, Jeng and Wexler [95] reported on a boundary element method employing parametric representation of surface/contour and source and an automatic algorithm to address Green's function singularities over arbitrarily shaped geometries. The Galerkin method was used to approximate the boundary integral equation.

In 1979, Daffe and Olson [96] made use of the BEM based on Green's theorem for axisymmetric electrostatic field problems to solve Laplace's equation. The BEM was also approximated by the point-matching method with constant elements. In 1982, Schneider [92] presented a boundary integral equation for the single-component magnetic vector potential of infinite domain problems in two dimensions. The BEM was approximated by the point-matching method. In 1984, Kagami and Fukai [97] applied the BEM to electromagnetic waveguide discontinuities, multimedia problems, and scattering problems. They formulated the scalar Helmholtz wave equation with a Hankel function of the second kind as the kernel of the integral equation.

In the following procedure, the governing equation is transformed into integral equations on the boundaries of the domains of interest. These equations are discretized into a number of elemental segments over the boundaries. The unknown field variables are assumed to vary as specified local shape functions. Substituting these shape functions into the integral equations, a set of linear equations is obtained in terms of nodal unknown field variables. After enforcing the appropriate interface conditions, a linear system of equations is solved. The main advantages of the BEM for wave applications are that

- The open boundary problem is easily modeled.
- Numerical accuracy is generally greater than that of finite element method.
- The order of the system of equations is less than that of the finite element method.
- The unknowns are only on the boundaries. If additional field values are of interest, they can be calculated.

The main disadvantages of the boundary element method are that

- It is more difficult to model complex geometry.
- It is not easy to represent nonlinear and/or nonhomogeneous materials.
- The coefficient matrix is full and positive definiteness is not assured.

Starting from the vector equivalent of Green's second identity, Stratton's formula [35] is obtained using the following procedure:

$$\int_\Omega (\mathbf{Q} \cdot \nabla \times \nabla \times \mathbf{P} - \mathbf{P} \cdot \nabla \times \nabla \times \mathbf{Q})d\Omega = \oint_\Gamma (\mathbf{P} \times \nabla \times \mathbf{Q} - \mathbf{Q} \times \nabla \times \mathbf{P}) \cdot d\Gamma$$
(9.38)

Let $\mathbf{P} = \mathbf{E}$, $\mathbf{Q} = \phi\mathbf{a}$, where $\mathbf{a}$ is the unit vector and $\phi = e^{ikr}/r$ for the three-dimensional case. We then have

$$\nabla \times \nabla \times \mathbf{Q} = \mathbf{a}k^2\phi + \nabla(\mathbf{a}\phi)$$
(9.39)

$$\nabla \times \nabla \times \mathbf{P} = k^2\mathbf{E} + j\omega\mu\mathbf{J}_e - \nabla \times \mathbf{J}_m$$
(9.40)

Using these equations, the value of $\mathbf{E}$ at any interior point of Ω in Figure 9.2 is obtained as [98]

$$\mathbf{E}(x', y', z') = \frac{1}{4\pi} \int_\Omega \left(j\omega\mu\mathbf{J}_e\phi - \mathbf{J}_m \times \nabla\phi + \frac{1}{\epsilon}\rho\nabla\phi\right) d\Omega$$

$$- \frac{1}{4\pi} \oint_\Gamma [j\omega\mu(\mathbf{n} \times \mathbf{H})\phi + (\mathbf{n} \times \mathbf{E}) \times \nabla\phi + (\mathbf{n} \cdot \mathbf{E})\nabla\phi] d\Gamma$$
(9.41)

If the region Ω contains no charges or current within its interior or on its boundary Γ, the electric field at an interior point is

$$\mathbf{E}(x', y', z') = -\frac{1}{4\pi} \oint_\Gamma \{j\omega\mu(\mathbf{n} \times \mathbf{H})\phi + (\mathbf{n} \times \mathbf{E}) \times \nabla\phi + (\mathbf{n} \cdot \mathbf{E})\nabla\phi\} d\Gamma \quad (9.42)$$

For magnetic field problems

$$\mathbf{H}(x', y', z') = \frac{1}{4\pi} \oint_\Gamma \{j\omega\epsilon(\mathbf{n} \times \mathbf{E})\phi + (\mathbf{n} \times \mathbf{H}) \times \nabla\phi + (\mathbf{n} \cdot \mathbf{H})\nabla\phi\} d\Gamma \quad (9.43)$$

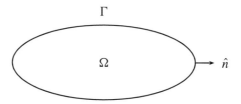

Figure 9.2 Boundary Element Region and Surface

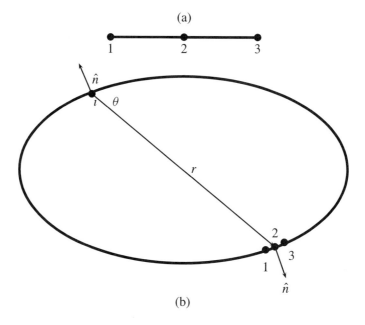

Figure 9.3 (a) Second-Order Line Element, (b) Boundary Element Integration

High-Frequency Problems with Finite Elements

Because **H** is assumed to have only a z component for the TM case, the last term of equation (9.43) goes to zero. Hence this equation becomes

$$H_i = \oint_\Gamma \left(j\omega \epsilon G \cdot E_t - \frac{\partial G}{\partial n} \cdot H \right) d\Gamma \tag{9.44}$$

where G is the corresponding Green's function for the 2D scalar wave equation

$$G = \frac{j}{4} H_0^{(2)}(k\mathbf{r}) \tag{9.45}$$

$H_0^{(2)}$ is the Hankel function of the second kind and of zero order, and $k = \omega/c$ is the wave number in free space. In equation (9.44), the subscript i refers to an arbitrary point in the region Ω, and $\frac{\partial G}{\partial n}$ is the outward normal derivative of G on the boundary. When the point i is on the boundary Γ, a singularity occurs in the Green's function as $r \to 0$. This case needs special consideration. As demonstrated in Chapter 7, we integrate around a circular arc centered at the singular point and allow the radius of the arc to shrink to zero.

Using this integration path, equation (9.44) is rewritten as follows:

$$\begin{aligned} H_i \; + \; &\lim_{\epsilon \to 0} \int_{\Gamma_1} \frac{\partial G}{\partial n} \cdot H \, d\Gamma + \lim_{\epsilon \to 0} \int_{\Gamma_2} \frac{\partial G}{\partial n} \cdot H \, d\Gamma \\ = \; &j\omega\epsilon \left[\lim_{\epsilon \to 0} \int_{\Gamma_1} G \cdot E_t \, d\Gamma + \lim_{\epsilon \to 0} \int_{\Gamma_2} G \cdot E_t \, d\Gamma \right] \end{aligned} \tag{9.46}$$

The integration over the boundary Γ_2 is estimated as follows:

$$\begin{aligned} \lim_{\epsilon \to 0} \int_{\Gamma_2} \frac{\partial G}{\partial n} \cdot H \, d\Gamma &= \lim_{\epsilon \to 0} \int_{\Gamma_2} \frac{j}{4} k H_1^{(2)}(k\epsilon) \cdot H \, d\Gamma \\ &= \lim_{\epsilon \to 0} \frac{j}{4} k H_1^{(2)}(k\epsilon) \epsilon \theta_i \cdot H_i \\ &= \frac{j}{4} k \theta_i \lim_{\epsilon \to 0} \epsilon \left(\frac{k\epsilon}{2} - j\left(-\frac{2}{\pi} \cdot \frac{1}{k\epsilon}\right) \right) \cdot H_i \\ &= -\frac{\theta_i}{2\pi} H_i \end{aligned} \tag{9.47}$$

$$\begin{aligned}
\lim_{\epsilon \to 0} \int_{\Gamma_2} G\, E_t d\Gamma &= \lim_{\epsilon \to 0} \int_{\Gamma_2} \left(\frac{j}{4} H_0^{(2)}(k\epsilon) \right) \cdot E_t d\Gamma \\
&= \lim_{\epsilon \to 0} \left[-\frac{j}{4}\left(1 - j\frac{2}{\pi}(\ln k\epsilon + \gamma - \ln 2)\right) \epsilon\theta \cdot E_{t_i} \right] \\
&= 0
\end{aligned} \qquad (9.48)$$

where $\gamma = 0.5772\ldots$ is Euler's number.

From equations (9.46), (9.47), and (9.48), the following equation is derived:

$$\alpha_i H_i + \lim_{\epsilon \to 0} \int_{\Gamma_1} \frac{\partial G}{\partial n} \cdot H d\Gamma = j\omega\epsilon \left[\lim_{\epsilon \to 0} \int_{\Gamma_1} G \cdot E_t d\Gamma \right] \qquad (9.49)$$

where

$$\alpha_i = 1 - \frac{\theta_i}{2\pi} \qquad (9.50)$$

Equation (9.49) can be rewritten as

$$\alpha_i H_i + \int_\Gamma \left(\frac{\partial G}{\partial n} \cdot H - j\omega\epsilon G \cdot E_t \right) d\Gamma = 0 \qquad (9.51)$$

where

- α_i is the integrated value of the singularity as given by equation (9.50).
- θ_i is the interior angle at node i.
- $\frac{\partial G}{\partial n}$ is the outward normal derivative of G

$$\frac{\partial G}{\partial n} = -\frac{jk}{4} H_1^{(2)}(k\mathbf{r}) \cos\beta \qquad (9.52)$$

- $H_1^{(2)}$ is the Hankel function of the second kind, order one.
- $\cos \beta$ is the direction cosine.

The surface integral of equation (9.51) is evaluated approximately using the shape functions.

$$\alpha_i H_i - \sum_{e=1}^{N}\left[\int_{\Gamma_e} \frac{\partial G}{\partial n} \cdot N(x,y) d\Gamma\right] \cdot \{H\}$$

$$= \sum_{e=1}^{N} j\omega\epsilon \left[\int_{\Gamma_e} G \cdot N(x,y) d\Gamma\right] \cdot \{E_t\} \qquad (9.53)$$

The assembled global system matrix of equation (9.53) can be obtained in the form

$$[U]\{H\} = [V]\{E_t\} \qquad (9.54)$$

where

$$U_{ik} = \begin{cases} \alpha_i & \text{if } i = k \\ \sum_{g=1}^{\text{Gauss points}} \frac{\partial G}{\partial n} \cdot N_k \cdot w_g \cos \beta_g & \text{otherwise} \end{cases} \qquad (9.55)$$

and

$$V_{ik} = j\omega\epsilon \sum_{g=1}^{\text{Gauss points}} G \cdot N_k \cdot w_g \qquad (9.56)$$

where w_g is a weight associated with the integration point, g, and β_g is the direction cosine.

TE Polarization

For TE polarization, we again consider the region shown in Figure 9.2. Because **E** has only a z component, the last term of equation (9.42) goes to zero. This equation becomes

$$E_i = \oint_\Gamma \left(j\omega\mu G \cdot H_t - \frac{\partial G}{\partial n} \cdot H \right) d\Gamma \tag{9.57}$$

where G is the corresponding Green's function for the 2D scalar wave equation. Using the same integration path as for the TM case, equation (9.57) becomes

$$\begin{aligned} E_i &+ \lim_{\epsilon \to 0} \int_{\Gamma_1} \frac{\partial G}{\partial n} \cdot E \, d\Gamma + \lim_{\epsilon \to 0} \int_{\Gamma_2} \frac{\partial G}{\partial n} \cdot E \, d\Gamma \\ &= j\omega\epsilon \left[\lim_{\epsilon \to 0} \int_{\Gamma_1} G \cdot H_t d\Gamma + \lim_{\epsilon \to 0} \int_{\Gamma_2} G \cdot H_t d\Gamma \right] \end{aligned} \tag{9.58}$$

Equation (9.58) then becomes

$$\alpha_i E_i + \int_\Gamma \left(\frac{\partial G}{\partial n} \cdot E - j\omega\mu G \cdot H_t \right) d\Gamma = 0 \tag{9.59}$$

In equation (9.60), the surface integral of equation (9.59) is evaluated approximately using the shape functions.

$$\begin{aligned} \alpha_i E_i &- \sum_{e=1}^N \left[\int_{\Gamma_e} \frac{\partial G}{\partial n} \cdot N(x, y) \, d\Gamma \right] \cdot \{E\} \\ &= \sum_{e=1}^N j\omega\mu \left[\int_{\Gamma_e} G \cdot N(x, y) \, d\Gamma \right] \cdot \{H_t\} \end{aligned} \tag{9.60}$$

The global system matrix of equation (9.60) can be rewritten in the familiar form of

$$[U]\{E\} - [V]\{H_t\} = \{0\} \tag{9.61}$$

where

$$U_{ik} = \begin{cases} \alpha_i & \text{if } i = k \\ \sum_{g=1}^{\text{Gauss points}} \frac{\partial G}{\partial n} \cdot N_k \cdot w_g \cos \beta_g & \text{otherwise} \end{cases} \tag{9.62}$$

and

$$V_{ik} = j\omega\mu \sum_{g=1}^{\text{Gauss points}} G \cdot N_k \cdot w_g \tag{9.63}$$

9.4 IMPLEMENTATION OF THE HYBRID METHOD (HEM)

As we discussed, both the FEM and BEM have distinct advantages and disadvantages. Fortunately, a particular disadvantage of one method is complemented by a corresponding advantage of the other. Finite elements are used to represent the geometry, material properties, and the region containing the source structure. The boundary elements are used to represent the exterior boundary region and to evaluate field variables in the exterior region. The advantage of the HEM lies in the fact that it takes advantages of both methods. However, the assembled system matrix is nonsymmetric, so a special technique is needed to solve it effectively. The finite element equations (9.19),

$$[S]\{H\} - [T]\{E_t\} = \{F\} \tag{9.64}$$

for TM polarization are rearranged by partitioning the matrix:

$$\begin{pmatrix} S_{ii} & S_{ib} \\ S_{bi} & S_{bb} \end{pmatrix} \begin{pmatrix} H_i \\ H_b \end{pmatrix} + \begin{pmatrix} 0 & 0 \\ 0 & T \end{pmatrix} \begin{pmatrix} 0 \\ E_t \end{pmatrix} = \begin{pmatrix} F \\ 0 \end{pmatrix} \quad (9.65)$$

where

- The matrix $[S_{ii}]$ is the stiffness matrix for interior nodes. Its order is the same as the number of interior nodes.
- The matrix $[S_{bb}]$ is the stiffness matrix for boundary nodes. Its order is the number of the boundary nodes.
- The matrices $[S_{ib}]$ and $[S_{bi}]$ represent the coupling between the two regions.
- The subscripts i and b represent the interior and boundary nodes, respectively.
- The term $\{F\}$ describes the forcing functions in the finite element region.

The corresponding boundary element equation, equation (9.54), is

$$[U]\{H_b\} = [V]\{E_t\} \quad (9.66)$$

This equation contains two unknown quantities, the magnetic field and its normal derivative on the surface. In the FE regions, there is only one interior unknown, the magnetic field H. The boundary surface has two unknowns, H and E_t. Assuming the materials are the same, when the finite element and boundary element regions are coupled, the magnetic field and its normal derivative must be continuous across the boundary. We therefore use a free space layer between interior objects and the exterior boundary surfaces as shown in Figure 9.4.

With a free space interface, therefore, the finite element and boundary element equations can be directly combined by adding the two sets of equations (9.65) and (9.66):

$$\begin{Bmatrix} S_{ii} & S_{ib} & 0 \\ S_{bi} & S_{bb} & T \\ 0 & U & V \end{Bmatrix} \begin{pmatrix} H_i \\ H_b \\ E_t \end{pmatrix} = \begin{pmatrix} F \\ 0 \\ 0 \end{pmatrix} \quad (9.67)$$

There are three ways to solve the system matrix equation (9.67). In this equation there are two unknowns (the magnetic field H and the normal derivative of H). The first

High-Frequency Problems with Finite Elements

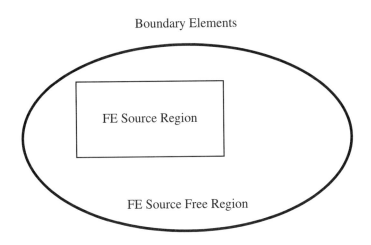

Figure 9.4 Domain of the Hybrid Element Method

method is to solve equation (9.67) directly in terms of H and E_t without modifying the matrix, which is complex, nonsymmetric, and sparse.

The second method is to eliminate one of the unknowns and to solve the reduced system matrix. In this case, the normal derivative, E_t, can be described in terms of the magnetic field, H, from equation (9.66),

$$\{E_t\} = [V]^{-1}[U]\{H\} \qquad (9.68)$$

Substitution of equation (9.68) into equation (9.19) leads to

$$[S]\{H\} - [T][V]^{-1}[U]\{H\} = \{F\} \qquad (9.69)$$

or

$$[A]\{H\} = \{F\} \qquad (9.70)$$

where

$$[A] = [S] - [T][V]^{-1}[U] \qquad (9.71)$$

The only unknown quantity is H, but the inversion of matrix $[V]$ and a matrix multiplication are required. The system matrix is still nonsymmetric and sparse.

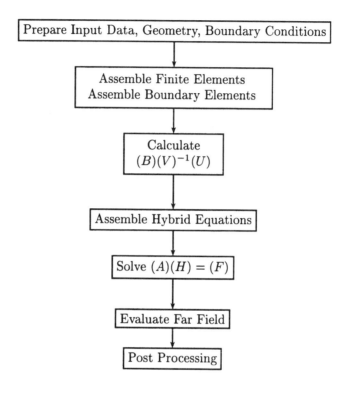

Figure 9.5 Flowchart of the Hybrid Program

A third method is to symmetrize equation (9.71) by using the average contribution of the boundary element matrix, as in the following equation.

$$[A] = [S] - \frac{1}{2}\{[T][V]^{-1}[U] + [T][V]^{-1}[U]\} \qquad (9.72)$$

This is, of course, an approximation, but one which has been found acceptable [98]. A flowchart of the hybrid program is shown in Figure 9.5.

9.5 EVALUATION OF THE FAR-FIELD

In order to check the numerical results, we present here two analytical solutions for radiation from a parallel-plate waveguide. One is an exact solution calculated by the Wiener-Hopf method and the other is a solution obtained by the geometrical theory of diffraction. These analytical methods are limited to simple geometries. A numerical method for calculating the far-field is formulated using the pre-calculated boundary values which come from the HEM program. This numerical result will be compared with analytical solutions for the far-field pattern of a parallel-plate waveguide.

9.5.1 Analytical Solutions

Wiener–Hopf Method

Collin and Zucker [99] introduced a Wiener–Hopf method to solve radiation problems of open waveguides. They assumed that

- The waveguide is semi-infinitely long.
- The waveguide has infinitely thin walls.
- The cross sections coincide with a single constant-coordinate curve.
- The waveguide has infinite conductivity.

Starting from the well-known Helmholtz wave equation, a general solution is obtained by enforcing boundary conditions. Using the Fourier transform method and an asymptotic expansion, the normalized radiation pattern, equation (9.73), is obtained for TE mode propagation.

$$E_y = e^{-(k_0 a/2)} \left(\frac{1+\sin\theta}{2}\right)^{1/2} \left(\frac{\cos(k_0 a \cos\theta)}{1-(\frac{4a}{\lambda_0})^2 \cos^2\theta}\right)^{1/2} \qquad (9.73)$$

The far-zone normalized electric field pattern is given by

$$E_y = \frac{\sin\theta \cos(k_0 a \cos\theta)}{1-(\frac{4a}{\lambda_0^2}) \cos^2\theta} \qquad (9.74)$$

Figure 9.6 shows the normalized far-field pattern based on equation (9.74). On increasing the frequency, the beam width becomes narrow as expected.

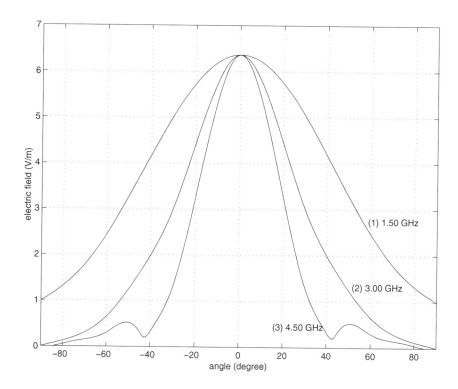

Figure 9.6 Normalized Far-Field Patterns of the Closed-Form Solutions as a Function of Frequency

Solutions Using the Geometrical Theory of Diffraction

One of the simplest problems using the method of the geometrical theory of diffraction (GTD) is evaluating the radiated field from the open-ended parallel plate waveguide. Initially, each plate is considered as an isolated half-plane as shown in Figure 9.7.

When a plane wave is normally incident upon a half-plane, two field components have to be considered. One is the geometrical optics field, and the other is the diffraction field, which is the part of the scattered field. The total field is a combination of these two components; the geometrical optics fields and the edge diffracted fields from the upper and lower plate. For the TEM case, if the field point $P(\rho, \phi)$ is taken to be in the far field, then only the edge-diffracted field components will contribute. Taking the

High-Frequency Problems with Finite Elements 473

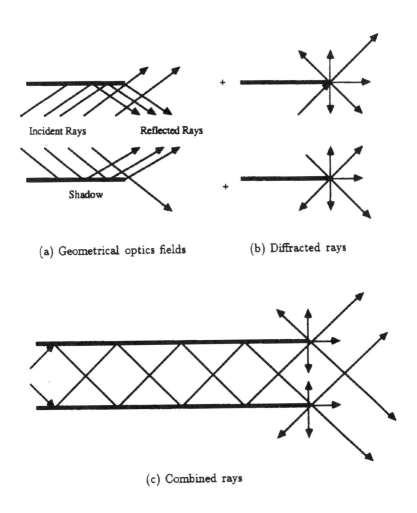

Figure 9.7 Interpretation of the Geometrical Theory of Diffraction (after Shin [98]).

asymptotic form and adding the two fields from each edge, the electric radiation field, E_ϕ, is obtained in the forward region as

$$E_\phi = 4jka E_{in} \cos\frac{\phi}{2} \frac{\sin u}{u} \frac{e^{-jk\rho}}{\sqrt{8j\pi k\rho}} \tag{9.75}$$

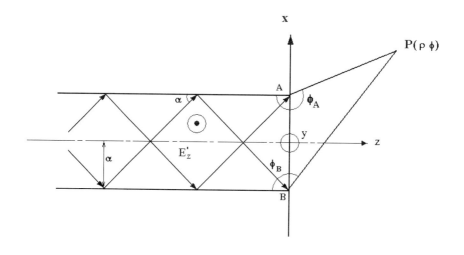

Figure 9.8 TE_1 Mode Propagation in Parallel Plate Waveguide

If the parallel plate waveguide is supporting the TE_1 mode, then this mode can be decomposed into two plane waves propagating between the two plates as shown in Figure 9.8. Adding these two diffracted far fields by the two incident waves, the radiation field, E_y, for $\frac{-\pi}{2} < \phi < \frac{\pi}{2}$ is

$$E_y = 8E_{in} \frac{\cos u \cos \frac{1}{2}\phi \sin \frac{1}{2}\alpha}{\cos \alpha - \cos \phi} \frac{e^{-jk\rho}}{\sqrt{8j\pi k\rho}} \qquad (9.76)$$

Figure 9.9 shows a far-field pattern for the TE_1 mode as a function of frequency. On increasing the frequency, the beam width becomes narrower as in the case of the exact solution.

High-Frequency Problems with Finite Elements

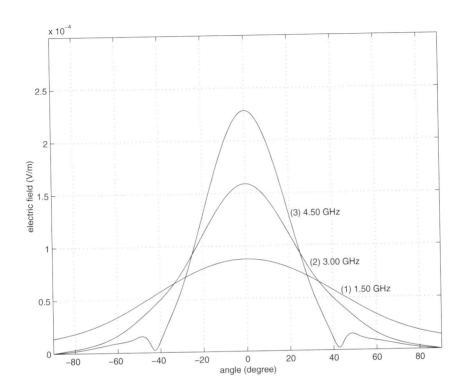

Figure 9.9 Far-Field Patterns of GTD Solutions as a Function of Frequency.

9.5.2 Hybrid Element Method Solutions

The far field, H, will be evaluated by using the calculated boundary values (H_z and E_t):

$$H = \frac{1}{2\pi} \int_\Gamma \left(j\omega\epsilon G \cdot E_t - \frac{\partial G}{\partial n} \cdot H \right) d\Gamma \qquad (9.77)$$

where

- H is the magnetic field at an observation point (far field).
- H is the calculated magnetic field on the boundary.

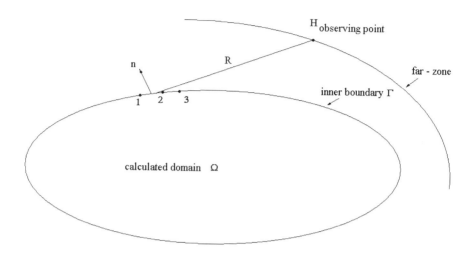

Figure 9.10 Evaluation of the Far Field

- E_t is the tangential component of electric field on the boundary.
- G is the Green's function.
- $\frac{\partial G}{\partial n}$ is the outward normal derivative of the Green's function.

The surface integral of equation (9.77) is approximated by using the second-order interpolation function and ten-point Gauss integration.

$$H = \sum_{e=1}^{n} \frac{1}{2\pi} \left(j\omega\epsilon\, N \cdot E_t \cdot G_k \cdot w_k \cdot \frac{l}{2} - N \cdot H \frac{dG}{dr} \cdot \frac{l}{2} \cdot \cos\beta \right) \quad (9.78)$$

where

- N is the second-order interpolation function
- w_k is the weighting function at a Gauss integration point
- $\cos\beta$ is the direction cosine of $\frac{\partial G}{\partial n}$.

High-Frequency Problems with Finite Elements 477

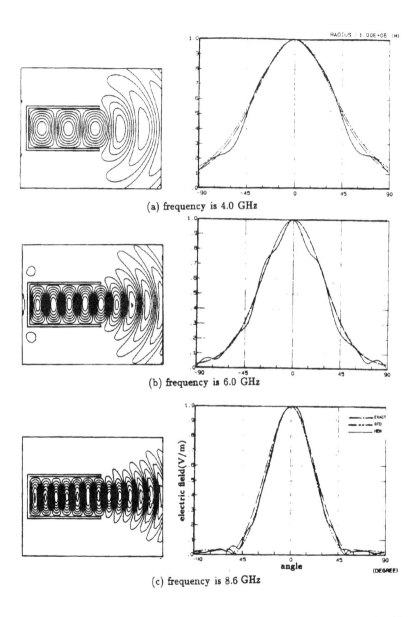

Figure 9.11 Near-Field and the Corresponding Far-Field Patterns of the TE_1 Mode with Analytical Solutions (after Shin[98]).

In order to verify the solution of the hybrid element method, the TE_1 mode is calculated as a function of frequency. As shown in Figure 9.11, the contour lines are constant electric field lines and the graphs on the right side show the corresponding far-field patterns of the exact solution, GTD solution, and HEM solution, respectively. The two methods have different assumptions and different approximating procedures. For instance,

1. The waveguide is infinitely thin in the analytical methods and finite (1.0 cm in this example) in the HEM analysis.
2. The waveguide is semi-infinitely long in the analytical methods and finite (13 cm in this example) in the HEM analysis.
3. The approximating procedures use the Fourier transform method and an asymptotic expansion in the analytical solutions while the FEM and boundary integral equations are used in the HEM analysis.

Even though the analytical (closed form and GTD) and numerical (HEM) methods are quite different in approach, the results obtained from both are in close agreement.

The Poynting vector in the far-field zone can be calculated using H as follows.

$$\mathbf{P} = \mathbf{E} \times \mathbf{H}^* = E_y H_z^* \mathbf{x} - E_x H_z^* \mathbf{y} \tag{9.79}$$

where

$$E_x \propto \frac{\partial H_z}{\partial y} \tag{9.80}$$

$$E_y \propto \frac{\partial H_z}{\partial x} \tag{9.81}$$

As the radius of the observation point, R, goes to infinity, the Hankel function has the asymptotic behavior

$$H_0^{(2)}(kR) \approx \sqrt{\frac{2i}{\pi k r_q}} e^{-jkR} \tag{9.82}$$

High-Frequency Problems with Finite Elements

where $R = r_q - r_m$ and $r_q \gg r_m$. Equation (9.82) becomes

$$H_0^{(2)}(k(r_q - r_m)) \approx \sqrt{\frac{2j}{\pi k r_q}} e^{-jkr_q} e^{jkr_m} \tag{9.83}$$

When R goes to infinity,

$$H_1^{(2)}(kR) = j H_0^{(2)}(kR) \tag{9.84}$$

The boundary integral equation becomes

$$H = \frac{1}{2\pi} \int_{S_m} \left(\frac{k}{4} H_m \frac{dr}{dn} - \frac{j}{4} \frac{\partial H_m}{\partial n} \sqrt{\frac{2j}{\pi k r_q}} \right) e^{-jk\, r_q} e^{jk\, r_m}\, dS \tag{9.85}$$

The $e^{-jk\, r_q}$ is a constant phase term that can be eliminated. Equation (9.85) can be used to evaluate the far-field pattern.

In a far zone of free space, **E** can be evaluated using the characteristic impedance, η, as shown in the following equation

$$\mathbf{E} = \eta\, \mathbf{H} \tag{9.86}$$

In addition, the *directive gain* of each case will be evaluated using far-field values. A property of an antenna indicating how effectively its radiation pattern concentrates its power in a desired direction is known as its directive gain. The directive gain of a given antenna is defined as the ratio of the power density, p_{av}, radiated in the desired direction (r, θ) at some distance r, to the total power (P_{av}) radiated by the antenna averaged over the surrounding sphere of area $4\pi r^2$:

$$D(r, \theta) = \frac{p_{av}(r, \theta)}{P_{av}} \tag{9.87}$$

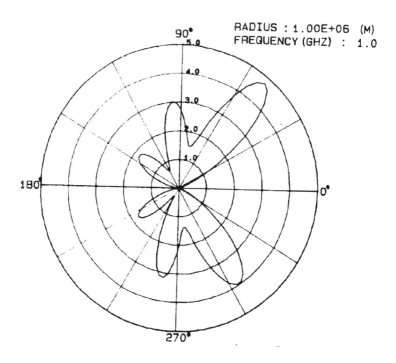

Figure 9.12 Polar Plot of a Far-Field Pattern for the TM_1 Mode at 1.414 GHz (after Shin [98]).

where

$$P_{\text{av}} = \frac{1}{4\pi r^2} \oint_s p_{\text{av}} \cdot ds. \tag{9.88}$$

Figure 9.12 shows a polar plot of the far-field pattern of the parallel plate waveguide at a frequency of 1.414 GHz. The observation point is 1000 km away. The plot shows the magnitude of the Poynting vector. The pattern is symmetric with respect to the z-axis and the main lobes are apparent.

9.6 SCATTERING PROBLEMS

There is an interest in solving electromagnetic scattering problems by numerical methods because analytical methods can only be used for a few simple shapes such as cylinders, wedges, half-planes, and strip lines. Also, as we have seen, the numerical methods can be used to solve scattering problems with complex geometries and material properties via direct treatment of Maxwell's equations. In this section, the hybrid method and the finite element method with absorbing boundary conditions are employed to solve scattering problems with various shapes. The numerical results are compared with those of analytical solutions where possible.

9.6.1 Formulation Using the Hybrid Element Method

TM Polarization

The hybrid finite element–boundary element method described earlier can be applied to solve a two-dimensional electromagnetic scattering problem (see Figure 9.13) by adding an incident wave, H_{in}, to the boundary integral equation, as follows. For the TM case,

$$H_{in} + \alpha_i H_i + \int_\Gamma \left(\frac{\partial G}{\partial n} \cdot H - j\omega\epsilon G \cdot E_t \right) d\Gamma = 0 \quad (9.89)$$

Using a second-order interpolation function and Gauss integration, the matrix form of equation (9.89) becomes

$$[U]\{H\} - [V]\{E_t\} = \{H_{in}\} \quad (9.90)$$

where the matrices $[U]$ and $[V]$ have been defined previously. Equation (9.90) is solved for E_t as

$$\{E_t\} = [V]^{-1}[U]\{H\} - [V]^{-1}\{H_{in}\} \quad (9.91)$$

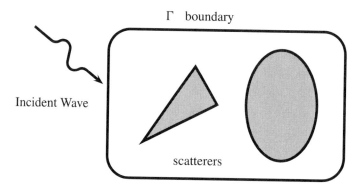

Figure 9.13 Scattering Problem

The second term on the right-hand side of equation (9.91) will be the forcing function of the hybrid system matrix, which is combined with the finite element equations,

$$[S]\{H\} - [T]\{E_t\} = \{F\}$$

as follows: The FE equation is

$$\begin{pmatrix} S_{ii} & S_{ib} \\ S_{bi} & S_{bb} \end{pmatrix} \begin{pmatrix} H_i \\ H_b \end{pmatrix} + \begin{pmatrix} 0 & 0 \\ 0 & T \end{pmatrix} \begin{pmatrix} 0 \\ E_t \end{pmatrix} = \begin{pmatrix} 0 \\ 0 \end{pmatrix} \qquad (9.92)$$

where the $[S]$ and $[T]$ matrices have been defined in the previous section. The combined matrix equation is:

$$\begin{pmatrix} S_{ii} & S_{ib} \\ S_{bi} & S_{bb}^c \end{pmatrix} \begin{pmatrix} H_i \\ H_b \end{pmatrix} = \begin{pmatrix} 0 \\ F^c \end{pmatrix} \qquad (9.93)$$

where

$$[S_{bb}^c] = [S_{bb}] - [T][V]^{-1}[U] \qquad (9.94)$$
$$\{F^c\} = [T][V]^{-1}\{H_{\text{in}}\} \qquad (9.95)$$

High-Frequency Problems with Finite Elements

As a source on the boundary, an incident plane wave, H_{in}, is described as,

$$H_{in} = H_0\, e^{-j(k_z z + k_x x)} \tag{9.96}$$

where $k_z = k \cos \psi_0$, $k_x = k \sin \psi_0$, and ψ_0 is the incident angle to the normal. In case of normal incidence, the incident wave, H_{in}, is dependent only on z,

$$H_{in} = H_0 e^{-jkz} \tag{9.97}$$

Using equation (9.96) or (9.97), the incident field on the boundary nodes is calculated. Then the forcing function $\{F^c\}$ can be obtained using equation (9.95).

TE Polarization

As described earlier, using the hybrid finite element–boundary element method, electromagnetic scattering problems for TE polarization can also be formulated by adding an incident wave, E_{in}, to the boundary integral equation as follows

$$E_{in} + \alpha_i E_i + \int_\Gamma \left(\frac{\partial G}{\partial n} \cdot H - j\omega\mu G \cdot E_t \right) d\Gamma = 0 \tag{9.98}$$

where

$$E_{in} = E_0 e^{-j(k_x x + k_z z)} \tag{9.99}$$

Using the second-order interpolation function and Gauss integration, the matrix form of equation (9.98) becomes

$$[U]\{E\} - [V]\{H_t\} = \{E_{in}\} \tag{9.100}$$

Equation (9.100) is solved for H_t as

$$\{H_t\} = [V]^{-1}[U]\{E\} - [V]^{-1}\{E_{in}\} \tag{9.101}$$

The second term on the right-hand side of equation (9.101) will be the forcing function of the hybrid system matrix, which is combined with the FE equations as was done for the TM polarization case. The FE equation is

$$\begin{pmatrix} S_{ii} & S_{ib} \\ S_{bi} & S_{bb} \end{pmatrix} \begin{pmatrix} E_i \\ E_b \end{pmatrix} + \begin{pmatrix} 0 & 0 \\ 0 & T \end{pmatrix} \begin{pmatrix} 0 \\ H_t \end{pmatrix} = \begin{pmatrix} 0 \\ 0 \end{pmatrix} \quad (9.102)$$

The combined matrix equation is

$$\begin{pmatrix} S_{ii} & S_{ib} \\ S_{bi} & S_{bb}^c \end{pmatrix} \begin{pmatrix} E_i \\ E_b \end{pmatrix} = \begin{pmatrix} 0 \\ F^c \end{pmatrix} \quad (9.103)$$

where

$$[S_{bb}^c] = [S_{bb}] - [T][V]^{-1}[U] \quad (9.104)$$
$$\{F^c\} = [T][V]^{-1}\{E_{\text{in}}\} \quad (9.105)$$

9.6.2 Finite Element Formulation with Absorbing Boundary Conditions

Using the second-order Engquist–Majda absorbing boundary condition, a finite element formulation for scattering problems will now be derived. The total field satisfies the Helmholtz equation

$$\nabla \cdot \frac{1}{j\omega\epsilon} \nabla \Phi + j\omega\mu \Phi = 0 \quad (9.106)$$

Multiplying equation (9.106) by a weighting function, W and integrating by parts over the boundary Γ, a weak form of the two-dimensional Galerkin weighted residual method for equation (9.106) is obtained as

$$\int_\Omega \left(\nabla W \cdot \frac{1}{j\omega\epsilon} \nabla \Phi^t + j\omega\mu W \cdot \Phi^t \right) d\Omega - \frac{1}{j\omega\epsilon} \oint_\Gamma W \cdot \frac{\partial \Phi^t}{\partial n} d\Gamma = 0 \quad (9.107)$$

High-Frequency Problems with Finite Elements

The second term on the right-hand side of equation (9.107) can be replaced with the absorbing boundary condition [100].

$$\frac{\partial \Phi^s}{\partial n} = -jk\Phi^s - \frac{j}{2k}\frac{\partial^2 \Phi^s}{\partial \tau^2} \tag{9.108}$$

In order to apply the Galerkin method, equation (9.108) is multiplied by the weighting function, W.

$$W\frac{\partial \Phi^s}{\partial n} = -W\left(jk\Phi^s + \frac{j}{2k}\frac{\partial^2 \Phi^s}{\partial \tau^2}\right) \tag{9.109}$$

We then integrate equation (9.109) over the boundary Γ.

$$\oint_\Gamma W \frac{\partial \Phi}{\partial n} d\Gamma = -\oint_\Gamma W \left(jk\Phi + \frac{j}{2k}\frac{\partial^2 \Phi}{\partial \tau^2}\right) d\Gamma \tag{9.110}$$

In order to solve scattering problems, both the scattered field, Φ^s, and total field, Φ^t, have to be considered.

$$\Phi^s = \Phi^t - \Phi^{in} \tag{9.111}$$

where Φ^{in} is the incident wave

$$\Phi^{in} = \Phi_0 e^{-j(k_x x + k_z z)} \tag{9.112}$$

Equation (9.110) becomes

$$\oint_\Gamma W \left(\frac{\partial \Phi^s}{\partial n} + \frac{\partial \Phi^{in}}{\partial n}\right) d\Gamma =$$
$$-\oint_\Gamma W \left(jk(\Phi^t - \Phi^{in}) + \frac{j}{2k}\frac{\partial^2(\Phi^t - \Phi^{in})}{\partial \tau^2} - \frac{\partial \Phi^{in}}{\partial n}\right) d\Gamma \tag{9.113}$$

Integrating the right-hand side of equation (9.113),

$$-\oint_\Gamma \left(jkW \cdot (\Phi^t - \Phi^{in}) + \frac{j}{2k}\frac{\partial W}{\partial \tau} \cdot \frac{\partial(\Phi^t - \Phi^{in})}{\partial \tau}\right) d\Gamma + \oint_\Gamma W \cdot \frac{\partial \Phi^{in}}{\partial n} d\Gamma \quad (9.114)$$

Equation (9.114) can be replaced by the second term on the right-hand side of equation (9.107) as

$$\int_\Omega \left(\nabla W \cdot \frac{1}{j\omega\epsilon}\nabla\Phi^t + j\omega\mu W \Phi^t\right) d\Omega$$
$$+ \frac{1}{j\omega\epsilon}\oint_\Gamma \left(jkW \cdot (\Phi^t - \Phi^{in}) - \frac{j}{2k}\frac{\partial W}{\partial \tau} \cdot \frac{\partial(\Phi^t - \Phi^{in})}{\partial \tau}\right) d\Gamma$$
$$= \frac{1}{j\omega\epsilon}\oint_\Gamma W \cdot \frac{\partial \Phi^{in}}{\partial n} d\Gamma \quad (9.115)$$

Rearranging equation (9.115) for the TM case ($\Phi^t = H$), the finite element formulation is obtained as

$$\int_\Omega \left(\nabla W \cdot \frac{1}{j\omega\epsilon}\nabla H + j\omega\mu W \cdot H\right) d\Omega$$
$$+ \frac{1}{j\omega\epsilon}\oint_\Gamma \left(jkW \cdot H - \frac{j}{2k}\frac{\partial W}{\partial \tau} \cdot \frac{\partial H}{\partial \tau}\right) d\Gamma$$
$$= \frac{1}{j\omega\epsilon}\oint_\Gamma \left(jkW \cdot H^{in} - \frac{j}{2k}\frac{\partial W}{\partial \tau} \cdot \frac{\partial H^{in}}{\partial \tau} + W \cdot \frac{\partial H^{in}}{\partial n}\right) d\Gamma \quad (9.116)$$

Using a second-order shape function, equation (9.116) becomes the matrix equation

$$[S]\{H\} + [C]\{H_b\} + [D]\{H_b\} = +[C]\{H_b^{in}\} + [D]\{H_b^{in}\} + [B]\left(\frac{\partial H^{in}}{\partial n}\right) \quad (9.117)$$

where the $[S]$, $[C]$ and $[D]$ matrices have been previously been defined and $[B]$ is :

$$[B]\left(\frac{\partial H^{in}}{\partial n}\right) = \frac{k}{\omega\epsilon}\oint_\Gamma W \cdot \frac{\partial H^{in}}{\partial n} d\Gamma$$
$$= \frac{k}{\omega\epsilon}\sum_{\text{elements}}\int_{\Gamma_e} N^T \cdot N d\Gamma \left(\frac{\partial H^{in}}{\partial n}\right) \quad (9.118)$$

High-Frequency Problems with Finite Elements

where $\{H_b^{in}\}$ is the incident H-field on the boundary and $\{\frac{\partial H^{in}}{\partial n}\}$ is the normal component of the incident field on the boundary.

The TE Case Using Absorbing Boundary Conditions

For TE polarization, the finite element formulation becomes

$$\int_\Omega \left(\nabla W \cdot \frac{1}{j\omega\mu} \nabla E + j\omega\epsilon W \cdot E\right) d\Omega$$
$$+ \frac{1}{j\omega\mu} \oint_\Gamma \left(jkW \cdot E - \frac{j}{2k} \frac{\partial W}{\partial \tau} \cdot \frac{\partial E}{\partial \tau}\right) d\Gamma$$
$$= \frac{1}{j\omega\mu} \oint_\Gamma \left(jkW \cdot E^{in} - \frac{j}{2k} \frac{\partial W}{\partial \tau} \cdot \frac{\partial E^{in}}{\partial \tau} + W \cdot \frac{\partial E^{in}}{\partial n}\right) d\Gamma \quad (9.119)$$

Using a second-order shape function, as was done for the TM case, equation (9.119) becomes a set of matrix equations

$$[S]\{E\} + [C]\{E_b\} + [D]\{E_b\} = +[C]\{E_b^{in}\} + [D]\{E_b^{in}\} + [B]\{\frac{\partial E^{in}}{\partial n}\} \quad (9.120)$$

where the $[S]$, $[C]$, and $[D]$ matrices have been previously defined and $[B]$ is

$$[B]\left(\frac{\partial E^{in}}{\partial n}\right) = \frac{k}{\omega\mu} \oint_\Gamma W \cdot \frac{\partial E^{in}}{\partial n} d\Gamma$$
$$= \frac{k}{\omega\mu} \sum_{element} \int_{\Gamma_e} N^T \cdot N d\Gamma \left(\frac{\partial E^{in}}{\partial n}\right) \quad (9.121)$$

where $\{E_b^{in}\}$ represents an incident electric field on the boundary and $\{\frac{\partial E^{in}}{\partial n}\}$ represents the normal component of the incident field on the boundary.

9.7 NUMERICAL EXAMPLES

Several examples of scattering problems will now be used to illustrate the method. Both the hybrid element method and the absorbing boundary condition approach have been used.

9.7.1 Free Space

First a region of free space is solved (i.e., there is no scatterer in the region). Figure 9.14a shows the geometry of the free space window whose size is $3\lambda \times 4\lambda$ at a frequency of 2 GHz. It is assumed that a plane wave is incident normal to the boundary. Figure 9.14b represents constant magnetic field lines that are found by the hybrid element method for the case of TM polarization. In Figure 9.14c, the corresponding fields are solved by the absorbing boundary condition method. The analytic solution in this case is the incident wave itself,

$$H_{in} = \frac{1}{2} e^{-jkx} \tag{9.122}$$

The result found by the absorbing boundary condition method is somewhat more accurate than the result from the hybrid element method. The reason is that the absorbing boundary conditions are very accurate for normal incident waves.

9.7.2 Scattering Cylinder

One of the standard geometries for a 2D scattering problem is a cylinder. Figures 9.15a and b show equifield lines for the total magnetic field for a conducting cylinder in free space whose radius is 0.5λ at the applied frequency of 2 GHz. The window size is $3\lambda \times 4\lambda$. A plane wave is normally incident to the boundary. Figure 9.15a shows equifield lines for the total magnetic field found by the hybrid element method and b shows the corresponding field lines solved by the absorbing boundary condition method. Figure 9.15c is a polar plot of the magnitude of the total field along a circle of radius $1.5\ \lambda$ for each method.

The total fields include the incident field and the scattered field. We will solve directly for the scattered field. The total field, Φ^t, satisfies the Helmholtz equation

$$\nabla^2 \Phi^t + k^2 \Phi^t = 0 \tag{9.123}$$

High-Frequency Problems with Finite Elements

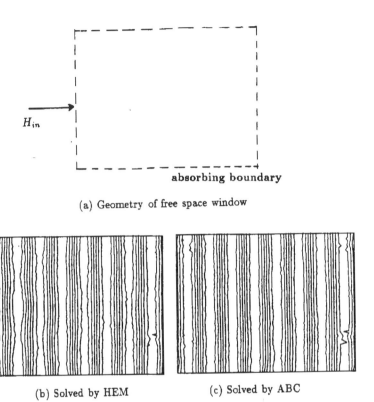

Figure 9.14 Geometry and Free Space Window for Scattering Problems (after Shin [98]).

and the incident field, Φ^{in}, also satisfies the Helmholtz equation

$$\nabla^2 \Phi^{in} + k_0^2 \Phi^{in} = 0 \qquad (9.124)$$

where k_0 is a free space wave number and k is the wave number in any material. The scattered field can be defined in terms of the total and incident fields as

$$\Phi^s = \Phi^t - \Phi^{in} \qquad (9.125)$$

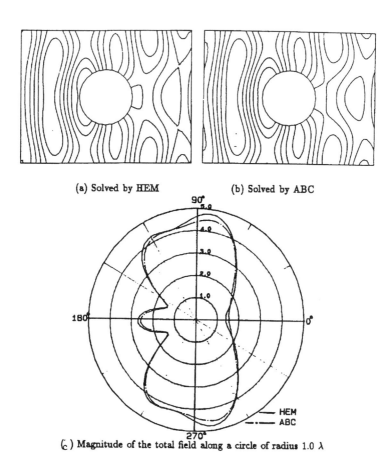

(a) Solved by HEM (b) Solved by ABC

(c) Magnitude of the total field along a circle of radius 1.0 λ

Figure 9.15 Total Near-Field Pattern of a Conducting Cylinder with Incident Plane Wave (after Shin [98]).

Subtracting equation (9.124) from equation (9.123),

$$\nabla^2 \Phi^s + k^2 \Phi^s = (k_0^2 - k^2)\Phi^{\text{in}} \tag{9.126}$$

An alternative is to subtract the known incident field from the calculated total field. Figure 9.16a shows the scattered field obtained using this method at two instants, $\omega t = 0$ and $\omega t = \pi/2$.

High-Frequency Problems with Finite Elements

(a) Scattered field lines at $\omega t = 0$

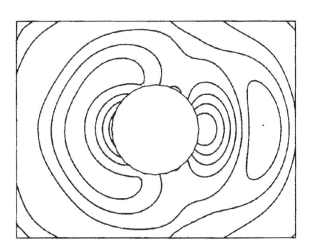

(b) Scattered field lines at $\omega t = \pi/2$

Figure 9.16 Scattered Field for the Conducting Cylinder (after Shin [98]).

9.7.3 Scattering Wedge

Another standard geometry for scattering problems is the wedge-shaped scatterer. A conducting wedge is shown in Figure 9.17a. The size is $2\lambda \times 3\lambda$ and the window size is $5\lambda \times 8\lambda$. A plane wave travels along the positive x-axis and its frequency is 2.0 GHz. Figure 9.17b shows constant magnetic field lines for this case. Figure 9.17c shows constant magnetic field lines for a dielectric wedge with a relative dielectric constant of 2.0 at 2.0 GHz. As expected, the wavelength shortens in the dielectric region.

Another example can be used to compare the calculated result with the analytical solution. Figure 9.18 shows the geometry for a problem with a conducting wedge. The incident field is a plane wave at an angle of 78.75° with respect to the x-axis. The window size is $8\lambda \times 7\lambda$. The length of the wedge is 3.5λ and the closed angle of the wedge is 24.5°. We distinguish the incident region, reflection region, and shadow region. Figure 9.19 shows the magnitude of the total field along a circle whose radius is 3λ and whose origin is the vertex of the wedge on the x-axis. The solid line represents the amplitude of the magnetic field found by the finite element method with absorbing boundary conditions and the broken line represents the corresponding H-field found by the analytical method. An analytical solution is described by Bouman and Senior [101]. For high-frequency applications, a convenient decomposition of the field is

$$H_z = H_z^{g.o.} + H_z^d \tag{9.127}$$

where $H_z^{g.o.}$ and H_z^d are the geometrical optics and diffracted fields, respectively. The geometrical optics fields are

$$H_z^{g.o.} = \sum_{n_1} e^{jk\rho \cos \alpha_{n_1}} + \sum_{n_2} e^{jk\rho \cos \alpha_{n_2}} \tag{9.128}$$

where

$$\alpha_{n_1} = \pi - \varphi + \varphi_0 - 2n_1 \nu \pi \tag{9.129}$$

$$\alpha_{n_2} = \pi - \varphi - \varphi_0 + 2\pi - 2n_1 \nu \pi \tag{9.130}$$

$$\nu \pi = 2\pi - 2\psi, \tag{9.131}$$

and ψ_0 is the incident angle of the plane wave. The summations in equation (9.128) extend over all integers n_1 and n_2 satisfying the inequalities

$$|\varphi - \varphi_0 + 2n_1 \nu \pi| < \pi \tag{9.132}$$

$$|\varphi - \varphi_0 - 2\psi + 2n_1 \nu \pi| < \pi \tag{9.133}$$

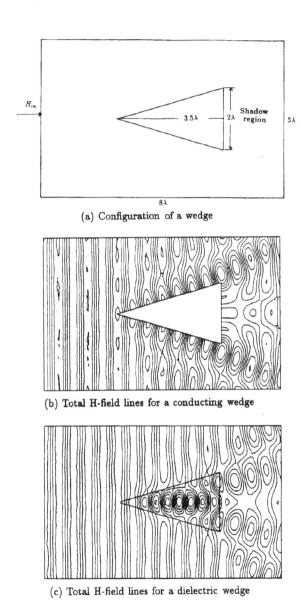

Figure 9.17 Geometry and Constant Magnetic Field Lines for a Wedge (after Shin [98])

and ψ represents the closed angle of the wedge. The diffracted field H_z^d can be approximated using the leading term of the asymptotic expansion,

$$H_z^d = \frac{\exp j(k\rho + \pi/4)}{\sqrt{2\pi k\rho}} \frac{1}{\nu} \sin\frac{\pi}{\nu} \left(\left(\cos\frac{\pi}{\nu} - \cos\frac{\pi - \pi_0}{\nu}\right)^{-1} \right.$$
$$\left. + \left(\cos\frac{\pi}{\nu} + \cos\frac{2\pi - \varphi - \varphi_0}{\nu}\right)^{-1} \right) \quad (9.134)$$

On the geometrical optics boundaries (reflection boundaries from the upper face and the lower face boundary of the geometrical shadow), the total magnetic field has a different asymptotic representation. It is difficult to compare the numerical solutions with the analytical solutions quantitatively, because the two methods have different assumptions and a different solution process. For example, in the analytical solution, the wedge is assumed to be infinitely long, but in the numerical solutions the size of wedge is finite. However, as shown in Figure 9.19, the scattered fields have generally the same pattern in the refraction and shadow regions. Because of the different lengths of the wedge, there is disagreement in the reflection region.

Finally, we show an airplane-shaped scatterer [102]. This problem illustrates the advantages of numerical methods in which complicated boundaries can be easily represented.

9.8 THREE DIMENSIONAL FEM FORMULATION FOR THE ELECTRIC FIELD

We begin with the time harmonic form of Maxwell's equations which are

$$\nabla \times \vec{E} = -j\omega\mu\vec{H} \quad (9.135)$$

$$\nabla \times \vec{H} = \vec{J} + j\omega\epsilon\vec{E} \quad (9.136)$$

High-Frequency Problems with Finite Elements 495

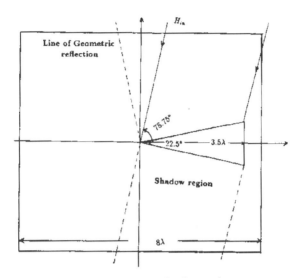

(a) Geometry for a conducting wedge

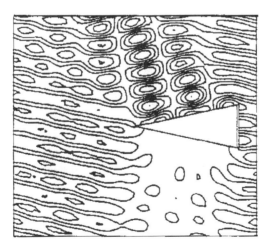

(b) Total H-field pattern of a conducting wedge scatter

Figure 9.18 Geometry and Constant Magnetic Field Pattern with an Incident Wave at 78.75° (after Shin [98])

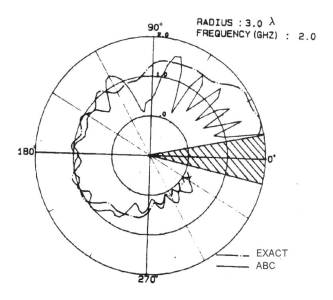

Figure 9.19 Total Field Pattern along a Circle Around the Wedge (after Shin [98])

Figure 9.20 Wave Scattering From an Airplane [102]

High-Frequency Problems with Finite Elements

From these

$$\frac{1}{\mu}\nabla \times \vec{E} = -j\omega\vec{H} \tag{9.137}$$

$$\nabla \times \left(\frac{1}{\mu}\nabla \times \vec{E}\right) = \nabla \times \left(-j\omega\vec{H}\right) = -j\omega\nabla \times \vec{H} \tag{9.138}$$

$$\nabla \times \left(\frac{1}{\mu}\nabla \times \vec{E}\right) = -j\omega\vec{J} + \omega^2\epsilon\vec{E} \tag{9.139}$$

Using

$$\mu = \mu_0\mu_r \tag{9.140}$$

$$\epsilon = \epsilon_0\epsilon_r \tag{9.141}$$

and defining the wave number

$$k_0^2 = \omega^2\epsilon_0\mu_0 \tag{9.142}$$

we have

$$\omega^2 = k_0^2/\epsilon_0\mu_0 \tag{9.143}$$

Substituting

$$\frac{1}{\mu_0}\nabla \times \left(\frac{1}{\mu_r}\nabla \times \vec{E}\right) = -j\omega\frac{\mu_0}{\mu_0}\vec{J} + \frac{k_0^2\epsilon_0\epsilon_r}{\epsilon_0\mu_0}\vec{E} \tag{9.144}$$

we obtain

$$\nabla \times \left(\frac{1}{\mu_r}\nabla \times \vec{E}\right) - k_0^2 \epsilon_r \vec{E} = -j\omega\mu_0 \vec{J} \quad (9.145)$$

Applying Galerkin's method, we multiply by the shape (weighting) function, integrate, and set the resulting integral to zero.

$$\iiint N_i^T \left[\nabla \times \frac{1}{\mu_r}\nabla \times \vec{E} - k_0^2 \epsilon_r \vec{E}\right] dV = -j\omega\mu_0 \iiint N_i^T \vec{J} dV \quad (9.146)$$

Consider the first term.

$$\iiint N_i^T \left[\nabla \times \frac{1}{\mu_r}\nabla \times \vec{E}\right] dV \quad (9.147)$$

We use the vector identity

$$\nabla \times \left(\alpha \vec{F}\right) = (\nabla \alpha) \times \left(\vec{F}\right) + \alpha \left(\nabla \times \vec{F}\right) \quad (9.148)$$

In our case let $\alpha = N_i^T$ and $\vec{F} = \frac{1}{\mu_r}\nabla \times \vec{E}$. This gives

$$\iiint N_i^T \left[\nabla \times \frac{1}{\mu_r}\nabla \times \vec{E}\right] dV = \iiint \nabla \times \left(N_i^T \frac{1}{\mu_r}\nabla \times \vec{E}\right) dV$$
$$- \iiint \nabla N_i^T \times \left(\frac{1}{\mu}\nabla \times \vec{E}\right) dV \quad (9.149)$$

Using

$$\iiint \nabla \times \vec{F} dV = \iint \hat{n} \times \vec{F} dS \quad (9.150)$$

we have for the first term of equation (9.149)

$$\iint \hat{n} \times \left(N_i^T \frac{1}{\mu_r} \nabla \times \vec{E} \right) dS = \iiint \left(\nabla N_i^T \times \frac{1}{\mu_r} \nabla \times \vec{E} \right) dV \quad (9.151)$$

This first term which is evaluated only on the boundary of the problem, will be used in the implementation of the absorbing boundary conditions. The Galerkin form of the equation is now

$$-\iiint \left(\nabla N_i^T \times \frac{1}{\mu_r} \nabla \times \vec{E} \right) dV - \iiint N_i^T k_0^2 \epsilon_r \vec{E} dV$$
$$+ \iint \hat{n} \times \left(N_i^T \frac{1}{\mu_r} \nabla \times \vec{E} \right) dS = -j\omega\mu_0 \iiint N_i^T J dV \quad (9.152)$$

This is the so-called curl–curl form of the equation. It has been shown that with the proper penalty function the spurious modes can be eliminated and further that this is equivalent to a Laplacian formulation. The penalty function (with a unity coefficient) is

$$-\nabla \left(\nabla \cdot \epsilon_r \vec{E} \right) \quad (9.153)$$

This choice is reasonable because $\nabla \cdot \vec{D} = 0$ in free space. In other words, we are constraining the divergence in a weak sense. In Galerkin form

$$-\iiint N_i^T \nabla \left(\nabla \cdot \epsilon_r \vec{E} \right) dV \quad (9.154)$$

Using the identity

$$\nabla (\alpha\beta) = \alpha \nabla \beta + \beta \nabla \alpha \quad (9.155)$$

with

$$\alpha = N_i^T, \quad \beta = \nabla \cdot \epsilon_r \vec{E} \quad (9.156)$$

we get

$$-\iiint N_i^T \nabla \left(\nabla \cdot \epsilon_r \vec{E}\right) dV$$
$$= \iiint \left(\nabla \cdot \epsilon_r \vec{E}\right) \left(\nabla N_i^T\right) dV - \iiint \nabla \left(N_i^T \nabla \cdot \epsilon_r \vec{E}\right) dV \quad (9.157)$$

Using the divergence theorem on the last term, the penalty function becomes

$$\iiint \left(\nabla \cdot \epsilon_r \vec{E}\right) \left(\nabla N_i^T\right) dV - \iint N_i^T \left(\nabla \cdot \epsilon_r E\right) dS \quad (9.158)$$

The entire expression is now

$$-\iiint \left(\nabla N_i^T \times \frac{1}{\mu_r} \nabla \times \vec{E}\right) dV - \iint N_i^T k_0^2 \epsilon_r \vec{E} dV$$
$$+ \iint \hat{n} \times \left[N_i^T \frac{1}{\mu_r} \nabla \times \vec{E}\right] dS + \iiint \left(\nabla \cdot \epsilon_r \vec{E}\right) \left(\nabla N_i^T\right) dV$$
$$- \iint N_i^T \left(\nabla \cdot \epsilon_r \vec{E}\right) dS = -j\omega\mu_0 \iint N_i^T \vec{J} dV \quad (9.159)$$

In three-dimensional Cartesian coordinates

$$\vec{E} = \left\{ \begin{array}{c} E_x \\ E_y \\ E_z \end{array} \right\} \quad (9.160)$$

High-Frequency Problems with Finite Elements

$$\nabla \times \vec{E} = \left\{ \begin{array}{l} \frac{\partial E_z}{\partial y} - \frac{\partial E_y}{\partial z} \\ \frac{\partial E_x}{\partial z} - \frac{\partial E_z}{\partial x} \\ \frac{\partial E_y}{\partial x} - \frac{\partial E_x}{\partial y} \end{array} \right\} \quad (9.161)$$

$$\nabla N_i^T = \left\{ \begin{array}{l} \frac{\partial N_i^T}{\partial x} \\ \frac{\partial N_i^T}{\partial y} \\ \frac{\partial N_i^T}{\partial z} \end{array} \right\} \quad (9.162)$$

Considering the first term in equation (9.159),

$$\nabla N_i^T \times \nabla \times \vec{E} = \left[\begin{array}{l} \frac{\partial N_i^T}{\partial y} \left(\frac{\partial E_y}{\partial x} - \frac{\partial E_x}{\partial y} \right) - \frac{\partial N_i^T}{\partial z} \left(\frac{\partial E_x}{\partial z} - \frac{\partial E_z}{\partial x} \right) \\ \frac{\partial N_i^T}{\partial z} \left(\frac{\partial E_z}{\partial y} - \frac{\partial E_y}{\partial z} \right) - \frac{\partial N_i^T}{\partial x} \left(\frac{\partial E_y}{\partial x} - \frac{\partial E_x}{\partial y} \right) \\ \frac{\partial N_i^T}{\partial x} \left(\frac{\partial E_x}{\partial z} - \frac{\partial E_z}{\partial x} \right) - \frac{\partial N_i^T}{\partial y} \left(\frac{\partial E_z}{\partial y} - \frac{\partial E_y}{\partial z} \right) \end{array} \right] \quad (9.163)$$

Writing $\vec{E}$ in terms of the shape functions,

$$E_x = \sum_i N_i E_{xi}, \quad \text{etc.} \quad (9.164)$$

The derivatives become, for example,

$$\frac{\partial E_x}{\partial y} = \sum_i \frac{\partial N_i}{\partial y} E_{xi}, \quad \text{etc.} \quad (9.165)$$

The first integral of (9.159) is

$$= \int \begin{bmatrix} \left(\frac{\partial N_i^T}{\partial y}\frac{\partial N_i}{\partial y} + \frac{\partial N_i^T}{\partial z}\frac{\partial N_i}{\partial z}\right) & \left(-\frac{\partial N_i^T}{\partial y}\frac{\partial N_i}{\partial x}\right) & \left(-\frac{\partial N_i^T}{\partial z}\frac{\partial N_i}{\partial x}\right) \\ \left(-\frac{\partial N_i^T}{\partial x}\frac{\partial N_i}{\partial y}\right) & \left(\frac{\partial N_i^T}{\partial z}\frac{\partial N_i}{\partial x} + \frac{\partial N_i}{\partial z}\frac{\partial N_i^T}{\partial x}\right) & \left(-\frac{\partial N_i^T}{\partial z}\frac{\partial N_i}{\partial y}\right) \\ \left(-\frac{\partial N_i^T}{\partial x}\frac{\partial N_i}{\partial z}\right) & \left(-\frac{\partial N_i^T}{\partial y}\frac{\partial N_i}{\partial z}\right) & \left(\frac{\partial N_i^T}{\partial y}\frac{\partial N_i}{\partial y} + \frac{\partial N_i^T}{\partial x}\frac{\partial N_i}{\partial x}\right) \end{bmatrix}$$

$$\times \int \left(\nabla N_i^T \times \frac{1}{\mu_r} \nabla \times \vec{E}\right) dV \frac{1}{\mu_r} \times \begin{bmatrix} E_{xi} \\ E_{yi} \\ E_{zi} \end{bmatrix} dV \qquad (9.166)$$

The second integral in equation (9.159) is

$$-\iiint N_i^T k_o^2 \epsilon_r \vec{E}\, dV = -k_o^2 \epsilon_r \iiint \begin{bmatrix} N_i^T N_i & 0 & 0 \\ 0 & N_i^T N_i & 0 \\ 0 & 0 & N_i^T N_i \end{bmatrix} \begin{bmatrix} E_{xi} \\ E_{yi} \\ E_{zi} \end{bmatrix} dV$$

(9.167)

For the penalty function

$$\iiint \left(\nabla \cdot \epsilon_r \vec{E}\right) \nabla N_i^T dV - \iint N_i^T \left(\nabla \cdot \epsilon_r \vec{E}\right) dS \qquad (9.168)$$

For a constant ϵ_r we have

$$\epsilon_r \nabla \cdot \vec{E} = \epsilon_r \left\{ \frac{\partial E_x}{\partial x} + \frac{\partial E_y}{\partial y} + \frac{\partial E_z}{\partial z} \right\} \qquad (9.169)$$

and

$$\nabla N_i^T = \begin{Bmatrix} \partial N_i/\partial x \\ \partial N_i/\partial y \\ \partial N_i/\partial z \end{Bmatrix} \qquad (9.170)$$

The penalty function terms are then

$$\epsilon_r \iiint \begin{bmatrix} \frac{\partial N_i^T}{\partial x}\frac{\partial N_i}{\partial x} & \frac{\partial N_i^T}{\partial x}\frac{\partial N_i}{\partial y} & \frac{\partial N_i^T}{\partial x}\frac{\partial N_i}{\partial x} \\ & \frac{\partial N_i^T}{\partial y}\frac{\partial N_i}{\partial y} & \frac{\partial N_i^T}{\partial y}\frac{\partial N_i}{\partial z} \\ & & \frac{\partial N_i^T}{\partial z}\frac{\partial N_i}{\partial z} \end{bmatrix} \begin{Bmatrix} E_{xi} \\ E_{yi} \\ E_{zi} \end{Bmatrix} dV \qquad (9.171)$$

and

$$\epsilon_r \iint \begin{bmatrix} N_i^T \frac{\partial N_i}{\partial x} & N_i^T \frac{\partial N_i}{\partial y} & N_i^T \frac{\partial N_i}{\partial z} \\ N_i^T \frac{\partial N_i}{\partial x} & N_i^T \frac{\partial N_i}{\partial y} & N_i^T \frac{\partial N_i}{\partial z} \\ N_i^T \frac{\partial N_i}{\partial x} & N_i^T \frac{\partial N_i}{\partial y} & N_i^T \frac{\partial N_i}{\partial z} \end{bmatrix} \begin{Bmatrix} E_{xi} \\ E_{yi} \\ E_{zi} \end{Bmatrix} dS \qquad (9.172)$$

Treatment of the surface integral term is postponed until later. The finite element implementation shown here is for first-order isoparametric brick elements. For first-order bricks, the formulation would be simpler using direct integration, but the use of isoparametric elements in the program structure makes it more easily adapted to high-order and curved elements. The parent element in Figure 9.21 is a cube with $-1 \leq \mu_1 \leq 1$, $-1 \leq \mu_2 \leq 1$, and $-1 \leq \mu_3 \leq 1$. The point (0,0,0) is in the center. There are eight nodes located on the vertices and their shape function are as follows:

$$N_i = \frac{(1+p\mu_1)(1+q\mu_2)(1+r\mu_3)}{8} \qquad (9.173)$$

where $i = 1, \ldots 8$ and $p = +1$ for nodes 2, 3, 6, and 7, $p = -1$ for nodes 1, 4, 5, and 8, $q = +1$ for nodes 3, 4, 7, and 8, $q = -1$ for nodes 1, 2, 5, and 6, $r = +1$ for nodes 5, 6, 7, and 8, and $r = -1$ for nodes 1, 2, 3, and 4.

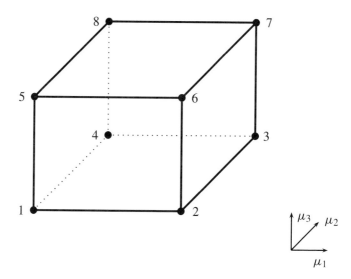

Figure 9.21 3D Brick Element

These have the usual property of shape functions in that they are equal to 1 at their corresponding node and 0 at all other nodes.

In evaluating the integrals we will also need the derivative of the shape functions with respect to the local variables. These are as follows:

$$\frac{\partial N_i}{\partial \mu_1} = p(1 + q\mu_2)(1 + r\mu_3) \tag{9.174}$$

$$\frac{\partial N_i}{\partial \mu_2} = q(1 + p\mu_1)(1 + r\mu_3) \tag{9.175}$$

$$\frac{\partial N_i}{\partial \mu_3} = r(1 + p\mu_1)(1 + q\mu_2) \tag{9.176}$$

The Jacobian matrix is defined in the transformation:

$$\begin{bmatrix} \frac{\partial N_i}{\partial \mu_1} \\ \frac{\partial N_i}{\partial \mu_2} \\ \frac{\partial N_i}{\partial \mu_3} \end{bmatrix} = \begin{bmatrix} \frac{\partial x}{\partial \mu_1} & \frac{\partial y}{\partial \mu_1} & \frac{\partial z}{\partial \mu_1} \\ \frac{\partial x}{\partial \mu_2} & \frac{\partial y}{\partial \mu_2} & \frac{\partial z}{\partial \mu_2} \\ \frac{\partial x}{\partial \mu_3} & \frac{\partial y}{\partial \mu_3} & \frac{\partial z}{\partial \mu_3} \end{bmatrix} \begin{bmatrix} \frac{\partial N_i}{\partial x} \\ \frac{\partial N_i}{\partial y} \\ \frac{\partial N_i}{\partial z} \end{bmatrix} \tag{9.177}$$

where the global coordinates are written in terms of the shape functions and nodal coordinates:

$$x = \sum_{i=1}^{8} N_i x_i \tag{9.178}$$

$$y = \sum_{i=1}^{8} N_i y_i \tag{9.179}$$

$$z = \sum_{i=1}^{8} N_i z_i \tag{9.180}$$

Because we have found the derivatives of the shape function, we can write the components of the Jacobian matrix.

$$[J] = \frac{1}{8} \begin{bmatrix} a & b & c \\ d & e & f \\ g & h & i \end{bmatrix} \tag{9.181}$$

where

$$\begin{aligned} a = &-(1-\mu_2)(1-\mu_3)x_1 + (1-\mu_2)(1-\mu_3)x_2 \\ &+(1+\mu_2)(1-\mu_3)x_3 - (1+\mu_2)(1-\mu_3)x_4 \\ &-(1-\mu_2)(1+\mu_3)x_5 + (1-\mu_2)(1+\mu_3)x_6 \\ &+(1+\mu_2)(1+\mu_3)x_7 - (1+\mu_2)(1+\mu_3)x_8 \end{aligned} \tag{9.182}$$

Note that b is the same as a with y_i instead of x_i and c is the same as a with z_i instead of x_i.

$$
\begin{aligned}
d = & -(1-\mu_1)(1-\mu_3)x_1 - (1+\mu_1)(1-\mu_3)x_2 \\
& +(1+\mu_1)(1-\mu_3)x_3 + (1-\mu_1)(1-\mu_3)x_4 \\
& -(1-\mu_1)(1+\mu_3)x_5 - (1+\mu_1)(1+\mu_3)x_6 \\
& +(1+\mu_1)(1+\mu_3)x_7 + (1-\mu_1)(1+\mu_3)x_8
\end{aligned} \tag{9.183}
$$

As before, e is the same as d with y_i instead of x_i and f is the same as d with z_i instead of x_i.

$$
\begin{aligned}
g = & -(1-\mu_1)(1-\mu_2)x_1 - (1+\mu_1)(1-\mu_2)x_2 \\
& -(1+\mu_1)(1+\mu_2)x_3 - (1-\mu_1)(1+\mu_2)x_4 \\
& +(1-\mu_1)(1-\mu_2)x_5 + (1+\mu_1)(1-\mu_2)x_6 \\
& +(1+\mu_1)(1+\mu_2)x_7 + (1-\mu_1)(1+\mu_2)x_8
\end{aligned} \tag{9.184}
$$

We have the relations that h is the same as g with y_i instead of x_i and i is the same as g with z_i instead of x_i. We now compute the inverse of J. The determinant is

$$
|J| = \frac{1}{8^3}[a(ei-hf) - d(bi-ch) + g(bf-ec)] = \frac{|J'|}{8^3} \tag{9.185}
$$

This gives

$$
[J]^{-1} = \frac{1}{8^2|J|}\begin{bmatrix} ei-hf & ch-bi & bg-ec \\ fg-di & ai-cg & cd-af \\ dh-eg & bg-ah & ae-bd \end{bmatrix} \tag{9.186}
$$

or

$$
[J]^{-1} = \frac{8}{|J'|}\begin{bmatrix} A & B & C \\ D & E & F \\ G & H & I \end{bmatrix} \tag{9.187}
$$

so

$$
\begin{bmatrix} \frac{\partial N_i}{\partial x} \\ \frac{\partial N_i}{\partial y} \\ \frac{\partial N_i}{\partial z} \end{bmatrix} = [J]^{-1} \begin{bmatrix} \frac{\partial N_i}{\partial \mu_1} \\ \frac{\partial N_i}{\partial \mu_2} \\ \frac{\partial N_i}{\partial \mu_3} \end{bmatrix} \tag{9.188}
$$

$$\frac{\partial N_1}{\partial x} = \frac{1}{|J'|}[A(1+\mu_2)(1+\mu_3) + B(1+\mu_1)(1+\mu_3) + C(1+\mu_1)(1+\mu_3)]$$
(9.189)

$$\frac{\partial N_2}{\partial x} = \frac{1}{|J'|}[-A(1+\mu_2)(1+\mu_3) + B(1-\mu_1)(1+\mu_3) + C(1-\mu_1)(1+\mu_2)]$$
(9.190)

$$\frac{\partial N_3}{\partial x} = \frac{1}{|J'|}[-A(1-\mu_2)(1+\mu_3) - B(1-\mu_1)(1+\mu_3) + C(1-\mu_1)(1-\mu_2)]$$
(9.191)

$$\frac{\partial N_4}{\partial x} = \frac{1}{|J'|}[A(1-\mu_2)(1+\mu_3) - B(1+\mu_1)(1+\mu_3) + C(1+\mu_1)(1-\mu_2)]$$
(9.192)

$$\frac{\partial N_5}{\partial x} = \frac{1}{|J'|}[A(1+\mu_2)(1-\mu_3) + B(1+\mu_1)(1-\mu_3) - C(1+\mu_1)(1+\mu_2)]$$
(9.193)

$$\frac{\partial N_6}{\partial x} = \frac{1}{|J'|}[-A(1+\mu_2)(1-\mu_3) + B(1-\mu_1)(1-\mu_3) - C(1-\mu_1)(1+\mu_2)]$$
(9.194)

$$\frac{\partial N_7}{\partial x} = \frac{1}{|J'|}[-A(1-\mu_2)(1-\mu_3) - B(1-\mu_1)(1-\mu_3) - C(1-\mu_1)(1-\mu_2)]$$
(9.195)

$$\frac{\partial N_8}{\partial x} = \frac{1}{|J'|}[A(1-\mu_2)(1-\mu_3) - B(1+\mu_1)(1-\mu_3) - C(1+\mu_1)(1-\mu_2)]$$
(9.196)

To find $\frac{\partial N_i}{\partial y}$ replace A by D, B by E and C by F. To find $\frac{\partial N_i}{\partial z}$ replace A by G, B by H and C by I.

We integrate these equations in the local coordinate system using Gauss quadrature. We have terms of the form:

$$\iiint f\left(E, \frac{\partial E}{\partial x}, \frac{\partial E}{\partial y}, \frac{\partial E}{\partial z}\right) dx\, dy\, dz \qquad (9.197)$$

In local coordinates we have

$$dx\, dy\, dz = |J| d\mu_1 d\mu_2 d\mu_3 \qquad (9.198)$$

For example, a transformed term becomes

$$\iiint \frac{1}{\mu}\left(\frac{\partial N^T}{\partial y}\frac{\partial N}{\partial y} + \frac{\partial N^T}{\partial z}\frac{\partial N}{\partial z}\right) dx\, dy\, dz$$
$$= \iiint \frac{1}{\mu}\left(\frac{\partial N^T}{\partial y}\frac{\partial N}{\partial y} + \frac{\partial N^T}{\partial z}\frac{\partial N}{\partial z}\right) |J| d\mu_1 d\mu_2 d\mu_3 \qquad (9.199)$$

9.8.1 Application of the Absorbing Boundary Condition

The surface integral term will be modified with the vector Engquist–Majda boundary condition. Consider the surface S_{x+} in Figure 9.22.

The surface integral is

$$\iint \hat{n} \times \left(N_i^T \frac{1}{\mu_r} \nabla \times \vec{E}\right) dS \qquad (9.200)$$

Evaluating the curl:

$$\nabla \times \vec{E} = \begin{bmatrix} \frac{\partial E_z}{\partial y} - \frac{\partial E_y}{\partial z} \\ \frac{\partial E_x}{\partial z} - \frac{\partial E_z}{\partial x} \\ \frac{\partial E_y}{\partial x} - \frac{\partial E_x}{\partial y} \end{bmatrix} \qquad (9.201)$$

High-Frequency Problems with Finite Elements

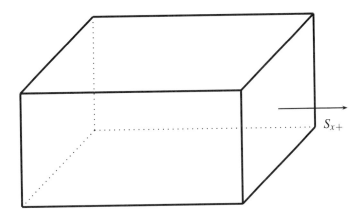

Figure 9.22 3D Region and Surface

Taking the cross product with the normal vector

$$\hat{n} = \begin{bmatrix} n_x \\ 0 \\ 0 \end{bmatrix} \tag{9.202}$$

$$\hat{n} \times \nabla \times \vec{E} = \begin{bmatrix} 0 \\ n_x \left(\frac{\partial E_x}{\partial y} - \frac{\partial E_y}{\partial x} \right) \\ n_x \left(\frac{\partial E_x}{\partial z} - \frac{\partial E_z}{\partial x} \right) \end{bmatrix} \tag{9.203}$$

+++ Note that if we keep the boundaries lined up with the Cartesian axes, then $n_x = 1$. The derivatives $\frac{\partial E_y}{\partial x}$ and $\frac{\partial E_z}{\partial x}$ are now replaced by the relationship

$$\frac{\partial E_y}{\partial x} = jk_0 E_y + \frac{j}{2k_0} \nabla_{yz}^2 E_y \tag{9.204}$$

$$\frac{\partial E_z}{\partial x} = jk_0 E_z + \frac{j}{2k_0}\nabla_{yz}^2 E_z \tag{9.205}$$

First consider the y component.

$$\frac{1}{\mu_r}\iint n_x N_i^T \left(\frac{\partial E_x}{\partial y} - \frac{\partial E_y}{\partial x}\right) dS = \frac{1}{\mu_r}\iint n_x N_i^T \frac{\partial E_x}{\partial y} dS$$
$$- \frac{1}{\mu_r}\iint jk_0 N_i^T E_y dS - \frac{1}{\mu_r}\iint \frac{j}{2k_0} N_i^T \nabla_{yz}^2 E_y dS \tag{9.206}$$

In terms of the shape functions, the first term on the right of (9.206) is

$$\frac{1}{\mu_r}\iint n_x N_i^T \frac{\partial E_x}{\partial y} dS = \frac{1}{\mu_r}\iint \Sigma n_x N_i^T \frac{\partial N_i}{\partial y} E_{xi} dS \tag{9.207}$$

The second term on the right becomes

$$-\frac{1}{\mu_r}\iint jk_0 N_i^T E_y dS = -\frac{jk_0}{\mu_r}\iint \Sigma N_i^T N_i E_{yi} dS \tag{9.208}$$

The third term can be integrated by parts.

$$\frac{-j}{2\mu_r k_0}\iint N_i^T \left[\frac{\partial}{\partial y}\left(\frac{\partial E_y}{\partial y}\right) + \frac{\partial}{\partial z}\left(\frac{\partial E_y}{\partial z}\right)\right] dS$$
$$= \frac{j}{2\mu_r k_0}\iint \Sigma \left[\frac{\partial N_i^T}{\partial y}\frac{\partial N_i}{\partial y} E_{yi} + \frac{\partial N_i^T}{\partial z}\frac{\partial N_i}{\partial z} E_{yi}\right] dS$$
$$- \frac{j}{2\mu_r k_0}\int \Sigma \left(N_i^T \frac{\partial N_i}{\partial y} E_y + N_i^T \frac{\partial N_i}{\partial z} E_y\right) dC \tag{9.209}$$

High-Frequency Problems with Finite Elements

For the z component

$$\frac{1}{\mu_r}\iint n_x N_i^T \left(\frac{\partial E_x}{\partial z} - \frac{\partial E_z}{\partial x}\right) dS = \frac{1}{\mu_r}\iint \Sigma n_x N_i^T \frac{\partial N_i}{\partial z} E_{xi} dS$$

$$- \frac{jk_0}{\mu_r}\iint \Sigma N_i^T N_i E_{zi} dS + \frac{j}{2\mu_r k_0}\iint \Sigma \left(\frac{\partial N_i^T}{\partial y}\frac{\partial N_i}{\partial y} E_{zi} + \frac{\partial N_i^T}{\partial z}\frac{\partial N_i}{\partial z} E_{zi}\right) dS$$

$$- \frac{j}{2\mu_r k_0}\int \Sigma \left(N_i^T \frac{\partial N_i}{\partial z} E_z + N_i^T \frac{\partial N_i}{\partial y} E_z\right) dC \tag{9.210}$$

It has been show that the line integral contribution is very small [103], so ignoring the line integral term, the integrand becomes

$$\frac{1}{\mu_r}\begin{bmatrix} 0 & 0 & 0 \\ n_x N_i^T \frac{\partial N_i}{\partial y} & -jk_0 N_i^T N_i + \frac{j}{2k_0}\left(\frac{\partial N_i^T}{\partial y}\frac{\partial N_i}{\partial y} + \frac{\partial N_i^T}{\partial z}\frac{\partial N_i}{\partial z}\right) & 0 \\ n_x N_i^T \frac{\partial N_i}{\partial z} & 0 & -jk_0 N_i^T N_i + \frac{j}{2k_0}\left(\frac{\partial N_i^T}{\partial y}\frac{\partial N_i}{\partial y} + \frac{\partial N_i^T}{\partial z}\frac{\partial N_i}{\partial z}\right) \end{bmatrix} \tag{9.211}$$

Similarly, for a surface S_y we have

$$\frac{1}{\mu_r} \times$$

$$\begin{bmatrix} -jk_0 N_i^T N_i & \frac{j}{2k_0}\left[\frac{\partial N_i^T}{\partial x}\frac{\partial N_i}{\partial x} + \frac{\partial N_i^T}{\partial x}\frac{\partial N_i}{\partial z}\right] & n_y N_i^T \frac{\partial N_i}{\partial z} & 0 \\ 0 & 0 & 0 \\ 0 & N_i^T N_i + \frac{\partial N_i}{\partial z} - jk_0 N_i^T N_i & \frac{j}{2k_0}\left(\frac{\partial N_i^T}{\partial x}\frac{\partial N_i}{\partial x} + \frac{\partial N_i^T}{\partial z}\right) \end{bmatrix}$$

$$\tag{9.212}$$

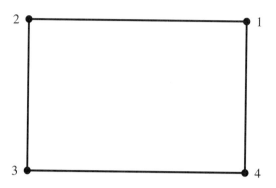

Figure 9.23 3D Surface for Absorbing Boundary

For the surface S_z

$$\frac{1}{\mu_r} \times$$

$$\begin{bmatrix} -jk_0 N_i^T N_i + \frac{j}{2k_0}(\frac{\partial N_i^T}{\partial x}\frac{\partial N_i}{\partial x} + \frac{\partial N_i^T}{\partial y}\frac{\partial N_i}{\partial y}) & 0 & n_z N_i^T \frac{\partial N_i}{\partial x} \\ 0 & -jk_0 N_i^T N_i + \frac{j}{2k_0}(\frac{\partial N_i^T}{\partial x}\frac{\partial N_i}{\partial x} + \frac{\partial N_i^T}{\partial y}\frac{\partial N_i}{\partial y}) & n_z N_i^T \frac{\partial N_i}{\partial y} \\ 0 & 0 & 0 \end{bmatrix}$$

(9.213)

Note that these matrices are not symmetric. To evaluate these integrals, we consider the first-order rectangle that is the facet of the brick element in common with the absorbing boundary.

The shape functions are

$$N_i = \frac{(1 + p\mu_1)(1 + q\mu_2)}{4} \tag{9.214}$$

where $p = +1$ for nodes 1 and 4, $p = -1$ for nodes 2 and 3, $q = +1$ for nodes 1 and 2, and $q = -1$ for nodes 3 and 4.

The derivatives of the shape functions are

$$\frac{\partial N_i}{\partial \mu_1} = \frac{p(1+q\mu_2)}{4} \qquad (9.215)$$

and

$$\frac{\partial N_i}{\partial \mu_2} = \frac{q(1+p\mu_1)}{4} \qquad (9.216)$$

The Jacobian is

$$[J] = \begin{bmatrix} \frac{\partial x}{\partial \mu_1} & \frac{\partial y}{\partial \mu_1} & \frac{\partial z}{\partial \mu_1} \\ \frac{\partial x}{\partial \mu_2} & \frac{\partial y}{\partial \mu_2} & \frac{\partial z}{\partial \mu_2} \\ n_x & n_y & n_y \end{bmatrix} = \begin{bmatrix} a & b & c \\ d & e & f \\ n_x & n_y & n_z \end{bmatrix} \qquad (9.217)$$

We also have

$$\vec{S} = \begin{bmatrix} \frac{\partial x}{\partial \mu_2} \\ \frac{\partial y}{\partial \mu_2} \\ \frac{\partial z}{\partial \mu_2} \end{bmatrix} \qquad (9.218)$$

The normal vector is found by

$$\hat{n} = \hat{t} \times \hat{s} = \begin{bmatrix} n_x \\ n_y \\ n_z \end{bmatrix} \qquad (9.219)$$

To evaluate the Jacobian

$$\frac{\partial x}{\partial \mu_1} = \frac{1}{4}[(x_1(1+\mu_2) - x_2(1+\mu_2) - x_3(1-\mu_2) + x_4(1-\mu_2)] \quad (9.220)$$

$$\frac{\partial y}{\partial \mu_1} = \frac{1}{4}[(y_1(1+\mu_2) - y_2(1+\mu_2) - y_3(1-\mu_2) + y_4(1-\mu_2)] \quad (9.221)$$

$$\frac{\partial z}{\partial \mu_1} = \frac{1}{4}[(z_1(1+\mu_2) - z_2(1+\mu_2) - z_3(1-\mu_2) + z_4(1-\mu_2)] \quad (9.222)$$

$$\frac{\partial x}{\partial \mu_2} = \frac{1}{4}[(x_1(1+\mu_1) + x_2(1-\mu_1) - x_3(1-\mu_1) - x_4(1+\mu_1)] \quad (9.223)$$

$$\frac{\partial y}{\partial \mu_2} = \frac{1}{4}[(y_1(1+\mu_1) + y_2(1-\mu_1) - y_3(1-\mu_1) - y_4(1+\mu_1)] \quad (9.224)$$

$$\frac{\partial z}{\partial \mu_2} = \frac{1}{4}[(z_1(1+\mu_1) + z_2(1-\mu_1) - z_3(1-\mu_1) - z_4(1+\mu_1)] \quad (9.225)$$

If we again define

$$J^{-1} = \frac{1}{|J|} \begin{bmatrix} A & B & C \\ D & E & F \\ G & H & I \end{bmatrix} \quad (9.226)$$

$$\begin{pmatrix} \frac{\partial N_i}{\partial x} \\ \frac{\partial N_i}{\partial y} \\ \frac{\partial N_i}{\partial z} \end{pmatrix} = J^{-1} \begin{pmatrix} \frac{\partial N_i}{\partial \mu_1} \\ \frac{\partial N_i}{\partial \mu_2} \\ \frac{\partial N_i}{\partial \mu_3} \end{pmatrix} \quad (9.227)$$

(for example, $A = en_z - fn_y$). Then

$$\frac{\partial N_1}{\partial x} = \frac{1}{4|J|}[A(1+\mu_2) + B(1+\mu_1)] \qquad (9.228)$$

$$\frac{\partial N_2}{\partial x} = \frac{1}{4|J|}[-A(1+\mu_2) + B(1-\mu_1)] \qquad (9.229)$$

$$\frac{\partial N_3}{\partial x} = \frac{1}{4|J|}[-A(1-\mu_2) - B(1-\mu_1)] \qquad (9.230)$$

$$\frac{\partial N_4}{\partial x} = \frac{1}{4|J|}[A(1-\mu_2) - B(1+\mu_1)] \qquad (9.231)$$

To find $\frac{\partial N_i}{\partial y}$ replace A by D and B by E. To find $\frac{\partial N_i}{\partial z}$ replace A by G and B by H.

TE and TM Problems

The formulation is the same whether we solve for $\vec{E}$ or $\vec{H}$. The differences lie in the excitation and the interface conditions on the conducting wall.

(a) For the TE cases we use a current sheet $J_s = \hat{z} \times (H^+ - H^-)$. Because

$$\hat{z} = \begin{bmatrix} 0 \\ 0 \\ 1 \end{bmatrix} \qquad (9.232)$$

$$J_s = \begin{bmatrix} -H_y \\ H_x \\ 0 \end{bmatrix} \qquad (9.233)$$

Writing the field in cylindrical coordinates

$$H_r = \frac{-j\beta}{k_c}(A \sin n\phi + B \cos n\phi) J_n'(k_c r) \tag{9.234}$$

$$H_\phi = \frac{-j\beta n}{k_c^2 p}(A \cos n\phi - B \sin n\phi) J_n(k_c r) \tag{9.235}$$

On the conductor $H_r = 0$ so $k_c = \frac{x_{np}'}{a}$, where a is the radius of the waveguide and x_{np}' are the zeros of the derivative of the Bessel function $J_n(x)$. A and B are constants defined by the user. For example, for the TE_{11} we have $x_{11}' = 1.841$, for TE_{21}, $x_{21}' = 3.054$, etc. To get back to rectangular coordinates we have $H_x = H_r \cos\phi - H_\phi \sin\phi$ and $H_y = H_1 \sin\phi + H_\phi \cos\phi$.

(b) For the TM cases

$$\vec{J}_m = \left(\vec{E}^+ - \vec{E}^-\right) \times \hat{z} \tag{9.236}$$

In cylindrical coordinates

$$E_r = \frac{-j\beta}{k_c}[A \sin n\phi + B \cos n\phi] J_n'(k_c r) \tag{9.237}$$

and

$$E_\phi = \frac{-j\beta}{k_c}[A \cos n\phi - B \sin n\phi] J_n(k_c r) \tag{9.238}$$

On the conductor $E_\phi = 0$ so that $k_c = \frac{x_{np}}{a}$, where x_{np} are the zeros of the Bessel function $J_n(x)$. For example, for TM_{01}, $x_{01} = 2.405$, for TM_{11}, $x_{11} = 3.832$, etc. When solving for $\vec{E}$, we require that the tangential components are zero on a conductor boundary, that is,

$$\hat{n} \times E = 0 \tag{9.239}$$

When solving for $\vec{H}$ we require that the normal component is zero on a conductor boundary, that is,

$$\hat{n} \cdot \hat{H} = 0. \tag{9.240}$$

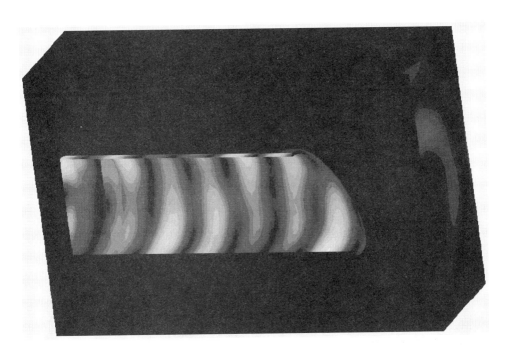

Figure 9.24   3D Solution of Waveguide with Absorbing Boundary Conditions

9.9   EXAMPLE

To illustrate the formulation, we solve the problem of a Vlasov waveguide in three dimensions[102]. Figure 9.24 shows the field on a symmetric cross section through the center of the waveguide. Figure 9.25 shows the two dimensional solution.

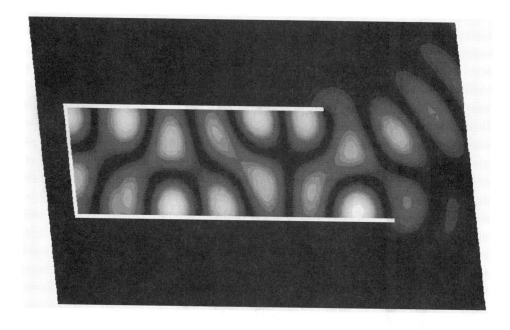

Figure 9.25 2D Solution of Waveguide with Absorbing Boundary Conditions

10
LOW-FREQUENCY APPLICATIONS

10.1 TIME DOMAIN MODELING OF ELECTROMECHANICAL DEVICES

We will now consider the transient (time domain) behavior of electromechanical devices that are activated by a voltage or current source, such as motors or actuators. The voltage (current) source for such devices may be time dependent. One cannot specify *a priori* the numerical value of the current density in all of the conductive source regions of the device. The finite element procedures described previously required current density as a known input to the analysis. In addition, it may be necessary to attach external circuit components between the voltage or current sources and the region to be modeled by finite elements. The lumped components may represent the internal impedance of the voltage or current source, or they may be used to approximate the effects of the parts of the device that are outside the region modeled by finite elements. The corresponding transient circuit equations must be coupled to the transient finite element field equations. Moreover, in the case of the motor or actuator, there are movable components. Magnetic forces in part determine the positions of these components, and the positions, in turn, affect the magnetic field within the device. Provisions must be made for the transient modeling of such a coupled electromechanical system.

A method for the proper coupling of transient fields, circuits, and motion must be such that: (1) only terminal voltage (or total terminal current) applied to the device is required as a known input quantity, and total terminal current (terminal voltage) is calculated as an unknown; (2) the transient external circuit equations that represent the electrical sources and circuit components are coupled to the finite element field equations; and (3) equations for motion are coupled to the finite element field equations.

10.1.1 Electromagnetic and Mechanical Theory

The electromagnetic field theory, electric circuit theory, and basic mechanical motion theory, upon which the proposed method is based, are summarized in the following sections [104][105] [106][107][108].

Time-Dependent Magnetic Diffusion Equation

We begin with the two-dimensional diffusion equation,

$$\nabla \times \nu \nabla \times A = J \tag{10.1}$$

where J and A are assumed to be z-directed and independent of z. Because these vectors have only one component, they can be treated as scalars.

We can speak of the current density, J, as having three parts: one due to the applied source, another due to the induced electric field produced by time-varying magnetic flux, and the third due to motion-induced or "speed" voltage. Therefore,

$$J = \sigma \frac{V_b}{\ell} - \sigma \frac{\partial A}{\partial t} + \sigma v \times B \tag{10.2}$$

where

$$\begin{aligned}
\sigma &= \text{electrical conductivity} \\
\ell &= \text{length of problem in } z\text{-direction (for 2D problems)} \\
V_b &= \text{voltage applied to the finite element region} \\
v &= \text{velocity of the conductor with respect to } B
\end{aligned}$$

It is assumed here that the gradient of V_b is z-directed and has a magnitude of V_b/ℓ. (Figure 10.1 illustrates the physical meaning of the V_b term.)

The first term on the right of equation (10.2) represents the current density due to the applied source, the second term represents the induced current density, and the third term represents current density produced by the speed voltage. Note that the distinction among the three parts is purely mathematical; the different components cannot be separated by experiment.

Low-Frequency Applications

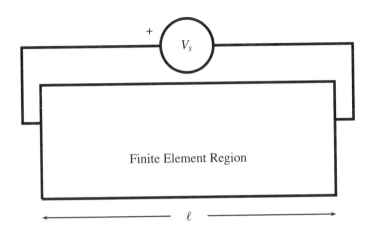

Figure 10.1 Voltage Applied to the Finite Element Region

The time-dependent magnetic diffusion equation is then

$$\nabla \times \nu \nabla \times A = \sigma \frac{V_b}{\ell} - \sigma \frac{\partial A}{\partial t} + \sigma v \times \nabla \times A \qquad (10.3)$$

By employing a frame of reference that is fixed with respect to the component under consideration,[1] the relative velocity v becomes zero and the diffusion equation simplifies to

$$\nabla \times \nu \nabla \times A = \sigma \frac{V_b}{\ell} - \sigma \frac{\partial A}{\partial t} \qquad (10.4)$$

Total Conductor Current Equation

In order to couple the circuit and field equations, it is necessary to calculate the total current flowing in each conductor in the problem.

---

[1] In finite element analysis, such a reference frame is effected by fixing the mesh to the surface of the moving component and moving/deforming only those elements that lie in the air surrounding the component. For a more detailed explanation, see Istfan [109].

From the previous section, the expression for current density is known to be

$$J = \sigma \frac{V_b}{\ell} - \sigma \frac{\partial A}{\partial t} \quad (10.5)$$

The total current in the conductor is then found by integrating equation (10.5) over the cross section of the conductor:

$$I = \iint_{\text{conductor}} \left(\sigma \frac{V_b}{\ell} - \sigma \frac{\partial A}{\partial t} \right) dx\, dy \quad (10.6)$$

Circuit Equations

The total conductor current is coupled to the voltage or current source through the lumped resistance and inductance, R_{ext} and L_{ext}. The time-dependent circuit equations that describe this coupling are now derived.

The most basic component of a circuit in the finite element region is a *bar*. A bar is defined as a single conductive region of length ℓ in the z-direction. Bars may be connected in series to form *coils*, and coils may be connected together in parallel.

Series Connection of Bars to Form a Coil

A number of bars $(b_1, b_2, \ldots, b_n)$ may be connected together in series to form a *coil*. All bars in a coil carry the same total current, but successive bars carry the current in opposite directions. The leads of the coil are brought out of the finite element region and connected to a voltage (or current) source. The resistance and reactance of the leads are modeled with lumped resistance, R_{ext}, and inductance, L_{ext}, as shown in Figure 10.2.

V_c is the voltage applied to the external terminals of the coil, and V_t is the "terminal voltage" of the finite element region. V_t is the sum of the voltages across the bars that constitute the coil, that is, for coil c,

$$V_{t,c} = \sum_{b \in c} d_b V_b \quad (10.7)$$

Low-Frequency Applications

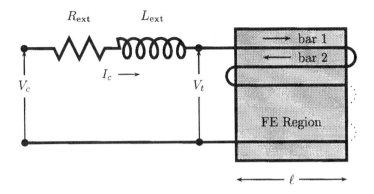

Figure 10.2 Series Connection of Bars to Form a Coil

where d_b is $+1$ or -1 and represents the polarity of bar b. (In Figure 10.2, bar 1 is "positive," bar 2 is "negative," etc., and V_b is as shown in Figure 10.1.) Therefore,

$$V_c = \{d_b\}_c^T \{V_b\}_c + L_{ext} \frac{dI_c}{dt} + R_{ext} I_c \tag{10.8}$$

This important equation serves to couple the finite element region, represented by V_t, to the external circuits and sources, represented by R_{ext}, L_{ext}, and V_c.

Parallel Connection of Coils

Sometimes a set of coils is connected in parallel, forming a single circuit at the external terminals of the device. The general case of p coils connected in parallel to a source V_s with internal resistance R_s and inductance L_s is illustrated in Figure 10.3.

The governing R–L equation for this configuration is

$$V_s = R_s \sum_{Coils} I_c + L_s \sum_{Coils} \frac{dI_c}{dt} + V_c \tag{10.9}$$

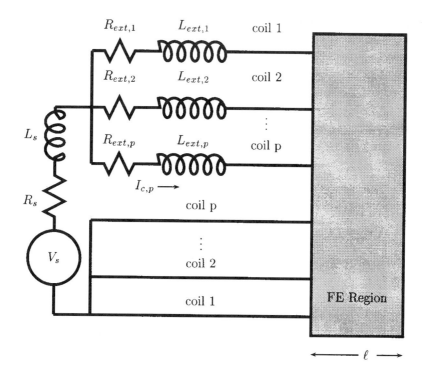

Figure 10.3 Parallel Connection of p Coils

or, in matrix form,

$$V_s = R_s \{1\}^T \{I\}_c + L_s \{1\}^T \left\{\frac{dI}{dt}\right\}_c + V_c \qquad (10.10)$$

where $\{1\}$ is a column vector of dimension p whose entries are all 1. In the global system of equations, there is one equation of this form for each set of parallel-connected coils.

Equations of Motion

The basic equations of motion are as follows:

$$m\frac{dv}{dt} + \lambda v = F_{em} - F_{ext} \qquad (10.11)$$

$$v = \frac{dx}{dt} \qquad (10.12)$$

where

$$\begin{aligned}
m &= \text{mass} \\
v &= \text{velocity} \\
x &= \text{position} \\
\lambda &= \text{damping coefficient} \\
F_{em} &= \text{electromagnetic force} \\
F_{ext} &= \text{externally applied mechanical force (load)}
\end{aligned}$$

All of these pertain to the moving part.

The electromagnetic force, F_{em}, may be written in terms of magnetic field quantities, specifically, magnetic vector potential A. The magnetic force is proportional to the square of flux density:

$$F_{em} \propto B^2 \qquad (10.13)$$

meaning that

$$F_{em} \propto (\nabla \times A)^2 \qquad (10.14)$$

Thus, electromagnetic force is a nonlinear (quadratic) function of vector potential.

Summary of Equations

The equations necessary for the transient coupling of field, circuit, and motion equations are summarized next:

Field Equation

$$\nabla \times \nu \nabla \times A = \sigma \frac{V_b}{\ell} - \sigma \frac{\partial A}{\partial t}$$

Total Current Equation

$$I = \iint_{\text{conductor}} \left(\sigma \frac{V_b}{\ell} - \sigma \frac{\partial A}{\partial t} \right) dx\, dy$$

Series Bar–Coil Equation

$$V_c = \{d_b\}_c^T \{V_b\}_c + L_{\text{ext}} \frac{dI_c}{dt} + R_{\text{ext}} I_c$$

Parallel Coil Equation

$$V_s = R_s \{1\}^T \{I\}_c + L_s \{1\}^T \left\{\frac{dI}{dt}\right\}_c + V_c$$

Mechanical Acceleration Equation

$$m \frac{dv}{dt} + \lambda v = F_{\text{em}} - F_{\text{ext}}$$

Mechanical Velocity Equation

$$v = \frac{dx}{dt}$$

Note that the first three equations are coupled by the voltage applied to the finite element region, V_b; the second, third, and fourth equations are coupled by the conductor currents; and the first, second, and fifth equations are coupled by magnetic vector potential A (A is implicit in F_{em} in the fifth equation).

10.1.2 Galerkin Formulation

The portion of the problem to be analyzed with finite elements must be discretized in space, that is, meshed. The Galerkin method is used to approximate the field and current equations in discretized space.

Formulation of the Field Equation

The magnetic diffusion equation was derived as

$$\nabla \times \nu \nabla \times A = \sigma \frac{V_b}{\ell} - \sigma \frac{\partial A}{\partial t} \qquad (10.15)$$

For the 2D case, current density, J, and vector potential, A, are assumed to be invariant in the z-direction, so equation (10.15) becomes

$$\frac{\partial}{\partial x}\left(\nu \frac{\partial A}{\partial x}\right) + \frac{\partial}{\partial y}\left(\nu \frac{\partial A}{\partial y}\right) + \sigma \frac{V_b}{\ell} - \sigma \frac{\partial A}{\partial t} = 0 \qquad (10.16)$$

Equation (10.16) is to be solved for A. Applying the method of weighted residuals for an approximate solution, $\hat{A}$, and weighting function W, leads to

$$\iint \nu \left[\frac{\partial W}{\partial x}\frac{\partial \hat{A}}{\partial x} + \frac{\partial W}{\partial y}\frac{\partial \hat{A}}{\partial y} \right] dx\,dy$$
$$+ \iint \sigma W \frac{\partial \hat{A}}{\partial t} dx\,dy - \iint \sigma W \frac{V_b}{\ell} dx\,dy = 0 \qquad (10.17)$$

Now, for the case of first-order, triangular finite elements,

$$\hat{A} = \sum_{i=1}^{3} N_i A_i(t) \qquad (10.18)$$

where the N_i are the shape functions and the A_i are approximations to the vector potentials at the nodes of the mesh. The Galerkin method of weighted residuals uses the shape functions as the weighting functions, W_i:

$$W_i = N_i, \quad i = 1, 2, 3 \tag{10.19}$$

Substituting equations (10.18) and (10.19) into equation (10.17), we obtain

$$\sum_e \iint \left\{ \nu \left[\frac{\partial N_i}{\partial x} \frac{\partial \left(\sum_{j=1}^{3} N_j A_j \right)}{\partial x} + \frac{\partial N_i}{\partial y} \frac{\partial \left(\sum_{j=1}^{3} N_j A_j \right)}{\partial y} \right] \right. \\ \left. + \sigma \left[N_i \sum N_j \frac{\partial A_j}{\partial t} - N_i \frac{V_b}{\ell} \right] \right\} dx\,dy = 0 \tag{10.20}$$

Alternatively, in matrix form,

$$\sum_e \left[\nu [S]_e \{A(t)\}_e + \sigma [T]_e \left\{ \frac{\partial A(t)}{\partial t} \right\}_e - \sigma [Q]_e \frac{V_b}{\ell} \right] = \{0\} \tag{10.21}$$

where $\sum_e$ indicates summation over all of the finite elements and the subscript e means that the matrices refer to a particular element. Typical entries in these matrices are

$$S_{e,ij} = \iint \left(\frac{\partial N_i}{\partial x} \frac{\partial N_j}{\partial x} + \frac{\partial N_i}{\partial y} \frac{\partial N_j}{\partial y} \right) dx\,dy = \frac{b_i b_j + c_i c_j}{4\Delta_e}$$

$$T_{e,ij} = \iint N_i N_j \, dx\,dy = \begin{cases} \dfrac{\Delta_e}{6} & i = j \\ \dfrac{\Delta_e}{12} & i \neq j \end{cases}$$

$$Q_{e,ij} = \iint N_i \, dx\,dy = \frac{\Delta_e}{3}$$

Low-Frequency Applications

where Δ_e is the area of the element (assuming counterclockwise ordering of the vertices) and the b's and c's are the first-order elemental geometric coefficients.

Now, are the $[S]$, $[T]$, and $[Q]$ matrices time dependent? For elements that deform with time, all three matrices are clearly time dependent. Recall, however, that only the elements in air (or free space) are permitted to deform. The conductivity, σ, in such elements is zero, which nullifies the contribution of the $[T]$ and $[Q]$ matrices in equation (10.21). Therefore, only the time dependence of the $[S]$ matrix requires consideration.

The final Galerkin form of the diffusion equation is then

$$\sum_e \left[\nu [S(t)]_e \{A(t)\}_e + \sigma [T]_e \left\{ \frac{\partial A(t)}{\partial t} \right\}_e - \sigma \{Q\}_e \frac{V_b}{\ell} \right] = 0 \quad (10.22)$$

or

$$\nu [S]\{A(t)\} + \sigma [T] \left\{ \frac{\partial A(t)}{\partial t} \right\} - \sigma \{Q\} \frac{V_b}{\ell} = 0 \quad (10.23)$$

Formulation of the Current Equation

The total current in one element, e, is

$$I_e = \iint_e \sigma \frac{V_b}{\ell} \, dx \, dy - \iint_e \sigma \frac{\partial A}{\partial t} \, dx \, dy \quad (10.24)$$

The integrand of the first term is a constant, so the integration yields

$$\iint_e \sigma \frac{V_b}{\ell} \, dx \, dy = \sigma \frac{V_b}{\ell} \Delta_e \quad (10.25)$$

For the second term of equation (10.24), recall that

$$\frac{\partial A}{\partial t} = \sum_{i=1}^{3} N_i \frac{\partial A_i}{\partial t} \quad (10.26)$$

for first-order elements. Therefore,

$$\iint_e \sigma \frac{\partial A}{\partial t} dx\, dy = \iint_e \sigma \sum_{i=1}^{3} N_i \frac{\partial A_i}{\partial t} dx\, dy \tag{10.27}$$

or, in matrix form,

$$\iint_e \sigma \frac{\partial A}{\partial t} dx\, dy = \sigma \, \{Q\}_e^T \left\{ \frac{\partial A}{\partial t} \right\}_e \tag{10.28}$$

where $\{Q\}$ is defined as before. Thus, the total current in one element is found by substituting equations (10.25) and (10.28) into equation (10.24):

$$I_e = \sigma \frac{V_b}{\ell} \Delta_e - \sigma \, \{Q\}_e^T \left\{ \frac{\partial A}{\partial t} \right\}_e \tag{10.29}$$

Now we consider a bar. Recall that a bar is defined as a conductor of length ℓ. The total current in a bar, b, is the summation of current in the elements that make up the cross section of the bar:

$$I_b = \sum_{e \in b} \left[\sigma \frac{V_b}{\ell} \Delta_e - \sigma \, \{Q\}_e^T \left\{ \frac{\partial A}{\partial t} \right\}_e \right] \tag{10.30}$$

or

$$I_b = \sigma_b \frac{V_b}{\ell} \Delta_b - \sigma_b \, \{Q\}_b^T \left\{ \frac{\partial A}{\partial t} \right\}_b \tag{10.31}$$

where Δ_b is the cross-sectional area of the bar, $\{Q\}_b$ and $\left\{ \frac{\partial A}{\partial t} \right\}_b$ are m-dimensional vectors, and m is the number of mesh nodes on the cross section of the bar.

Low-Frequency Applications

Next, consider an even number of bars, $(b_1, b_2 \ldots, b_n)$ that are connected together in series to form a *coil*. All bars in a coil carry the same current, but successive bars carry the current in opposite directions. Equation (10.31) is modified as follows for a coil:

$$[D]\{I_b\}_c = \left[\frac{\sigma}{\ell}\Delta_b\right]\{V_b\}_c - [\sigma Q]_c^T \left\{\frac{\partial A}{\partial t}\right\}_c \quad (10.32)$$

where

$[D]$ = a diagonal matrix with entries of $+1$ or -1, indicating the polarity of each bar in the coil

$\{I_b\}_c = \{I_{b1}, I_{b2}, \ldots, I_{bn}\}^T$

$\left[\frac{\sigma}{\ell}\Delta_b\right]$ = a diagonal matrix whose ith entry is $\left(\frac{\sigma}{\ell}\Delta_b\right)_{c,ii} = \frac{\sigma_{bi}}{\ell}\Delta_{bi}$

$[\sigma Q]_c^T = \{\sigma_{b1}\{Q\}_{b1}, \sigma_{b2}\{Q\}_{b2}, \ldots, \sigma_{bn}\{Q\}_{bn}\}^T$

$\{V_b\}_c = \{V_{b1}, V_{b2}, \ldots, V_{bn}\}^T$

Note that the entries of $\{I_b\}_c$ are all equal, because the bars are connected in series.

10.1.3 Time Discretization

In this section, the field equation, total current equation, circuit equations, and the equations of motion are discretized in the time domain. The method of time discretization used is based on the following equation:

$$\beta\left\{\frac{\partial A}{\partial t}\right\}^{t+\Delta t} + (1-\beta)\left\{\frac{\partial A}{\partial t}\right\}^t = \frac{\{A\}^{t+\Delta t} - \{A\}^t}{\Delta t} \quad (10.33)$$

The value of the constant β determines whether the algorithm is of the forward difference type ($\beta = 0$), backward difference type ($\beta = 1$), or some intermediate type ($0 < \beta < 1$). If $\beta = \frac{1}{2}$, we have the Crank–Nicholson method.

The goal is to solve for $\{A\}^{t+\Delta t}$. The derivatives $\left\{\frac{\partial A}{\partial t}\right\}^{t+\Delta t}$ and $\left\{\frac{\partial A}{\partial t}\right\}^t$ are unknown.

Time Discretization of the Diffusion Equation

Recall the Galerkin form of the diffusion equation[2]:

$$\nu [S]\{A(t)\} + \sigma [T]\left\{\frac{\partial A(t)}{\partial t}\right\} - \sigma \{Q\}\frac{V_b}{\ell} = 0 \tag{10.34}$$

We now split equation (10.34) into two equations as follows:

$$\beta\sigma [T]\left\{\frac{\partial A}{\partial t}\right\}^{t+\Delta t} = -\beta\nu [S]\{A\}^{t+\Delta t} + \beta\frac{\sigma}{\ell}\{Q\} V_b^{t+\Delta t} \tag{10.35}$$

$$(1-\beta)\sigma [T]\left\{\frac{\partial A}{\partial t}\right\}^{t} = -(1-\beta)\nu [S]\{A\}^{t} + (1-\beta)\frac{\sigma}{\ell}\{Q\} V_b^{t} \tag{10.36}$$

Substitute equations (10.35) and (10.36) into equation (10.33) and obtain

$$-\beta\nu [S]\{A\}^{t+\Delta t} + \beta\frac{\sigma}{\ell}\{Q\} V_b^{t+\Delta t} - (1-\beta)\nu [S]\{A\}^{t}$$
$$+\nu(1-\beta)\frac{\sigma}{\ell}\{Q\} V_b^{t} = \sigma [T]\frac{\{A\}^{t+\Delta t} - \{A\}^{t}}{\Delta t} \tag{10.37}$$

Rearranging terms to isolate the $t + \Delta t$ terms yields:

$$\left[\nu [S] + \frac{\sigma [T]}{\beta\Delta t}\right]\{A\}^{t+\Delta t} - \frac{\sigma}{\ell}\{Q\} V_b^{t+\Delta t}$$
$$= \left[\frac{\sigma [T]}{\beta\Delta t} - \frac{1-\beta}{\beta}\nu [S]\right]\{A\}^{t} + \frac{\sigma}{\ell}\{Q\}\frac{1-\beta}{\beta} V_b^{t} \tag{10.38}$$

Time Discretization of the Current Equation

The time discretization of the total current equation is performed in a similar fashion. The current equation for one coil, consisting of n series-connected bars, was given as

$$[D]\{I_b\}_c = \left[\frac{\sigma}{\ell}\Delta_b\right]\{V_b\}_c - [\sigma Q]_c^T \left\{\frac{\partial A}{\partial t}\right\}_c \tag{10.39}$$

---

[2] Consideration of the time dependence of the $[S]$ matrix will be deferred to a later section.

We now split equation (10.39) into two equations as follows:

$$\beta [\sigma Q]_c^T \left\{\frac{\partial A}{\partial t}\right\}_c^{t+\Delta t} = \beta \left[\frac{\sigma}{\ell}\Delta_b\right]\{V_b\}_c^{t+\Delta t} - \beta [D]\{I_b\}_c^{t+\Delta t} \tag{10.40}$$

$$(1-\beta)[\sigma Q]_c^T \left\{\frac{\partial A}{\partial t}\right\}_c^{t} = (1-\beta)\left[\frac{\sigma}{\ell}\Delta_b\right]\{V_b\}_c^{t} - (1-\beta)[D]\{I_b\}_c^{t} \tag{10.41}$$

Substituting equations (10.40) and (10.41) into equation (10.33), we obtain

$$\beta\left[\frac{\sigma}{\ell}\Delta_b\right]\{V_b\}_c^{t+\Delta t} - \beta[D]\{I_b\}_c^{t+\Delta t} + (1-\beta)\left[\frac{\sigma}{\ell}\Delta_b\right]\{V_b\}_c^{t}$$
$$-(1-\beta)[D]\{I_b\}_c^{t} = [\sigma Q]_c^T \frac{\{A\}^{t+\Delta t} - \{A\}^{t}}{\Delta t} \tag{10.42}$$

Rearranging terms to isolate $\{A\}^{t+\Delta t}$, $V_b^{t+\Delta t}$, and $\{I_b\}_c^{t+\Delta t}$ yields

$$-[\sigma Q]_c^T \{A\}^{t+\Delta t} + \beta \Delta t \left[\frac{\sigma}{\ell}\Delta_b\right]\{V_b\}_c^{t+\Delta t} - \beta \Delta t [D]\{I_b\}_c^{t+\Delta t}$$
$$= -[\sigma Q]_c^T \{A\}^t - (1-\beta)\Delta t \left[\frac{\sigma}{\ell}\Delta_b\right]\{V_b\}_c^t + (1-\beta)\Delta t [D]\{I_b\}_c^t \tag{10.43}$$

Equation (10.43) is the time discretization of the current equation. The global system of equations will contain one equation of this form for each series-connected coil in the problem.

Time Discretization of the Circuit Equations

We will now find the equations describing the series and parallel circuit connections, respectively.

Time Discretization of the Series Bar–Coil Equation

The circuit equations, like the diffusion and current equations, are discretized according to equation (10.33):

$$\beta \frac{dI^{t+\Delta t}}{dt} + (1-\beta)\frac{dI^t}{dt} = \frac{I^{t+\Delta t} - I^t}{\Delta t} \tag{10.44}$$

Now take the equation for the series connection of bars forming a coil, equation (10.8), and split it into two equations as follows:

$$\beta L_{\text{ext}} \frac{dI^{t+\Delta t}}{dt} = \beta V_c^{t+\Delta t} - \beta \{D\}_c^T \{V_b\}_c^{t+\Delta t} - \beta R_{\text{ext}} I_c^{t+\Delta t} \tag{10.45}$$

$$(1-\beta)L_{\text{ext}} \frac{dI^t}{dt} = (1-\beta)V_c^t - (1-\beta)\{D\}_c^T \{V_b\}_c^t - (1-\beta)R_{\text{ext}} I_c^t \tag{10.46}$$

Now substitute equations (10.45) and (10.46) into equation (10.44) to obtain

$$\beta V_c^{t+\Delta t} - \beta \{D\}_c^T \{V_b\}_c^{t+\Delta t} - \beta R_{\text{ext}} I_c^{t+\Delta t} + (1-\beta)V_c^t$$
$$-(1-\beta)\{D\}_c^T \{V_b\}_c^t - (1-\beta)R_{\text{ext}} I_c^t = L_{\text{ext}} \frac{I^{t+\Delta t} - I^t}{\Delta t} \tag{10.47}$$

Rearranging terms to isolate the $t + \Delta t$ terms,

$$-\beta \Delta t \{D\}_c^T \{V_b\}_c^{t+\Delta t} - (L_{\text{ext}} + \beta \Delta t R_{\text{ext}})I_c^{t+\Delta t} + \beta \Delta t V_c^{t+\Delta t}$$
$$= (1-\beta)\Delta t \{D\}_c^T \{V_b\}_c^t - [L_{\text{ext}} - (1-\beta)\Delta t R_{\text{ext}}] I_c^t - (1-\beta)\Delta t V_c^t \tag{10.48}$$

Equation (10.48) is the time-discretized form of the equation that joins bars in series to form coils and brings the coil leads out of the finite element region to external terminals, through R_{ext} and L_{ext}. In the global system, there is one equation of this form for each coil.

Time Discretization of the Parallel Circuit Equation

The parallel circuit equation is time discretized in the same fashion, using equation (10.33):

$$\beta \left\{ \frac{dI}{dt} \right\}^{t+\Delta t} + (1-\beta) \left\{ \frac{dI}{dt} \right\}^{t} = \frac{\{I\}^{t+\Delta t} - \{I\}^{t}}{\Delta t} \tag{10.49}$$

We take the equation for the parallel connection of coils, equation (10.10), and split it into two equations as follows:

$$\beta L_s \{1\}^T \left\{ \frac{dI}{dt} \right\}^{t+\Delta t} = \beta V_s^{t+\Delta t} - \beta R_s \{1\}^T \{I\}_c^{t+\Delta t} - \beta \{V\}_c^{t+\Delta t} \tag{10.50}$$

$$(1-\beta) L_s \{1\}^T \left\{ \frac{dI}{dt} \right\}^{t} = (1-\beta) V_s^{t} - (1-\beta) R_s \{1\}^T \{I\}_c^{t}$$
$$- (1-\beta) \{V\}_c^{t} \tag{10.51}$$

Now substitute equations (10.50) and (10.51) into equation (10.49) to obtain

$$\beta V_s^{t+\Delta t} - \beta R_s \{1\}^T \{I\}_c^{t+\Delta t}$$
$$- \beta \{V\}_c^{t+\Delta t} + (1-\beta) V_s^{t}$$
$$- (1-\beta) R_s \{1\}^T \{I\}_c^{t} - (1-\beta) \{V\}_c^{t}$$
$$= L_s \{1\}^T \frac{\{I\}^{t+\Delta t} - \{I\}^{t}}{\Delta t} \tag{10.52}$$

Rearranging terms to isolate the $t + \Delta t$ terms,

$$-\{1\}^T (L_s + \beta \Delta t R_s) \{I\}_c^{t+\Delta t} - \beta \Delta t V_c^{t+\Delta t}$$
$$= -\beta \Delta t V_s^{t+\Delta t} - (1-\beta) \Delta t V_s^{t} - \{1\}^T [L_s - (1-\beta) \Delta t R_s] \{I\}_c^{t}$$
$$+ (1-\beta) \Delta t V_c^{t} \tag{10.53}$$

where $V_s^{t+\Delta t}$ is a known quantity (the external source voltage). Equation (10.53) is the time-discretized form of the equation that joins coils in parallel to form circuits. In the global system, there is one equation of this form for each set of coils connected in parallel.

Time Discretization of the Mechanical Equations

In this section we perform the time discretization of the equations of motion.

Time Discretization of the Acceleration Equation

The acceleration equation is time discretized using equation (10.33):

$$\beta m \frac{dv^{t+\Delta t}}{dt} + (1-\beta)m \frac{dv^t}{dt} = m \frac{v^{t+\Delta t} - v^t}{\Delta t} \quad (10.54)$$

We now take equation (10.11), the acceleration equation, and split it into two equations, as follows:

$$\beta m \left(\frac{dv}{dt}\right)^{t+\Delta t} = -\beta \lambda v^{t+\Delta t} + \beta F_{em}^{t+\Delta t} - \beta F_{ext}^{t+\Delta t} \quad (10.55)$$

$$(1-\beta)m \left(\frac{dv}{dt}\right)^t = -(1-\beta)\lambda v^t + (1-\beta)F_{em}^t - (1-\beta)F_{ext}^t \quad (10.56)$$

Now substitute equations (10.55) and (10.56) into equation (10.54) to obtain

$$\left(\beta \lambda + \frac{m}{\Delta t}\right) v^{t+\Delta t} - \beta F_{em}^{t+\Delta t} = -\beta F_{ext}^{t+\Delta t}$$
$$+ \left[\frac{m}{\Delta t} - (1-\beta)\lambda\right] v^t + (1-\beta)F_{em}^t - (1-\beta)F_{ext}^t \quad (10.57)$$

Equation (10.57) is the time-discretized form of the acceleration equation. There is one equation of this form for each moving object in the system.

Time Discretization of the Velocity Equation

The velocity equation is also time discretized using equation (10.33):

$$\beta \frac{dx^{t+\Delta t}}{dt} + (1-\beta)\frac{dx^t}{dt} = \frac{x^{t+\Delta t} - x^t}{\Delta t} \quad (10.58)$$

Low-Frequency Applications

We take equation (10.12), the acceleration equation, and split it into two equations, as follows:

$$\beta \left(\frac{dx}{dt}\right)^{t+\Delta t} = \beta v^{t+\Delta t} \tag{10.59}$$

$$(1-\beta) \left(\frac{dx}{dt}\right)^{t} = (1-\beta) v^{t} \tag{10.60}$$

Now substitute equations (10.59) and (10.60) into equation (10.58) to obtain

$$\beta v^{t+\Delta t} - \frac{1}{\Delta t} x^{t+\Delta t} = -(1-\beta) v^{t} - \frac{1}{\Delta t} x^{t} \tag{10.61}$$

Equation (10.61) is the time-discretized form of the velocity equation. There is one equation of this form for each movable object in the system.

10.1.4 Linearization

The field equation and the acceleration equations are nonlinear functions of vector potential, A, and/or component displacement, x. These equations must be linearized before they can be combined with the other equations of the system in a global matrix equation. The linearization of the field and acceleration equations is described briefly in the following sections.

Linearization of the Field Equation

The field equation,

$$\nabla \times \nu \nabla \times A = \sigma \frac{V_b}{\ell} - \sigma \frac{\partial A}{\partial t} \tag{10.62}$$

is nonlinear in A in cases where the reluctivity, ν, is a function of flux density, B (and hence of vector potential, A).

To linearize this equation we begin with equation (10.38):

$$\left[\nu [S] + \frac{\sigma [T]}{\beta \Delta t}\right]\{A\}^{t+\Delta t} - \frac{\sigma}{\ell}\{Q\} V_b^{t+\Delta t}$$

$$= \left[\frac{\sigma [T]}{\beta \Delta t} - \frac{1-\beta}{\beta}\nu [S]\right]\{A\}^t + \frac{\sigma}{\ell}\{Q\}\frac{1-\beta}{\beta}V_b^t$$

We now apply the Newton–Raphson procedure to this equation, introduce the time dependence of the $[S]$ matrix, and obtain

$$\left[[G] + \frac{\sigma [T]}{\beta \Delta t}\right]\{\Delta A\}_{k+1}^{t+\Delta t} - \frac{\sigma}{\ell}\{Q\}\Delta V_{b,k+1}^{t+\Delta t}$$

$$= -\left[\nu_k^{t+\Delta t}[S]_k^{t+\Delta t} + \frac{\sigma [T]}{\beta \Delta t}\right]\{A\}_k^{t+\Delta t}$$

$$+ \left[\frac{\sigma [T]}{\beta \Delta t} - \frac{1-\beta}{\beta}\nu^t [S]^t\right]\{A\}^t + \frac{\sigma}{\ell}\{Q\}\frac{1-\beta}{\beta}V_b^t \qquad (10.63)$$

where

$$[G] = \nu_k^{t+\Delta t}[S]_k^{t+\Delta t} + \frac{2}{\Delta}\left(\frac{\partial \nu}{\partial B^2}\right)\left([S]_k^{t+\Delta t}\{A\}_k^{t+\Delta t}\right)\left([S]_k^{t+\Delta t}\{A\}_k^{t+\Delta t}\right)^T$$

Note that for linear materials, $\partial \nu / \partial B^2 = 0$, and equation (10.63) reduces to equation (10.38).

The field equation is also a nonlinear function of component displacement, x, when moving components are present. Using the Newton–Raphson method to linearize with respect to x, equation (10.63) yields:

$$\left[[G] + \frac{\sigma [T]}{\beta \Delta t}\right]\{\Delta A\}_{k+1}^{t+\Delta t} - \frac{\sigma}{\ell}\{Q\}\Delta V_{b,k+1}^{t+\Delta t}$$

$$+ \nu_k^{t+\Delta t}\left[\frac{\partial S}{\partial x}\right]\{A\}_k^{t+\Delta t}\{\Delta x\}_{k+1}^{t+\Delta t}$$

$$= -\left[\nu_k^{t+\Delta t}[S]_k^{t+\Delta t} + \frac{\sigma [T]}{\beta \Delta t}\right]\{A\}_k^{t+\Delta t}$$

$$+ \left[\frac{\sigma [T]}{\beta \Delta t} - \frac{1-\beta}{\beta}\nu^t [S]^t\right]\{A\}^t + \frac{\sigma}{\ell}\{Q\}\frac{1-\beta}{\beta}V_b^t \qquad (10.64)$$

(Differentiation of the [S] matrix is addressed in the next section.) Equation (10.64) is the linearized time-discretized diffusion equation.

Linearization of the Acceleration Equation

In this section, the acceleration equation will be linearized and combined with the velocity equation.

The acceleration equation,

$$m\frac{dv}{dt} + \lambda v = F_{em} - F_{ext} \tag{10.65}$$

is nonlinear with respect to vector potential, A, because the electromagnetic force, F_{em}, is a quadratic function of A. The Newton–Raphson technique will be used to linearize the acceleration equation with respect to the vector potential at the mesh nodes.

Recall the time-discretized form of the acceleration equation:

$$\left(\beta\lambda + \frac{m}{\Delta t}\right)v^{t+\Delta t} - \beta F_{em}^{t+\Delta t}$$
$$= -\beta F_{ext}^{t+\Delta t} + \left[\frac{m}{\Delta t} - (1-\beta)\lambda\right]v^t + (1-\beta)F_{em}^t - (1-\beta)F_{ext}^t \tag{10.66}$$

Now apply the Newton–Raphson procedure to this equation and obtain

$$\left(\beta\lambda + \frac{m}{\Delta t}\right)\Delta v_{k+1} - \beta\left\{\frac{\partial F_{em}}{\partial A}\right\}\{\Delta A\}_{k+1} - \beta\frac{\partial F_{em}}{\partial x}\Delta x_{k+1}$$
$$= -\beta F_{ext}^{t+\Delta t} + \left[\frac{m}{\Delta t} - (1-\beta)\lambda\right]v^t + (1-\beta)F_{em}^t - (1-\beta)F_{ext}^t$$
$$- \left(\frac{m}{\Delta t} + \beta\lambda\right)v_k^{t+\Delta t} + \beta F_{em,k}^{t+\Delta t} \tag{10.67}$$

Now, how are the derivatives $\left\{\frac{\partial F_{em}}{\partial A}\right\}$ and $\frac{\partial F_{em}}{\partial x}$ computed? First note that the electromagnetic force acting on a component, according to the method of virtual work, is

$$F_{em} = -\left.\frac{\partial W_{mag}}{\partial x}\right|_{\text{constant magnetic flux}} \tag{10.68}$$

where W_{mag} is the stored magnetic energy of the system and ∂x represents "virtual motion" of the component under consideration.

The magnetic energy can be written as

$$W_{mag} = \frac{1}{2} v\ell \{A\}^T [S] \{A\} \tag{10.69}$$

Substitution of equation (10.69) into equation (10.68) yields

$$F_{em} = -\frac{1}{2} v\ell \{A\}^T \left[\frac{\partial S}{\partial x}\right] \{A\} \tag{10.70}$$

Differentiation of the stiffness matrix with respect to x (or y) is straightforward, because the entries of the stiffness matrix are simple functions of x and y. Numerical values are given to the derivatives by assigning a factor between 0 and 1 to each mesh node. Nodes that are displaced (virtually, in the case of equation (10.70), or physically, in the case of equation (10.64)) directly by object motion, such as the nodes fixed to the surface of a moving object, are assigned a factor of 1. Nodes that are unaffected by motion, such as those attached to fixed objects, are assigned a factor of 0. Intermediate factors between 0 and 1 may be assigned to nodes in the air surrounding the moving object in such a way that the object motion is "absorbed" by the finite elements in the air region.

From equation (10.70), then, we define new variables C and U as follows:

$$\frac{\partial F_{em}}{\partial A} = -v\ell \{A\}^T \left[\frac{\partial S}{\partial x}\right] = -\{C\}^T \ell \tag{10.71}$$

$$\frac{\partial F_{em}}{\partial x} = -\frac{1}{2} v\ell \{A\}^T \left[\frac{\partial^2 S}{\partial x^2}\right] \{A\} = -U \tag{10.72}$$

Low-Frequency Applications

Substituting equations (10.71) and (10.72) into equation (10.67), we obtain

$$-\left(\beta\lambda + \frac{m}{\Delta t}\right)\Delta v_{k+1} + \beta\ell\{C\}^T\{\Delta A\}_{k+1} + \beta U \Delta x_{k+1} =$$
$$-\beta F_{\text{ext}}^{t+\Delta t} + \left[\frac{m}{\Delta t} - (1-\beta)\lambda\right]v^t$$
$$+(1-\beta)F_{\text{em}}^t - (1-\beta)F_{\text{ext}}^t - \left(\beta\lambda + \frac{m}{\Delta t}\right)v_k^{t+\Delta t} + \beta F_{\text{em}}^{t+\Delta t} \quad (10.73)$$

Next, we take equation (10.61), the velocity equation, and put it in Newton–Raphson form.

$$\beta \Delta v_{k+1} - \frac{1}{\Delta t}\Delta x_{k+1} = -\beta v_{k+1}^{t+\Delta t} + \frac{1}{\Delta t}x_k^{t+\Delta t} - (1-\beta)v^t - \frac{1}{\Delta t}x^t$$

or

$$\Delta v_{k+1} = \frac{1}{\beta \Delta t}\Delta x_{k+1} - v_k^{t+\Delta t} + \frac{1}{\beta \Delta t}x_k^{t+\Delta t} - \frac{1-\beta}{\beta}v^t - \frac{1}{\beta \Delta t}x^t \quad (10.74)$$

Now substitute equation (10.74) into equation (10.73) to obtain

$$\left(\beta\lambda + \frac{m}{\Delta t}\right)\left[\frac{1}{\beta \Delta t}\Delta x_{k+1} - v_k^{t+\Delta t} + \frac{1}{\beta \Delta t}x_k^{t+\Delta t} - \frac{1-\beta}{\beta}v^t - \frac{1}{\beta \Delta t}x^t\right]$$
$$+\beta\ell\{C\}^T\{A\}_{k+1} + \beta U \Delta x_{k+1} = -\beta F_{\text{ext}}^{t+\Delta t} + \left[\frac{m}{\Delta t} - (1-\beta)\lambda\right]v^t$$
$$+(1-\beta)F_{\text{em}}^t - (1-\beta)F_{\text{ext}}^t - \left(\frac{m}{\Delta t} + \beta\lambda\right)v_k^{t+\Delta t} + \beta F_{\text{em},k}^{t+\Delta t}$$

Note that $\beta F_{\text{em}}^{t+\Delta t} = -\beta \frac{1}{2}\ell\{C\}^T\{A\}_k^{t+\Delta t}$, and rearrange terms to yield

$$\beta\ell\{C\}^T\{\Delta A\}_k^{t+\Delta t} + \left(\frac{\lambda}{\Delta t} + \frac{m}{\beta(\Delta t)^2} + \beta U\right)\Delta x_{k+1}$$
$$= -\frac{1}{2}\ell\{C\}^T\{A\}_k^{t+\Delta t} + \left(-\frac{\lambda}{\Delta t} - \frac{m}{\beta(\Delta t)^2}\right)x_k^{t+\Delta t} + \frac{m}{\beta \Delta t}v^t$$
$$+ \left(\frac{\lambda}{\Delta t} + \frac{m}{\beta(\Delta t)^2}\right)x^t + (1-\beta)F_{\text{em}}^t - (1-\beta)F_{\text{ext}}^t - \beta F_{\text{ext}}^{t+\Delta t} \quad (10.75)$$

10.1.5 Global System of Equations

The field, circuit, and mechanical equations are now available in a discretized and linearized form. It remains to assemble these matrix equations into a global system of equations describing the entire problem.

Summary of the Global Equations

The six discretized and linearized system equations are summarized next. Note that all equations have been placed in Newton–Raphson form.

Field Equation *(one per mesh node)*

$$\left[[G] + \frac{\sigma [T]}{\beta \Delta t} \right] \{\Delta A\}_{k+1}^{t+\Delta t} - \frac{\sigma}{\ell} \{Q\} \Delta V_{b,k+1}^{t+\Delta t} + v_k^{t+\Delta t} \left[\frac{\partial S}{\partial x} \right] \{A\}_k^{t+\Delta t} \{\Delta x\}_{k+1}^{t+\Delta t}$$

$$= -\left[v_k^{t+\Delta t} [S]_k^{t+\Delta t} + \frac{\sigma [T]}{\beta \Delta t} \right] \{A\}_k^{t+\Delta t} + \left[\frac{\sigma [T]}{\beta \Delta t} - \frac{1-\beta}{\beta} v^t [S]^t \right] \{A\}^t$$

$$+ \frac{\sigma}{\ell} \{Q\} \frac{1-\beta}{\beta} V_b^t \qquad (10.76)$$

Total Current Equation *(one per coil)*

$$-[\sigma Q]_c^T \{\Delta A\}_{k+1}^{t+\Delta t} + \beta \Delta t \left[\frac{\sigma}{\ell} \Delta_b \right] \{\Delta V_b\}_{c,k+1}^{t+\Delta t} - \beta \Delta t [D] \{\Delta I_b\}_{c,k+1}^{t+\Delta t}$$

$$= \ell \{C\} [\sigma Q]_c^T \{A\}_k^{t+\Delta t} - \beta \Delta t \left[\frac{\sigma}{\ell} \Delta_b \right] \{V_b\}_{c,k}^{t+\Delta t} + \beta \Delta t [D] \{I_b\}_{c,k}^{t+\Delta t}$$

$$- [\sigma Q]_c^T \{A\}^t - (1-\beta) \Delta t \left[\frac{\sigma}{\ell} \Delta_b \right] \{V_b\}_c^t + (1-\beta) \Delta t [D] \{I_b\}_c^t \qquad (10.77)$$

Series Bar-Coil Equation *(one per coil)*

$$-\beta \Delta t \{D\}_c^T \{\Delta V_b\}_{c,k+1}^{t+\Delta t} - (L_{\text{ext}} + \beta \Delta t R_{\text{ext}}) \Delta I_{c,k+1}^{t+\Delta t} + \beta \Delta t \Delta V_{c,k+1}^{t+\Delta t}$$

$$= \beta \Delta t \{D\}_c^T \{V_b\}_{c,k}^{t+\Delta t} + (L_{\text{ext}} + \beta \Delta t R_{\text{ext}}) I_{c,k}^{t+\Delta t} - \beta \Delta t V_{c,k}^{t+\Delta t}$$

$$+ (1-\beta) \Delta t \{D\}_c^T \{V_b\}_c^t$$

$$- [L_{\text{ext}} - (1-\beta) \Delta t R_{\text{ext}}] I_c^t - (1-\beta) \Delta t V_c^t \qquad (10.78)$$

Low-Frequency Applications

Parallel Coil Equation *(one per set of parallel-connected coils)*

$$\{1\}^T \beta \Delta t \{\Delta I\}_{c,k+1}^{t+\Delta t} + \frac{(\beta \Delta t)^2}{L_s + \beta \Delta t R_s} \Delta V_{c,k+1}^{t+\Delta t}$$

$$= -\{1\}^T \beta \Delta t \{I\}_{c,k}^{t+\Delta t} - \frac{(\beta \Delta t)^2}{L_s + \beta \Delta t R_s} V_{c,k}^{t+\Delta t} + \frac{(\beta \Delta t)^2}{L_s + \beta \Delta t R_s} V_s^{t+\Delta t}$$

$$+ \frac{\beta \Delta t}{L_s + \beta \Delta t R_s} (1 - \beta) \Delta t V_s^t$$

$$+ \{1\}^T \frac{\beta \Delta t}{L_s + \beta \Delta t R_s} [L_s - (1 - \beta) \Delta t R_s] \{I\}_c^t$$

$$- \frac{\beta \Delta t}{L_s + \beta \Delta t R_s} (1 - \beta) \Delta t V_c^t \quad (10.79)$$

Motion Equation *(one per movable component)*

$$\ell \{C\}^T \{\Delta A\}_{k+1}^{t+\Delta t} + \left(\frac{\lambda}{\beta \Delta t} + \frac{m}{(\beta \Delta t)^2} + U \right) \Delta x_{k+1}^{t+\Delta t}$$

$$= -\frac{1}{2\beta} \ell \{C\}^T \{A\}_k^{t+\Delta t} + \left(-\frac{\beta \lambda}{\Delta t} - \frac{m}{(\beta \Delta t)^2} \right) x_k^{t+\Delta t} + \frac{m}{\beta^2 \Delta t} v^t$$

$$+ \left(\frac{\lambda}{\beta \Delta t} + \frac{m}{(\beta \Delta t)^2} \right) x^t + \frac{1-\beta}{\beta} F_{em}^t - \frac{1-\beta}{\beta} F_{ext}^t - F_{ext}^{t+\Delta t} \quad (10.80)$$

Assembly of the Global System of Equations

From the preceding summary, we see that there are five vector unknowns:

$\{\Delta A\}_{k+1}^{t+\Delta t}$:   change in vector potential of each node

$\{\Delta V_b\}_{k+1}^{t+\Delta t}$:   change in voltage across each bar

$\{\Delta I\}_{c,k+1}^{t+\Delta t}$:   change in current in each coil

$\{\Delta V_c\}_{k+1}^{t+\Delta t}$:   change in terminal voltage of each set of parallel-connected coils

$\{\Delta x\}_{k+1}^{t+\Delta t}$:   change in position of each movable component

The global system matrix equation may then be set up in the form

$$[M]\{f\} = \{N\} \tag{10.81}$$

or, in expanded form,

$$\begin{bmatrix} M_{1,1} & M_{1,2} & & & M_{1,5} \\ M_{1,2}^T & M_{2,2} & M_{2,3} & & \\ & M_{2,3}^T & M_{3,3} & M_{3,4} & \\ & & M_{3,4}^T & M_{4,4} & \\ M_{1,5}^T & & & & M_{5,5} \end{bmatrix} \begin{Bmatrix} \{\Delta A\} \\ \{\Delta V_b\} \\ \{\Delta I\} \\ \{\Delta V_c\} \\ \{\Delta x\} \end{Bmatrix}_{k+1}^{t+\Delta t} = \begin{Bmatrix} \{N_1\} \\ \{N_2\} \\ \{N_3\} \\ \{N_4\} \\ \{N_5\} \end{Bmatrix} \tag{10.82}$$

This global system of equations must now be solved for $\{f\}$.

System Solution

In this section, the characteristics of the global system of equations will be discussed with respect to some common solution algorithms.

The square matrix, $[M]$, is symmetric and generally sparse (the mechanical motion equations are somewhat dense, but they are usually vastly outnumbered by the sparse finite element field equations) and can therefore be stored in an efficient manner.

$[M]$ is not positive definite, however, because

$$M_{3,3} = -(L_{\text{ext}} + \beta \Delta t R_{\text{ext}})$$

which is always negative, while all other diagonal entries in $[M]$ are always positive. This is of no concern if Gaussian elimination is to be used to solve the system, but it precludes the use of solvers that perform Choleski decompositions, such as the incomplete Choleski conjugate gradient (ICCG) method.[3]

It is possible, by adding another equation to the system, to force $[M]$ to be both symmetric and positive definite. First take the third equation:

$$M_{3,2}\{\Delta V_b\} + M_{3,3}\{\Delta I\} + M_{3,4}\{\Delta V_c\} = \{N_3\} \qquad (10.83)$$

Now split equation (10.83) into two equations, as follows:

$$M_{3,2}\{\Delta V_b\} - M_{3,3}\{\Delta I\}$$
$$+ M_{3,4}\{\Delta V_c\} + 2M_{3,3}\{\Delta I_{\text{aux}}\} = \{N_3\} \qquad (10.84)$$
$$2M_{3,3}\{\Delta I\} - 2M_{3,3}\{\Delta I\}_{\text{aux}} = \{0\} \qquad (10.85)$$

With the addition of the new equation the global system components become

---
[3] The Choleski decomposition involves taking the square root of the diagonal elements. When the negative-signed diagonal elements are present, the square root cannot be taken.

$$\begin{bmatrix} M_{1,1} & M_{1,2} & & & & M_{1,5} \\ M_{1,2}^T & M_{2,2} & M_{2,3} & & & \\ & M_{2,3}^T & -M_{3,3} & 2M_{3,3} & M_{3,4} & \\ & & 2M_{3,3}^T & -2M_{3,3} & & \\ & & M_{3,4}^T & & M_{4,4} & \\ M_{1,5}^T & & & & & M_{5,5} \end{bmatrix} \begin{Bmatrix} \{\Delta A\} \\ \{\Delta V_b\} \\ \{\Delta I\} \\ \{\Delta I_{\text{aux}}\} \\ \{\Delta V_c\} \\ \{\Delta x\} \end{Bmatrix}_{k+1}^{t+\Delta t}$$

$$= \begin{Bmatrix} \{N_1\} \\ \{N_2\} \\ \{N_3\} \\ \{0\} \\ \{N_4\} \\ \{N_5\} \end{Bmatrix} \quad (10.86)$$

where

$$\begin{aligned} M_{1,1} &= \ell\left[[G] + \frac{\sigma[T]}{\beta\Delta t}\right] \\ M_{1,2} &= -\sigma\{Q\} \\ M_{1,5} &= \ell\{C\}^T \\ M_{2,2} &= \beta\Delta t\left[\frac{\sigma}{\ell}\Delta_b\right] \\ M_{2,3} &= -\beta\Delta t\,[D] \\ M_{3,3} &= -(L_{\text{ext}} + \beta\Delta t\,R_{\text{ext}}) \\ M_{3,4} &= \beta\Delta t \\ M_{4,4} &= \frac{(\beta\Delta t)^2}{L_s + \beta\Delta t\,R_s} \\ M_{5,5} &= \frac{\lambda}{\beta\Delta t} + \frac{m}{(\beta\Delta t)^2} + U \end{aligned}$$

Low-Frequency Applications

and

$$\{N_1\} = -\ell \left[v_k^{t+\Delta t} [S]_k^{t+\Delta t} + \frac{\sigma [T]}{\beta \Delta t} \right] \{A\}_k^{t+\Delta t}$$

$$+ \ell \left[\frac{\sigma [T]}{\beta \Delta t} - \frac{1-\beta}{\beta} v^t [S]^t \right] \{A\}^t + \sigma \{Q\} \frac{1-\beta}{\beta} V_b^t$$

$$\{N_2\} = [\sigma Q]_c^T \{A\}_k^{t+\Delta t} - \beta \Delta t \left[\frac{\sigma}{\ell} \Delta_b \right] \{V_b\}_{c,k}^{t+\Delta t} + \beta \Delta t [D] \{I_b\}_{c,k}^{t+\Delta t}$$

$$- [\sigma Q]_c^T \{A\}^t - (1-\beta) \Delta t \left[\frac{\sigma}{\ell} \Delta_b \right] \{V_b\}_c^t + (1-\beta) \Delta t [D] \{I_b\}_c^t$$

$$\{N_3\} = \beta \Delta t \{D\}_c^T \{V_b\}_{c,k}^{t+\Delta t} + (L_{\text{ext}} + \beta \Delta t R_{\text{ext}}) I_{c,k}^{t+\Delta t} - \beta \Delta t V_{c,k}^{t+\Delta t}$$

$$+ (1-\beta) \Delta t \{D\}_c^T \{V_b\}_c^t - [L_{\text{ext}} - (1-\beta) \Delta t R_{\text{ext}}] I_c^t - (1-\beta) \Delta t V_c^t$$

$$\{N_4\} = -\{1\}^T \beta \Delta t \{I\}_{c,k}^{t+\Delta t} - \frac{(\beta \Delta t)^2}{L_s + \beta \Delta t R_s} V_{c,k}^{t+\Delta t} + \frac{(\beta \Delta t)^2}{L_s + \beta \Delta t R_s} V_s^{t+\Delta t}$$

$$+ \frac{\beta \Delta t}{L_s + \beta \Delta t R_s} (1-\beta) \Delta t V_s^t + \{1\}^T \frac{\beta \Delta t}{L_s + \beta \Delta t R_s}$$

$$\times [L_s - (1-\beta) \Delta t R_s] \{I\}_c^t - \frac{\beta \Delta t}{L_s + \beta \Delta t R_s} (1-\beta) \Delta t V_c^t$$

$$\{N_5\} = -\frac{1}{2\beta} \ell \{C\}^T \{A\}_k^{t+\Delta t} + \left(-\frac{\beta \lambda}{\Delta t} - \frac{m}{(\beta \Delta t)^2} \right) x_k^{t+\Delta t} + \frac{m}{\beta^2 \Delta t} v^t$$

$$+ \left(\frac{\lambda}{\beta \Delta t} + \frac{m}{(\beta \Delta t)^2} \right) x^t + \frac{1-\beta}{\beta} F_{\text{em}}^t - \frac{1-\beta}{\beta} F_{\text{ext}}^t - F_{\text{ext}}^{t+\Delta t}$$

The global system matrix is now sparse, symmetric, and positive definite, and the system may be solved by any of a variety of methods.

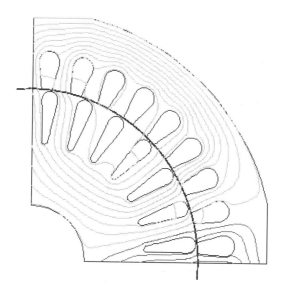

Figure 10.4 Equipotential Plot at One Instant

10.1.6 Example

The preceding formulation has been applied to a number of coupled electromechanical problems. An example illustrates the coupling of fields, circuits, and motion.

Squirrel Cage Induction Motor

We consider a three-phase squirrel cage induction motor. The motor is rated at 5 hp. The parameters of the motor are listed in Table 10.1. In this example the entire motor winding is represented. The inputs are the instantaneous voltages at the three terminals. The currents in the windings are unknown. The rotor is free to turn. Each rotor bar is represented as an independent circuit connected to an end ring that has a constant resistance and inductance. The mesh in the air gap may or may not be remeshed at each time step, depending on the distortion of the elements. In any case, the remeshing is done so that the number of nodes and elements remains the same. The sequence of plots in Figures 10.4, 10.5, and 10.6 shows the motor operating at full load at various positions in a cycle.

Low-Frequency Applications

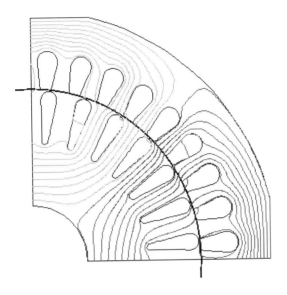

Figure 10.5 Plot at a Later Time

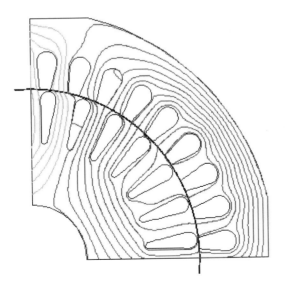

Figure 10.6 Plot at a Still Later Time

Table 10.1 Electrical and Mechanical Parameters of 5-HP Induction Motor

	Parameter	Value	Units
Voltage source	Phase voltage	220	V
	Frequency	50	Hz
	Resistance/phase	0.13	Ω
	Inductance/phase	0.02	mH
Winding coils	DC resistance	0.951	Ω
	Lead inductance	4.87	mH
	Winding type	Double layer	
	Pole pitch	6	slots
	Number of turns	160	turns
Rotor end ring	Interbar resistance	2.10	$\mu\Omega$
	Interbar inductance	0.04	μH
Rotor load	Inertia	$6.191 \cdot 10^{-3}$	kg·m^2
Rotor bars	Electric conductivity	$4.90 \cdot 10^7$	℧/m

10.2 MODELING OF FLOW ELECTRIFICATION IN INSULATING TUBES

The streaming electrification phenomena due to the flow of low-conductivity liquids through pipes, filters, etc. was initially recognized and studied in the petroleum industry [110] [111]. Investigation of the same phenomena in electric power apparatus dates from the 1970's when the failures of some forced oil cooled transformers were found to be triggered by static electrification [112] [113] [114] initiated by the flow of liquid coolant.

Streaming electrification occurs when a low-conductivity liquid flows over a solid surface and entrains the diffused part of the charge double layer [115], while the solid phase retains the corresponding countercharge on its surface. The accumulation of unipolar charges on an insulating or isolated part of the system causes a voltage buildup, which,

Low-Frequency Applications 551

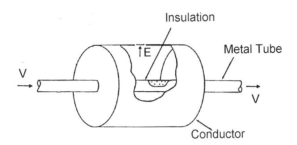

Figure 10.7 Experimental Arrangement of the Model after Gasworth [116]

above some threshold level, may cause electrical discharges and result in dielectric failure in the case of power apparatus.

The previous work on streaming electrification (mainly experimental) has been primarily devoted to steady state phenomena. The temporal behavior of flow-induced electrification will now be discussed.

10.2.1 Description of the Physical Model

Figure 10.7 shows the physical model to be studied. This model was originally used in the experimental work done by Gasworth [116] in support of electrification studies in Freon-filled power equipment. The insulating tube lies on the axis of a grounded capped cylindrical conducting annulus and is connected to grounded metallic pipe sections on both ends. The metallic pipe section located upstream generates the convection current that drives the charging process. It is sufficiently long to ensure fully developed turbulent flow.

The insulating tube is made of Tefzel, a type of Teflon (polytetrafluoroethylene) developed by Dupont. The liquid is Dupont's refrigerant Freon TF (CCl_2F–$CClF_2$).

The model geometry is axisymmetric with respect to the axis of the insulating tube. The related parameters summarized in Table 10.2 are those of run 60 of the experimental model reported in [116].

The external charging mode is defined as the case in which the dominant contribution to charge generation is in the metallic pipes upstream of the insulating section; any local charge generation inside the insulating tube is neglected. In addition, the ion migration

Table 10.2 Parameters for the Capped Cylinder Case

Symbol	Description	Value	Unit
a	Tube radius	1.3	mm
R	Outer conductor radius	75	mm
L	Tube length	300	mm
$\frac{\sigma}{\ell}$	Liquid bulk conductivity	47.0	pS/m
σ_e	Effective conductivity	52.8	pS/m
σ_Ω	Conductivity of domain Ω	0	pS/m
$\epsilon_\Omega/\epsilon_0$	Permittivity of domain Ω	1	
ϵ_w/ϵ_0	Permittivity of tube wall	2.6	
R_y	Reynolds number	12,000	
δ_e	Molecular Debye length	20.0	μm
δ_f	Diffusion sublayer length	5.0	μm
$I(0, t)$	Tube inlet current	-1.2	nA

is regarded as the dominant mechanism for the surface charge accumulation process. Therefore, the insulating tube remains a passive charge collecting and leakage site.

The following assumptions are made in the modeling:

1. The flow is turbulent and the velocity profile is fully developed throughout the pipe.
2. The Debye length exceeds the thickness of the laminar sublayer or $\delta_e \gg \delta_f$.
3. The bulk of the charge is spread out and carried by the turbulent core region, where the mean liquid velocity is practically the nominal velocity: $\bar{\rho}_v(z, t)\pi a^2 \gg \rho_w 2\pi a \delta_f$.
4. The migration component of the radial current density within the diffusion sublayer dominates the diffusion component. For the cases where $\frac{LV}{D_e}$ is large, axial diffusion is neglected as well.

Low-Frequency Applications 553

5. The volume charge density in the fluid, which arises from a small imbalance of the positive and negative ion concentrations, is very small. This implies that the liquid conductivity, σ_l, is uniform and independent of the charge density throughout the fluid. It remains close to its nominal value, σ_0, where the net charge density vanishes.

6. The bulk conductivity of the insulating wall is so small that the bulk charge relaxation time ($\tau_w = \frac{\epsilon_w}{\sigma_w}$) of the tube is long compared with the transient charging time. This means that the charge precipitated from the flow is stored at the liquid–solid interface as a surface charge density, ρ_s, that can leak away only by conduction through the liquid or along the interface.

7. The space surrounding the insulating tube is perfectly insulating

8. The axial electric field is practically uniform over the tube cross section. This is true if the contribution of the local charges is not significant and the axial field arises from distant sources (image charges on the external conductors) distributed over a length scale larger than a tube radius.

9. The surface charge accumulation is due to upstream current, $I(0, t)$, which is assumed constant or slowly varying on the scale of the liquid residence time in the tube for the examples shown. This is not a limitation of the proposed modeling technique, which can easily handle a time variation of the influent charge.

10.2.2 Electric Potential Calculation in Ω Region

The primary objective of this modeling is the prediction of the temporal and spatial evolution of the charge distribution along the insulating tube and the electric stress within the insulating tube wall, shown as $\mathbf{E}_r^w$ in the cross-sectional view of Figure 10.8. When the thickness of the tube wall is regarded as small compared with the inner diameter a shown in Figure 10.9, the field computations can be achieved conveniently if the potential distributions in the axisymmetric domain Ω are calculated first. Domain Ω represents the region surrounding the insulating tube and bounded by the capped cylinder conducting annulus. Once the radial field $\mathbf{E}_r^\Omega$ on the Γ_4 boundary is calculated, the radial field within the wall, $\mathbf{E}_r^w$, follows from Gauss' law as

$$\mathbf{E}_r{}^w(z, t) = \frac{\epsilon_\Omega}{\epsilon_w} \mathbf{E}_r^\Omega(a, z, t) \qquad (10.87)$$

where ϵ_Ω and ϵ_w are the permittivities of region Ω and the tube wall, respectively.

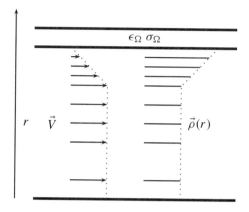

Figure 10.8 Insulating Tube Definitions With Charge and Velocity

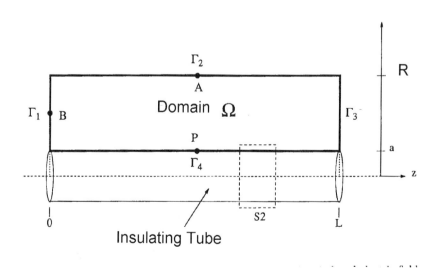

Figure 10.9 Axisymmetric Domain Used for the Analysis of the Flow-Induced Field

Low-Frequency Applications 555

Based on the model arrangement shown in Figure 10.9), the material properties and the assumptions previously described, the electric potential ϕ in domain Ω is formulated [117] as summarized here:

$$\phi = \phi\,(r, z, t) \tag{10.88}$$

$$\nabla^2 \phi = 0 \quad \text{on} \quad \Omega \tag{10.89}$$

$$\phi = 0 \quad \text{on} \quad \Gamma_1, \Gamma_2, \Gamma_3 \tag{10.90}$$

$$\frac{\partial \phi}{\partial r} + c_1 \frac{\partial^2 \phi}{\partial z^2} = -c_2\,\lambda(z, t) \quad \text{on} \quad \Gamma_4 \tag{10.91}$$

$$\frac{\partial \lambda(z, t)}{\partial t} = c_3 \frac{\partial^2 \phi}{\partial z^2} + c_4\,e^{-\frac{z}{d}} \tag{10.92}$$

where the symbols are defined in the list of symbols.

10.2.3 Solution

The quantities ϕ and λ are both functions of time (t) and space (r, z). The solution method combines the Laplace transform and finite element method. The major steps of the method are as follows:

Step 1: The Laplace transform method is used to transform all time-dependent equations from the (r, z, t) domain to algebraic equations in the (r, z, s) domain.

Step 2: The transformed potential variable, $\bar{\phi}(r, z, s)$, is expanded in terms of a power series of s^{-1}, where the coefficients of the series terms denoted by $\psi(r, z)$ are functions of the space variables (r, z).

Step 3: The finite element method is used to solve for spatial coefficients $\psi(r, z)$ over predefined nodal points throughout the domain of the problem for an appropriate number of iterations.

Step 4: After performing inverse Laplace transforms, the computed values of $\psi_k(rz)$ for $k = 1, 2, \ldots$ are used to obtain the transient time scale, the nodal potentials $\phi(r, z, t)$, and the equivalent line charge density along the tube for any given time. Electrical stresses can then be calculated from the electrostatic potentials obtained.

10.2.4 Laplace Transform Formulation

Assuming $\bar{\lambda}(z, s)$ and $\bar{\phi}(r, z, s)$ to be the Laplace transform of $\lambda(z, t)$ and $\phi(r, z, t)$, respectively, after some manipulations we get the following set of equations:

$$\bar{\phi} = \bar{\phi}(r, z, s) \tag{10.93}$$

$$\nabla^2 \bar{\phi} = 0 \quad \text{on} \quad \Omega \tag{10.94}$$

$$\bar{\phi} = 0 \quad \text{on} \quad \Gamma_1, \Gamma_2, \Gamma_3 \tag{10.95}$$

$$\frac{\partial \bar{\phi}}{\partial r} + c_1 \frac{\partial^2 \bar{\phi}}{\partial z^2} = -c_2 \bar{\lambda}(z, s) \quad \text{on} \quad \Gamma_4 \tag{10.96}$$

and

$$\bar{\lambda}(z, s) = \gamma^2 c_4 e^{\frac{-z}{d}} + \gamma c_3 \frac{\partial^2 \bar{\phi}}{\partial z^2} + \gamma \lambda(z, 0) \quad \text{on} \quad \Gamma_4 \tag{10.97}$$

where $\gamma = s^{-1}$.

10.2.5 Introduction of an Auxiliary Variable

The s-dependent $\bar{\phi}(r, z, s)$ is expanded as a power series in $\gamma = s^{-1}$:

$$\bar{\phi}(r, z, s) \equiv \sum_{k=0}^{\infty} \psi_k(rz) \gamma^k \tag{10.98}$$

Notice that the series coefficients, $\psi_k(rz)$, for $k = 1, 2, \ldots$, are functions of the space variables (r and z). Because the transformed potential variable $\bar{\phi}(r, z, s)$ satisfies Laplace's equation on Ω, equation (10.94), and because $\gamma^k \neq 0$ for $k = 0, 1, 2 \ldots$, it is clear that $\psi_k(rz)$ satisfies the following set of equations and boundary conditions:

$$\nabla^2 \psi_k(r, z) = 0 \tag{10.99}$$

$$\psi_k(r, z) = 0 \quad \text{on} \quad \Gamma_1, \Gamma_2, \Gamma_3 \tag{10.100}$$

Low-Frequency Applications

The boundary condition for $\psi_k(rz)$ on Γ_4 comes from equation (10.96) by substituting for $\bar{\phi}$ its value from equation (10.98), expanding and rearranging the resulting equation in ascending powers of γ, and equating the coefficients of equal powers of γ:

$$\frac{\partial \psi_k}{\partial r} = -c_1 \frac{\partial^2 \psi_k}{\partial z^2} + f \qquad (10.101)$$

where

$$f = -c_{24} e^{-\frac{z}{d}} \quad \text{for } k = 2 \qquad (10.102)$$

$$f = -c_{23} \frac{\partial^2 \psi_{k-1}}{\partial z^2} \quad \text{for } k \geq 3 \qquad (10.103)$$

Equation (10.101) for $k = 0$ to ∞ gives the boundary conditions on Γ_4 required for the solution of the corresponding variable $psi_k(rz)$ over the domain Ω. When the initial equivalent line charge density along the pipe $\lambda(z, 0)$ is zero, ψ_0 and ψ_1 both have zero value throughout Ω, making no contribution to the final results. For $k = 2$, the right-hand side $(-c_{24} e^{-\frac{z}{d}}) = (\frac{-I(0,t)}{2\pi a \epsilon_\Omega \tau_l V} e^{-\frac{z}{d}})$, which is a function of the influent current as well as pipe and fluid parameters, works as the driving force for the entire charging process.

Finally, based on the preceding formulations, we solve for the auxiliary variable $\psi_k(rz)$ over the axisymmetric domain Ω for a finite number of iterations. $\psi_k(rz)$ is governed by Laplace's equation and satisfies the boundary conditions as shown by equations (10.100) and (10.101).

10.2.6 Finite Element Formulation

The finite element method is used to solve for the sequence of auxiliary variables, $\psi_k(rz)$, over the axisymmetric domain Ω shown in Figure 10.9. At each iteration the unknown $\psi_k(rz)$ satisfies Laplace's equation (10.99) and the boundary conditions of equations (10.100) and (10.101). The Galerkin weighted residual method (see Chapter 4) is used to find an approximate solution by transforming the original PDE problem to the following system of linear algebraic equations:

$$\mathbf{S}\,\psi_k = \mathbf{F}_k \qquad (10.104)$$

where $\mathbf{S}$ is the stiffness or coefficient matrix, ψ_k is the unknown vector, and $\mathbf{F}$ is the load or forcing vector. As inferred from the preceding formulations, the solution for ψ_0 and ψ_1 is always zero. When solving for ψ_2, the right-hand side vector **F**, as given by equation (10.102), is physically related to the influent charge or streaming current to the tube. When solving for ψ_k with $k \geq 3$, the right-hand side vector **F** is computed from equation (10.103), where the solution from the previous iteration ψ_{k-1} is used. The stiffness matrix, **S**, is constructed, decomposed, and saved only once. For any further iterations only the right-hand side must be updated.

10.2.7 Computation of Charge, Potential, and Time Scale

Once the sequence of the auxiliary variable, $\psi_k(rz)$ is found using an adequate number of iterations, an inverse Laplace transformation is performed. After some mathematical manipulations, the necessary expressions for the electric potential $\phi(r, z, t)$ over the entire domain, Ω, and the line charge density $\lambda(z, t)$ along the tube at any given time are obtained as follows:

$$\phi(r, z, t) = \sum_{k=1}^{\infty} \psi_{k+1}(rz) \frac{t^k}{k!} \tag{10.105}$$

$$\lambda(z, t) = c_4 t e^{-\frac{z}{d}} + c_3 \sum_{k=2}^{\infty} \frac{t^k}{k!} \frac{\partial^2 \psi_k}{\partial z^2} + \lambda(z, 0) \tag{10.106}$$

where $\lambda(z, 0)$ is the initial value of the equivalent line charge density. As for the time scale of the charging transient, τ, in the present study our solution for $\psi_k(rz)$, appears exponential in time for time sufficiently large. Thus, a numerical scheme can be adopted in which the coefficient of the exponential dependence at large t can be extracted [117]. To be specific, when looking at the sequence of functions $\psi_k(rz)$, as k gets larger, we observe that:

$$\psi_{k+1} \approx -\beta \psi_k \tag{10.107}$$

where the positive constant β is the reciprocal of the relaxation time scale, τ. Therefore,

$$\tau = |\frac{1}{\beta}| \approx |\frac{\psi_k}{\psi_{k+1}}| \tag{10.108}$$

This approximation gets better as k increases. We can find the value of β numerically by monitoring how the ratio of the maximum values of ψ_{k+1} and ψ_k changes as k increases. When this ratio is constant to about three decimal places, we take it as β.

Low-Frequency Applications

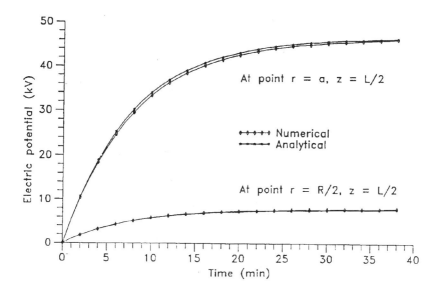

Figure 10.10 Numerical–Analytical Temporal Evolution of Potential $\phi(a, L/2, t)$

10.2.8 Numerical versus Analytical Results

The method is applied to the physical model of Figure 10.9 for the purpose of checking its validity and accuracy. Table 10.2 contains the related parameters. For test purposes the exponential term $c_4 e^{-\frac{z}{a}}$ in the forcing function of equation (10.92) is replaced by the sine function $c_4 \sin(\omega z)$ where $\omega = \pi/L$. With this simple, if nonphysical, substitution we can get an exact analytical solution to our original problem represented by equations (10.88)-(10.92) by using the separation of variables technique (see [117] for details).

The same problem is solved using the finite element method. The sequence of auxiliary variables, ψ_k, is calculated for 15 iterations. Figure 10.10 shows close agreement between the analytical and numerical values of the potential at an arbitrary point of domain Ω. The Laplace finite element method has been verified with test data as well. For further details see [118]

List of Symbols

ϕ : scalar potential

r : radial coordinate

z : axial coordinate

t : time

a : inner radius of the insulating tube

R : radius of the outer conductor

$I(0, t)$: tube inlet current

E : electric field

L : length of the insulating tube

R_y : Reynolds number

δ_e : molecular Debye length

δ_f : diffusion sublayer length

λ : equivalent line charge density along the tube

$c_1 = \frac{\epsilon_l \, a}{2 \, \epsilon_\Omega}$

$c_2 = \frac{1}{2\pi a \, \epsilon_\Omega}$

$c_3 = \pi a^2 \, \sigma_e$

$c_4 = \frac{\pi a^2}{\tau_l} \, \bar{\rho}_v(0, t)$

d : development length ($\equiv \tau_l \, V$)

V : flow velocity

τ_l : charge relaxation time in liquid ($\equiv \epsilon_l / \frac{\sigma}{\ell}$)

ϵ_l : permittivity of the liquid

$\frac{\sigma}{\ell}$: conductivity of the liquid

ϵ_Ω : permittivity of the Ω region

σ_Ω : conductivity of the Ω region

σ_e : effective axial conductivity

$\rho_v(0, t)$: influent volume charge density at tube inlet

10.3 COUPLED FINITE ELEMENT AND FOURIER TRANSFORM METHOD FOR TRANSIENT SCALAR FIELD PROBLEMS

Many transient problems occur in low frequency electromagnetic field applications. These have been traditionally solved by lumped parameter equivalent circuits for which the parameter values are obtained either by closed form analysis or by numerical methods. In either case, parameter evaluation is carried out at steady state, and their time and field dependence are either ignored or accounted for by empirical methods. The simplicity of these methods have proved to be useful especially in system studies where the behavior of each individual component may not be critical to the prediction of the overall performance of the system. However, for device design applications or for the detailed study of a particular phenomenon in physical problems, lumped circuit analysis has been ineffective.

A rigorous numerical modeling by finite differences, finite elements, and integral equation methods has been proposed by researchers for such applications. Time-stepping methods have been found to be useful, specifically for nonlinear problems, but they are expensive in terms of computer implementation and often encounter problems of convergence and numerical stability.

For linear problems, where the effect of the geometry of the device or the physical analog are of importance, simpler but robust and economical methods of solution have been proposed. Some of these are by time-harmonic solutions, combined numerical field solution and transform methods and others. In the previous section we have presented a combined finite element and Laplace transform method for evaluating the transient electrification phenomena in dielectrics due to flow induced charges, but excluding electrical conductivity of the dielectric medium. A method based on combining closed form or numerical solutions with discrete Fourier transform was proposed by Lawrenson and Miller [119], and demonstrated on eddy current applications.

In this section, a method of combining the finite element method for two-dimensional scalar problems with the Fourier Integral Transform is described. The method is first illustrated by application to a simple L-R circuit excited by a step input of voltage at the terminals. It is then applied to detailed modeling of static electrification phenomena in dielectric media with finite electrical conductivity.

10.3.1 Transient Analysis by Fourier Integral Transform Method

The solution to the transient field problem is obtained by first evaluating its frequency response at a number of discrete frequencies and combining these with the Fourier integral transform method. We can best illustrate this by application to a simple L-R circuit and its transient response to a voltage step function.

The equation to be solved is

$$v = Ri + L\frac{di}{dt} \tag{10.109}$$

where i is the instantaneous current, L and R are inductance and resistance of the circuit, v is the applied voltage, and t is time.

Now $v = v \cdot u(t)$, where $u(t)$ is the unit step function.

First, we shall convert equation (10.109) into a frequency domain relationship at a given frequency ω, such that

$$v = Ri + j\omega Li \tag{10.110}$$

where $i = |i|e^{j\omega t}$ and $v = |v|e^{j\omega t}$ are time-harmonic functions.

The Fourier transform of (10.110) will yield

$$\hat{i} = \frac{\hat{v}}{R + j\omega L} F(u(t)) \tag{10.111}$$

where $\hat{i}$ and $\hat{v}$ are Fourier transforms of i and v, respectively.

$$F(u(t)) = \left(\pi\delta(\omega) + \frac{1}{j\omega}\right) \tag{10.112}$$

and $\delta(\omega) = 1$ at $\omega = 0$ and $\delta(\omega) = 0$ for $\omega > 0$. Substituting (10.112) into (10.111), one obtains

$$\hat{i} = \frac{\hat{v}\left(\pi\delta(\omega) + \frac{1}{j\omega}\right)}{R + j\omega L} \tag{10.113}$$

Taking the inverse Fourier transform of (10.113)

$$i(t) = \frac{v}{\pi R}\int_0^\infty \left(\frac{1}{j\omega} - \frac{1}{R/L + j\omega}\right)e^{j\omega t}\,d\omega + \frac{v}{2\pi}\int_0^\infty \frac{\pi\delta(\omega)e^{j\omega t}}{R + j\omega L}\,d\omega \tag{10.114}$$

Because the time domain solution must be real, we shall consider only the real part of equation (10.114). Therefore,

$$i(t) = \frac{v}{\pi R} \int_0^\infty \left(\frac{\sin \omega t}{\omega} - \frac{R/L \cos \omega t}{\frac{R^2}{L^2} + \omega^2} - \frac{\omega \sin \omega t}{\frac{R^2}{L^2} + \omega^2} \right) d\omega + \frac{v}{2\pi} |_0^\infty \frac{\pi \delta(\omega)}{R} \quad (10.115)$$

Evaluating the integral in equation (10.115) term by term and adding the last term, called the DC term,

$$i(t) = \frac{v}{\pi R} \left(\frac{\pi}{2} - \frac{\pi}{2} e^{-\frac{Rt}{L}} - \frac{\pi}{2} e^{-\frac{Rt}{L}} \right) + \frac{v}{2R} \quad (10.116)$$

or

$$i(t) = \frac{v}{R} \left(1 - e^{-\frac{Rt}{L}} \right) \quad (10.117)$$

Equation (10.117) is the classical solution to (10.109). It should be noted that in equation (10.111),

$$\frac{v}{R + j\omega L}$$

represents the solution at the frequency ω. We could rewrite equation (10.111) as

$$\hat{i} = s(\omega) F(u(t)) \quad (10.118)$$

where $s(\omega)$ is the frequency domain solution; $F(u(t))$ is the Fourier transform of the excitation function.

The last term in equation (10.115), the DC term, is evaluated separately. The integrations in (10.115) excluding the DC term were also performed using Simpson's rule as follows:

$$i(t) = \frac{v}{\pi R} \left(f(\omega_0) + 4 f(\omega_{1,3,\ldots,n-2}) + \ldots f(\omega_n) \right) \frac{\Delta \omega}{3} \quad (10.119)$$

where $f(\omega)$ is the integral in equation (10.115) and $\Delta \omega$ is the angular frequency interval. The results obtained by evaluating equation (10.119) were compared with those of equation (10.117) with excellent agreement as seen in Figure 10.11.

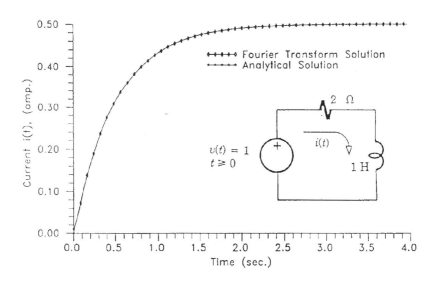

Figure 10.11 Current in a R-L Circuit with Applied Step Voltage. Fourier Transform Solutions versus Analytical Solution

10.3.2 Application to the Electrification Problem in Dielectric Media

The Fourier transform method described above was applied to the evaluation of electrification phenomena in dielectric media. Figure 10.12 shows the insulating tube with transformer coolant oil flowing through it. As before, the electrostatic charge generated due to oil flow accumulates on the insulating cylinder and causes a voltage buildup. For a fuller description of the phenomenon, the reader is referred to references [117],[120], and [121]. In this application, we shall assume that the dielectric has finite electrical conductivity, sufficient to drive a resistive current in addition to the capacitive current. The potential will be a complex quantity which will vary both in magnitude and in phase throughout the structure. Therefore, the numerical computational method needs to take into account the nature of the complex potentials and the resulting field distribution. The complex potentials obtained by the numerical method at different frequencies can then be used to obtain the transient response of the system by coupling them with the Fourier transform method.

The governing equations in the frequency domain described by Tahani *et al.*, [118] can be summarized as follows:

$$\nabla \cdot (\sigma + j\omega\epsilon)\nabla\phi = 0 \tag{10.120}$$

Low-Frequency Applications

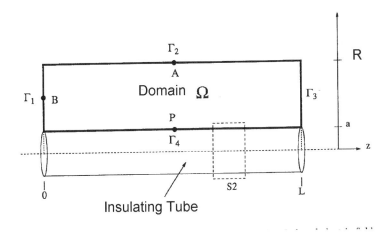

Figure 10.12 Axisymmetric Model of the Insulating Tube

With boundary conditions $\phi = 0$ on Γ_1, Γ_2 and Γ_3. On Γ_4

$$\frac{\partial \phi}{\partial r} + C_1 \frac{\partial^2 \phi(a,z,t)}{\partial z^2} - C_2 \lambda(z,t) \tag{10.121}$$

and

$$\frac{\partial \lambda(z,t)}{\partial t} + C_3 \frac{\partial^2 \phi(a,z,t)}{\partial z^2} + C_4 e^{-z/d} \tag{10.122}$$

where ϕ is the scalar potential (volts), r is the radius, λ is the equivalent line charge density on the tube wall, d is the development length, a is the radius of the tube (the thickness of the tube is neglected), ϵ_Ω is the permittivity of the dielectric region, v is the velocity of the fluid, $C_1 = \frac{\epsilon_1 a}{2\epsilon_\Omega}$, $C_2 = \frac{1}{2\pi a \epsilon_\Omega}$, $\rho_v(0,t)$ is the influent charge density at the inlet σ_e is the electrical conductivity, $C_3 = \pi a^2 \sigma_e$, $C_4 = \frac{\pi a^2 \rho_v(0,t)}{\Gamma_\ell}$, $d = \Gamma_\ell V$, σ_ℓ is the oil conductivity, σ_Ω is the conductivity of the Ω region. Assuming the ϕ, ρ_v and λ are time harmonic; $\phi = |\phi|e^{j\omega t}$, $\rho_v = |\rho_v|e^{j\omega t}$ and $\lambda = |\lambda|e^{j\omega t}$.

We can rewrite equations (10.121) and (10.122) as

$$e^{j\omega t}\left(\frac{\partial |\phi|}{\partial r} + C_1 \frac{\partial^2 |\phi|}{\partial z^2}\right) = -C_2|\lambda|e^{j\omega t} \qquad (10.123)$$

$$j\omega|\lambda|e^{j\omega t} + C_3 \frac{\partial^2 |\phi|e^{j\omega t}}{\partial z^2} + C_4 e^{j\omega t} e^{-z/d} \qquad (10.124)$$

Combining equations (10.123) and (10.124), eliminating $e^{j\omega t}$ and substituting $\frac{\partial \phi}{\partial n} = -\frac{\partial \phi}{\partial r}$, one obtains the boundary condition on Γ_4 as

$$\frac{\partial phi}{\partial n} = (C_1 + \frac{C_2 C_3}{j\omega})\frac{\partial^2 \phi}{\partial z^2} + \frac{C_2 C_4}{j\omega} e^{-z/d} \qquad (10.125)$$

Finite Element Formulation and Discretization

The Galerkin weighted residual procedure applied to the complex scalar Poisson equation (10.120) yields

$$-\int_v (\sigma + j\omega\epsilon)\nabla W \cdot \nabla \phi \, dv + \oint_\Gamma (\sigma + j\omega\epsilon)W \frac{\partial \phi}{\partial n} \, d\Gamma \qquad (10.126)$$

Substituting for $\frac{\partial \phi}{\partial n}$ from equation (10.125) into equation (10.126), the finite element expression for ϕ is obtained for the axisymmetric geometry of Figure 10.12 as

$$-\int_v (\sigma + j\omega\epsilon)\nabla W \cdot \nabla \phi r \, dr dz -$$
$$\int_0^L (\sigma + j\omega\epsilon)(C_1 + \frac{c_2 C_3}{j\omega})\frac{\partial W}{\partial z} \cdot \frac{\partial \phi}{\partial z} \, dz$$
$$\int_0^L W(\sigma + j\omega\epsilon)\frac{C_2 C_4}{j\omega} a e^{-z/d} \, dz = 0 \qquad (10.127)$$

where W is the weighting or shape function and L is the length of the tube.

Low-Frequency Applications

The Ω region of Figure 10.12 was discretized by second order triangular finite elements and the coefficient matrix and forming function for each triangular element were obtained from the usual finite element procedure. For the sake of completeness, the element matrices are in included in Appendix C.1.

Equation (10.127) can be formally written in matrix form as

$$(P)(\phi) = (R) \qquad (10.128)$$

where P is the coefficient matrix including the contributions of volume and line elements, R is the forcing function from the Γ_4 boundary condition.

Transient Solution by Fourier Integral Transform

As in the case of the R-L circuit, solutions were obtained for equation (10.128) at several discrete frequencies. In order to obtain an estimate of the range of frequencies at which the solution is required, the modulus of the solution

$$|s(\omega)| = |\phi| \qquad (10.129)$$

is plotted versus ω as shown in Figure 10.13.

A final value of frequency f_n and the frequency step Δf were found. This resulted in the total number of frequency steps $N = f_n/\Delta f$. Unlike in the discrete Fourier Transform method, wherein the time step is rigidly coupled to the frequency interval by the relationship

$$\frac{1}{\Delta f \Delta t} = N \qquad (10.130)$$

in the Fourier integral transform method the time step is decoupled from the frequency interval. This provides greater flexibility in computation.

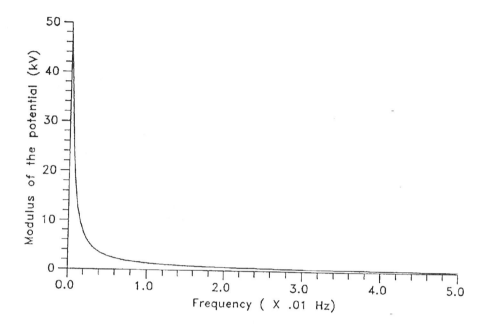

Figure 10.13 Modulus of the Frequency Domain Solution versus Frequency

The next computational step is to combine the frequency domain solution $s(\omega)$ with the Fourier transform of the forcing function, in this case a step input. Thus, as before, from equation (10.112)

$$F(u(t)) = \left(\pi\delta(\omega) + \frac{1}{j\omega}\right) \qquad (10.131)$$

then the time response of the solution is obtained as

$$\phi(t) = \int_0^\infty s(\omega)\left(\pi\delta(\omega) + \frac{1}{j\omega}\right)e^{j\omega t}\,d\omega \qquad (10.132)$$

The DC term in equation (10.132) is removed as before and its solution is obtained as

$$\frac{1}{2\pi}|_0^\infty \pi\delta(\omega)s(\omega) = \frac{s(0)}{2} \qquad (10.133)$$

Low-Frequency Applications

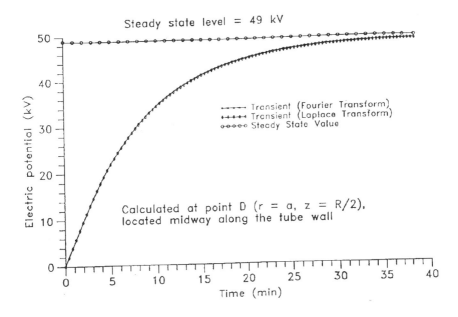

Figure 10.14 Comparison of Fourier and Laplace Transform Methods in Insulating Tube

The rest of the integral in equation (10.132) is evaluated by Simpson's rule using the following expression

$$\phi(t) = \frac{\Delta\omega}{3\pi}\left[f(\omega_0) + 4f(\omega_{1,3,\ldots,n-1}) + 2f(\omega_{2,4,\ldots,n-2}) + f(\omega_n)\right] \quad (10.134)$$

where $f(\omega) = \frac{s(\omega)}{j\omega}$. The total solution is then obtained by adding the DC term to equation (10.134). The time response of the scalar potential to the step input of voltage was evaluated and illustrated in Figure 10.14, assuming zero conductivity of the dielectric in the Ω region. In the same figure, the solution obtained from the Laplace transform method described in the previous section is also plotted for comparison. The agreement is good between the two solutions.

Figure 10.15 shows the plot of the solution as a function of time for various values of conductivity of the dielectric. It is apparent that the solution is exceedingly sensitive to conductivity, and after a threshold value of σ, the solution for the potential or voltage is a very low value such as in a short circuit.

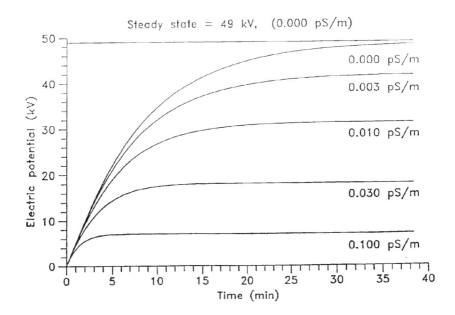

Figure 10.15 Fourier Transform Solution with Different Dielectric Conductivities

10.4 AXIPERIODIC ANALYSIS

In certain geometries, the fields are three dimensional, but in one of the dimensions the fields are periodic. In this case we can, by the use of special shape functions, use a two-dimensional analysis in which this variation is included. The result is that by solving the problem in one plane we have the full three-dimensional representation of the fields [75]. An example of a turbine generator end region analysis will illustrate the procedure. First we consider the scalar potential formulation and representation of the currents [122].

In Figure 10.16 we have region Ω_1 containing the currents J_s; the region Ω_2, which is current free; the boundary Γ_{12} separating Ω_1 and Ω_2; and the boundary Γ_2, which is the exterior.

The defining equations are

- In region Ω_1

$$\begin{aligned} \nabla \cdot (\mu_1 H_1) &= 0 \text{ and} \\ \nabla \times H_1 &= J_s \end{aligned} \qquad (10.135)$$

Low-Frequency Applications

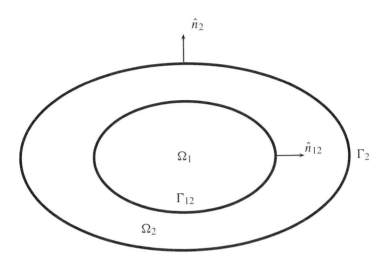

Figure 10.16 Regions and Boundary in Axiperiodic Problem

- In region Ω_2

$$\nabla \cdot (\mu_2 H_2) = 0 \text{ and}$$
$$\nabla \times H_2 = 0 \tag{10.136}$$

- Along boundary Γ_{12}

$$n_{12} \times H_1 = n_{12} \times H_2 \tag{10.137}$$

- Across boundary Γ_{12}

$$n_{12} \cdot \mu_1 H_1 = n_{12} \cdot \mu_2 H_2 \tag{10.138}$$

- On boundary Γ_2

$$n_2 \times H_2 = 0 \text{ or}$$
$$n_2 \cdot \mu_2 H_2 = 0 \tag{10.139}$$

In region Ω_1 we break H_1 into two components, H_0 and H_s, where H_0 is the field due to the source currents alone.

$$H_1 = H_0 + H_s \tag{10.140}$$

where

$$\begin{aligned} \nabla \times H_0 &= J_s \quad \text{in } \Omega_1 \\ n_{12} \times H_0 &= 0 \quad \text{along } \Gamma_{12} \end{aligned} \tag{10.141}$$

This last relation comes from the boundary condition at the magnetic shell (see Carpenter [123]), where there is no tangential component of the field. Therefore, from $\nabla \times H_1 = J_s$ and $\nabla \times H_0 = J_s$ we conclude that

$$\begin{aligned} \nabla \times (H_1 - H_0) &= 0 \quad \text{or} \\ \nabla \times H_s &= 0 \end{aligned} \tag{10.142}$$

so we may write

$$H_s = -\nabla \phi_1 \tag{10.143}$$

In region Ω_2 we have $\nabla \times H_2 = 0$ so that

$$H_2 = -\nabla \phi_2 \tag{10.144}$$

Therefore, to summarize the equations describing the magnetostatic field:

- In region Ω_1

$$\begin{aligned} \nabla \cdot \mu_1 (H_0 - \nabla \phi_1) &= 0 \\ H_1 &= H_0 - \nabla \phi_1 \end{aligned} \tag{10.145}$$

Low-Frequency Applications

- In region Ω_2

$$\nabla \cdot \mu_2(-\nabla\phi_2) = 0$$
$$H_2 = -\nabla\phi_2 \qquad (10.146)$$

- Along boundary Γ_{12}

$$n_{12} \times (H_0 - \nabla\phi_1) = n_{12} \times (-\nabla\phi_2) \qquad (10.147)$$

- Across boundary Γ_{12}

$$n_{12} \cdot \mu_1(H_0 - \nabla\phi_1) = n_{12} \cdot \mu_2(-\nabla\phi_2) \qquad (10.148)$$

- On Γ_2

$$\phi_2 = 0 \text{ or}$$
$$\frac{\partial \phi_2}{\partial n} = 0 \qquad (10.149)$$

We note that since H_0 is found by solving for the field in region Ω_1 imbedded in an infinitely permeable region,

$$n_{12} \times H_0 = 0 \qquad (10.150)$$

and therefore equation (10.147) is consistent.

The Galerkin form of the equations is

$$\int_{\Omega_1} \nabla w \cdot \mu_1(-\nabla\phi_1 + H_0) dv - \int_{\Gamma_{12}} w\mu_1(H_0 - \nabla\phi_1) \cdot n_{12} \, d\Gamma_{12} = 0 \qquad (10.151)$$

and

$$\int_{\Omega_2} \nabla w \cdot \mu_2(-\nabla\phi_2) dv + \int_{\Gamma_{12}} w\mu_2(-\nabla\phi_2) \cdot n_{12} \, d\Gamma - \int_{\Gamma_2} w\mu_2(-\nabla\phi_2) \cdot n_2 \, d\Gamma = 0$$
$$(10.152)$$

From equations (10.151) and (10.152) we obtain

$$\int_{\Omega_1+\Omega_2} \nabla w \cdot \mu(H_0 - \nabla\phi)\,dv + \oint_{\Gamma_{12}} -wn_{12} \cdot (\mu_1(H_0 - \nabla\phi_1) + \mu_2\nabla\phi_2)\,d\Gamma = 0 \tag{10.153}$$

The surface integral in equation (10.153) is zero due to the continuity of the normal component of the flux density on the boundary.

We are then left with

$$\int_{\Omega_1+\Omega_2} \nabla w \cdot \mu\nabla\phi\,dv = \int_{\Omega_1} \nabla w \cdot \mu_1 H_0\,dv \tag{10.154}$$

By our assumption of axiperiodicity, we can represent the current source as a harmonic expansion in the θ direction. Therefore

$$J_s(r,\theta,z) = \sum_{n=-\infty}^{\infty} J_s^n(r,z)e^{-jnp\theta} \tag{10.155}$$

where

$$J_s^n(r,\theta,z) = \frac{p}{2\pi}\int_0^{2\pi} J_s(r,\theta,z)e^{jnp\theta}\,d\theta \tag{10.156}$$

Each term in the series has zero divergence of the current and we can write H_0^n as

$$\nabla \times H_0^n = J_s e^{-jnp\theta} \tag{10.157}$$

Low-Frequency Applications

We also note that along the interface Γ_{12} we have $n_{12} \times H_0^n = 0$.

Equation (10.154) becomes

$$\sum_{n=-\infty}^{\infty} \int_{\Omega} (\nabla w e^{jnp\theta} \cdot \mu \nabla (w e^{-jnp\theta} \phi^n) \, dv)$$
$$= \sum_{n=-\infty}^{\infty} \int_{\Omega_1} (\nabla (w e^{jnp\theta} \cdot \mu H_0^n) \, dv \qquad (10.158)$$

Because the problem is linear we can solve for each of the harmonics independently.

$$\int_{\Omega} \nabla w e^{jnp\theta} \cdot \mu \nabla w e^{-jnp\theta} \phi^n \, dv$$
$$= \int_{\Omega_1} \nabla w e^{jnp\theta} \cdot \mu H_0^n \, dv \qquad (10.159)$$

The potential is solved for in the (r, z) plane with the exponential variation in the θ direction implied. The potential is then

$$\phi^n = \sum_{i=1}^{N} w_i(r, z) e^{-jnp\theta} \phi_i^n \qquad (10.160)$$

where ϕ_i is the nodal potential, N is the number of nodes, and $w_i(r, z) e^{-jp\theta}$ is the shape function.

Using this expression for potential, we obtain N equations for N unknowns. For $1 \le k \le N$ we have

$$\sum_{i=1}^{N} (\int_{\Omega} (\nabla w_k e^{jnp\theta} \cdot \mu \nabla w_i e^{-jnp\theta} \, dv) \phi_i^n = \int_{\Omega_1} \nabla (w_k e^{jnp\theta} \cdot \mu H_0^n) \, dv \qquad (10.161)$$

Expanding the gradient in cylindrical coordinates

$$\nabla(w_k e^{jp\theta}) = \frac{\partial w_k}{\partial r} e^{jp\theta} \hat{a}_r + \frac{1}{r} jp w_k e^{jp\theta} \hat{a}_\theta + \frac{\partial w_k}{\partial z} e^{jp\theta} \hat{a}_z \qquad (10.162)$$

Using (10.162), equation (10.161) becomes for $1 \leq k \leq N$

$$\sum_{i=1}^{N} \left(\int_\Omega \mu \left(\frac{\partial w_k}{\partial r} \frac{\partial w_i}{\partial r} + \frac{p^2}{r^2} w_k w_i + \frac{\partial w_k}{\partial z} \frac{\partial w_i}{\partial z} \right) dv \right) \phi_i \right)$$
$$= \int_{\Omega_1} \left(\mu_1 \frac{\partial w_k}{\partial r} e^{jp\theta} H_{0r} + \frac{1}{r} jp w_k e^{jp\theta} H_{0\theta} + \frac{\partial w_k}{\partial z} e^{jp\theta} H_{0z} \right) dv \qquad (10.163)$$

To solve equation (10.163) we now must find H_0.

10.4.1 MVP Formulation to Find H_0

We seek the solution for the source field in a magnetic shell. We will assume that the three components of the current density are known. (Refer to Figure 10.16.) The boundary Γ_{12} of Ω_1 is broken into two parts, Γ_{12H} and Γ_{12B}. The Γ_{12H} boundary is the interface between the conductor region and the other regions of the problem, on which we assume that $\hat{n}_{12} \times H_0 = 0$. The Γ_{12B} boundary will be an antisymmetry for the current sources. We will be able to model, for example, half of the machine in this case. We therefore have the following conditions to meet:

- In region Ω_1

$$\nabla \times \nu \nabla \times A = J_s \qquad (10.164)$$

- On the magnetic boundary surrounding region Ω_1.

$$n \times \nu \nabla \times A = 0 \qquad (10.165)$$

- On the symmetry surface

$$n \cdot \nabla \times A = 0 \qquad (10.166)$$

To ensure the uniqueness of the solution, we introduce a penalty function $-\nabla(\nu\nabla \cdot A)$ into equation (10.164) which now becomes

$$\nabla \times \nu\nabla \times A - \nabla\nu\nabla \cdot A = J_s \tag{10.167}$$

We also require that on the symmetry surface $\nu\nabla \cdot A = 0$.

Gauge Condition

To see that this leads to the Coulomb gauge we take the divergence of equation (10.167)

$$\nabla \cdot (\nabla \times \nu\nabla \times A - \nabla\nu\nabla \cdot A - J_s) = 0 \tag{10.168}$$

Because $\nabla \cdot J_s = 0$ and $\nabla \cdot \nabla \times A = 0$, we are left with Poisson's equation for the scalar variable $\nu\nabla \cdot A$.

$$\nabla^2(\nu\nabla \cdot A) = 0 \tag{10.169}$$

We also know the normal component of equation (10.167) along the surface Γ_{12}.

$$n \cdot (\nabla \times \nu\nabla \times A - \nabla\nu\nabla \cdot A - J_s) = 0 \tag{10.170}$$

from which we find (because $J_s \cdot n = 0$)

$$n \cdot \nabla \times \nu\nabla \times A = \frac{\partial}{\partial n}(\nu\nabla \cdot A) \tag{10.171}$$

We also have

$$n \cdot \nabla \times \nu\nabla \times A = -\nabla \cdot (n \times \nu\nabla \times A) = -\nabla \cdot (n \times H_0) = 0 \tag{10.172}$$

on Γ_{12}.

So we have

$$\begin{aligned}\nabla^2(\nu\nabla\cdot A) &= 0 \text{ in } \Omega_1 \\ \frac{\partial}{\partial n}(\nu\nabla\cdot A) &= 0 \text{ on } \Gamma_{12H} \\ \nu\nabla\cdot A &= 0 \text{ on } \Gamma_{12B}\end{aligned} \qquad (10.173)$$

This ensures the Coulomb gauge.

Uniqueness of A

As we now have $\nabla \cdot A = 0$, we assume that the problem has two solutions, A_1 and A_2. We call their difference δA.

$$\nabla \times \delta A = 0 \qquad (10.174)$$

This leads to the conclusion that δA can be expressed as the gradient of a scalar, ϕ. Because we also have

$$\nabla \cdot A = 0 \qquad (10.175)$$

we conclude that

$$\nabla^2 \phi = 0 \qquad (10.176)$$

The boundary conditions are $n \cdot A = 0$ or $\frac{\partial \phi}{\partial n} = 0$ on Γ_{12H} and $n \times A = 0$ or $n \times \nabla\phi = 0$ on Γ_{12B} or ϕ is constant on the boundary.

Low-Frequency Applications

Therefore ϕ satisfies

$$\nabla^2 \phi = 0 \text{ in } \Omega_1$$
$$\frac{\partial \phi}{\partial n} = 0 \text{ on } \Gamma_{12H} \tag{10.177}$$
$$\phi = \text{constant on } \Gamma_{12B} \tag{10.178}$$

So we see that ϕ is constant in Ω_1 and this means that $A_1 = A_2$.

Summary of Equations

The final set of equation is then

$$\begin{aligned}
\nabla \times \nu \nabla \times A &= J_s \text{ in } \Omega_1 \\
n \cdot A &= 0 \text{ on } \Gamma_{12H} \\
n \times A &= 0 \text{ on } \Gamma_{12B} \\
n \times (\nabla \times A) &= 0 \text{ on } \Gamma_{12B}
\end{aligned} \tag{10.179}$$

10.4.2 Finite Element Formulation

The Galerkin form of equation (10.167) is

$$\int_\Omega W_i (\nabla \times \nu \nabla \times A - \nabla \nu \nabla \cdot A - J_s) \, d\Omega = 0 \tag{10.180}$$

where W_i are the vector weighting functions.

We find A in an element by interpolation of

$$A = \sum_{j=1}^{N} W_j A_j \tag{10.181}$$

where

$$W_i = \begin{pmatrix} \xi_i & 0 & 0 \\ 0 & \xi_i & 0 \\ 0 & 0 & \xi_i \end{pmatrix} \tag{10.182}$$

Integrating by parts gives

$$\int_{\Omega_1} ((\nabla \times W_i)v \cdot (\nabla \times A) + (\nabla \cdot W_i)v(\nabla \cdot A) - W_i \cdot J_s) \, d\Omega$$
$$- \int_{\Gamma_{12}} ((n \times W_i)v \cdot (\nabla \times A) + (n \cdot W_i)v\nabla \cdot A) \, d\Gamma = 0 \quad (10.183)$$

The surface integrals vanish as follows:

For the first term

$$\int_{\Gamma_{12}} (n \times W_i) v \nabla \times A \, d\Gamma = -\int_{\Gamma_{12}} (n \times H_0) \cdot W_i \, d\Gamma \quad (10.184)$$

Because we have $n \times H_0 = 0$ on Γ_{12H} and $n \times W_i = 0$ on Γ_{12B} the integral is equal to zero.

For the second term

$$\int_{\Gamma} (n \cdot W_i) v (\nabla \cdot A) \, d\Gamma = 0 \quad (10.185)$$

On Γ_{12H}, because $n \cdot A = 0$, we have $n \cdot W_i = 0$. Also, because $v\nabla \cdot A = 0$ on Γ_{12B} the integral is zero.

To simplify the equations we make the following change of variables:

$$\begin{aligned} A_r &= \frac{1}{r} A_r^* \\ A_\theta &= \frac{1}{r} A_\theta^* \\ A_z &= A_z^* \end{aligned} \quad (10.186)$$

or

$$A^* = PA \quad (10.187)$$

where

$$P = \begin{pmatrix} r & 0 & 0 \\ 0 & r & 0 \\ 0 & 0 & r \end{pmatrix} \quad (10.188)$$

Low-Frequency Applications

Using the weighting functions

$$A = \sum_{j=1}^{N} W_j P^{-1} A_j^* \tag{10.189}$$

Using the new variable, the system of equations becomes (for $1 \leq i \leq N$)

$$\sum_{j=1}^{N} \nu \int_{\Omega_1} (P)^{-1} \left(\frac{(S)}{(\nabla \xi_i^T \cdot \nabla \xi_j) I d_3 - (\nabla_j \nabla_i^T) + (\nabla \xi_i \nabla_j^T)} \right) (P)^{-1} A_j^* r \, dr \, d\theta \, dz$$

$$= \int_{\Omega_1} (P)^{-1} \xi_i J_s \, r \, dr \, d\theta \, dz \tag{10.190}$$

We assume that for the space harmonics A has the form

$$A = \sum_{j=1}^{N} \begin{pmatrix} \frac{1}{r} & 0 & 0 \\ 0 & \frac{1}{r} & 0 \\ 0 & 0 & 1 \end{pmatrix} \xi'_j(r,z) e^{-jp\theta} \begin{pmatrix} A_{jr}^* \\ A_{j\theta}^* \\ A_{jz}^* \end{pmatrix} \tag{10.191}$$

where the components of A^* are complex and $\xi'_j(r,z)e^{-jp\theta} = \xi_j(r,z,\theta)$ with ξ' being real.

We now integrate equation (10.190) from 0 to 2π, which eliminates the θ variable, meaning that we must solve the problem only in the (r, z) plane.

The system is then (at $\theta = 0$) for $1 \leq i \leq N$

$$\sum_{j=1}^{N} 2\pi \nu \int_{\Omega_1} (P)^{-1} (S)(P)^{-1} \begin{pmatrix} A_{jr}^* \\ A_{j\theta}^* \\ A_{jz}^* \end{pmatrix} r \, dr \, dz = 2\pi \int_{\Omega_1} \begin{pmatrix} \frac{1}{r}(\xi'_i J_{sr}) \\ \frac{1}{r}\xi'_i J_{s\theta} \\ \xi'_i J_{sz} \end{pmatrix} r \, dr \, dz$$

$$\tag{10.192}$$

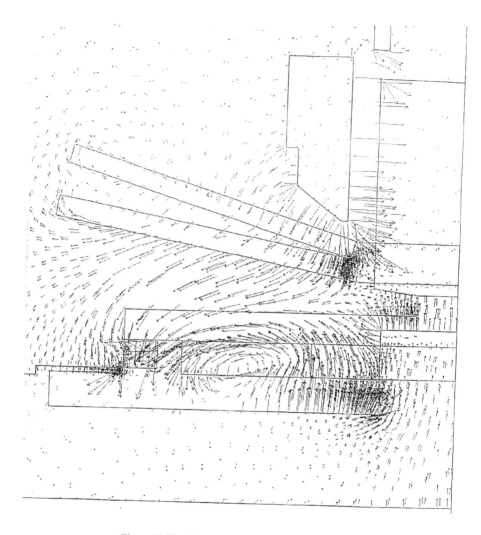

Figure 10.17 H-Field Vectors from Axiperiodic Solution

Example

The method has been applied to the end winding of a turbine generator. The arrows of Figure 10.17 shows the direction of the field vectors with both the rotor and the stator excited. For further information on the application of this method to electric machines see [75].

Low-Frequency Applications

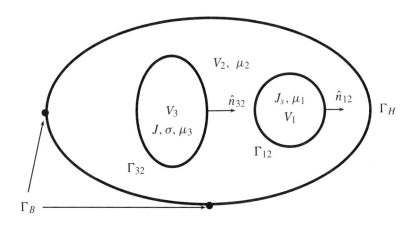

Figure 10.18 Regions and Boundary in Eddy Current Problem

10.4.3 Magnetodynamic $T - \Omega$ Formulation

Consider the problem of computation of eddy current in the region shown in Figure 10.18 [124].

Here V_1 is a region with source current J_s but no induced current. Region V_2 is a region with no current and region V_3 is a region with eddy currents. In region 3 we have

$$\nabla \times T = J \qquad (10.193)$$

As $\nabla \times H = J$ we can write

$$H = T - \nabla \Omega \qquad (10.194)$$

where Ω is the magnetic scalar potential.

The equation relating T and Ω in region 3 is

$$\nabla \times \frac{1}{\sigma}(\nabla \times T) + \mu_3 \frac{\partial}{\partial t}(T - \nabla \Omega_3) = 0 \qquad (10.195)$$

Uniqueness of T and Ω

To ensure the uniqueness of T we add a term $-\nabla \frac{1}{\sigma} \nabla \cdot T$ (Coulomb gauge) to equation (10.195) which now becomes

$$\nabla \times \frac{1}{\sigma}(\nabla \times T) - \nabla \frac{1}{\sigma} \nabla \cdot T + \mu_3 \frac{\partial}{\partial t}(T - \nabla \Omega_3) = 0 \qquad (10.196)$$

Taking the divergence of (10.196)

$$\nabla \cdot (\nabla \times \frac{1}{\sigma}(\nabla \times T)) - \nabla^2 \frac{1}{\sigma} \nabla \cdot T + \frac{\partial}{\partial t} \nabla \cdot (\mu_3(T - \nabla \Omega_3)) = 0 \qquad (10.197)$$

The first term of (10.197) is zero for any vector. The third term is zero since $\nabla \cdot B = 0$. So the scalar $\frac{1}{\sigma} \nabla \cdot T$ satisfies Laplace's equation in region V_3.

$$\nabla^2 \frac{1}{\sigma} \nabla \cdot T = 0 \qquad (10.198)$$

If we set the Dirichlet boundary condition $\frac{1}{\sigma} \nabla \cdot T = 0$ on Γ_{32}, then from equation (10.198) we conclude $\frac{1}{\sigma} \nabla \cdot T = 0$ in V_3. This satisfies the gauge condition in the conductor.

Now assume that there are two solutions to the problem, T_1 and T_2. Call the difference $\delta T = T_1 - T_2$. Then because $\nabla \times \delta T = 0$ we can write δT as the gradient of a scalar. So

$$\delta T = \nabla \phi \qquad (10.199)$$

Low-Frequency Applications

Because $\nabla \cdot \delta T = 0$ we know that $\nabla^2 \phi = 0$ in V_3. As we require that $n_{32} \times T = 0$ on the boundary Γ_{32}, $n_{32} \times \nabla \phi = 0$ as well. So ϕ must be constant on the boundary and therefore in V_3. In this case $T_1 = T_2$. We must also enforce the condition that $\nabla \cdot B = 0$, which is not implied by equation (10.196). This becomes

$$\nabla \cdot \mu_3 (T - \nabla \phi) = 0 \qquad (10.200)$$

In a similar way we must enforce the condition that the current density normal to the surface of the conductor is zero.

In the regions with no induced current, we have only one unknown, the magnetic field $\vec{H}$. We separate $\vec{H}$ into two components

$$H = H_0 + H_s \qquad (10.201)$$

The component H_0 satisfies the relationships

$$\nabla \times H_0 = J_s, \quad \text{in } V_1$$
$$n_{12} \times H_0 = 0 \quad \text{along } \Gamma_{12} \qquad (10.202)$$

H_0 is the field that would be produced by the current J_s if this current were imbedded in a region surrounded by infinitely permeable iron. This is the so-called *magnetic shell*. From $\nabla \times H = J_s$ and $\nabla \times H_0 = J_s$ we deduce that

$$\nabla \times (H - H_0) = 0 \qquad (10.203)$$

or

$$\nabla \times H_s = 0 \qquad (10.204)$$

In this case H_s can be written as the gradient of a scalar so

$$H_s = -\nabla \Omega \qquad (10.205)$$

In the nonconducting regions we must satisfy

$$\nabla \cdot \mu_1(-\nabla \Omega_1) = -\nabla \cdot \mu_1 H_0 \text{ in } V_1$$
$$\nabla \cdot \mu_2(-\nabla \Omega_2) = 0 \text{ in } V_2 \quad (10.206)$$

We discretize these variables as follows:

$$\Omega = \sum_{i=1}^{N} w_i \Omega_i \quad (10.207)$$

$$T = \sum_{i=1}^{N} W_i T_i \quad (10.208)$$

where w_i is the weighting function at node i and

$$W_i = \begin{pmatrix} w_i & 0 & 0 \\ 0 & w_i & 0 \\ 0 & 0 & w_i \end{pmatrix} \quad (10.209)$$

is the vector weighting function at node i.

The Galerkin form of the differential equations is

$$\int_{V_1} w(\nabla \cdot \mu_1(H_0 - \nabla \Omega_1)) \, dv = 0 \quad (10.210)$$

$$\int_{V_2} w(\nabla \cdot \mu_2(-\nabla \Omega_2)) \, dv = 0 \quad (10.211)$$

$$\int_{V_3} w(\nabla \cdot \mu_3(T - \nabla \Omega_3)) \, dv = 0 \quad (10.212)$$

$$\int_{V_3} [W \cdot \nabla \times \frac{1}{\sigma}(\nabla \times T) - W \cdot \nabla \frac{1}{\sigma}(\nabla \cdot T) + W \cdot \frac{1}{\mu_3} \frac{\partial}{\partial t}(T - \nabla \Omega_3)] \, dv = 0 \quad (10.213)$$

Low-Frequency Applications

We now integrate these equations by parts to obtain

$$-\int_{V_1} \nabla w \cdot \mu_1 (H_0 - \nabla \Omega_1) \, dv + \oint_{\Gamma_{12}} w \mu_1 (H_0 - \nabla \Omega_1) \cdot \hat{n}_{12} \, d\Gamma_{12} \qquad (10.214)$$

$$-\int_{V_2} \nabla w \cdot \mu_2 (-\nabla \Omega_2) \, dv + \oint_{\Gamma_{21}} w \mu_2 (-\nabla \Omega_2) \cdot \hat{n}_{21} \, d\Gamma_{12}$$
$$+ \oint_{\Gamma_{23}} w \mu_2 (-\nabla \Omega_2) \cdot \hat{n}_{23} \, d\Gamma_{23} + \oint_{\Gamma} w \mu_2 (-\nabla \Omega_2) \cdot \hat{n} \, d\Gamma = 0 \qquad (10.215)$$

$$-\int_{V_3} \nabla w \cdot \mu_3 (T - \nabla \Omega_3) \, dv + \oint_{\Gamma_{32}} w \mu_3 (T - \nabla \Omega_3) \cdot \hat{n}_{32} \, d\Gamma_{32} \qquad (10.216)$$

$$-\int_{V_3} \left(\nabla \times W \cdot \frac{1}{\sigma} \nabla \cdot T + W \cdot \mu_3 \frac{\partial}{\partial t}(T - \nabla \Omega_3) \right) dv$$
$$- \oint_{\Gamma_{32}} W \cdot \hat{n}_{32} \frac{1}{\sigma} \nabla \cdot T \, d\Gamma_{32}$$
$$+ \oint_{\Gamma_{32}} W \cdot \hat{n}_{32} \times \left(\frac{1}{\sigma} \nabla \times T \right) d\Gamma_{32} = 0 \qquad (10.217)$$

The reader may verify that the interface conditions on normal B and J and tangential H are satisfied on all boundaries in these integral expressions.

We now must solve

$$-\int_{V_1} \nabla w \cdot \mu_1 \nabla \Omega_1 \, dv = -\int_{V_1} \nabla w \cdot \mu_1 H_0 \, dv \qquad (10.218)$$

$$-\int_{V_2} \nabla w \cdot \mu_2 \nabla \Omega_1 \, dv = 0 \qquad (10.219)$$

$$-\int_{V_3} \nabla w \cdot \mu_3 (T - \nabla \Omega_3) \, dv = 0 \qquad (10.220)$$

$$-\int_{V_3} \left(\nabla \times W \cdot \frac{1}{\sigma} \nabla \times T + \nabla \cdot W \frac{1}{\sigma} \nabla \cdot T + W \cdot \mu_3 \frac{\partial}{\partial t}(T - \Omega_3) \right) dv = 0$$

(10.221)

with boundary conditions $\hat{n}_{12} \times H_0 = 0$ on Γ_{12}, $\hat{n}_{32} \times T = 0$ on Γ_{32}, and $\Omega_2 = 0$ on Γ_H.

The matrix form of the equations is as follows:

In region V_1

$$\sum_{j=1}^{N} \left(\int_{V_1} -\nabla w_i \cdot \mu \nabla w_j \, dv \right) dv_j = \int_{V_1} -\nabla w_i \cdot \mu H_0 \, dv \qquad (10.222)$$

We evaluate these integrals using the methods presented in Chapters 5 and 6 and with the aid of the integration formulas in the Appendices. For region 1 we obtain the matrix equation

$$(A)\Omega = B \qquad (10.223)$$

For region V_2

$$\sum_{j=1}^{N} \left(\int_{V_2} -\nabla w_i \cdot \mu \nabla w_j \, dv \right) dv_j = 0 \qquad (10.224)$$

This gives

$$(C)\Omega = 0 \qquad (10.225)$$

In region V_3

$$\sum_{j=1}^{N} \left(\int_{V_3} -\nabla w_i \cdot \mu W_j \, dv \right) T_j - \sum_{j=1}^{N} \left(\int_{V_3} \nabla w_i \cdot \mu \nabla w_j \, dv \right) \Omega_j = 0 \qquad (10.226)$$

This gives
$$(D)T + (E)\Omega = 0 \tag{10.227}$$
and in region V_3

$$\sum_{j=1}^{N} \left(\int_{V_3} \left(\nabla \times W_i^T \frac{1}{\sigma} \nabla \times W_j + \nabla \cdot W_i^T \frac{1}{\sigma} \nabla \cdot W_j \right) dv \right) T_j$$
$$+ \sum_{j=1}^{N} \left(\int_{V_3} W_i \cdot \mu W_j \, dv \right) \frac{\partial T_j}{\partial t} - \sum_{j=1}^{N} \left(\int_{V_3} W_i \cdot \mu \nabla w_j \, dv \right) \frac{\partial \Omega_j}{\partial t} = 0 \tag{10.228}$$

In matrix form
$$(F)T + (G)\frac{\partial T}{\partial t} + (H)\frac{\partial \Omega}{\partial t} = 0 \tag{10.229}$$

In the steady-state sinusoidal formulation the w_i and w_js are complex. We have

$$\begin{aligned} w_i(r,\theta,z) &= w_i(r,z)e^{jp\theta} \\ w_j(r,\theta,z) &= w_j(r,z)e^{-jp\theta} \end{aligned} \tag{10.230}$$

This formulation has been successfully implemented as described in [124].

11
SOLUTION OF EQUATIONS

11.1 INTRODUCTION

Numerical field computation normally results in a large set of linear algebraic equations. This set of equations is then solved for the field or potential. The different formulations result in different matrices. These may be large or small, sparse or dense, possibly banded, symmetric or nonsymmetric, in rare cases even singular. The systems are often ill conditioned, even for physically realistic problems. A single technique for solving linear algebraic equations is therefore not sufficient. In this section we will consider a number of so-called *direct* methods that the authors have found particularly useful and efficient in the solution of electromagnetics problems. In direct methods we can (at least in principle) compute the number of steps or operations involved in the solution *a priori*. This finite set of operations would give the exact result were it not for round-off.

We will begin with a short review of matrix operations and notation. Consider the system of equations

$$\begin{aligned} a_{11}x_1 + a_{12}x_2 + \cdots + a_{1m}x_n &= y_1 \\ a_{21}x_1 + a_{22}x_2 + \cdots + a_{2n}x_n &= y_2 \\ &\vdots \\ a_{m1}x_1 + a_{m2}x_2 + \cdots + a_{m1}x_n &= y_m \end{aligned} \quad (11.1)$$

In matrix notation we represent this as

$$Ax = y \qquad (11.2)$$

where

$$A = \begin{pmatrix} a_{11} & a_{12} & \dots \\ a_{21} & a_{22} & \dots \\ a_{31} & \vdots & \dots \\ \vdots & \vdots & \dots \end{pmatrix} \qquad (11.3)$$

is an $m \times n$ (rows $\times$ columns) matrix

$$x = \begin{pmatrix} x_1 \\ \vdots \\ x_n \end{pmatrix} \qquad (11.4)$$

is an n vector and

$$y = \begin{pmatrix} y_1 \\ \vdots \\ y_m \end{pmatrix} \qquad (11.5)$$

is an m vector.

We will be using a number of matrix operations. Some of these are:

- Multiplication by a scalar

$$\alpha A = \begin{pmatrix} \alpha a_{11} & \alpha a_{12} & \dots \\ \alpha a_{21} & \alpha a_{22} & \dots \\ \vdots & \vdots & \vdots \end{pmatrix} \qquad (11.6)$$

Solution of Equations

- Addition and subtraction

$$A + B = \begin{pmatrix} a_{11} + b_{11} & a_{12} + b_{12} & \cdots \\ a_{21} + b_{21} & a_{22} + b_{22} & \cdots \\ \vdots & \vdots & \vdots \end{pmatrix} \quad (11.7)$$

- Multiplication by a matrix

$$AB \quad (11.8)$$

where A is $m \times n$, B is $n \times p$. The number of multiplications is $m \times n \times p$. The number of additions is $(n-1) \times m \times p$. The resulting matrix is $m \times p$. It is important to note that multiplication is not commutative. If A and B are square then $AB \neq BA$ in general.

- Inversion

$$AA^{-1} = I \quad (11.9)$$

- 5. Transposition

$$A^T = \begin{pmatrix} a_{11} & a_{21} & a_{31} \\ a_{12} & a_{22} & \cdots \\ \vdots & \vdots & \ddots \\ a_{13} & & \end{pmatrix} \quad (11.10)$$

- Conjugate transpose

$$A^* = \begin{pmatrix} a_{11}^* & a_{21}^* & \cdots \\ a_{12}^* & a_{22}^* & \vdots \\ \vdots & \vdots & \cdots \end{pmatrix} \quad (11.11)$$

We will be concerned with a number of special matrices:

- Square matrix: $m = n$
- Symmetric matrix: $A = A^T$
- Skew symmetric matrix: $A = -A^T$
- Hermetian matrix: $A = A^*$
- Diagonal matrix: $a_{ij} = 0$ if $i \neq j$
- Unit matrix or identity matrix: $a_{ij} = 0$, if $i \neq j$, $a_{ii} = 1$
- Upper triangular matrix: $a_{ij} = 0$ for $i > j$
- Lower triangular matrix: $a_{ij} = 0$ for $i < j$

We also note that the sum or product of two upper (lower) triangular matrices is upper (lower) triangular. The inverse of an upper (lower) triangular matrix is also upper (lower) triangular.

Matrix manipulations can be expressed in terms of simple operations known as elementary row (column) operations. We will be concerned with three such operations,

- Row (column) i and row (column) j are interchanged.
- Row (column) i is multiplied by a scalar.
- Row (column) i is replaced by itself plus k times row (column) j.

An elementary row (column) operation $r(c)$ on a matrix A can be accomplished by premultiplying (postmultiplying) by an identity matrix on which $r(c)$ has been performed. Therefore

$$r(A) = r(I)A$$
$$c(A) = Ac(I) \tag{11.12}$$

Solution of Equations

11.2 DIRECT METHODS

Consider a system of linear equations in matrix form

$$Ax = y \tag{11.13}$$

where A is $n \times n$ and x and y are $n \times 1$. The problem is that given y, the forcing vector, we want to solve for x. This problem arises in the analysis of linear systems such as electrical networks and in the numerical solution of differential equations. We may also be required to solve the system of equations in each iteration step of a nonlinear solution method, as when considering magnetic saturation, or we may have to solve the system of equations in each incremental step of a dynamic or transient problem such as in the diffusion problem.

In principle, the system $Ax = y$ has a simple solution

$$x = A^{-1}y \tag{11.14}$$

where A can also be written in the compact form

$$A^{-1} = \text{adj}\,(A)/|A| \tag{11.15}$$

where the ijth element of adj(A) is -1^{i+1} (jith minor of A). Although this may look simple and is actually very straightforward, let us consider it from the computational point of view. To evaluate the determinant of an $n \times n$ matrix requires $(n-1)n!$ multiplications. For large values of n this is clearly a very computationally intensive formulation and as a result is never used for reasonably sized problems.

11.2.1 Gauss Elimination

The standard Gauss elimination method is still one of the most popular and most efficient methods of solving a linear system of equations. If there are no special properties of the matrix to exploit (sparsity, bandedness, symmetry, etc.) then Gauss elimination is still the method of choice. We will describe Gauss elimination by columns, although elimination by rows works equally well and in the same number of steps. Consider a system of equations

$$Ax = b \tag{11.16}$$

In the Gauss elimination method we perform elimination one column at a time in order to obtain a triangular system. We then back-substitute to find the solution.

Writing out the components of A, we have

$$\begin{pmatrix} a_{11} & a_{12} & a_{13} & \cdots \\ a_{21} & a_{22} & a_{23} & \cdots \\ a_{31} & a_{32} & a_{33} & \cdots \\ \vdots & & \vdots & \end{pmatrix} \tag{11.17}$$

To modify A such that we have zeros below the diagonal, proceed as follows: Let's say we wish to place zeros in column 1 below the diagonal. We begin with the second equation. If we multiply the first equation by $\frac{a_{21}}{a_{11}}$ and subtract this from the second equation, then we must have a zero in the (2,1) position of A. Similarly, for the third equation we multiply the first equation by $\frac{a_{31}}{a_{11}}$ and subtract it from the third equation. We must also perform the same operation on the right-hand side vector y. The algorithm is then as follows

$$a_{jk} = a_{jk} - \frac{a_{ji}}{a_{ii}} a_{ik}, \quad \begin{aligned} k &= 1+1, \ldots, n \\ j &= i+1, \ldots, n \end{aligned} \tag{11.18}$$

and

$$y_j = y_j - \frac{a_{ji}}{a_{ii}}, \quad j = i+1, \ldots, n \tag{11.19}$$

The "=" sign in equation (11.18) has the meaning *is replaced by*. Note that all of the diagonals must be nonzero in order to apply equation (11.18). If this is not the case we must *pivot*, as will be explained later. If the matrix is nonsingular we will always find a nonzero pivot.

Solution of Equations

When the elimination process is completed, the resulting system matrix is upper triangular. Therefore the last equation contains only one unknown and can be solved for directly.

$$x_n = \frac{y_n}{a_{nn}} \tag{11.20}$$

All other unknowns are found (in order from the nth equation to the 1st) by back-substitution using

$$x_i = \frac{y_i - \sum_{j-i+1}^{n} a_{ij} x_j}{a_{ii}} \quad i = n-1, n-2, \ldots 1 \tag{11.21}$$

Pivoting

Most of the systems that we will be studying will be positive definite and well posed. However, the accuracy of the solution may be improved by partial pivoting, that is, by interchanging equations during elimination such that for any i, the diagonal element a_{ii} is always the largest element in column i of the remaining sub-matrix to be eliminated. This is because the elimination factor a_{ji}/a_{ii} is always less than or equal to one, thus reducing the round-off error. If the original diagonal element is zero, partial pivoting must be applied. If no nonzero element can be found in column i, then the matrix is singular and no unique solution exists.

For full pivoting the entire submatrix is searched for the largest element, which is then positioned into the diagonal. If this involves a column interchange, then the unknowns must be renumbered. This may be necessary for a near-singular set of equations.

Symmetry

A symmetric matrix has the property of remaining symmetric after each reduction step.

$$a_{jk} = a_{jk} - \frac{a_{ji} a_{ik}}{a_{ii}} \tag{11.22}$$

$$a_{kj} = a_{kj} - \frac{a_{ki} a_{ij}}{a_{ii}} = a_{jk} - \frac{a_{ik} a_{ji}}{a_{ii}} = a_{jk} \tag{11.23}$$

Therefore, only the elements in the upper triangle need to be modified and the algorithm becomes

$$a_{jk} = a_{jk} - \frac{a_{ij}}{a_{ii}} a_{ik}, \quad k = j, j+1, ..., n \quad (11.24)$$
$$j = i+1, ..., n \quad (11.25)$$

In this case, only a little more than half of the operations need to be performed.

11.3 LU DECOMPOSITION

Any nonsingular matrix, A, can be factored into a product

$$A = LU \quad (11.26)$$

The matrices L and U are lower and upper triangular, respectively, and usually one or the other has 1's on the diagonal. These are called unit lower triangular and unit upper triangular matrices, respectively. We will illustrate the case of the unit lower triangular factorization and develop the algorithm as follows:

First consider the nonsingular (possibly nonsymmetric and complex) matrix

$$A = \begin{pmatrix} a_{11} & a_{12} & \cdots & a_{1n} \\ a_{21} & a_{22} & \cdots & a_{2n} \\ \vdots & \vdots & \vdots & \vdots \\ a_{n1} & a_{n2} & \cdots & a_{nn} \end{pmatrix} \quad (11.27)$$

Solution of Equations

By looking at the decomposition component by component, we have (for the unit lower triangular decomposition)

$$\begin{pmatrix} a_{11} & a_{12} & \cdots & a_{1n} \\ a_{21} & a_{22} & \cdots & a_{2n} \\ \vdots & \vdots & \vdots & \vdots \\ a_{n1} & a_{n2} & \cdots & a_{nn} \end{pmatrix} = \begin{pmatrix} 1 & 0 & \cdots & 0 \\ l_{21} & 1 & \cdots & 0 \\ \vdots & \vdots & \vdots & \vdots \\ l_{n1} & l_{n2} & \cdots & 1 \end{pmatrix} \begin{pmatrix} u_{11} & u_{12} & \cdots & u_{1n} \\ 0 & u_{22} & \cdots & u_{2n} \\ \vdots & \vdots & \vdots & \vdots \\ 0 & 0 & 0 & u_{nn} \end{pmatrix} \tag{11.28}$$

The first row of L is known, and by multiplying the first row of L by the first column of U we see that $u_{11} = a_{11}$. In fact, because there is only one nonzero term in the first column of U, we see that the first column of L is found as

$$l_{i1} = \frac{a_{i1}}{u_{11}}, \quad i = 2, 3, \ldots, n \tag{11.29}$$

We can then proceed to find the first row of U, and so forth. The LU decomposition can be expressed as follows:

$$A = \begin{bmatrix} A^{(1)}_{11} & a^{(1)T}_{1r} \\ a^{(1)}_{c1} & A^{(1)} \end{bmatrix} = \begin{bmatrix} A^{(1)}_{11} & 0 \\ a^{(1)}_{c1} & I \end{bmatrix} \begin{bmatrix} 1 & a^{(1)T}_{1r}/a^{(1)}_{11} \\ 0 & A^{(1)} - \frac{a^{(1)}_{c1} a^{(1)T}_{1r}}{a_{11}} \end{bmatrix} \tag{11.30}$$

where $a^{(1)}_{c1}$ is a column vector $((n-1) \times 1)$, $a^{(1)T}_{1r}$ is a row vector $(1 \times (n-1))$ and A^1 is a matrix $((n-1) \times (n-1))$.

Then we LU decompose

$$\begin{bmatrix} A^{(1)} - \frac{a^{(1)}_{c1} a^{(1)T}_{1r}}{a_{11}} \end{bmatrix} = \begin{bmatrix} a^{(2)}_{22} & a^{(2)T}_{2r} \\ a^{(2)}_{22} & A^{(2)}_{2} \end{bmatrix}$$

$$= \begin{bmatrix} a^{(2)}_{22} & 0 \\ a^{(2)}_{c2} & I \end{bmatrix} \begin{bmatrix} 1 & a^{(2)T}_{2r}/a^{(2)}_{22} \\ 0 & A^{(2)}_{2} - \frac{a^{(2)}_{22} a^{(2)T}_{2r}}{a^{(2)}_{22}} \end{bmatrix} \tag{11.31}$$

$$A = \begin{bmatrix} a_{11}^{(1)} & 0 & 0 & \cdots \\ a_{21}^{(1)} & a_{22}^{(2)} & & \cdots \\ a_{31}^{(1)} & a_{32}^{(2)} & 1 & \\ \vdots & \vdots & & \end{bmatrix} \begin{bmatrix} 1 & a_{21}^{(1)} & a_{13}^{(1)}/a_{11}^{(1)} & \cdots \\ 0 & 1 & a_{2r}^{(22)}/a_{22} & \cdots \\ & & A_2^{(2)} - \dfrac{a_{c2}^{(2)} a_{2r}^{(2)T}}{a_{22}^{(2)}} \end{bmatrix}$$

We therefore have the recurrence formula

$$\begin{aligned} \ell_{ij} &= a_{ij} - \sum_{k=1}^{j-1} \ell_{ik} u_{kj} & i \geq j \\ u_{ij} &= \frac{1}{\ell_{ii}} \left(a_{ij} - \sum_{k=1}^{i-1} \ell_{ik} u_{kj} \right) & i < j \end{aligned} \quad (11.32)$$

In the computer implementation we can store the composite $[L/U]$ matrix F as follows:

$$\begin{aligned} f_{ij} &= \ell_{ij} \text{ if } i \geq j \\ &= u_{ij} \text{ if } i < j \end{aligned} \quad (11.33)$$

Because the diagonal elements of U are all 1's, they need not be stored. This composite matrix $F = [L/U]$ is sometimes called the table of factors for A. Although A may be sparse, we are not assured that F will be sparse. We will address this problem when we talk about optimal ordering.

To summarize, there are two steps in the solution.

- Factorization

$$A = LU$$

Solution of Equations

- **Substitution**

$$Lz = y$$
$$Ux = z$$

Other computational advantages of the method are as follows:

- (1) If we want to solve $A^T x = y$ and are given $A = LU$ then $A^T = (LU)^T = U^T L^T$, where U^T is a lower triangular matrix L_1 and L^T is an upper triangular matrix U_1 so that $A^T = L_1 U_1$. We can then solve

$$L_1 z = y$$
$$U_1 x = z \quad (11.34)$$

- (2) If we want to solve

$$A \begin{bmatrix} x_1 \\ x_2 \end{bmatrix} = \begin{bmatrix} y_1 \\ y_2 \end{bmatrix} \quad (11.35)$$

for x_1 y_2 (given y_1, x_2)

$$A = LU = \begin{bmatrix} L_{11} & 0 \\ L_{21} & L_{22} \end{bmatrix} \begin{bmatrix} U_{11} & U_{12} \\ 0 & U_{22} \end{bmatrix}$$

$$\begin{bmatrix} L_{11} & 0 \\ L_{21} & L_{22} \end{bmatrix} \begin{bmatrix} U_{11} & U_{12} \\ 0 & U_{22} \end{bmatrix} \begin{bmatrix} x_1 \\ x_2 \end{bmatrix} = \begin{bmatrix} y_1 \\ y_2 \end{bmatrix} \quad (11.36)$$

$$L_{11}U_{11}x_1 + L_{11}U_{12}x_2 = y_1$$
$$L_{11}U_{11}x_1 = y_1 - L_{11}U_{12}x_2$$
$$L_{21}U_{11}x_1 + (L_{21}U_{12} + L_{22}U_{22})x_2 = y_2 \quad (11.37)$$

The first equation can be solved for x and the equations are already in the desired LU factorized form. Substituting into the second equation, we get y_2.

- (3) We can solve the system

$$A^T \begin{bmatrix} x_1 \\ x_2 \end{bmatrix} = \begin{bmatrix} y_1 \\ y_2 \end{bmatrix} \text{ for } x_1, y_1 \text{ (given } y_1, x_2)$$

using a combination of (1) and (2).

- (4) We may want to solve the same system for different values of input (right-hand side). If we save the decomposition, we need only to perform the forward and back substitutions to find the solution.

Variations of the the LU Decomposition

We have not proved it, but the decomposition $A = LU$ where L is unit lower triangular and U is upper triangular is unique. Further, both L and U are nonsingular. That being so, there are variations of the LU decomposition that may prove useful. For example, the 1's could be on the diagonal of the upper triangular matrix instead of the lower triangular matrix. The algorithm for this possibility is a simple variation of the one presented before and is left to the reader. Two other variations follow:

The LDU' Decomposition

In this process the matrix A is factored into a unit lower triangular matrix L, a diagonal matrix, D, and a unit upper triangular matrix U'. This possibility follows from the fact that because U is upper triangular and nonsingular, then $u_{ii} \neq 0$, $i = 1, ..., n$. Let D be the diagonal matrix made of the diagonal elements of U. Then D^{-1} exists. Its elements are simply $\frac{1}{u_{ii}}$. Let $U' = D^{-1}U$. The matrix U' is upper triangular. As $A = LU$, then $A = LDD^{-1}U = LDU'$.

The LDL^T Decomposition

If A is symmetric, then the decomposition can be made in the more convenient form $A = LDL^T$, where L is unit lower triangular and D is a diagonal matrix. We see this as follows: We have shown that A can be expressed as $A = LDU'$. In this case $A = A^T = (LDU')^T = U'^T D^T L^T$. Here U'^T is unit lower triangular, D^T is diagonal, and L^T is upper triangular. As $U^T D^T L^T$ is a decomposition of A and this decomposition is unique, we deduce that $U'^T = L^T$. If A is positive definite, then the elements of D are positive.

Solution of Equations

We will now show that Gauss elimination, triangular factorization, and LU decomposition involve the same operations. We will use the following example [125]:

$$\begin{aligned} 2x_1 + x_2 + 3x_3 &= 6 \\ 2x_1 + 3x_2 + 4x_3 &= 9 \\ 3x_1 + 4x_2 + 7x_3 &= 14 \end{aligned} \qquad (11.38)$$

In matrix notation

$$A = \begin{pmatrix} 2 & 1 & 3 \\ 2 & 3 & 4 \\ 3 & 4 & 7 \end{pmatrix} \qquad (11.39)$$

Each step of the Gaussian elimination procedure is what we have defined as an elementary row operation. Each corresponds to the premultiplication of A by an elementary matrix. We can therefore express the Gaussian elimination as follows:

$$\begin{bmatrix} 1 & -1/2 & -3/2 \\ 0 & 1 & 0 \\ 0 & 0 & 1 \end{bmatrix} \begin{bmatrix} 1 & 0 & 0 \\ 0 & 1 & -1/2 \\ 0 & 0 & 1 \end{bmatrix} \begin{bmatrix} 1 & 0 & 0 \\ 0 & 1 & 0 \\ 0 & 0 & 4/5 \end{bmatrix}$$
$$\times \begin{bmatrix} 1 & 0 & 0 \\ 0 & 1/2 & 0 \\ 0 & -5/4 & 1 \end{bmatrix} \begin{bmatrix} 1/2 & 0 & 0 \\ -1 & 1 & 0 \\ -3/2 & 0 & 1 \end{bmatrix} \begin{bmatrix} 2 & 1 & 3 \\ 2 & 3 & 4 \\ 3 & 4 & 7 \end{bmatrix} \qquad (11.40)$$

This procedure will transform the original matrix into an identity matrix. Consider the last two matrices. Premultiplying the original matrix by the row vector [1/2, 0, 0] divides the first row by 2, thus putting a 1 in the (1,1) place. Premultiplying by [-1, 1, 0] subtracts the second row from the first, putting a 0 in the (2,1) position. Premultiplying by the third row, [-3/2, 0, 1], puts a zero in the (3,1) place. The reader can confirm that the process results in the identity matrix. Note that all of the elementary matrices are triangular, that is,

$$U_1 U_2 L_3 L_2 L_1 A = I \qquad (11.41)$$

Because the product of upper (lower) triangular matrices is an upper (lower) triangular matrix, we can write

$$U_\pi L_\pi A = I \tag{11.42}$$

Here the subscript π is used to indicate a product of matrices. It is also true that the inverse of an upper (lower) triangular matrix is an upper (lower) triangular matrix. Therefore

$$A = LU, \quad L = L_\pi^{-1}, \quad U = U_\pi^{-1} \tag{11.43}$$

The solution by Gauss elimination involves two stages: (1.) forward elimination (premultiplication by L_π) brings the system into an upper triangular form ($U = L_\pi A$) and (2.) back-substitution (premultiplication by U_π) transforms the system into an identity matrix ($I = U_\pi L_\pi A$).

Returning to $Ax = y$,

$$
\begin{array}{ll}
Ax = y & Ax = y \\
L_\pi Ax = L_\pi y & A = LU \\
Ux = z & Lz = y \\
U_\pi U_x = U_\pi z & Lz = y \\
x = U^{-1} z & Ux = z \\
\text{Gauss elimination} & LU \text{ decomposition}
\end{array}
\tag{11.44}
$$

Therefore the two methods are equivalent.

Solution of Equations

11.4 CHOLESKY DECOMPOSITION

Any symmetric positive definite matrix can be decomposed into a product

$$A = L \cdot L^T \tag{11.45}$$

where L is a lower triangular matrix. To see this, we write A in a partitioned form

$$A = \begin{pmatrix} a_{11} & b^T \\ b & A_2 \end{pmatrix} \tag{11.46}$$

From the positive definite property of A we know that $a_{11} > 0$. We see that A can be factored as

$$A = \begin{pmatrix} \sqrt{a_{11}} & 0^T \\ \frac{b}{\sqrt{a_{11}}} & I \end{pmatrix} \begin{pmatrix} 1 & 0^T \\ 0 & A^{(2)} \end{pmatrix} \begin{pmatrix} \sqrt{a_{11}} & \frac{b}{\sqrt{a_{11}}} \\ 0 & I \end{pmatrix} \tag{11.47}$$

The $(n-1) \times (n-1)$ matrix $A^{(2)}$ is given by

$$A^{(2)} = A_2 - \frac{b \cdot b^T}{a_{11}} \tag{11.48}$$

$A^{(2)}$ is symmetric (from the form of equation (11.48)) and is also positive definite. To see this, let us define a vector

$$x = \begin{pmatrix} x_1 \\ x' \end{pmatrix} \tag{11.49}$$

Here x' can be any nonzero vector of $(n-1)$ components. Using equation (11.46) and the fact that A is positive definite, we find

$$x^T A x = a_{11} x_1^2 + 2 x_1 x'^T b + x'^T A^{(2)} x' > 0 \tag{11.50}$$

If we set

$$x_1 = -\frac{x'^T b}{a_{11}} \tag{11.51}$$

then

$$x'^T A^{(2)} x' > 0 \tag{11.52}$$

Because x' is any nonzero vector, $A^{(2)}$ is positive definite. This means we can factor $A^{(2)}$ in the same way and continue the process until we have a matrix of order 1. We then obtain

$$A = L_1 L_2 ... L_n U_n ... U_2 U_1 \tag{11.53}$$

where L_k is the transpose of U_k. Multiplying out the factors in equation (11.53) gives equation (11.45).

Algorithm for Cholesky Decomposition

To compute the Cholesky factors of a matrix we can use the following algorithm:

1. Set $L_{11} = \sqrt{a_{11}}$.

2. In each row i, compute the off-diagonals as

$$l_{ik} = (a_{ik} - \sum_{j=1}^{k-1} l_{ij} l_{kj}) / l_{kk}$$

3. Compute the diagonal element as

$$l_{ii} = \sqrt{a_{ii} - \sum_{j=1}^{i-1} l_{ij}^2}$$

Solution of Equations

Numerical Example

Consider the 6 × 6 system

$$A = \begin{bmatrix} 7 & -2 & -3 & 0 & -1 & 0 \\ -2 & 8 & 0 & 0 & -1 & 0 \\ -3 & 0 & 4 & -1 & 0 & 0 \\ 0 & 0 & -1 & 5 & 0 & -2 \\ -1 & -1 & 0 & 0 & 4 & 0 \\ 0 & 0 & 0 & -2 & 0 & 6 \end{bmatrix} \quad (11.54)$$

The matrix is symmetric and positive definite. In the Cholesky decomposition the l_{11} term is the square root of a_{11} or $\sqrt{7} = 2.646$. To find l_{21} we note that the product $l_{11}l_{21} = -2$. Because we know $l_{11} = 2.646$, we now find $l_{21} = -0.756$. Similarly, the first column of L (and therefore the first row of L^T) is found by dividing the first column of A by $\sqrt{7}$. To find l_{22} we note that $l_{11}l_{21} + l_{22}^2 = 8$. The only unknown here is l_{22}. We now find $l_{22} = \sqrt{a_{22} - l_{11}l_{12}} = 2.726$. We leave it as an exercise to find the remaining elements of the decomposition as

$$L = \begin{bmatrix} 2.646 & 0 & 0 & 0 & 0 & 0 \\ -0.756 & 2.726 & 0 & 0 & 0 & 0 \\ -1.134 & -0.314 & 1.617 & 0 & 0 & 0 \\ 0 & 0 & -0.618 & 2.149 & 0 & 0 \\ -0.378 & -0.472 & -0.357 & -0.103 & 1.870 & 0 \\ 0 & 0 & 0 & -0.931 & -0.051 & 2.265 \end{bmatrix} \quad (11.55)$$

11.5 SPARSE MATRIX TECHNIQUES

Most matrices resulting from finite difference or finite element formulations are sparse. Although the number of equations may be large, each equation will have relatively few terms. We exploit this sparse structure by, wherever possible, not storing the zeros and not operating on them.

We have seen that Gauss elimination, LDU decomposition, Cholesky decomposition, and related methods all involve essentially the same steps. The inverse or decomposition of a sparse matrix is not usually sparse. This is because elimination or decomposition results in *fill-ins*. Although it is usually not possible to avoid creating fill-ins, it is possible to minimize the number of fill-ins by renumbering or reordering schemes.

11.5.1 Storage Methods

As we will be dealing with matrices in which most of the elements are zero, we take advantage of the sparse structure by (1.) not storing the zeros and (2.) not operating on the zeros. To accomplish the first goal we use what is known as compact storage techniques. There are many variations of these techniques, and we will look at three popular methods. These methods vary in efficiency and in complexity, and the one to choose will depend on the specific problem and the method of matrix solution.

The first method is known as a *linked list*. It is very efficient in that it stores only the nonzeros. It is also well adapted to either direct or iterative solvers. For direct solvers we must deal with *fill-ins* and the linked list approach allows us to modify the matrix very easily. The method is however more complicated to program. The second method, which uses three arrays, is simpler and also stores only the nonzeros. Fill-ins are more difficult to accommodate but the simplicity of programming makes up for this disadvantage. The third method is band storage, and this method, although less efficient in that it stores some zeros, is very popular for direct solvers. The method is usually used in conjunction with a bandwidth reduction scheme that compresses the nonzero matrix elements around the diagonal.

Linked List Structure

This involves storing the matrix row by row. Three pieces of information are required.

- 1. The starting address of each row
- 2. The diagonal elements
- 3. The nonzero off-diagonal elements

Each nonzero element a_{ij} is stored as an item in its row i. An item consists of the following three attributes, which completely specify the matrix element:

- 1. The column number j.
- 2. The numerical value of the element.
- 3. The location of the next element in the row.

For more information on linked lists see [126].

Solution of Equations

In FORTRAN this procedure could be implemented as follows:

- 1. Create five arrays IA, AD, JA, AN, KA.
- 2. Array IA is the starting address $k = IA(i)$, which means that the first off-diagonal nonzero in row i starts at $JA(k)$.
- 3. AD represents the diagonal elements.
- 4. $JA(k)$ gives the column number j.
- 5. $AN(k)$ gives the value of a_{ij}.
- 6. $KA(k)$ gives the location of the next element in the row.

Pointer Method

In this method the structure of A is described by three one-dimensional arrays IA, JA, and AN. The n rows are stored contiguously in compact form. The nonzero matrix elements are stored in the array AN. The array JA is an integer array that gives the column numbers corresponding to the nonzero elements in AN. The array IA is another integer array that gives the location of the first element in each row. The dimension of this array is therefore equal to the number of rows. Thus, in the following example if we consider row 3, we see that the third element of IA is 8 so the third row begins with the eighth element of AN which is -3. The column number is given by the eighth element of JA which is 1. Therefore the entire matrix can be recreated by one real (or complex) array and two integer arrays. This reduces the storage considerably. For more information on this scheme see [127].

Example

We use the previous example matrix on which we performed Cholesky decomposition.

$$A = \begin{bmatrix} 7 & -2 & -3 & 0 & -1 & 0 \\ -2 & 8 & 0 & 0 & -1 & 0 \\ -3 & 0 & 4 & -1 & 0 & 0 \\ 0 & 0 & -1 & 5 & 0 & -2 \\ -1 & -1 & 0 & 0 & 4 & 0 \\ 0 & 0 & 0 & -2 & 0 & 6 \end{bmatrix} \qquad (11.56)$$

The three arrays are

$$
\begin{array}{ccc}
IA & JA & AN \\
1 & 1 & 7 \\
  & 2 & -2 \\
  & 3 & -3 \\
  & 5 & -1 \\
5 & 1 & -2 \\
  & 2 & 8 \\
  & 5 & -1 \\
8 & 1 & -3 \\
  & 3 & 4 \\
  & 4 & -1 \\
11 & 3 & -1 \\
  & 4 & 5 \\
  & 6 & -2 \\
14 & 1 & -1 \\
  & 2 & -1 \\
  & 5 & 4 \\
17 & 4 & -2 \\
  & 6 & 6 \\
\end{array}
\qquad (11.57)
$$

11.5.2 Optimal Ordering

During the Gaussian elimination process, we call an element that changes from a zero to a nonzero a *fill-in*. Sparse matrix techniques are aimed at reducing the fill-ins. We save time by not having to compute these and we save memory by not storing them. Some of the most successful techniques involve renumbering or reordering the equations. Although no one has yet come up with a method that guarantees a minimum number of fill-ins, there are a number of methods that may be called near-optimal and that greatly reduce the computation.

As an example, consider the system graph of a 10-node problem shown in Figure 11.1.

This results in a matrix with structure illustrated in Figure 11.2.

If factorization or decomposition proceeds in the usual order, most of the zero elements will fill in during the process. We will be left with the matrix shown in Figure 11.3.

However, if we reorder the rows and columns as indicated in Figure 11.4 then the decomposition produces only two fill-ins as shown in Figure 11.5

Solution of Equations

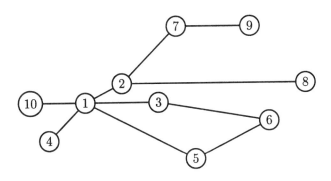

Figure 11.1 Connection Graph for 10-Node System

	1	2	3	4	5	6	7	8	9	10
1	×	×	×	×	×					×
2	×	×					×	×		
3	×		×			×				
4	×			×						
5	×				×	×				
6			×		×	×				
7		×					×		×	
8		×						×		
9							×		×	
10	×									×

Figure 11.2 A Matrix of Order 10 Showing Non-Zeros

In the kth step of decomposition we are working on the kth row and can write the partially decomposed system as

$$A^{(k)} = \begin{pmatrix} a_{kk}^{(k)} & 0 \\ a_{k+1,k}^{(k)} & \\ \vdots & I \end{pmatrix} \begin{pmatrix} 1 & \dfrac{a_{k,k+1}^{(k)}}{a_{kk}^{(k)}} & \cdots \\ 0 & A_{k+1}^{(k+1)} \end{pmatrix} \qquad (11.58)$$

	1	2	3	4	5	6	7	8	9	10
1	×	×	×	×	×					×
2	×	×	F	F	F		×	×		F
3	×	F	×	F	F	×	F	F		F
4	×	F	F	×	F	F	F	F		F
5	×	F	F	F	×	×	F	F		F
6			×	F	×	×	F	F		F
7		×	F	F	F	F	×	F	×	F
8		×	F	F	F	F	F	×	F	F
9							×	F	×	F
10	×	F	F	F	F	F	F	F	F	×

Figure 11.3   The Matrix of Order 10 after Factorization

	8	9	4	10	7	2	6	3	5	1
8	×				×					
9		×			×					
4			×							×
10				×						×
7		×			×	×				
2	×				×	×				×
6							×	×	×	
3							×	×		×
5							×		×	×
1			×	×		×		×	×	×

Figure 11.4   The Matrix of Order 10 Reordered

where the ijth element of $A_{k+1}^{(k+1)}$ is given by

$$a_{ij}^{(k+1)} = a_{ij}^{(k)} - \frac{a_{ik}^{(k)} a_{kj}^{(k)}}{a_{kk}^{(k)}} \qquad (11.59)$$

Solution of Equations

	8	9	4	10	7	2	6	3	5	1
8	×				×					
9		×			×					
4			×							×
10				×						×
7	×				×	×				
2					×	×				×
6							×	×	×	
3							×	×	F	×
5							×	F	×	×
1			×	×		×		×	×	×

Figure 11.5 The Matrix of Order 10 Reordered Showing Fill-ins

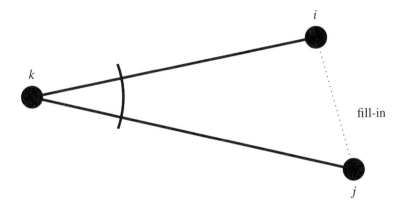

Figure 11.6 Generation of a Fill-in (dotted) by Eliminating Node k

Therefore a fill-in is introduced if $a_{ij}^{(k)} = 0$ and $a_{ij}^{(k+1)} \neq 0$. The latter condition occurs only if $a_{ik}^{(k)} \neq 0$ and $a_{kj}^{(k)} \neq 0$. This is illustrated in Figure 11.6. Here node k is connected to nodes i and j. But there is no ij connection. Elimination of the kth column produces the ij connection.

It seems reasonable then to bring the row and column that have the fewest nonzero terms among the remaining $n - (k - 1)$ rows to be the kth row. This is the scheme suggested by Tinney and Walker [125].

11.5.3 BAND AND PROFILE STORAGE

The previous methods store only the nonzero elements of the matrix. Two methods that are simpler to program but not quite as efficient are band storage and profile storage. The two methods take advantage of the fact that the matrices will be sparse and work best if the nonzero elements are clustered tightly around the diagonal. During the elimination or factorization process, we consider keeping all of the terms, including the zeros, from the first non-zero term in the equation to the diagonal. This assumes that the matrix is symmetric. If the leftmost non-zero term in row i occurs in column $n(i)$, the half-bandwidth M is defined as

$$M = 1 + \max_i [i - n(i)]$$

In the band storage we store M elements for each row. In row k the first entry would be column $k - M + 1$, the last element would be the diagonal of row k. For example, consider the following matrix (the ordering of which we will obtain by a bandwidth reduction technique in the next section).

	1	2	3	4	5	6	7	8	9	10
1	12	-2	-3							
2	-2	8	-1	-4						
3	-3	-1	10	-5	-1					
4		-4	-5	16	-4					
5			-1	-4	18	-1	-6	-2		
6					-1	7		-3		
7					-6		8		-1	
8					-2	-3		17	-5	-3
9							-1	-5	10	-2
10								-3	-2	9

Solution of Equations

$$\begin{array}{cccc} 0 & 0 & 0 & 12 \\ 0 & 0 & -2 & 8 \\ 0 & -3 & -1 & 10 \\ & -4 & -5 & 16 \\ 0 & -1 & -4 & 18 \\ & & -1 & 7 \\ & -6 & & 8 \\ -2 & -3 & & 17 \\ & -1 & -5 & 10 \\ & -3 & -2 & 9 \end{array}$$

This scheme uses 40 memory locations instead of 55 for the lower triangle of the full matrix. Note that this includes some leading zeros in the first equation which are not actually in the matrix. As the system becomes larger the savings increase.

Profile or skyline storage works in a similar way. Instead of finding the largest bandwidth in the system and using that for each row, the skyline storage method uses the bandwidth of each equation. The preceding example would become

$$PROF(A) = [12, -2, 8, -3, -1, 10, -4, -5, 16, -1, -4, 18, -1, 7,$$
$$-6, 0, 8, -2, -3, 0, 17, -1, -5, 10, -3, -2, 9] \quad (11.60)$$

Notice that any zeros that occur after the first nonzero are stored. However, leading zeros are never stored, as they sometimes are in the profile method. We need another piece of information here, the location of the leftmost zero, $n(i)$.

$$LEFT(A) = [1, 1, 1, 2, 3, 5, 5, 5, 7, 8]$$

The storage in this case is 27. These methods are illustrated for our symmetric 10×10 matrix in Figures 11.7 and 11.8.

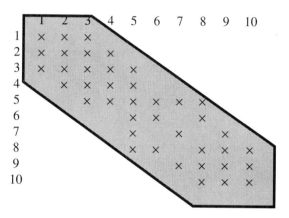

Figure 11.7 Band Storage

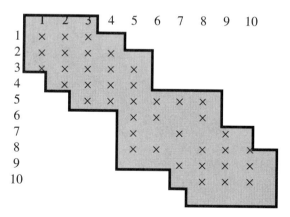

Figure 11.8 Profile Storage

Solution of Equations 617

If we use these methods, we can modify the Cholesky decomposition algorithm to take into account the fact that we are not storing the whole matrix. The change occurs in the limits of the summations. The new formulas become:

For the band solver,

$$L_{ik} = (a_{ik} - \sum_{j=J}^{k-1} L_{ij} L_{kj})/L_{kk}$$

$$L_{ii} = \sqrt{a_{ii} - \sum_{j=J}^{i-1} L_{ij}^2} \quad \text{where } J = i - M \quad (11.61)$$

For the skyline solver,

$$J = \max[n(i), n(k)]$$

Node Numbering

When using band or profile storage, node numbering becomes important. Most node numbering schemes are variations of the one proposed by Cuthill and McKee [128]. A similar method is found in Gibbs *et al.* [129]. The best method for band storage is not necessarily the best method for profile storage; however, the methods seem to improve the performance of either solver. The intent of the method is to number the nodes so that the bandwidth is a (near) minimum. To do this we would like to keep the nodes that are connected to each other, close to each other numerically; we do not want node 1 connected to node 10,000 as this results in a large bandwidth and many stored zeros. These methods use the concept of "levels." An example will illustrate. We choose a starting node as level 0, or $L(0)$. $L(i)$ is made of all variables connected to $L(i-1)$ that have not been assigned a level yet. The object is to make the number $i - j$ as small as possible. To do this we number all members of $L(i)$ before numbering $L(i+1)$.

The method involves essentially two steps:

- (1.) Choose the node with the fewest connections as the starting node.

- (2.) Within each level, number the nodes connected to the fewest members of the next level first. This is because if we wait as long as possible to number a node that is strongly connected to members of the next level, its node number will be higher and therefore closer to the node numbers of the next level.

It should also be noted that a scheme which gives a greater number of levels is preferred. The ordering of the graph using numbers is better than the numbering of the graph using letters. For the numbered graph in Figure 11.9 we pick the lower left node as the starting node. Table 11.1 shows the node, the level, and the number of connections to the next (higher) level. This is the Cuthill–McKee solution.

NODE	1	2	3	4	5	6	7	8	9	10
LEVEL	0	1	1	2	2	3	3	3	4	4
CONNECTIONS	2	1	2	0	3	0	1	2	0	

Table 11.1 Level Structure and Node Numbers

1. Select a starting node.

2. Form level 1 as the set of all nodes connected to level 0.

3. Form level 2 as the set of all nodes connected to level 1 not yet in another level.

4. Repeat until all nodes are in a level.

5. Begin with level 0 and number the nodes in each level before proceeding to the next level. Within each level number the nodes so that the nodes with the fewest connections to the next level are numbered first and the ones with the most connections are numbered last.

Figure 11.10 shows the original matrix, without renumbering. Figure 11.11 shows the nonzeros after the renumbering. We see that the bandwidth has been reduced.

The final bandwidth depends on the starting or root node, Reference [127] describes algorithms for finding good choices of starting nodes. Generally speaking, a starting node resulting in a larger number of levels is best.

Reverse Cuthhill–McKee

Although the Cuthill–McKee algorithm is effective in reducing bandwidth, it has been found that reversing the numbering of the Cuthill–McKee algorithm may decrease the profile storage requirements. It has been proven [127] that the reverse numbering will never produce a larger profile. Because this is a simple operation, it is often performed.

Solution of Equations

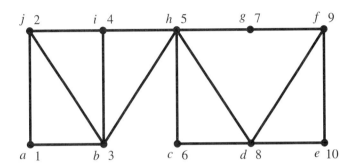

Figure 11.9 Cuthill–McKee Numbering

	a	b	c	d	e	f	g	h	i	j
a	×	×								×
b	×	×	×					×	×	×
c		×	×	×				×		
d			×	×	×	×		×		
e				×	×	×				
f				×	×	×	×			
g						×	×	×		
h		×	×	×			×	×	×	
i								×	×	×
j	×	×							×	×

Figure 11.10 Nonzeros with Original Numbering

We can see this by referring to Figures 11.12 and 11.13. In the Figure 11.12 we have used the Cuthill–McKee algorithm to renumber the nine-node system taking the leftmost node as the starting node. The storage for the half bandwidth is 21. By reversing the order of the nodes in Figure 11.13 we have reduced the storage to 20.

	1	2	3	4	5	6	7	8	9	10
1	×	×	×							
2	×	×	×	×						
3	×	×	×	×	×					
4		×	×	×	×					
5			×	×	×	×	×	×		
6					×	×		×		
7					×		×		×	
8					×	×		×	×	×
9							×	×	×	×
10								×	×	×

Figure 11.11 Nonzeros with Cuthill–McKee

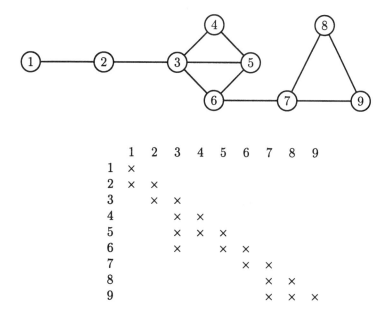

	1	2	3	4	5	6	7	8	9
1	×								
2	×	×							
3		×	×						
4			×	×					
5			×	×	×				
6			×		×	×			
7						×	×		
8							×	×	
9							×	×	×

Figure 11.12 Cuthill–McKee Numbering

Solution of Equations

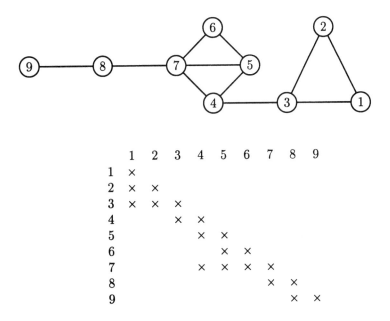

Figure 11.13 Reverse Cuthill–McKee Numbering

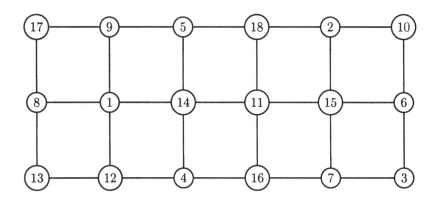

Figure 11.14 Random Ordering of 18-Node Problem

11.5.4 Nested Dissection Reordering

The purpose of the nested dissection method is to reorder the equations and variables of a symmetric matrix to attempt to minimize fill-ins during the solution process. This reordering is equivalent to renumbering the nodes of its associated graph or network.

By reducing fill-ins, we minimize the number of operations performed and therefore the time to reach the solution. We also minimize storage requirements by using a storage scheme that stores only nonzero elements.

The method is well suited for matrix problems arising in finite element and finite difference applications due to system sparsity and symmetry. We will first examine how the method works for a regular $m \times n$ grid, and then we will study the more complicated general case (irregular grids).

Nested Dissection for a Regular m × n Grid

A matrix with a rectangular grid is not a typical problem, but it does illustrate how the renumbering takes place and how it works. The example in Figure 11.14 shows a 3×6 grid with its nodes randomly numbered and its corresponding connection matrix. If the matrix were to be factored with this ordering, 34 fill-ins would be generated in the lower Choleski factor as shown in Figure 11.16.

Solution of Equations 623

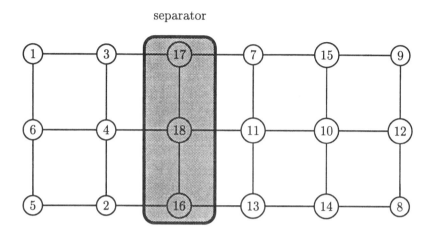

Figure 11.15 First Dissection of 18-Node Problem

The grid is *dissected* by breaking it into two grids of approximately equal size using a separator. Renumbering then proceeds by renumbering all the nodes of one subgrid first, then continuing with the nodes of the second subgrid and finally numbering the nodes of the separator, as shown in Figure 11.15. The separator should be chosen to contain as few nodes as possible.

It can be seen that its corresponding matrix contains four distinct areas: Area 1 corresponds to the first subgrid matrix, area 2 to the second subgrid matrix, area 3 to the separator, and area 4 is a zero-filled block. The key observation is that area 4 will remain zero during factorization, whereas areas 1, 2, and 3 may or may not generate fill-ins. Figure 11.17 shows the matrix structure after the first dissection.

The same method can now be applied to each subgrid, which in turn will separate into two smaller subgrids (see Figure 11.18) to which the method is applied again, and so on, until the subgrids can be divided no further. The effect of this is to create more and more zero block areas in which no factorization fill-ins can occur. This process is performed until the final ordering in Figure 11.19 is reached.

Symbolic factorization of the reordered matrix shows that fill-ins were reduced to 23. While this may or may not be the absolute minimum number of fill-ins obtainable with a given ordering, this reordering method gets very close to the minimum degree ordering with little effort.

	1	2	3	4	5	6	7	8	9	10	11	12	13	14	15	16	17	18
1	×							×	×			×		×				
2		×								×					×			×
3			×			×	×											
4				×								×		×		×		
5					×				×					×				×
6			×			×				×					×			
7			×			F	×								×	×		
8	×							×					×				×	
9	×				×			F	×								×	
10		×				×	F			×								
11											×				×	×	×	×
12	×			×				F	F			×	×					
13								×	F			×	×					
14	×			×	×			F	F		×	F	F	×				
15		×				×	×		F		×			F	×			
16			×				×		F		×	F	F	F	F	×		
17								×	×			F	F	F	F	F	×	
18		×		×					F	F	×	F	F	F	F	F	F	×

Figure 11.16 Random Numbering of 18-Node Problem Showing Fill-ins in Lower Triangle

Solution of Equations

	1	2	3	4	5	6	7	8	9	10	11	12	13	14	15	16	17	18
1	×																	
2		×																
3	×		×															
4		×	×	×														
5		×		F	×													
6	×		F	×	×	×												
7							×											
8								×										
9									×									
10										×								
11							×			×	×							
12								×	×	×	F	×						
13										×	F	×						
14								×		×	F	F	×	×				
15						×		×	×	×	F	F	F	F	×			
16	×		F	F	F								×	F	F	×		
17		×	F	F	F		×				F	F	F	F	F	F	×	
18			×	F	F						×	F	F	F	F	×	×	×

Figure 11.17 First Dissection Numbering of 18-Node Problem Showing Fill-ins in Lower Triangle

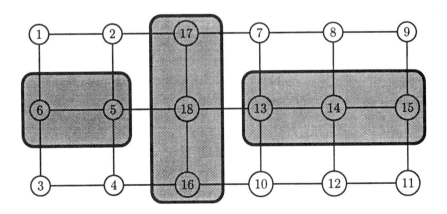

Figure 11.18 Second Dissection of 18-Node Problem

	1	2	3	4	5	6	7	8	9	10	11	12	13	14	15	16	17	18
1	×																	
2	×	×																
3			×															
4			×	×														
5		×		×	×													
6	×		×	F	×	×												
7							×											
8							×	×										
9								×	×									
10										×								
11											×							
12										×	×	×						
13							×	F	F	×		F	×					
14								×	F			×	×	×				
15									×		×	F	F	×	×			
16			×	F	F					×		F	F	F	F	×		
17		×		F	F	×	F	F				F	F	F	F	×	×	
18				×	F							×	F	F	×	×	×	×

Figure 11.19 Nested Dissection Numbering of 18-Node Problem Showing Fill-ins in Lower Triangle

Nested Dissection for the General Case

It is easy to see how a separator can be chosen for a rectangular grid but with an irregular grid choosing separators is not as obvious. In this case we want to generate a long level structure from the grid and then pick a middle level as the separator.

Generating a Long Level Structure

In order to generate a level structure a starting node must be chosen. Experience has proved that the best starting nodes for several ordering algorithms are *peripheral* nodes. This is true in this case, as well as in the reverse Cuthill-McKee algorithm. A peripheral node is one of either of the two nodes that are separated by the longest path, the "path" being the shortest distance between them. In a large problem, it becomes impractical to find a peripheral node, so a *pseudoperipheral* node is found that gives a "good" starting

Solution of Equations

point to generate a long, slender level structure. Once a starting node has been selected, the level structure is generated and the separator is found by picking the middle level [127]. From here on, the reordering is the same as in the regular grid case, except that every time we want to find a new separator for any subgrid, we must find its peripheral node, generate its level structure, and pick a middle level as the separator.

Algorithm for Finding a Pseudoperipheral Node

This algorithm does not guarantee the finding of a peripheral node, but the node found will have a high degree of *eccentricity*, in other words, it will have a long path between itself and another node in the grid, the "path" being the shortest distance between the two nodes. High eccentricity guarantees long level structures.

The steps involved in finding a pseudoperipheral node are as follows:

- Step 1 — Initialization: Choose an arbitrary node r in the grid.
- Step 2 — Generate a level structure rooted at node r.
- Step 3 — Choose a node x in the last level that is of minimum degree (has the lowest number of connections to other nodes).
- Step 4 — Generate a level structure rooted at x.
- Step 5 — If the number of levels with root x is larger than the number of levels with root r, then x becomes r, and go to step 3.
- Step 6 – Finished: Node x is a pseudoperipheral node.

11.6 THE PRECONDITIONED CONJUGATE GRADIENT METHOD

Since the 1970s, the preconditioned conjugate gradient method has been extensively used to solve sparse matrix equations resulting from the discretization of boundary value problems. This method combines useful aspects of iterative methods with the desirable characteristics of direct methods and is a modification of the well-known, but little used, conjugate gradient method.

The method of conjugate gradients appears to have been first proposed by Hestenes and Stiefel [130] in 1952 and was later discussed by Forsythe [131]. A useful bibliography of iterative and gradient solutions can be found in the work of Engeli *et al.* [132]. A clear description of this technique with illustrative examples was presented by Fox [58]. Manteuffel [133] discusses the preconditioned conjugate gradient method with shifted Cholesky factorization at great length. Silvester [48] summarized the method for engineering applications.

The conjugate gradient method is applicable to symmetric positive definite matrices and is executed in a semi-iterative fashion stepwise. This technique, however, is guaranteed to converge in N steps for a matrix equation of order N, provided that round-off errors are not generated in the process. Similar to iterative methods, this scheme yields an improved estimate of the solution vector at each step, while permitting restarting at any point during the process. Because the original matrix is preserved throughout the computation such that the zeros in the matrix are not replaced by fill-ins (as would be the case with Gauss elimination) the conjugate gradient method results in memory economy.

We shall discuss in this section the method of conjugate gradients and briefly describe the preconditioning modifications that have been introduced to render the technique practical in terms of speed of execution and attractive from a computer storage point of view.

Consider the solution of a set of linear equations

$$Ax = b \tag{11.62}$$

where A is a positive definite symmetric matrix ($N \times N$), x is the unknown solution vector ($N \times 1$), b is the source or forcing function vector (Nx1), and N is the order of the matrix. The quadratic form, which on minimization yields the solution to equation (11.62) is given by

$$F = \frac{1}{2} x^T A x - b^T x \tag{11.63}$$

It is evident that the negative gradient of F yields the residual vector r, such that

$$-\nabla F = b - Ax = r \tag{11.64}$$

Solution of Equations

Because A is positive definite, for any arbitrary nonzero vector W, the following relationship will hold.

$$W_i^T A W_j > 0, \text{ for } i = j \tag{11.65}$$

We can also find a set of N vectors orthogonal with respect to A, or *conjugate,* such that

$$W_i^T A W_j = 0, \text{ for } i \neq j \tag{11.66}$$

and from (11.65)

$$W_i^T A W_j \neq 0, \text{ for } i = j \tag{11.67}$$

These vectors, as we shall show later on, are independent vectors and span the space in which any solution of x of the matrix equation (11.62) must lie. Therefore, any possible solution to (11.62) must be representable as a linear combination of these vectors W, such that

$$x = C_i W_i \tag{11.68}$$

where the vectors W can be termed the direction or basis vectors. If we can determine the coefficients C, the solution to equation (11.62) can be found exactly. Multiplying equation (11.62) by the ith vector of W

$$W_i^T A x = W_i^T b \tag{11.69}$$

Substituting for x from equation (11.68) in equation (11.69), and using the conjugacy relationship (A-orthogonal property) of equations (11.66) and (11.67), we have

$$C_i W_i^T A W_i = W_i^T b \tag{11.70}$$

because C is a constant.

Therefore,
$$C_i = \frac{W_i^T b}{W_i^T A W_i} \quad (11.71)$$

Thus the problem of solving the matrix equation (11.62) may be viewed as one of generating a set of conjugate or A-orthogonal vectors in an efficient and economical manner. In the conjugate gradient method, the direction vectors are generated one at a time, so there is no need to store a large number of them. Beginning with an arbitrary starting vector $x = x_o$, the first basis vector is chosen as the residual such that

$$W_1 = r_1 = b - Ax_1 \quad (11.72)$$

The estimated solution to (11.62) is now improved by adding a vector parallel to the basis vector W. Thus, we have

$$x_{i+1} = x_i + \alpha_i W_i \quad (11.73)$$

We see that the residual component parallel to this basis vector is reduced to zero. We must choose α from equation (11.73) so as to minimize the quadratic (11.63). Therefore, by substituting x_{i+1} from equation (11.73) in place of x in equation (11.63) and setting its derivative with respect to α to zero,

$$\frac{\partial F}{\partial \alpha} = 0 = \frac{\partial}{\partial \alpha}\left(\frac{1}{2}x_{i+1}^T A x_{i+1} - b_{i+1}^T x\right) \quad (11.74)$$

or

$$\frac{\partial F}{\partial \alpha} = 0 = \frac{\partial}{\partial \alpha}\left(\frac{1}{2}(x_i + \alpha_i W_i)^T A(x_i + \alpha_i W_i) - b^T(x_i + \alpha_i W_i)\right) \quad (11.75)$$

Performing the required algebraic steps and using (11.72), equation (11.75) becomes

$$\frac{\partial F}{\partial \alpha} = \alpha_i W_i^T A W_i + W_i^T (Ax_i - b) = \alpha_i W_i^T A W_i - W_i^T r_i = 0 \quad (11.76)$$

Solution of Equations

yielding

$$\alpha_i = \frac{W_i^T r_i}{W_i^T A W_i} \tag{11.77}$$

Let us examine the vector product of the present basis vector W and the new residual r_{i+1}. Using equations (11.64) and (11.73) appropriately, we have

$$W_i^T r_{i+1} = W_i^T (b - A x_i - \alpha_i A W_i) = W_i^T r_i - \alpha_i W_i^T A W_i \tag{11.78}$$

Substituting the relation of equation (11.76) in equation (11.78) for the minimization of F, one obtains

$$W_i^T r_{i+1} = W_i^T r_i - \alpha_i W_i^T A W_i = 0 \tag{11.79}$$

which yields the required value of α as in equation (11.77).

Many steps of this process may be carried out. But the successive residuals will remain orthogonal to all the previous vectors W that have already been used, because the ith step generates an approximate solution that is correct in its first i terms of the expansion (11.68). Thus after N steps, the correct solution will have been obtained and the process will terminate, provided round-offs do not introduce errors in the directions already considered. This will be discussed further later on.

We must now choose the basis vector W and this should be in the direction of the residual so as to reduce it. If we make the new W orthogonal to all the old ones, then we can determine the C's by multiplying equation (11.68) by each of the W's successively and by using the orthogonality relationships

$$\begin{aligned} W_i^T W_j &= 0, \text{ for } i \neq j \\ W_i^T W_j &\neq 0, \text{ for } i = j \end{aligned} \tag{11.80}$$

But the C's can be determined only by knowing the final value of x, which is not known *a priori*. We know Ax and not x. Therefore, if we make the new W's conjugate or A-orthogonal to each other by multiplying equation (11.68) by $W_i^T A$ instead of W_i^T, we have

$$W_i^T A x_i = C_i W_i^T A W_i \tag{11.81}$$

all other terms of $W_i^T A W_j$ vanishing for i not equal to j.

This would yield the value of C's as in equation (11.71), as before, so that

$$C_i = \frac{W_i^T A x_i}{W_i^T A W_i} = \frac{W_i^T b}{W_i^T A W_i} \tag{11.82}$$

In order to determine the C's, we need only know the matrix A, the right-hand vector b, and the basis vector W. In the conjugate gradient process, for the first step, the basis vector (direction vector) could be chosen more or less arbitrarily. Choosing it parallel to the residual merely ensures that the largest possible improvement will result in the first step. For succeeding steps, the same strategy may be followed, but it must be modified so as to ensure that the successive vectors W remain A-orthogonal. This may be illustrated by the following procedure.

Start from $x_0 = 0$, with the resulting residual

$$r_0 = b - Ax_0 = b \tag{11.83}$$

The residual is also the negative gradient of F at $x = x_o$. If, we make this the direction of W, then it is the method of steepest descent, and we go to the point using (11.73) such that

$$\begin{aligned} W_0 &= b \\ x_1 &= \alpha_0 W_0 \end{aligned} \tag{11.84}$$

Solution of Equations

We now compute the new residual

$$r_1 = b - Ax_1 \tag{11.85}$$

Then, r_1 would be the new direction of W_1, except that it is not conjugate to the first direction W_0, because

$$r_1^T r_0 = (b - Ax_1)^T b = W_1^T W_0 \tag{11.86}$$

What we need is that W_1 be A-orthogonal to W_0. To accomplish this, we must add to the residual vector, a component parallel to the previous W, such that

$$W_2 = r_2 + \beta_1 W_1 \tag{11.87}$$

We shall choose β_1 in equation (11.87), as will be shown later on, so as to terminate the iterative scheme in exactly N steps (never more than N steps) and to this end will set W_2 to be conjugate (or A-orthogonal) to W_1. Multiplying equation (11.87) by $W_1^T A$, we have

$$W_1^T A W_2 = W_1^T A r_2 + \beta_1 W_1^T A W_1 \tag{11.88}$$

Because A is symmetric and positive definite and W_2 is A-orthogonal to W_1

$$W_2^T A W_1 = r_2^T A W_1 + \beta_1 W_1^T A W_1 = 0 \tag{11.89}$$

and therefore,

$$\beta_1 = -\frac{r_2^T A W_1}{W_1^T A W_1} \tag{11.90}$$

Keeping successive vectors W mutually A-orthogonal suffices to generate a wholly orthogonal set, because successive residuals do not contain components parallel to previous vectors W. Now we move along the direction (11.87) to the new point

$$x_2 = x_1 + \alpha_1 W_1 \tag{11.91}$$

and so on. The cycle of calculations is as follows.

$$\begin{aligned}\text{(direction)} \quad W_{i+1} &= r_{i+1} + \beta_i W_i \\ \text{(new solution)} \quad x_{i+1} &= x_i + \alpha_i W_i \\ \text{(new residual)} \quad r_{i+1} &= b - A x_{i+1}\end{aligned} \tag{11.92}$$

and

$$\alpha_i = \frac{W_i^T r_i}{W_i^T A W_i} \tag{11.93}$$

$$\beta_i = -\frac{r_{i+1}^T A W_i}{W_i^T A W_i} \tag{11.94}$$

Several important results follow from the choices of α_i and β_i and the vectors x_i, W_i, and r_i. First, we shall determine the new residual in terms of the previous ones as follows.

$$r_{i+1} = b - A x_{i+1} = b - A x_i - \alpha_i A W_i = r_i - \alpha_i A W_i \tag{11.95}$$

Next we shall consider the vector product

$$r_{i+1}^T r_i = (r_i - \alpha_i W_i A)^T r_i \tag{11.96}$$

Solution of Equations

From (11.92)

$$r_i = W_i - \beta_{i-1} W_{i-1} \tag{11.97}$$

Substituting (11.97) into (11.96), we have

$$r_{i+1}^T r_i = (r_i - \alpha_i W_i A)^T (W_i - \beta_{i-1} W_{i-1}) = r_i^T W_i - \alpha_i W_i^T A W_i = 0 \tag{11.98}$$

all other terms vanishing by virtue of (11.79) and (11.67). Again, our choice of α makes r_{i+1} and r_i orthogonal.

We shall now examine the linear independence of the basis vectors and why the conjugate gradient process terminates exactly in N steps (or never more than in N steps).

If the N vectors W are linearly independent, then the following relationship will not be satisfied:

$$p_0 W_0 + p_1 W_1 + p_2 W_2 + \cdots + p_{N-1} W_{N-1} = 0 \tag{11.99}$$

unless every p is zero or there is some null vector W_i. If we multiply equation (11.99) by $W_i^T A$, all other terms will vanish except $p_i W_i^T A W_i$. However, since this term is also equal to zero by virtue of equation (11.99), p_i must necessarily be zero, making the vectors W linearly independent. Now if we consider the residual r_N, it will be orthogonal to every W in equation (11.99), thereby not requiring that every p be zero. This will violate the linear independence of W as shown before. Therefore, r_N must be identically equal to zero. Hence, the conjugate gradient process must terminate exactly in N steps or never more than in N steps.

Numerical Example

We are required to solve the following matrix equation by the conjugate gradient method.

$$\begin{pmatrix} 2 & 1 \\ 1 & 2 \end{pmatrix} \begin{pmatrix} x_1 \\ x_2 \end{pmatrix} = \begin{pmatrix} 2 \\ 1 \end{pmatrix} \tag{11.100}$$

Assuming a starting vector $x_0^T = [0, 0]$, we have

$$r_0^T = (b - Ax_0)^T = [2, 1] \tag{11.101}$$

We now compute

$$\alpha_0 = \frac{r_0^T W_0}{W_0^T A W_0} = \frac{5}{14} \tag{11.102}$$

From equation (11.82), we can evaluate

$$C_0 = \frac{W_0^T b}{W_0^T A W_0} = \frac{5}{14} \tag{11.103}$$

We can now find the new estimate of x as

$$x_1^T = (x_0 + \alpha_0 W_0^T) = \frac{5}{14}(2, 1) \tag{11.104}$$

The new residual r_1 is found from the relationship

$$r_1^T = (b - Ax_1)^T = \frac{3}{14}(1, -2) \tag{11.105}$$

Solution of Equations

k	x	r	α	β	W	Remarks
0	$\begin{pmatrix} 0 \\ 0 \end{pmatrix}$	$\begin{pmatrix} 2 \\ 1 \end{pmatrix}$	$\frac{5}{14}$	-	$\begin{pmatrix} 2 \\ 1 \end{pmatrix}$	Initial vector
1	$\frac{5}{14}\begin{pmatrix} 2 \\ 1 \end{pmatrix}$	$\frac{3}{14}\begin{pmatrix} 1 \\ -2 \end{pmatrix}$	$\frac{14}{15}$	$\frac{9}{14^2}$	$\frac{15}{14^2}\begin{pmatrix} 4 \\ -5 \end{pmatrix}$	
2	$\begin{pmatrix} 1 \\ 0 \end{pmatrix}$	$\begin{pmatrix} 0 \\ 0 \end{pmatrix}$	0	0	$\begin{pmatrix} 0 \\ 0 \end{pmatrix}$	Converged

Table 11.2 Summary of Conjugate Gradient Example

Next, we shall compute the value of β_0 from equation (11.90) as

$$\beta_0 = -\frac{r_1^T A_0 W_0}{W_0^T A W_0} = \frac{9}{14^2} \tag{11.106}$$

We can then find W_1 using the relation of equation (11.92) as

$$W_1^T = (r_1 + \beta_0 W_0)^T = \frac{15}{14^2}(4, -5) \tag{11.107}$$

The process is repeated to find the values of C_1, α_1, x_2, and r_2. The results are summarized in Tables 11.2 and 11.3. It is evident that the conjugate gradient process terminates exactly in two steps, excluding the starting step. The table also conclusively shows the orthogonality and A-orthogonality of the r and W vectors, respectively.

11.6.1 Preconditioning to Accelerate Convergence

In the previous section, we have seen that the conjugate gradient method converges in not more than N steps, because the direction vectors generated are mutually A-orthogonal and are linearly independent. As only nonzero terms of the matrix need

k	r	$r_i^T r_{i-1}$	W_i	$W_i^T A W_{i-1}$
0	$\begin{pmatrix} 2 \\ 1 \end{pmatrix}$	-	$\begin{pmatrix} 2 \\ 1 \end{pmatrix}$	-
1	$\frac{3}{14}\begin{pmatrix} 1 \\ -2 \end{pmatrix}$	0	$\frac{15}{14^2}\begin{pmatrix} 4 \\ -5 \end{pmatrix}$	0
2	$\begin{pmatrix} 0 \\ 0 \end{pmatrix}$	0	$\begin{pmatrix} 0 \\ 0 \end{pmatrix}$	0

Table 11.3 Summary of Conjugate Gradient Example

be stored, even this N-step convergence may be adequate for some cases in terms of both storage point and execution time. However, for large systems of matrix equations, attention needs to be directed at minimizing the number of steps for convergence. The rate at which the successive approximations of the unknown variables converge to the true solution depends on the eigenvalue spectrum of the matrix. If, for example, two eigenvalues are equal, and the direction vectors lie within the subspace spanned by the corresponding eigenvectors, excellent correction will be achieved for both direction vectors in their corresponding coordinate directions in the same iteration step. If several eigenvalues of the matrix are equal, fewer than N steps will be required for convergence. In fact, the number of iteration steps equals the number of distinct (unequal) eigenvalues. However, if the eigenvalues are close and clustered together, the conjugate gradient method converges rapidly within a small error. In the limiting case, if the matrix is an identity matrix, each of the eigenvalues being equal to 1, the method converges in one step. Our task is to find a preconditioning matrix that will modify the original matrix such that the new matrix so formed will be closer to the identity matrix, or at least that the eigenvalues of the new matrix are clustered together.

Let us consider the matrix equation (11.62), and operate on it by some other positive definite matrix B such that

$$BAB^T((B^T)^{-1}x) = Bb \tag{11.108}$$

We shall call the matrix B the preconditioning matrix. As a general rule, if the eigenvalues of B are closely clustered together, the conjugate gradient method applied to

Solution of Equations

(11.108) will converge faster than if it is applied to equation (11.62). We can choose the matrix B to be the inverse of A. Introducing this into equation (11.108), one obtains

$$A^{-1}A(A^T)^{-1}(A^T x) = A^{-1}b \qquad (11.109)$$

or

$$x = A^{-1}b \qquad (11.110)$$

The conjugate gradient iteration process converges in one step. Inverting matrix A is not practical, and one prefers to choose other simpler forms for matrix B to improve the performance of the method considerably.

One of the easy choices for B is the inverse of the matrix of the diagonal terms of A, and the following example illustrates the preconditioned conjugate gradient method.

Example

$$\begin{pmatrix} 9 & 1 & 1 & 1 \\ 1 & 8 & 1 & 1 \\ 1 & 1 & 7 & 1 \\ 1 & 1 & 1 & 6 \end{pmatrix} \begin{pmatrix} x_1 \\ x_2 \\ x_3 \\ x_4 \end{pmatrix} = \begin{pmatrix} 1 \\ 1 \\ 1 \\ 1 \end{pmatrix} \qquad (11.111)$$

We may consider the following inverse of the matrix of diagonal terms of A as the preconditioning matrix. Therefore,

$$B = \begin{pmatrix} \frac{1}{9} & & & \\ & \frac{1}{8} & & \\ & & \frac{1}{7} & \\ & & & \frac{1}{6} \end{pmatrix} \qquad (11.112)$$

Performing the matrix operations in equation (11.108), one obtains the preconditioned matrix equation as

$$\begin{pmatrix} \frac{1}{9} & \frac{1}{72} & \frac{1}{63} & \frac{1}{54} \\ \frac{1}{72} & \frac{1}{8} & \frac{1}{56} & \frac{1}{48} \\ \frac{1}{63} & \frac{1}{56} & \frac{1}{7} & \frac{1}{42} \\ \frac{1}{54} & \frac{1}{48} & \frac{1}{42} & \frac{1}{6} \end{pmatrix} \begin{pmatrix} z_1 \\ z_2 \\ z_3 \\ z_4 \end{pmatrix} = \begin{pmatrix} \frac{1}{9} \\ \frac{1}{8} \\ \frac{1}{7} \\ \frac{1}{6} \end{pmatrix} \qquad (11.113)$$

where

$$\begin{pmatrix} z_1 \\ z_2 \\ z_3 \\ z_4 \end{pmatrix} = B^{-1} x \qquad (11.114)$$

We then solve for the z's using the conjugate gradient procedure outlined before and obtain the solution for z and x in three steps as

$$\begin{pmatrix} z_1 \\ z_2 \\ z_3 \\ z_4 \end{pmatrix} = \begin{pmatrix} 0.8995 \\ 0.7950 \\ 0.7486 \\ 0.8947 \end{pmatrix}$$

$$\begin{pmatrix} x_1 \\ x_2 \\ x_3 \\ x_4 \end{pmatrix} = \begin{pmatrix} 0.09995 \\ 0.09937 \\ 0.10695 \\ 0.14910 \end{pmatrix} \qquad (11.115)$$

The exact solution for x obtained in four steps from matrix equation (11.111) is

$$\begin{pmatrix} x_1 \\ x_2 \\ x_3 \\ x_4 \end{pmatrix} = \begin{pmatrix} 0.100105 \\ 0.099055 \\ 0.107455 \\ 0.148757 \end{pmatrix} \qquad (11.116)$$

Solution of Equations

The root-mean-square error in (11.115) compared with 11.116 is 0.62%. and the r.m.s residual is of the order of 0.00056. Although, in this simple example, one could obtain the exact solution in four steps, it does signify the importance of preconditioning, particularly for large sparse matrices.

Incomplete Choleski Decomposition

Another method of accelerating convergence is to use a preconditioner based on Choleski factorization. In this method, the original matrix is decomposed into a lower triangular matrix consisting of the multipliers used in the Gauss elimination, a diagonal matrix comprising the diagonal terms resulting from forward modification by the Gauss process, and an upper triangular matrix that is the transpose of the lower triangular matrix. It is customary to further subdivide the diagonal matrix such that the square roots of the diagonal terms are associated with the lower triangular matrix and the remaining square roots are associated with the upper triangular matrix. We can illustrate this decomposition by the following example. Let us consider the matrix equation

$$\begin{pmatrix} 3 & 1 & 1 \\ 1 & 2 & 0 \\ 1 & 0 & 1 \end{pmatrix} \begin{pmatrix} x_1 \\ x_2 \\ x_3 \end{pmatrix} = \begin{pmatrix} 1 \\ 1 \\ 1 \end{pmatrix} \quad (11.117)$$

By standard Gaussian elimination, forward modification yields

$$\begin{pmatrix} 3 & 1 & 1 \\ \frac{1}{3} & \frac{5}{3} & \frac{-1}{3} \\ \frac{1}{3} & \frac{-1}{5} & \frac{3}{5} \end{pmatrix} \begin{pmatrix} x_1 \\ x_2 \\ x_3 \end{pmatrix} = \begin{pmatrix} 1 \\ \frac{2}{3} \\ \frac{4}{5} \end{pmatrix} \quad (11.118)$$

In equation (11.118), the entries in the lower triangle are the multipliers in the Gauss elimination process. Using these and the square roots of the diagonal terms, the lower triangular Choleski factor is obtained as

$$L = \begin{pmatrix} 1 & 0 & 0 \\ \frac{1}{3} & 1 & 0 \\ \frac{1}{3} & \frac{-1}{5} & 1 \end{pmatrix} \begin{pmatrix} \sqrt{3} & 0 & 0 \\ 0 & \sqrt{\frac{5}{3}} & 0 \\ 0 & 0 & \sqrt{\frac{3}{5}} \end{pmatrix} = \begin{pmatrix} \sqrt{3} & 0 & 0 \\ \sqrt{\frac{1}{3}} & \sqrt{\frac{5}{3}} & 0 \\ \sqrt{\frac{1}{3}} & -\sqrt{\frac{1}{15}} & \sqrt{\frac{3}{5}} \end{pmatrix} \quad (11.119)$$

The upper triangular Choleski factor L^T is the transpose of L and one can easily verify that
$$LL^T = A, \quad (11.120)$$
the original matrix.

The preconditioner is then chosen as the product of the inverses of L^T and L, so that
$$B = (L^T)^{-1}L^{-1} = A^{-1}, \quad (11.121)$$
the inverse of A.

Performing the standard arithmetic steps on L in equation (11.119) to find its inverse, there is
$$L^{-1} = \begin{pmatrix} \sqrt{\frac{1}{3}} & 0 & 0 \\ \frac{-1}{\sqrt{15}} & \frac{3}{\sqrt{15}} & 0 \\ \frac{-2}{\sqrt{15}} & \frac{1}{\sqrt{15}} & \sqrt{\frac{5}{3}} \end{pmatrix} \quad (11.122)$$
and
$$(L^T)^{-1} = \begin{pmatrix} \sqrt{\frac{1}{3}} & \frac{-1}{\sqrt{15}} & \frac{-2}{\sqrt{15}} \\ 0 & \frac{3}{\sqrt{15}} & \frac{1}{\sqrt{15}} \\ 0 & 0 & \sqrt{\frac{5}{3}} \end{pmatrix} \quad (11.123)$$

Substituting (11.122) and (11.123) in (11.121), the preconditioner is obtained as
$$B = (L^T)^{-1}L^{-1} = \begin{pmatrix} \frac{2}{3} & \frac{-1}{3} & \frac{-2}{3} \\ \frac{-1}{3} & \frac{2}{3} & \frac{1}{3} \\ \frac{-2}{3} & \frac{1}{3} & \frac{5}{3} \end{pmatrix} \quad (11.124)$$

Because B in (11.124) is the inverse of A, the solution is obtained in one step.

Solution of Equations

It is evident from equations (11.117), (11.122), and (11.123) that the Choleski factors L and L^T are denser than the original matrix, requiring more storage. Also, the computational work involved in the factorization and in determining the inverses may not be competitive with other methods. Nevertheless, the choice of this preconditioner is attractive from the point of view of accelerating convergence and has led to a family of preconditioned conjugate gradient methods based on incomplete Choleski decomposition. In this process, the sparsity structure of the Choleski factors is selected to be the same as that of matrix A. The decomposition itself is carried out in the standard manner, except that any element of the preconditioning matrix B corresponding to a zero entry in the original matrix A is set to zero without any computation being performed in finding the Choleski factors. These zero values are then used in the subsequent steps of the decomposition, so that the result is not the same as would be obtained by setting to zero certain terms in the correctly computed Choleski decomposition.

In the incomplete Choleski factorization, equation (11.119) is changed to

$$L = \begin{pmatrix} \sqrt{3} & 0 & 0 \\ \sqrt{\frac{1}{3}} & \sqrt{\frac{5}{3}} & 0 \\ \sqrt{\frac{1}{3}} & 0 & 1 \end{pmatrix} \qquad (11.125)$$

The corresponding preconditioner is obtained as

$$B = (L^T)^{-1} L^{-1} = \begin{pmatrix} \frac{23}{45} & \frac{-1}{5} & \frac{-1}{3} \\ \frac{-1}{5} & \frac{3}{5} & 0 \\ \frac{-1}{3} & 0 & 1 \end{pmatrix} \qquad (11.126)$$

The preconditioned matrix then becomes

$$C = BAB^T = \begin{pmatrix} \frac{58}{135} & \frac{-6}{45} & \frac{-7}{45} \\ \frac{-6}{45} & \frac{3}{5} & \frac{-1}{5} \\ \frac{-7}{45} & -\frac{1}{5} & \frac{2}{3} \end{pmatrix} \qquad (11.127)$$

The corresponding right-hand vector is obtained as

$$R = Bb = \begin{pmatrix} -\frac{1}{45} \\ \frac{2}{5} \\ \frac{2}{3} \end{pmatrix} \quad (11.128)$$

We then solve by the standard conjugate gradient procedure the new matrix equation

$$Cz = R \quad (11.129)$$

where

$$z = (B^T)^{-1}x \quad (11.130)$$

instead of the original matrix equation

$$Ax = b \quad (11.131)$$

and obtain

$$x = Bz \quad (11.132)$$

The preconditioned conjugate gradient method achieves economic storage and execution speeds that are highly competitive with those of nested dissection methods and other sparse matrix techniques. Choleski factorization is particularly useful for solving matrix equations resulting from finite element analysis. The method allows trading computer time for accuracy and allows good use to be made of approximate solutions obtained elsewhere by admitting them as initial iterates in the process. It is, therefore, applicable to linear and particularly nonlinear problems where N-R methods are

Solution of Equations 645

used for quasi-linearizing the nonlinear equations. Being semi-iterative, this method is also self-correcting against round-off errors. It is a strong contender for solving large systems of finite element equations.

11.7 GMRES

In the numerical solution of electromagnetic problems we frequently encounter systems of equations with nonsymmetric matrices. This nonsymmetry can be caused by a non-self-adjoint operator (for example, in the boundary element method), or by the application of certain boundary conditions, or by coupling equations. The direct methods, such as the LDU decomposition, can be easily modified to deal with nonsymmetric matrices. This of course means that the entire matrix (or at least the nonzeros) must be stored and operated on. The conjugate gradient method, discussed in the previous section, is valid only for symmetric systems. The conjugate gradient method has been generalized to deal with complex and/or nonsymmetric systems. This method is called the biconjugate gradient method. For nonsymmetric systems an incomplete LU decomposition can be used for preconditioning just as the incomplete Choleski decomposition was used for the conjugate gradient method. This method is not presented here. We now introduce a related method known as the *generalized method of minimum residuals* or GMRES [134] [135].

As in the conjugate gradient method we wish to solve the equation

$$Ax = b \qquad (11.133)$$

For an initial estimate of the solution we try x_0. This will give us a residual

$$r_0 = b - Ax_0 \qquad (11.134)$$

We now define a vector z by

$$z = x - x_0 \qquad (11.135)$$

Substituting into equation (11.133) gives

$$Az - r_0 = 0 \qquad (11.136)$$

Now if we can find z we have found x.

To find z, first consider a matrix, Z, made of columns v_i that span the vector space and are orthonormal. This means that

$$v_j^T v_j = 1 \qquad (11.137)$$
$$v_i^T v_j = 0; \; j \neq i \qquad (11.138)$$

To find the vectors, v_j we require that $Az_k - r_0$ is orthogonal to all of the columns in V.

$$v_i^T (Az_k - r_0) = 0; \; i = 1, 2, ..., k \qquad (11.139)$$

Therefore

$$V^T (Az_k - r_0) = 0 \qquad (11.140)$$

or

$$V^T Az_k - V^T r_0 = 0 \qquad (11.141)$$

Because z_k is an element of the vector space it can be expressed as the linear combination of the basis vectors, v_i.

Therefore

$$z_k = y_1 v_1 + y_2 v_2 + \cdots + y_k v_k = \sum_{i=1}^{k} y_i v_i \qquad (11.142)$$

This equation is the same as

$$z_k = V_k y \qquad (11.143)$$

Substituting (11.141) into (11.143) gives

$$V_k^T A V_k y - V_k^T r_0 = 0 \qquad (11.144)$$

We must now find y and the matrix of basis vectors V.

Solution of Equations

We start with the set of vectors $\{r_0, Ar_0, A^2r_0, ..., A^{k-1}r_0\}$.

We first choose

$$v_1 = \frac{r_0}{|r_0|} \tag{11.145}$$

Because r_0 is linearly dependent on v_1 and because v_1 is orthogonal to all other vectors v_i, $i = 2, ...k$ then

$$V_k^T r_0 = v_1^T r_0 e_1 + \cdots v_k^T r_0 e_k = v_1^T r_0 e_1 \tag{11.146}$$

Here e_i is a unit vector with 1 in the ith row and zeros elsewhere.

This gives

$$V_k^T r_0 = \frac{r_0^T}{|r_0|} r_0 e_1 = \frac{|r_0|^2}{|r_0|} e_1 = |r_0| e_1 \tag{11.147}$$

The second element is formed by a linear combination of v_1 and another vector orthogonal to v_1. Let us call that vector p_2. Now the projection of Av_1 on v_1 is

$$[(Av_1)^T v_1] v_1 \tag{11.148}$$

Therefore

$$Av_1 = [(Av_1)^T v_1] v_1 + p_2 \tag{11.149}$$

Solving for p_2 gives

$$p_2 = Av_1 - [(Av_1)^T v_1] v_1 \tag{11.150}$$

We find v_2 by normalizing p_2.

$$v_2 = \frac{p_2}{|p_2|} \tag{11.151}$$

To generalize the procedure to find v_k we note the following:

1. v_k is orthogonal to all previous v_j for $j =, 2, ..., k-1$.
2. v_k is a unit vector.
3. The kth element of our sequence $\{r_0, Ar_0, A^2 r_0, ..., A^{k-1} r_0\}$ is Av_{k-1}. This product can be expressed as the reflection of Av_{k-1} on each of the previous v_j and a vector p_k that is orthogonal to all previous v_k.

$$Av_{k-1} = [(Av_{k-1})^T v_1]v_1 + \vdots + [(Av_{k-1})^T v_{k-1}]v_{k-1} + p_k \tag{11.152}$$

This can be expressed as a summation,

$$Av_{k-1} = \sum_{i=1}^{k-1} [(Av_{k-1})^T v_i]v_i + p_k \tag{11.153}$$

so that

$$p_k = Av_{k-1} - \sum_{i=1}^{k-1} [(Av_{k-1})^T v_i]v_i \tag{11.154}$$

We now find v_k as

$$v_k = \frac{p_k}{|p_k|} \tag{11.155}$$

Solution of Equations

Our next step is to compute y. To do this we show that

$$H_k = V_k A^T V_k \tag{11.156}$$

is an upper Hessenberg matrix. Let us write this in component form

$$H_k = \begin{pmatrix} v_1^T \\ \vdots \\ v_k^T \end{pmatrix} \begin{pmatrix} A_1 \\ \vdots \\ A_n \end{pmatrix} (v_1 \ldots v_k) \tag{11.157}$$

Here A_i is the ith row of A. Multiplying, we get

$$\begin{aligned} H_k &= \begin{pmatrix} v_1^T \\ \vdots \\ v_k^T \end{pmatrix} \begin{pmatrix} A_1^T v_1 & \cdots & A_1^T v_k \\ \vdots & \ddots & \vdots \\ A_n^T v_1 & \cdots & A_n^T v_k \end{pmatrix} \\ &= \begin{pmatrix} v_1^T \\ \vdots \\ v_k^T \end{pmatrix} (Av_1 \ldots Av_k) \end{aligned} \tag{11.158}$$

Multiplying on the left gives

$$H_k = \begin{pmatrix} v_1^T A v_1 & \cdots & A_1^T v_k \\ \vdots & \ddots & \vdots \\ v_k^T A v_1 & \cdots & v_k A v_k \end{pmatrix} \tag{11.159}$$

That H_k is an upper Hessenberg matrix can be seen by examining an element of the subdiagonal

$$h_{j+1,j} = v_{j+1}^T A v_j \tag{11.160}$$

Using equation (11.153) we have

$$\begin{aligned} h_{j+1,j} &= v_{j+1}^T \left(\sum_{i=1}^{j} [(Av_j^T)v_i]v_i + p_{j+1} \right) \\ &= v_{j+1}^T \left(\sum_{i=1}^{j} \alpha_i v_i + p_{j+1} \right) \end{aligned} \quad (11.161)$$

where α_i is a scalar to be determined. We may now write

$$\begin{aligned} h_{j+1,j} &= v_{j+1}^T \left(\sum_{i=1}^{j} \alpha_i v_i + \alpha_{j+1} v_{j+1} \right) \\ &= v_{j+1}^T \left(\sum_{i=1}^{j+1} \alpha_i v_i \right) \end{aligned} \quad (11.162)$$

Because the v_i's are orthogonal unit vectors, we have

$$\begin{aligned} h_{j+1,j} &= \alpha_{j+1} \\ &= |p_{j+1}| \end{aligned} \quad (11.163)$$

Similarly, an element in the second subdiagonal is

$$\begin{aligned} h_{j+2,j} &= v_{j+2}T \left(\sum_{i=1}^{j+1} \alpha_i v_i \right) \\ &= 0 \end{aligned} \quad (11.164)$$

Solution of Equations

As v_{j+2} is orthogonal to all previous V_is. From this equation we see that all elements below the subdiagonal of H_k are zero.

From equation (11.160) we have

$$h_{j,k} = v_j^T A v_k \quad \text{for all } j = 1, \ldots k \tag{11.165}$$

We now can find the kth column of H_k. The orthogonality principle can now be written as

$$\begin{aligned} p_k &= A v_{k-1} - \sum_{i=1}^{k-1} \left(v_i^T A v_{k-1} \right) v_i \\ &= A v_{k-1} - \sum_{i=1}^{k-1} h_{i,k-1} v_i \end{aligned} \tag{11.166}$$

Once we have the kth column of H_k we can find p_{k+1} from equation (11.166). Let H^+ be the augmented H_k matrix with the last row containing $|p_{k+1}|$.

$$H^+ = \begin{pmatrix} v_1^T A v_1 & \ldots & & \ldots & v_1^T A v_k \\ |p_2| & v_2^T A v_2 & & \ldots & v_2^T A v_k \\ 0 & |p_3| & & \ldots & \ldots \\ 0 & \ldots & & \ldots & |p_{k+1}| \end{pmatrix} \tag{11.167}$$

We now must compute H_{k+1} from this augmented matrix. We have already found v_{k+1} and using equation (11.156)

$$V_{k+1}^T A V_k = \begin{pmatrix} v_1^T A v_1 & \ldots & v_1^T A v_k \\ \ldots & \ldots & \ldots \\ v_k^T A v_1 & \ldots & v_k^T A v_k \\ v_{k+1}^T A v_1 & \ldots & v_{k+1}^T A v_k \end{pmatrix} = H^+ \tag{11.168}$$

We now have

$$H_k y - |r_0|e_1 = 0 \tag{11.169}$$

Replacing H_k with the augmented matrix gives us the (rectangular) system

$$H^+ y - |r_0|e_1 \tag{11.170}$$

Because the system is overdetermined we find

$$\min_y |H^+ y - |r_0|e_1| \tag{11.171}$$

QR Factorization

We can factor this rectangular system as

$$H^+ = Q_k R_k \tag{11.172}$$

where Q_k is an orthonormal matrix and R_k is an upper triangular matrix. Since Q_k is orthonormal

$$Q_k^T = Q_k^{-1} \tag{11.173}$$

and

$$||Q_k|| = 1 \tag{11.174}$$

Solution of Equations

The minimization condition is

$$\begin{aligned}
\min_y |H^+ y - |r_0|e_1| &= \min_y |Q_k R_k y - |r_0|e_1| \\
&= \min_y |Q_k^T Q_k R_k y - Q_k |r_0|e_1| \\
&= \min_y |R_k y - g_k |r_0|e_1|
\end{aligned} \qquad (11.175)$$

where

$$g_k = |r_0| Q_k e_1 \qquad (11.176)$$

Equation (11.175) can be solved by solving

$$R_k y = g_k \qquad (11.177)$$

and because R_k is upper triangular this is done by a back-substitution. If we know y we know z_k and then can find x.

The problem is then to find the QR factors. Because H^+ is upper Hessenberg we can do this by a series of Givens rotations.

Givens Rotation

One Givens rotation will eliminate an element from a column. The decomposition can be written in the form

$$G_k ... G_1 H^+ = R_k \qquad (11.178)$$

where

$$Q_k^T = G_k...G_1 \qquad (11.179)$$

Therefore

$$g_k = G_k...G_1|r_0|e_1 \qquad (11.180)$$

and

$$g_0 = |r_0|e_1 \qquad (11.181)$$

The G matrices are of the form

$$G_i = \begin{pmatrix} 1 & 0 & \cdots & \cdots & \cdots & \cdots & \cdots & 0 \\ 0 & 1 & \cdots & \cdots & \cdots & \cdots & \cdots & 0 \\ 0 & 0 & \cdots & \cdots & \cdots & \cdots & \cdots & 0 \\ 0 & 0 & \cdots & c & s & \cdots & \cdots & 0 \\ 0 & 0 & \cdots & -s & c & \cdots & \cdots & 0 \\ 0 & 0 & \cdots & \cdots & \cdots & \cdots & \cdots & 0 \\ 0 & 0 & \cdots & \cdots & \cdots & \cdots & 1 & 0 \\ 0 & 0 & \cdots & \cdots & \cdots & \cdots & 0 & 1 \end{pmatrix} \qquad (11.182)$$

This is not the most general form of the matrix, but we will use a 2×2 submatrix at the ith and $(i+1)$st elements of the form

$$\begin{pmatrix} c & s \\ -s & c \end{pmatrix} \qquad (11.183)$$

In this case only the ith and $(i+1)$st elements of H_k will change when we perform the multiplication.

Solution of Equations

Showing only the region of the matrix we are working on

$$\begin{pmatrix} \cdots & \cdots & \cdots & \cdots \\ 0 & c & s & 0 \\ 0 & -s & c & 0 \\ \cdots & \cdots & \cdots & \cdots \end{pmatrix} \begin{pmatrix} \cdots & \cdots & \cdots & \cdots \\ \cdots & h_{i,j} & \cdots & \cdots \\ 0 & h_{i+1,j} & \cdots & \cdots \\ 0 & 0 & \cdots & \cdots \end{pmatrix}$$

$$= \begin{pmatrix} \cdots & \cdots & \cdots & \cdots \\ \cdots & ch_{i,j} + s_{i+1,i} & \cdots & \cdots \\ 0 & -sh_{i,j} + ch_{i+1,i} & \cdots & \cdots \\ 0 & 0 & \cdots & \cdots \end{pmatrix} \quad (11.184)$$

In order for the matrix to remain orthonormal we require

$$c^2 + s^2 = 1 \quad (11.185)$$

and

$$-sh_{i,j} + ch_{i+1,i} = 0 \quad (11.186)$$

These conditions are met if we set

$$s = \frac{h_{i+1,i}}{\sqrt{h_{k,k}^2 + h_{k+1,k}^2}}$$

$$c = -\frac{h_{i,i}}{\sqrt{h_{k,k}^2 + h_{k+1,k}^2}} \quad (11.187)$$

The first k elements of g_k are the elements of g_{k-1} rotated by G_k. So if we augment g_{k-1} by one element and set that element to zero we have

$$g_k = G_k g_{k-1} \quad (11.188)$$

So in practice we update g_{k-1} to form g_k rather than building a new vector in each iteration. Also, we do not need to compute y and z for each iteration to find the norm of the residual. After k iterations

$$|R_k y - g_k| = \sqrt{[(R_k y)_1 - (g_k)_1]^2 + \ldots + [(R_k y)_k - (g_k)_k]^2 + (g_k)_{k+1}^2} \quad (11.189)$$

Here the last term comes from the fact that the last row of R_k is zero. Therefore

$$\min_y |R_k y - g_k| = |(g_{k+1})_k| \quad (11.190)$$

We also see that

$$(g_k)_i = (g_{k-1})_i, \quad i = 1, \ldots, k-1 \quad (11.191)$$

and

$$\begin{aligned}(g_k)_{k+1} &= \min_y |R_k y - gk| \\ &= \min_y |H_k^+ y - |r_0| e_1| \end{aligned} \quad (11.192)$$

We can also write

$$(g_k)_{k+1} = |V_k^T (A z_k - r_0)| \quad (11.193)$$

and because V_k is orthonormal

$$(g_k)_{k+1} = |A z_k - r_0| \quad (11.194)$$

We then have

$$\begin{aligned}(g_k)_{k+1} &= |A(x_k - x_0) - r_0| \\ &= |A x_k - ((A x_0 + r_0)|\end{aligned} \quad (11.195)$$

Solution of Equations

Substituting

$$(g_k)_{k+1} = |Ax_k - b|$$
$$= |r_k| \qquad (11.196)$$

So $(g_k)_{k+1}$ is the magnitude of the residual vector after k iterations. Because the stopping criterion is based on $|r_k|$ we do not have to compute y or z until the final iteration.

We now summarize the entire procedure:

$$r_0 = b - Ax_0, \quad g_i = |r_0|$$

$$v_0 = \frac{r_0}{|r_0|}, \quad k = 1$$

$$\text{while} \quad \frac{|r_k|}{|b|} > \epsilon \quad \text{do}$$

$$\text{for} \quad i = 0, \ldots, k-1$$

$$H_{i,k} = (Av_{k-1}) \cdot v_i$$

$$p_{k+1} = Av_k - \sum_{i=1}^{k-1}[(Av_{k-1}) \cdot v_i]v_i$$

$$v_{k+1} = \frac{p_{k+1}}{|p_{k+1}|}$$

$$\text{for} \quad i = 1, \ldots, k-1$$

$$h_k = G_i h_k$$

$$\text{create} \quad G_k \quad \text{from} \quad h_k$$

$$g = G_k g$$

end while

$$\text{Solve for } y: \quad Hy = g$$

$$z = V_k y$$

$$x = x_0 + z$$

11.8 SOLUTION OF NONLINEAR EQUATIONS

In a large number of electromagnetics problems, the resulting set of equations are nonlinear. Often, this nonlinearity is expressible by means of a single analytical expression in powers of the sought solution, in a closed interval, or in terms of piecewise functions (usually a graph). Direct algebraic solutions are possible only in a few cases, such as when the degree of the resulting equations is not greater than 4 or for certain transcendental equations.

Another means of solving these nonlinear equations is by transformations in which the nonlinear equations are linearized. The Kirchhoff transformation for certain magnetostatic problems is an example. These are however limited to a few cases and in general, algebraic solutions cannot be found for most of the nonlinear problems. Numerical methods, on the other hand, are extremely well suited for these applications and yield good results, provided that the initial estimate of the solution is close to the actual one.

In this section a number of well-known numerical methods for one- and two-dimensional cases are discussed and are illustrated by examples. Existence, convergence, and uniqueness proofs are also presented for some examples. Finally, the procedure is extended to multidimensional cases for applications to nonlinear problems in electromagnetics by the finite element method.

Examples of Nonlinear Equations

The following are a few examples that illustrate the nature of the more common nonlinear problems met with in practice.

1. Find x such that
$$x^k = N \qquad (11.197)$$

2. Solve
$$a_0 x^n + a_1 x^{n-1} + \cdots + a_{n-1} + a_n = 0 \qquad (11.198)$$

3. If
$$y = f(x) \qquad (11.199)$$
is a continuous function of x such that $f(x_1) \cdot f(x_2) < 0$, find a value of x in (x_1, x_2) such that $f(x) = 0$.

4. Given values of M and e, find a value of x that satisfies

$$M = x - e \sin x \qquad (11.200)$$

(Kepler's equation)

In the foregoing, equations (11.197), (11.198) and (11.199) can be solved by algebraic formulae for certain conditions. Equation (11.200) cannot be solved by any explicit formula. Also, equation (11.198) can be solved directly only for $n \leq 4$, and equation (11.199) has an explicit solution only if $f(x)$ is a linear function of x. In this case at points x_1 and x_2 we have the relation

$$\frac{f(x) - f(x_1)}{x - x_1} = \frac{f(x_2) - f(x_1)}{x_2 - x_1} \qquad (11.201)$$

for which the solution is

$$x = \frac{x_1 f(x_2) - x_2 f(x_1)}{f(x_2) - f(x_1)} \qquad (11.202)$$

In view of the limitations of direct algebraic methods for the nonlinear equations just shown, iterative methods are used, some of which are now described.

11.8.1 Method of Successive Approximations (First Order Methods)

We shall first consider a single nonlinear equation $f(x) = 0$. Representative methods are

1. Successive bisection
2. Inverse linear interpolation
3. Functional iteration
4. Chord method

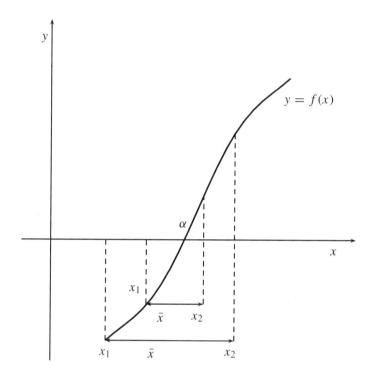

Figure 11.20 Method of Successive Bisection

Method of Successive Bisection

In this method, the interval (x_1, x_2) containing the root is bisected at $\bar{x} = \frac{(x_1 + x_2)}{2}$ and a determination is made as to which half-interval contains a root. Then that half-interval replaces (x_1, x_2) and this procedure is repeated. The process is continued until $||x_2 - x_1|| \leq \epsilon$. The geometric description of the method is illustrated in Figure 11.20.

Example

Let us consider the solution of the equation $x^2 = 2$. The iteration function is $f(x) = x^2 - 2$. It is necessary to find the root of the equation in the interval (x_1, x_2) such that $f(x_1) \cdot f(x_2) < 0$. The initial estimate of the interval is $x_1 = 0, x_2 = 1.8$, and the convergence criterion is taken to be $||x_{n+1} - x_n|| \leq 0.002$. The steps of the procedure are given in Table 11.4 and in the block diagram of Figure 11.21.

Solution of Equations

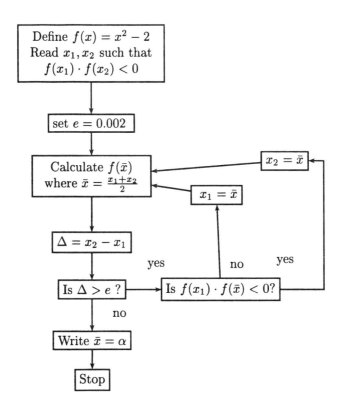

Figure 11.21 Block Diagram of Successive Bisection

Inverse Linear Interpolation

According to this method, $f(x)$ is approximated by the line segment (chord) through points $(x_1, f(x_1))$ and $(x_2, f(x_2))$ and this chord intersects the x axis at, say, x'. Then in this subinterval either (x_1, x') or (x', x_2) containing a root replaces (x_1, x_2) and the procedure is repeated. This procedure is continued until convergence is attained. The steps of the process are illustrated by solving the same equation, $x^2 = 2$. As before, the starting values are $x_1 = 1$ and $x_2 = 1.8$ so that $f(x_1) \cdot f(x_2) < 0$.

The inverse interpolation method is diagramatically represented in Figure 11.22. The method will converge provided the following Fourier conditions are satisfied.

$$\begin{aligned} f(x_1) \cdot f(x_2) &< 0 \\ f(x_1) \cdot f''(x_1) &> 0 \\ f(x'') \neq 0, \quad x_1 < x < x_2 & \end{aligned} \qquad (11.203)$$

Table 11.4 Example Problem

Iter. no.	x_1	x_2	$\bar{x}$	e
0	0	1.8	0.9	1.8
1	0.9	1.8	1.35	0.9
2	1.35	1.8	1.575	0.45
3	1.35	1.575	1.4625	0.225
4	1.35	1.4625	1.40625	0.112
5	1.40625	1.4625	1.434375	0.0562
6	1.40625	1.434375	1.4203125	0.0281
7	1.40625	1.4203125	1.41328	0.0140
8	1.41328	1.4203125	1.416802	0.0070
9	1.41328	1.416802	1.415041	0.0035
10	1.41328	1.415041	1.41416	0.0017

Table 11.5 Example Problem

Iter. no.	x_1	x_i	Δ
0	1	1.8	0.8
1	1	1.115	0.38
2	1	1.473	0.243
3	1	1.4043	0.0466
4	1	1.4159	0.0087
5	1	1.41392	0.0014
6	1	1.414	0.00005

Solution of Equations

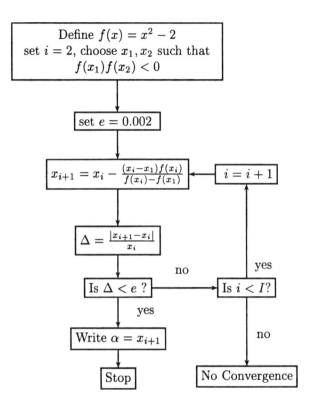

Figure 11.22 Block Diagram of Inverse Linear Interpolation

Functional Iteration Method

We shall now consider other methods of determining the roots of the equation $f(x) = 0$, where $f(x)$ and x are vectors of the same dimension k. When $k = 1$, we have a single equation, and if $k = n$ we have a system of n equations. If we now define a function

$$G(x) = x - f(x) \tag{11.204}$$

Then most of the iterative methods can be written in the form

$$x_{n+1} = G(x_n) \qquad (11.205)$$

for some suitable function G and initial approximation x_0. The convergence of the iteration process is assured if the mapping $G(x)$ carries a closed bounded set $S \subset C_k$ into itself and if the mapping is contracting, i.e., if

$$||G(x) - G(y)|| \leq M||x - y|| \qquad (11.206)$$

for some norm, for all (x, y) in S and for $M < 1$, known as the Lipschitz constant. Such an iterative scheme is sometimes called Picard iteration or the functional iteration method. This first-order unaccelerated iterative scheme is best illustrated by the flowchart of Figure 11.23.

Isaacson and Keller [136] have demonstrated the convergence of the iterative method and the existence of a unique solution in the form of a theorem as stated next.

Theorem 1:

Let $G(x)$ satisfy the Lipschitz condition

$$||G(x) - G(x')|| \leq \lambda ||x - x'| \qquad (11.207)$$

for all values of (x, x') in the closed interval

$$I = [x_0 - \rho, x_0 + \rho] \qquad (11.208)$$

where the Lipschitz constant satisfies

$$0 \leq \lambda < 1 \qquad (11.209)$$

Solution of Equations

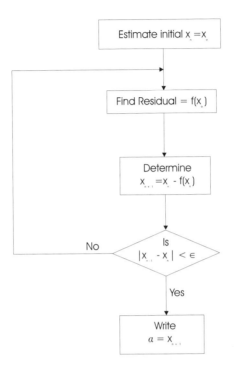

Figure 11.23 Functional Iteration Method

Let the initial estimate x_0 be such that

$$|x_0 - G(x_0)| \leq (1 - \lambda)\rho \tag{11.210}$$

Then the theorem states that

i. All the iterates x_n defined by the foregoing iterative sequence lie within the interval I, that is,

$$x_0 - \rho \leq x_n \leq x_0 + \rho \tag{11.211}$$

ii. (Existence) The iterates converge to some point, say

$$\lim_{n \to \infty} = \alpha \tag{11.212}$$

which is a root of the equation $x - G(x) = 0$.

iii. (Uniqueness) α is the only root in $[x_0 - \rho, x_0 + \rho]$.

Proof:

$x_i = G(x_0)$, we have by (11.209) and (11.210)

$$|x_0 - x_1| \leq (1 - \lambda)\rho \leq \rho \qquad (11.213)$$

and hence x_1 is in the interval (11.211). Then from (11.205), $|x_{n+1} - x_n| = |G(x_n) - G(x_{n-1})|$ and by the inductive assumption, x_n and x_{n-1} are in the interval (11.211). Thus by (11.207), the Lipschitz condition yields

$$\begin{aligned} |x_{n+1} - x_n| &\leq \lambda |x_n - x_{n-1}| \\ &\leq \lambda^2 |x_{n-1} - x_{n-2}| \\ &\leq \lambda^n |x_1 - x_0| \\ &\leq \lambda^n (1 - \lambda)\rho \end{aligned} \qquad (11.214)$$

Here we have used (11.205) and (11.206) recursively and then applied (11.213). However,

$$\begin{aligned} |x_{n+1} - x_0| &= |(x_{n+1} - x_n) + (x_n - x_{n-1}) + \cdots + (x_1 - x_0)| \\ &\leq |(x_{n+1} - x_n)| + |(x_n - x_{n-1})| + \cdots + |(x_1 - x_0)| \\ &\leq (\lambda^n + \lambda^{n-1} + \cdots + 1)(1 - \lambda)\rho \\ &= (1 - \lambda^{n+1})\rho \end{aligned} \qquad (11.215)$$

Convergence

We will first show that the sequence x_n is a Cauchy sequence. Thus for arbitrary positive integers m and p we consider

$$\begin{aligned} &|x_m - x_{m+p}| \\ &= |(x_m - x_{m-1}) + (x_{m-1} - x_{m-2}) + \cdots + (x_{m-p-1} - x_{m-p})| \\ &\leq |(x_m - x_{m+1}) + (x_{m-1} - x_{m+2}) + \cdots + (x_{m+p-1} - x_{m+p})| \\ &\leq (\lambda^m + \lambda^{m+1} + \cdots + \lambda^{m+p-1})(1 - \lambda)\rho \\ &\leq (1 - \lambda^p) p \lambda^m \end{aligned} \qquad (11.216)$$

Solution of Equations

Here we have used the inequalities (11.214). Given any $\epsilon > 0$, because λ in the interval $(0 \leq \lambda < 1)$ is fixed, we can find an integer $N(\epsilon)$ such that $|x_m - x_{m+p}| < \epsilon$ for all $m > N(\epsilon)$ and $p > 0$. Here N need only have a value such that $\lambda^n < \frac{\epsilon}{p}$. Hence the sequence $\{x_n\}$ is a Cauchy sequence and has a limit, say α, in I. As the function $G(x)$ is continuous in the interval I, the sequence $G(x_n)$ has the limit $G(\alpha)$ and by (11.205) this limit must also be α, that is, $\alpha = G(\alpha)$. Now $|x_n - \alpha| = |G(x_{n-1}) - G(\alpha)| \leq \lambda|x_{n-1} - \alpha|$ and hence $|x_n - \alpha| \leq \lambda^n |x_0 - \alpha| \leq \lambda^n \rho$.

Uniqueness

Let β be another root in $[x_0 - \rho, x_0 + \rho]$. Then because both α and β are in this interval, equation (11.207) holds and if $|\alpha - \beta| \neq 0$, we have $|\alpha - \beta| = |G(\alpha) - G(\beta)| \leq \lambda|\alpha - \beta| < |\alpha - \beta|$. This contradiction implies that $\alpha = \beta$ and hence the solution is unique.

11.8.2 Simple Iteration or the Chord Method

Once again we consider the solution of $f(x) = 0$ in some interval $a \leq x \leq b$. Let $\phi(x)$ be any function such that $0 < |\phi(x)| < \infty$ in $[a, b]$. Then the equation $x = G(x) = x - \phi(x) \cdot f(x)$ has roots that coincide with those of $f(x)$ in the interval $[a, b]$ and not others. The simplest choice for $\phi(x)$ is

$$\phi(x) = m \neq 0 \qquad (11.217)$$

If $f(x)$ is differentiable, we note that $G'(x) = 1 - m \cdot f'(x)$, and the scheme will be convergent in some interval provided m is chosen such that $0 < m \cdot f'(\alpha) < 2$. This is readily seen from Theorem 1, wherein the sequence $x_{n+1} = G(x_n)$ converges if

$$|G(x_1) - G(x_2)| \leq \lambda|x_1 - x_2| \qquad (11.218)$$

By the mean value theorem $|G(x_1) - G(x_2)|$ can be written as $G'\chi \cdot (x_1 - x_2)$, whence $|G'(\chi)| \leq \lambda < 1$ may serve as the Lipschitz constant. Hence if $G'(x) = 1 - m \cdot f'(x)$

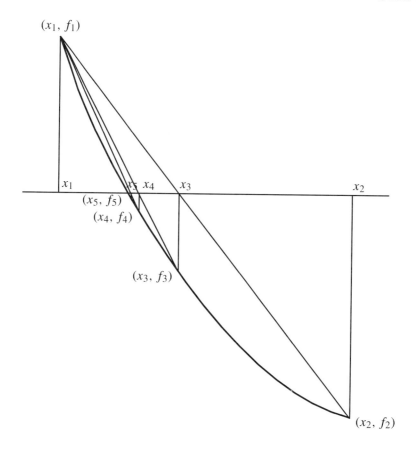

Figure 11.24 Method of Inverse Linear Interpolation

is always less than 1, then the sequence will be convergent. The iteration equation will be

$$x_{n+1} = x_n - m \cdot f(x_n) \qquad (11.219)$$

as illustrated in Figure 11.24.

The iterates described under the method of chords have geometric meaning as shown in Figure 11.25, in which the value x_{n+1} is the x intercept of the line with slope $\frac{1}{m}$ through $(x_n, f(x_n))$. The inequality implies that this slope should be taken between infinity and $\frac{1}{2}f'(\alpha)$, that is, half the slope of the tangent to the curve $y = f(x)$ at the root. Hence the name *chord* method.

Solution of Equations

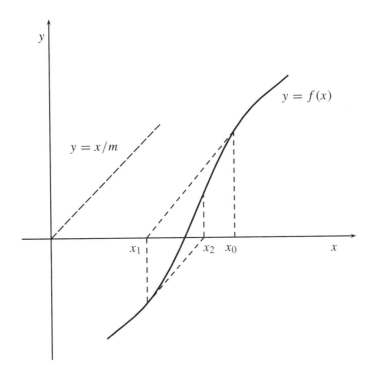

Figure 11.25 Chord Method

11.8.3 Second-Order Methods

Theorem 2: [136]

If $x = G(x)$ has a root at $x = \alpha$ and in the interval $|x - \alpha| < \rho$, $G(x)$ satisfies

$$|G(x) - G(\alpha)| \leq \lambda |x - \alpha| \qquad (11.220)$$

which is weaker than the general Lipschitz condition because α is fixed. It can be shown that with $\lambda < 1$ and for any x_0, we have that

1. All the iterates x_n lie in the interval $|x - \alpha| < \rho$.

2. The iterates converge to α.

3. The root α is unique in this interval.

Corollary:

If we now call $\frac{|G(x)-G(\alpha)|}{|x-\alpha|}$ as $|G'(x)| \leq \lambda < 1$, then the preceding conclusions follow. This can be proved by expanding $(\alpha - x_n)$ by the mean value theorem and using the results of Theorem 2. Also, because $|\alpha - x_{k+n}| \approx |G'(\alpha)|^n \cdot |\alpha - x_k|$, this would suggest that if $G'(\alpha) = 0$ the convergence of the sequence would be quite rapid.

In view of this corollary, if at $x = \alpha$, we have $G'(\alpha) = 0$ and if $G''(x)$ exists and is bounded in some interval $|\alpha - x| \leq \rho$, in which the conditions of the preceding theorem are satisfied, then for any x in this interval, by Taylor's theorem,

$$G(x) = G(\alpha) + 0 + \frac{(x-\alpha)^2}{2} \cdot G''(\chi) \tag{11.221}$$

where χ is some value between x and α. By using this result, we obtain for any iterate of the sequence $x_{n+1} = G(x_n)$, the expression

$$|x_n - \alpha| = |G(x_{n-1}) - G(\alpha)| = |\frac{1}{2}G''(\chi_{n-1})| \cdot |x_{n-1} - \alpha|^2_{n=1,2,3,\ldots} \tag{11.222}$$

Thus the error in any iterate is proportional to the square of the previous error and hence if $G''(\alpha) \neq 0$, this procedure will be called a second-order method.

Some examples of second-order methods are

1. Newton–Raphson method (also called Newton's method)
2. Regula falsi
3. Steffenson–Aitken method
4. Bailey's method
5. Accelerated convergence method

Solution of Equations

In the second-order methods, if the bound on $G''(x)$ is denoted by $|G''(x)| \leq 2M$ in the interval $|\alpha - x| < \rho$, then from (11.222) we have

$$|x_n - \alpha| \leq (M|x_0 - \alpha|)^{2^{n-1}} \cdot |x_0 - \alpha| \tag{11.223}$$

Thus if $M|x_0 - \alpha| < 1$, the second-order methods converge and reduce the initial error by at least 10^{-m} when

$$(M|x_0 - \alpha|)^{2^{n-1}} \approx 10^{-m} \tag{11.224}$$

11.8.4 Newton–Raphson Method

Suppose we are given an estimate x_i of a real root of the equation $f(x) = 0$, the equation of the line tangent to $f(x)$ at $x = x_i$ can be expressed as the linear Taylor polynomial

$$y(x) = f(x_i) + f'(x_i) \cdot (x_{i+1} - x_i) \tag{11.225}$$

Solving this equation for x_{i+1} and with $y(x) = 0$, we obtain

$$x_{i+1} = x_i - \frac{f(x_i)}{f'(x_i)} \tag{11.226}$$

which is the classical Newton–Raphson iterative method. In effect, we are obtaining a refined approximation x_{i+1} of a root α of $f(x) = 0$ by approximating the graph of $f(x)$ by the line tangent $f(x)$ at $x = x_i$. This is illustrated in Figure 11.26

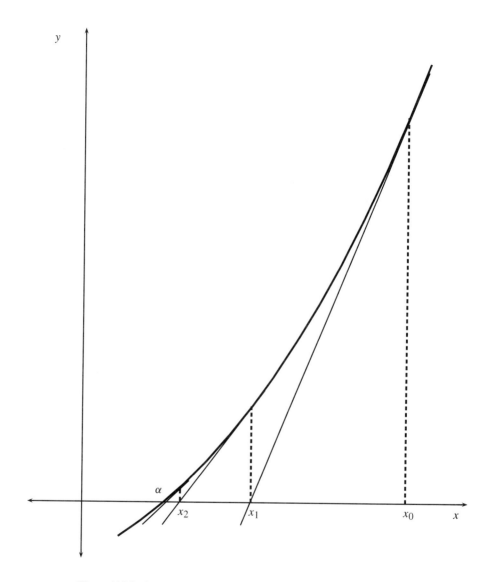

Figure 11.26 Graphical Interpretation of the Newton–Raphson Method

Solution of Equations

Existence and Convergence

We shall define an auxiliary iteration function

$$G(x) = x - \frac{f(x)}{f'(x)} \tag{11.227}$$

with

$$G'(x) = \frac{f(x) \cdot f''(x)}{|f'(x)|^2} \tag{11.228}$$

If α is a root of $f(x) = 0$ and if $f'(\alpha) \neq 0$ then we have

$$\begin{aligned} G(\alpha) &= \alpha \\ G'(\alpha) &= 0 \end{aligned} \tag{11.229}$$

We now have to show that the iterative sequence

$$x_{i+1} = x_i - \frac{f(x_i)}{f'(x_i)} \tag{11.230}$$

generates (from an initial value x_0), a convergent sequence $x_1, x_2, \ldots$ of successive approximation of a root α of the equation $f(x) = 0$.

Proof: Let λ be the largest magnitude of $G'(x)$ in the interval containing $x_0, x_1, x_2, \ldots$ and α. If $\lambda < 1$, then the sequence x_{i+1} converges to the root α of the equation $f(x) = 0$ and

$$G(x) = x - \frac{f(x)}{f'(x)} \tag{11.231}$$

From the iteration formula we have

$$x_{i+1} = G(x_i) \tag{11.232}$$

and

$$x_1 = 0 \tag{11.233}$$

Hence $(x1 - \alpha) = G(x_0) - G(\alpha)$, because $\alpha = G(\alpha)$ and by the mean value theorem

$$G(x_0) - G(\alpha) = (x_0 - \alpha)G'(\bar{(x)}_0) \tag{11.234}$$

where $\bar{x}_0 \in (x_0, \alpha)$. From the preceding two equations, we obtain

$$(x_1 - \alpha) = (x_0 - \alpha) \tag{11.235}$$

Taking absolute values, this equation reduces to the form

$$|x_1 - \alpha| = |x_0 - \alpha| \cdot |G'(\bar{x}_0)| \leq |x_0 - \alpha|\lambda \tag{11.236}$$

In the same manner we find

$$|x_2 - \alpha| = |x_1 - \alpha| \cdot |G'(\bar{x}_0)| \leq |x_1 - \alpha|\lambda \leq |x_0 - \alpha|^2 \tag{11.237}$$

Continuing, we obtain

$$|x_{i+1} - \alpha| \leq |x_0 - \alpha|\lambda^{i+1} \tag{11.238}$$

Solution of Equations

Where $\lambda < 1$, we see that

$$\lim_{i \to \infty} |x_{i+1} - \alpha| = 0 \qquad (11.239)$$

or

$$\lim_{i \to \infty} x_{i+1} = \alpha \qquad (11.240)$$

so that the sequence $\{x_{i+1}\}$ converges to the root α of the equation $f(x) = 0$.

Rate of Convergence of Newton's Method

Let the error term of the ith iterate be expressed as

$$\delta_i = x_i - \alpha \qquad (11.241)$$

By finding a relation between δ_{i+1} and δ_i we can estimate how rapidly (or how slowly) the algorithm converges to a root α of the equation $f(x) = 0$, provided the algorithm converges. Such a relation can be determined by expanding the iteration function $G(x)$ in a Taylor series about $x = \alpha$.

$$G(x) = G(\alpha) + G'(\alpha) \cdot (x - \alpha) + \frac{G''(\alpha) \cdot (x - \alpha)^2}{2} + \cdots \qquad (11.242)$$

For the Newton–Raphson algorithm, the iteration function $G(x)$ and its first two derivatives are

$$G(x) = x - \frac{f(x)}{f'(x)} \tag{11.243}$$

$$G'(x) = \frac{f(x) \cdot f''(x)}{(f'(x))^2} \tag{11.244}$$

$$G''(x) = \frac{((f'(x))^2(f(x) \cdot f''(x) + f'(x) \cdot f''(x)) - (f(x) \cdot f''(x))2f'(x)f''(x)}{f'(x)^4}$$

and

$$\begin{aligned} G(\alpha) &= \alpha \\ G'(\alpha) &= 0 \\ G''(\alpha) &= \frac{f''(\alpha)}{f'(\alpha)} \end{aligned} \tag{11.245}$$

Substituting the values in equation (11.245) into the series of equation (11.242) and truncating the series after terms of the second degree and evaluating at $x = x_i$, we obtain

$$G(x_i) = \alpha + \frac{f''(\alpha) \cdot (x_i - \alpha)^2}{2 f'(\alpha)} \tag{11.246}$$

Using the relation $x_{i+1} = G(x_i)$ and the definition of equation (11.241) we find that

$$\delta_{i+1} = \frac{f''(\alpha)\delta_i^2}{2 f'(\alpha)} \tag{11.247}$$

From this relation the rate of convergence of the Newton–Raphson method is said to be quadratic. A numerical example of a non-linear resistor current and voltage drop is solved by simple iteration and by the Newton–Raphson method.

Results of Iterative Methods Applied to a Nonlinear Resistor

Consider a nonlinear resistor whose $V - I$ and $\frac{\partial V}{\partial I}$ characteristics are as shown in Figure 11.27. It is required to find the voltage across the nonlinear resistor and the current by an iterative method. The circuit is shown in Figure 11.28. Let $R_x = 75 \, \Omega$ and $V = 100$ volts.

Solution of Equations

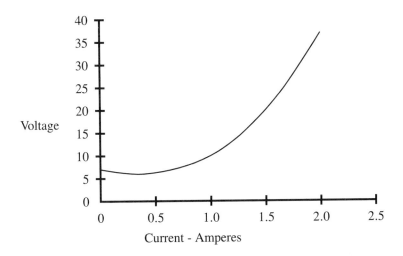

Figure 11.27 Nonlinear Resistor Characteristics

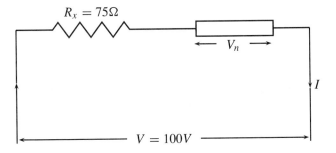

Figure 11.28 Circuit Diagram

Simple Iteration or Chord Method

If v_n^i is the ith value of v_n and v_n^{i+1} its value after the iteration, the iteration method (relaxation) can be stated as

$$v_n^{i+1} = v_n^i + K(v_n^{i+1} - v_n^i) \tag{11.248}$$

where K is the acceleration factor and is assumed to be between 0.1 and 0.2.

Also

$$v_n^{i+1} = V - I(v_n^i)R_x \tag{11.249}$$

We may therefore call the error

$$\epsilon(v_n^i) = V - I(v_n^i)R_x - v_n^i \tag{11.250}$$

Equation (11.250) may be rewritten as

$$v_n^{i+1} = v_n^i + K\epsilon(v_n^i) \tag{11.251}$$

The results are shown in Tables 11.6 and 11.7.

Newton–Raphson Method

$$v_n^{i+1} = v^i - \frac{\epsilon(v_n)}{\epsilon'(v_n)} \tag{11.252}$$

Solution of Equations

Table 11.6 Chord Method with $K = 0.1$

Iter. no.	v_n^i	I^i	$\epsilon(v_n)$
1	10	0.93	20.25
2	12.025	1.06	8.475
3	12.9225	1.12	3.0775
4	13.23025	1.14	1.76975
5	13.40723	1.145	0.71778
6	13.479	1.15	0.271
7	13.5061	1.151	0.1689
8	13.52379	1.152	0.07621
9	13.531411	1.1525	0.0311
10	13.5345	1.1526	0.0205

Table 11.7 Chord Method with $K = 0.2$

Iteration no.	v_n^i	I^i	$\epsilon(v_n)$
0	0	0	100
1	20	1.4	−25
2	15	1.2	−5
3	14	1.17	−1.75
4	13.65	1.16	−0.65
5	13.52	1.15	0.23
6	13.566	1.153	−0.041
7	13.558	1.1525	0.0045
8	13.5589	1.1523	−0.0189

Table 11.8 Newton Raphson–Iterations for the Example

Iter. no.	v_n^i	I^i	$\epsilon(v_n)$	$-\epsilon'(v_n)$	$1/\epsilon'(v_n)$	$-\epsilon''(v_n)E-3$
0	0	1.33	100	9.01754	0.111	0.7
1	11.1	1.0	13.9	5.42	0.1845	6.28
2	13.665	1.16	−0.665	4.56	0.219	4.00
3	13.519	1.15	0.281	4.60	0.2165	4.20
4	13.58	1.155	−0.20	4.58	0.218	4.01
5	13.5363	1.1525	0.0262	4.59	0.2178	4.15
6	13.5408	1.153	−0.0158	4.595	0.2173	4.1
7	13.53736	1.1528				

But

$$\epsilon(v_n) = V - [I(v_n)R_x + v_n] \tag{11.253}$$

$$\epsilon'(v_n) = -[\frac{\partial I(v_n)}{\partial v_n}R_x + 1]$$

$$v^{i+1} = v^i + \frac{V - [I(v_n^i)R_x + v_n^i]}{1 + \frac{\partial I(v_n^i)}{\partial v_n}R_x} \tag{11.254}$$

The results for the Newton–Raphson iterations are shown in Table 11.8.

11.8.5 Some Modifications of Newton's Methods

The evaluation of the derivative G' of the given iteration function G may not be a trivial problem in many practical situations, especially if G itself is the result of a complicated computation. Hence a variety of methods have been devised by different authors to obviate the need for calculating G'. Some of the methods are described next.

Whittaker's Method

In this method the derivative $G'(x_n)$ is simply replaced by a constant and the resulting iteration formula given as

$$x_{n+1} = x_n - \frac{G(x_n)}{m} \tag{11.255}$$

defines for a certain range of values of m a linearly convergent sequence, unless we happen to pick $m = G'(\alpha)$. If $m = 1$, the simple unaccelerated functional iteration results. If the estimate of m is good, convergence will be rapid. Further, in the initial stages of the Newton process, it is usually not necessary to recompute G' at each step.

Regula Falsi

Here the value of the derivative $f'(x_n)$ is approximated by the difference quotient

$$\frac{f(x_n) - f(x_{n-1})}{x_n - x_{n-1}} \tag{11.256}$$

formed with the two preceding approximations. The resulting iteration formula is given as

$$x_{n+1} = x_n - \frac{(x_n - x_{n-1})f(x_n)}{f(x_n) - f(x_{n-1})} \tag{11.257}$$

The algorithm suggested here is termed regula falsi, defined by a difference equation of order 2. Ostrowski [137] has shown that the degree of convergence of the method lies somewhere between that of Newton's method and that of the ordinary functional iteration method.

11.8.6 Muller's Method

Regula falsi can be obtained by approximating the graph of the function f by the straight line passing through the points $(x_{n-a}, G(x_{n-1}))$ and $(x_n, G(x_n))$. The point of intersection of this line with the x axis defines the new approximation x_{n+1}. Instead of approximating f by a linear function, more rapid convergence can be obtained by approximating f by a polynomial p of degree $k > 1$ coinciding with f at points $x_n, x_{n-1}, ..., x_{n-k}$ and determine x_{n+1} as one of the zeros of p. Muller's study reveals the the choice of $k = 2$ yields very satisfactory results. Because the construction of p depends on the theory of the interpolating polynomial, the matter will not be pursued any further here.

Newton's Method in the Case $f'(\alpha) = 0$

The Newton–Raphson method was derived under the assumption that $f'(x) \neq 0$, implying in particular that $f'(\alpha) \neq 0$. Let us now consider the general solution where

$$f(\alpha) = f'(\alpha) = ... f^{m-1}(\alpha) = 0 \tag{11.258}$$

with $f^m(\alpha) \neq 0$, where $m \geq 1$. If we set $x = \alpha + h$, the iteration function for Newton's process, namely

$$G(x) = x - \frac{f(x)}{f'(x)} \tag{11.259}$$

can be expanded in powers of h to give

$$G(\alpha + h) = \alpha + h - \frac{(m!)^{-1} f^m(\alpha) h^m + O(h^{m-1})}{(m-1)!^{-1} f^m(\alpha) h^{m-1} + O(h^m)} \tag{11.260}$$

or

$$G(\alpha + h) = \alpha + h - \frac{1 \cdot h}{m} + O(h^2) \tag{11.261}$$

From this we find

$$G'(\alpha) = \lim_{h \to 0} \frac{G(\alpha + h) - G(\alpha)}{h} = 1 - \frac{1}{m} \tag{11.262}$$

Solution of Equations

Thus if $m \neq 1$, and $G'(\alpha) \neq 0$, the condition for quadratic convergence proved earlier as $G'(\alpha) = 0$ is not satisfied. However, the preceding analysis shows how to modify the iteration function in order to achieve quadratic convergence. If we set

$$\begin{aligned} G(x) &= \alpha - m\frac{f(x)}{f'(x)}, \quad \text{for } x \neq \alpha \\ &= \alpha, \quad \text{for } x = \alpha \end{aligned} \qquad (11.263)$$

then a computation similar to the one performed before shows that $G'(\alpha) = 0$. By the theorem of quadratic convergence [138], the sequence defined by

$$x_{n+1} = x_n - \frac{mf(x_n)}{f'(x_n)}, \quad \text{for } n = 0, 1, \ldots \qquad (11.264)$$

converges to α quadratically provided that x_0 is sufficiently close to α. For $m = 1$, the algorithm reduces to the ordinary Newton process. Although only rarely, in practice, have we *a priori* knowledge of the fact $f'(x) = 0$ at a solution of $f(x) = 0$, the method has been used successfully in many cases (vide Greenspan [139]) and m is chosen in a heuristic fashion to lie somewhere between 1 and 2 [131].

11.8.7 Aitken's δ^2 Method (Arbitrary Order)

This procedure is usually presented as a means for accelerating the convergence of the ordinary functional iteration method, described earlier. If x_n is any number approximating a root of $f(x) = 0$ or $x = G(x)$ and if we define $x_{n+1} = G(x_n)$, then the measure of the errors in the two approximations x_n and x_{n+1} can be defined by

$$\epsilon_n = G(x_n) - x_n = x_{n+1} - x_n \qquad (11.265)$$

and

$$\epsilon_{n+1} = G(x_{n+1}) - x_{n+1} \qquad (11.266)$$

Because for each root, this error should vanish, that is,

$$\epsilon(\alpha) = G(\alpha) - \alpha = -\phi(\alpha)f(\alpha) = 0 \tag{11.267}$$

we may seek x_{n+1} by extrapolating the errors to zero. Here we define

$$x = G(x) = x - \phi(x)f(x) \tag{11.268}$$

That is the line segment joining the points (x_n, ϵ_n) and $(x_{n+1}, \epsilon_{n+1})$ extended to intersect the x axis and the point of intersection is taken as x_{n+1}. This yields the expression

$$x_{n+1} = \frac{x_n \epsilon_{n+1} - x_{n+1} \epsilon_n}{\epsilon_{n+1} - \epsilon_n} = F(x_n) \tag{11.269}$$

For computational use, this equation is rewritten as

$$x_{n+1} = x_n - \frac{\epsilon_n^2}{\epsilon_{n+1} - \epsilon_n} \tag{11.270}$$

Substituting for ϵ_n, ϵ_{n+1} from the previous relation, we have

$$x_n' = x_n - \frac{(x_{n+1} - x_n)^2}{x_{n+2} - 2x_{n+1} + x_n} \tag{11.271}$$

This accelerated functional iteration method is sometimes called Steffenson's method and was originally proposed by Aitken to convert any convergent sequence (no matter how generated) x_n into a more rapidly convergent scheme x_n'. It is seen that the denominator of equation (11.271) suggests the second difference notation δ^2.

Solution of Equations

Bailey's Method

Let x_i be an estimate of the real root of the equation $f(x) = 0$. The equation $f(x)$ at $x = x_i$ can be expressed as the quadratic Taylor polynomial

$$y(x) = f(x_i) + f'(x_i)(x - x_i) + \frac{f''(x_i)(x - x_i)^2}{2} + \cdots \tag{11.272}$$

The intersection of $y(x)$ with the x axis will then yield the required results. Hence we set $y(x) = 0$ at $x = x_{i+1}$ and obtain

$$0 = f(x_i) + f'(x_i)(x_{i+1} - x_i) + \frac{f''(x_i)}{2}(x_{i+1} - x_i)^2 \tag{11.273}$$

$$0 = f(x_i) + (x_{i+1} - x_i)\left[f'(x_i) + \frac{f''(x_i)(x_{i+1} - x_i)}{2}\right] \tag{11.274}$$

This equation can be rewritten in the form

$$x_{i+1} = x_i - \frac{f(x_i)}{f'(x_i) + \frac{f''(x_i)}{2}(x_{i+1} - x_i)} \tag{11.275}$$

Supposing now we let the coefficient of $\frac{f''(x_i)}{2}$ be computed by the Newton–Raphson iterative formula, $x_{i+1} = x_i - \frac{f(x_i)}{f'(x_i)}$, by substituting $-\frac{f(x_i)}{f'(x_i)}$ for $(x_{i+1} - x_i)$ in the term of the denominator of equation (11.275), the resulting formula will be

$$x_{i+1} = x_i - \frac{f(x_i)}{f'(x_i) - \frac{f(x_i)f''(x_i)}{2f'(x_i)}} \tag{11.276}$$

This is known as Bailey's iterative formula for computing a refined approximation x_{i+1} of root α from an approximation x_i. The proof of convergence of Bailey's method is identical to that of Newton's method.

Rate of Convergence of Bailey's Method

We define an iteration function

$$G(x) = x - \frac{f(x)}{f'(x) - \frac{f(x)f''(x)}{2f'(x)}} \tag{11.277}$$

and expand it as a Taylor series about $x = \alpha$ (where α is the root of $f(x) = 0$) as

$$G(x) = G(\alpha) + G'(\alpha)(x-\alpha) + \frac{G''(\alpha)}{2}(x-\alpha)^2 + \frac{G'''(\alpha)}{6}(x-\alpha)^3 + \cdots \tag{11.278}$$

It can be shown that $G(\alpha) = G''(\alpha) = 0$ and

$$G'''(\alpha) = \frac{3}{2}\left(\frac{f''(\alpha)}{f'(\alpha)}\right)^2 - \frac{f'''(\alpha)}{f'(\alpha)} \tag{11.279}$$

By substituting these values in the Taylor series and neglecting terms of order higher than 3, we have

$$G(x) - G(\alpha) \approx \lambda(x-\alpha)^3 \tag{11.280}$$

where $\lambda = G'''(\alpha)$. Evaluating at $x = x_{i+1}$ and using the relation $G(x_i) = x_{i+1}$ and $G(\alpha) = \alpha$, we find that

$$x_{i+1} - \alpha \approx \lambda(x_i - \alpha)^3 \tag{11.281}$$

From this relation, the rate of convergence is said to be cubic. Although the method converges faster than Newton's, the computation required may offset any gain due to speedier convergence.

Table 11.9 Comparison of Bailey's Method and the Newton–Raphson Method for the Example

Iteration	Bailey's Method		Newton–Raphson	
	x_i	x_{i+1}	x_i	x_{i+1}
0	1	1.4	1	1.5
1	1.4	1.4142	1.5	1.42
2			1.42	1.4167
3			1.4167	1.4142

Numerical Example

Table 11.9 shows the solution of $f(x) = (x^2 - 2)$ by Bailey's method. It is clearly seen that the convergence is rapid and certainly faster than that of the ordinary Newton-Raphson Method.

Systems of Nonlinear Equations

Let x be an n dimensional column vector with components $x_1, x_2, ... x_n$ and $G(x)$ an n-dimensional vector-valued function, that is, a column vector with components $G_1(x), G_2(x), ... G_n(x)$. Then the system to be solved is

$$x = G(x) \tag{11.282}$$

The solution or root is some vector, say α, with components $\alpha_1, \alpha_2, ..., \alpha_n$, which is, of course, some point in the n-dimensional space. Starting with a point

$$x^{(0)} = [x_1^{(0)}, x_2^{(0)}, ..., x_n^{(0)}]^T \tag{11.283}$$

the exact analogue of the functional iteration of the one dimensional case is

$$x^{n+1} = G(x^n), \text{ for } n = 0, 1, 2, \ldots \tag{11.284}$$

Theorem 1 applies except that vector norms are used in place of absolute values, which is restated in the following.

Theorem 3. If $G(x)$ satisfies the condition

$$||G(x) - G(y)|| \leq \lambda ||x - y|| \tag{11.285}$$

for all roots of x, y such that $||x - x^{(0)}||$, $||y - x^{(0)}|| \leq \rho$ with the Lipschitz constant satisfying $0 \leq \lambda < 1$. If the initial iterate $x^{(0)}$ satisfies $||G(x^{(0)}) - x^{(0)}|| \leq (1 - \lambda)\rho$ then

- i. All iterates of the sequence $x^{(n+1)} = G(x^n)$ satisfy

$$||x^n - x^{(0)}|| \leq \rho \tag{11.286}$$

- ii. The iterate converges to some vector, say

$$\lim_{n \to \infty} x^n = \alpha \tag{11.287}$$

which is a root of $x = G(x)$.

- iii. α is the only root of the preceding equation in $||x - x^{(0)}|| \leq \rho$. The proof is the same as for Theorem 1 for the functional iteration of a single equation.

Theorem 4. If $x = G(x)$ has a root α and if the components $G_i(x)$ have continuous first partial derivatives and satisfy

$$\left| \frac{\partial G_i(x)}{\partial x_j} \right| \leq \frac{\lambda}{n} \tag{11.288}$$

Solution of Equations

$\lambda < 1$, for all x in

$$||x - \alpha||_\infty \leq \rho \tag{11.289}$$

then

- i. For any $x^{(0)}$ satisfying equation (11.289) all the iterates of x^n of the sequence (11.284) also satisfies (11.289).

- ii. For any $x^{(0)}$ satisfying (11.289), the iterates (11.284) converge to the root α of equation (11.282), which is unique if $||x - \alpha||_\infty \leq \rho$.

Proof: For any two points x, y in (11.289), we have by Taylor's theorem

$$G_i(x) - G_i(y) = \sum_{j=1}^{n} \frac{\partial G_i(\chi')}{\partial x_j}(x_j - y_j), \text{ for } i = 1, 2, \ldots n \tag{11.290}$$

where χ' is a point in the open line segment joining x and y. Hence χ' is in (11.289).

From the infinity norm of equation (11.290) we have

$$|G_i(x) - G_i(y)| \leq (||x - y||_\infty \sum_{j=1}^{n} \left| \frac{\partial G_i(\chi')}{\partial x_j} \right|) \leq \lambda ||x - y||_\infty \tag{11.291}$$

Because the inequality holds for each i, we have

$$||G(x) - G(y)||_\infty \leq \lambda ||x - y||_\infty \tag{11.292}$$

and hence it is established that $G(x)$ is Lipschitz continuous in the domain $||x - \alpha||_\infty \leq \rho$ with respect to the indicated norm.

Also for any $x^{(0)}$ in (11.289)

$$||x^i - \alpha||_\infty = ||G(x^{(0)}) - G(\alpha)||_\infty \leq \lambda ||x^{(0)} - \alpha||_\infty \leq \lambda\rho \qquad (11.293)$$

and hence x^i is also in (11.289). Extending the argument, we obtain

$$\begin{aligned} ||x^n - \alpha||_\infty &= ||G(x^{n-1}) - G(\alpha)||_\infty \\ &\leq \lambda ||x^{n-1} - \alpha||_\infty \\ &\leq \lambda^n ||x^{(0)} - \alpha||_\infty \\ &\leq \lambda^n \rho \end{aligned} \qquad (11.294)$$

Therefore all x^n lie in (11.289). The convergence and uniqueness follow because $\lambda < 1$.

Corollary: If the function $G(x)$ is such that at a root the matrix

$$(G_{ij}(x)) = \left(\frac{\partial G_i(x)}{\partial x_j}\right) = 0 \qquad (11.295)$$

and these derivatives are continuous near the root, then

(a) $\qquad \left|\dfrac{\partial G_i(x)}{\partial x_j}\right| \leq \dfrac{\lambda}{n}$, for $\lambda < 1$

(b) $\qquad$ for all $||x - \alpha||_\infty < \rho \max_i \sum_{j=1}^n |G_{ij}(x)| \leq \lambda < 1$,

$$(11.296)$$

will be satisfied for some $\rho > 0$. If in addition, the second derivatives $\frac{\partial^2 G_i(x)}{\partial x_j \partial x_k}$ all exist in a neighborhood of the root, it is shown in Isaacson [136] that

$$||x^{(n)} - \alpha||_\infty \leq M ||x^{(n-1)} - \alpha||_\infty^2 \qquad (11.297)$$

Solution of Equations

where M is such that

$$\max_{i,j,k} \left| \frac{\partial^2 G_i(x)}{\partial x_j \partial x_k} \right| \leq \frac{2M}{n^2} \tag{11.298}$$

This shows that quadratic convergence can occur in solving a system of equations by iteration.

Explicit Iteration Schemes

In the general case, the system to be solved is of the form

$$f(x) = 0 \tag{11.299}$$

where $f(x) = [f_1(x), f_2(x), f_3(x), \ldots, f_n(x)]^T$ is an n-component vector. Such a system can be written in the form

$$x = G(x) \tag{11.300}$$

in a variety of ways. Hence we can examine the choice

$$G(x) = x - A(x) \cdot f(x) \tag{11.301}$$

where $A(x)$ is an nth order sequence matrix with components $a_{ij}(x)$. The equations (11.288) and (11.299) will have the same set of solutions if $A(x)$ is nonsingular (as in that case $A(x) \cdot f(x) = 0$ implies $f(x) = 0$).

The simplest choice for

$$A(x) = A \tag{11.302}$$

is a constant nonsingular matrix. If we introduce the matrix

$$J(x) = \frac{\partial f_i(x)}{\partial x_j} \qquad (11.303)$$

whose determinant is the Jacobian of the function $f_i(x)$, then from equations (11.301) and (11.303) we have by differentiation and substitution

$$F(x) = \left(\frac{\partial G_i(x)}{\partial x_j}\right) = I - AJ(x) \qquad (11.304)$$

By Theorem 2 or its corollary, the iteration determined by using

$$x^{n+1} = x^n - Af(x^n) \qquad (11.305)$$

will converge, for $x^{(0)}$ sufficiently close to α, if the elements of the matrix (11.304) are sufficiently small, for example, as in the case that $J(\alpha)$ is nonsingular, and A is approximated by the inverse of $J(\alpha)$. This procedure is the analog of the chord method.

11.8.8 Newton–Raphson Method for a System of Equations

This is an extension of the method already described for a single equation to a system of equations. Let us consider the following equations in two independent variables x and y.

$$\begin{aligned} f(x, y) &= 0 \\ g(x, y) &= 0 \end{aligned} \qquad (11.306)$$

If an initial estimate of the solution, namely (x_0, y_0), is available and this is incremented by changes δx, δy, then the functions can be expanded by Taylor's theorem as

$$f(x_0 + \delta x, y_0 + \delta y) = f(x_0, y_0) + f_x(x_0, y_0) \cdot \delta x + f_y(x_0, y_0) \cdot \delta y + \cdots \qquad (11.307)$$

Solution of Equations

$$g(x_0 + \delta x, y_0 + \delta y) = g(x_0, y_0) + g_x(x_0, y_0) \cdot \delta x + g_y(x_0, y_0) \cdot \delta y + \cdots \quad (11.308)$$

Here f_x, f_y, g_x, and g_y are the derivatives of f and g with respect to x and y. If we now truncate the series after terms of the first degree, we will obtain first-order approximations of the resulting changes in $f(x, y)$ and $g(x, y)$ as the total differentials

$$\begin{aligned} \delta f &= f_x(x_0, y_0) \cdot \delta x + f_y(x_0, y_0) \cdot \delta y \\ \delta g &= g_x(x_0, y_0) \cdot \delta x + g_y(x_0, y_0) \cdot \delta y \end{aligned} \quad (11.309)$$

A solution to the system (11.306) can be obtained by determining δx, δy satisfying the constraints

$$\begin{aligned} \delta f &= -f(x_0, y_0) \\ \delta g &= -g(x_0, y_0) \end{aligned} \quad (11.310)$$

Substituting the values of these constraints in equation (11.309) and solving the resulting set of linear equations, namely

$$\begin{aligned} -f(x_0, y_0) &= f_x(x_0, y_0) \cdot \delta x + f_y(x_0, y_0) \cdot \delta y \\ -g(x_0, y_0) &= g_x(x_0, y_0) \cdot \delta x + g_y(x_0, y_0) \cdot \delta y \end{aligned} \quad (11.311)$$

From equations (11.307) and (11.308), it is evident that if f and g are evaluated at $(x_0 + \delta x, y_0 + \delta y)$ and expressed in a linear Taylor expansion, the following relations hold.

$$\begin{aligned} &f(x_0 + \delta x, y_0 + \delta y) \\ &= f(x_0, y_0) + f_x(x_0, y_0) \cdot \delta x + f_y(x_0, y_0) \cdot \delta y = 0 \\ &g(x_0 + \delta x, y_0 + \delta y) \\ &= g(x_0, y_0) + g_x(x_0, y_0) \cdot \delta x + g_y(x_0, y_0) \cdot \delta y = 0 \end{aligned} \quad (11.312)$$

If these linear expansions are sufficiently accurate, then $(x_0 + \delta x, y_0 + \delta y)$ are fairly good approximations of the solution of equations (11.306). If $|\delta x| > \epsilon$ or if $|\delta y| > \epsilon$, where ϵ is a small positive quantity, it is necessary to replace x_0 by $x_0 + \delta x$ and y_0 by

$y_0 + \delta y$ and repeat the entire process. Usually a few iterates of the process will produce accurate values of the root, provided that the original estimates (x_0, y_0) are sufficiently close to the true solution.

Proof of Convergence

The proof [138] that follows is applicable to a problem in two variables but can be extended to more variables. Let (x_0, y_0) be the initial estimate of the solution so that the sequence of points (x_n, y_n) is given by

$$\begin{aligned} x_{n+1} &= \mathcal{F}(x_n, y_n) \\ y_{n+1} &= \mathcal{G}(x_n, y_n), \quad n = 0, 1, 2, \ldots \end{aligned} \quad (11.313)$$

where the functions $\mathcal{F}$ and $\mathcal{G}$ are defined by

$$\begin{aligned} \mathcal{F}(x, y) &= x + \delta(x, y) \\ \mathcal{G}(x, y) &= y + \epsilon(x, y) \end{aligned} \quad (11.314)$$

We assert that the sequence (x_n, y_n) converges quadratically to (α, t) for all (x_0, y_0) sufficiently close to (α, t). In order to verify this statement, it is sufficient to show that all elements of the Jacobian matrix of the iteration functions $\mathcal{F}$ and $\mathcal{G}$

$$J(x, y) = \begin{pmatrix} \mathcal{F}_x(x, y) & \mathcal{F}_y(x, y) \\ \mathcal{G}_x(x, y) & \mathcal{G}_y(x, y) \end{pmatrix} \quad (11.315)$$

are zero for $(x, y) = (\alpha, t)$. Omitting the arguments, we have

$$\begin{aligned} \mathcal{F}_x &= 1 + \delta_x \\ \mathcal{F}_y &= \delta_y \\ \mathcal{G}_x &= \epsilon_x \\ \mathcal{G}_y &= 1 + \epsilon_y \end{aligned} \quad (11.316)$$

Solution of Equations

The values of the derivatives δ_x, δ_y, ϵ_x, and ϵ_y are best determined from equation (11.314) by implicit differentiation. Differentiating the first of these equations

$$f(x, y) = 0$$
$$g(x, y) = 0 \qquad (11.317)$$

the Newton–Raphson scheme for the system is written as

$$f_x(x, y)\delta + g_y(x, y)\epsilon = -f(x, y)$$
$$g_x(x, y)\delta + g_y(x, y)\epsilon = -g(x, y) \qquad (11.318)$$

where δ is the increment in x and ϵ is the increment in y. Differentiating equation (11.318) with respect to x and y, we obtain

$$f_{xx}\delta + f_x\delta_x + f_{xy}\epsilon + f_y\epsilon_x = -f_x$$
$$f_{xy}\delta + f_x\delta_y + f_{yy}\epsilon + f_y\epsilon_y = -f_y$$
$$g_{xx}\delta + g_x\delta_x + g_{xy}\epsilon + g_y\epsilon_x = -g_x$$
$$g_{xy}\delta + g_x\delta_y + g_{yy}\epsilon + g_y\epsilon_y = -g_y \qquad (11.319)$$

We now set $(x, y) = (\alpha, t)$ and observe that $\delta(\alpha, t) = \epsilon(\alpha, t) = 0$. We thus obtain for $(x, y) = (\alpha, t)$

$$f_x\delta_x + f_y\epsilon_x = -f_x$$
$$g_x\delta_x + g_y\epsilon_x = -g_x$$
$$f_x\delta_y + f_y\epsilon_y = -f_y$$
$$g_x\delta_y + g_y\epsilon_y = -g_y \qquad (11.320)$$

The determinant of each of these two systems of linear equations is again the Jacobian determinant $D(\alpha, t)$ and hence is different from zero. We now find

Table 11.10 Newton–Raphson Iterations for Example

n	x_n	y_n
1	0.666666	0.583333
2	0.536240	0.5088490
3	0.5265620	0.5079319
4	0.5265226	0.5079197
5	0.5265226	0.5079197

$$\begin{aligned} \delta_x &= -1 \\ \delta_y &= 0 \\ \epsilon_x &= 0 \\ \epsilon_y &= -1 \end{aligned} \qquad (11.321)$$

According to equation (11.316), this implies that all elements of the matrix (J) of equation (11.315) are zero at the point (α, t) as desired.

Example

The results of the iterative solution by the Newton–Raphson method of the following equations are given in Table 11.10.

$$\begin{aligned} x - A\sin x - B\cos y &= 0 = f(x_n, y_n) \\ y - A\cos x + B\sin y &= 0 = g(x_n, y_n) \end{aligned} \qquad (11.322)$$

The much greater rapidity of convergence is evident from the example.

Solution of Equations

System of Equations for Which the Determinant of the Jacobian Is Zero

The Newton–Raphson method fails if the determinant D vanishes at or near a solution to the system. However, this gives us additional information about the problem as the vanishing of D indicates

- Multiple solutions, $f(x, y) = 0$ and $g(x, y) = 0$ are tangent to each other.
- Two or more solutions close together
- No solution in the neighborhood

Let us again consider the simple system of two equations in two unknowns, so that for the special case we have

$$D(x_i, y_i) = f_x(x_i, y_i) \cdot g_y(x_i, y_i) - f_y(x_i, y_i) g_x(x_i, y_i) = 0 \qquad (11.323)$$

Now the locus of points satisfying $D(x, y) = 0$ is the curve on which the loci of $f(x, y) = k_1$ and $g(x, y) = k_2$ have their common tangents or singular points. The procedure is to solve either of the two systems

$$\begin{aligned} D(x, y) &= 0 \\ f(x, y) &= 0 \end{aligned}$$

or

$$\begin{aligned} D(x, y) &= 0 \\ g(x, y) &= 0 \end{aligned} \qquad (11.324)$$

which can be done provided that one of the determinants

$$\begin{vmatrix} D_x & D_y \\ f_x & f_y \end{vmatrix} \quad \text{or} \quad \begin{vmatrix} D_x & D_y \\ g_x & g_y \end{vmatrix} \qquad (11.325)$$

does not vanish in the neighborhood of the point we are considering. Suppose we solve the first of these two systems and find the solution to be $x = a$ and $y = b$. We now calculate $g(a, b)$ and if this is zero, the curves are tangent at (a, b) and this point is said to be a double solution. If $g(a, b) = 0$, then either there is no solution or there are two solutions close together in the neighborhood of (a, b). Usually the case can be decided by graphing two functions, $f(x, y) = 0$ and $g(x, y) = 0$, and determining from the graph if the two functions intersect. Assuming that they do, the two functions can be expanded in a Taylor series about the point (x_0, y_0) and taking into account second-order terms we obtain

$$f(x_0, y_0) + f_x \Delta x + f_y \Delta y + \frac{1}{2} f_{xx}(\Delta x)^2 + f_{xy}\Delta x \Delta y + \frac{1}{2} f_{yy}(\Delta y)^2 = 0$$

$$g(x_0, y_0) + g_x \Delta x + g_y \Delta y + \frac{1}{2} g_{xx}(\Delta x)^2 + g_{xy}\Delta x \Delta y + \frac{1}{2} g_{yy}(\Delta y)^2 = 0 \quad (11.326)$$

where $\Delta x = x - x_0$ and $\Delta y = y - y_0$ and all partial derivatives are evaluated at (x_0, y_0). We then have a system of simultaneous quadratic equations, which we shall simplify somewhat. First we have $f(x_0, y_0) = 0$ and because $D = f_x g_y - f_y g_x = 0$ we obtain

$$\frac{g_y}{f_y} = \frac{g_x}{f_x} = k \quad (11.327)$$

or

$$g_x = k f_x$$

and

$$g_y = k f_y$$

Thus if we multiply the first equation of (11.326) by k and subtract it from the second and the linear terms are eliminated, we obtain

Solution of Equations

$$g + \frac{1}{2}(g_{xx} - kf_{xx})(\Delta x)^2 + (g_{xy} - kf_{xy})\Delta x \Delta y$$
$$+ \frac{1}{2}(g_{yy} - kf_{yy})(\Delta y)^2$$
$$= g + A(\Delta x)^2 + B\Delta x \Delta y + C(\Delta y)^2 = 0 \tag{11.328}$$

We shall now divide the first equation of (11.326) by f_x or f_y, whichever is larger. Let us suppose it is f_y. We obtain

$$\frac{f_x}{f_y}\Delta x + \Delta y + \frac{1}{2f_y}[f_{xx}(\Delta x)^2 + 2f_{xy}\Delta x \Delta y + f_{yy}(\Delta y)^2] = 0 \tag{11.329}$$

Divide this equation once more, this time by Δx, and solve for $\frac{\Delta y}{\Delta x} = m$ to obtain

$$\frac{\Delta y}{\Delta x} = m = -\frac{1}{f_y}[f_x + \frac{\Delta x}{2}(f_{xx} + 2f_{xy}m + f_{yy}m^2)] \tag{11.330}$$

Now consider equation (11.328) and write it in the form

$$g + (\Delta x)^2[A + B\frac{\Delta y}{\Delta x} + C(\frac{\Delta y}{\Delta x})^2] = 0 \tag{11.331}$$

and solve for Δx as

$$\Delta x = \pm\sqrt{\frac{-g}{A + Bm + Cm^2}} \tag{11.332}$$

The system of (11.330) and (11.331) may now be solved by the method of iteration. Let $m_0 = -\frac{f_x}{f_y}$. Using this value of m, solve for Δx using equation (11.332) and using m_0 and the new Δx, obtain a new m with equation (11.330). Eventually we have values for Δx and m from which we get approximations to the two solutions (x_1, y_1) and (x_2, y_2) by

$$\begin{aligned}\Delta y &= m\Delta x \\ x &= x_0 \pm \Delta x \\ y &= y_0 \pm \Delta y\end{aligned} \qquad (11.333)$$

From these values of x and y, D will not be zero and they may be improved by the usual method of successive approximation.

Aitken–Steffensen Method

Returning to the ordinary iteration methods of an arbitrary order, we may extend the Aitken–Steffensen formula to a system of nonlinear equations. If $\{x_n\}$ is a sequence of vectors generated by the algorithm

$$x_{n+1} = f(x_n) \qquad (11.334)$$

and α is the solution of $x = f(x)$, we can write

$$x_{n+1} - \alpha = J(x_n - \alpha), \text{ for } n = 0, 1, 2..., \qquad (11.335)$$

where $J = J(\alpha)$ is the Jacobian matrix of the function f taken at the solution α. The problem is to determine α from several consecutive iterates $x_n, x_{n-1}, x_{n-2}, \ldots$ although J is unknown.

Subtracting two consecutive equations (11.335) from each other, we find, using the symbol Δ to denote the forward differences,

$$\Delta x_{n+1} = J \Delta x_n, \text{ for } n = 0, 1, 2, \ldots \qquad (11.336)$$

We define X_n to be the matrix with the columns x_n and x_{n+1}, so that

$$X_n = (x_n, x_{n+1}) = \begin{pmatrix} x_n & x_{n+1} \\ y_n & y_{n+1} \end{pmatrix} \qquad (11.337)$$

Solution of Equations

Defining ΔX_n as the matrix of forward differences

$$\Delta X_n = \begin{pmatrix} (x_{n+1} - x_n) & (x_{n+2} - x_{n+1}) \\ (y_{n+1} - y_n) & (y_{n+2} - y_{n+1}) \end{pmatrix} \tag{11.338}$$

and from equation (11.336)

$$J \Delta X_n = \Delta X_{n+1}, \quad n = 0, 1, 2, \ldots \tag{11.339}$$

If the matrix is nonsingular, the solution for J will be $J = \Delta X_{n+1}(\Delta X_n)^{-1}$

Using equation (11.335), α is determined. If $(I - J)$ is nonsingular, we obtain

$$(I - J)\alpha = x_{n+1} - J x_n = (I - J)x_n + \Delta x_n \tag{11.340}$$

and therefore

$$\alpha = x_n + (I - J)^{-1} \Delta x_n \tag{11.341}$$

Substituting for J and after some algebraic manipulation, one obtains

$$(I - J)^{-1} = -\Delta X_n (\Delta^2 X_n)^{-1} \tag{11.342}$$

where

$$\Delta^2 X_n = (\Delta X_{n+1} - \Delta X_n). \tag{11.343}$$

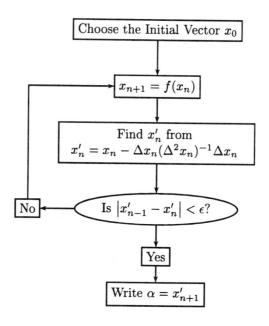

Figure 11.29 Flowchart Illustrating the Method

Finally, one obtains

$$\alpha = x_n - \Delta X_n (\Delta^2 X_n)^{-1} \Delta X_n \tag{11.344}$$

The iteration equation can be written as

$$x'_n = x_n - \Delta X_n (\Delta^2 X_n)^{-1} \Delta X_n \tag{11.345}$$

which yields a result closer to α than x_n, provided the matrix $\Delta^2 X_n$ is nonsingular. The algorithm is illustrated by the flowchart of Figure (11.29). The solution of the numerical problem

$$x = x^2 + y^2, \quad y = x^2 - y^2$$

is shown in Table 11.11.

Solution of Equations

Table 11.11 Iterations for Example

k	$x^{(k)}$	$y^{(k)}$	n	x_n	y_n
0	0.8	0.4	1	0.8	0.48
			2	0.80704	0.4096
			3	0.9253683	0.589824
1	0.7741243	0.419430	1	0.7751902	0.4233418
			2	0.7801424	0.4216974
			3	0.7864508	0.4367934
2	0.7718671	4196500	1	0.7718850	0.4196728
			2	0.7719317	0.4196813
			3	0.7720109	0.4197462
3	0.7718445	0.4196434	1	0.7718445	0.4196434

Bailey's Method for Higher Dimensional Problems

This is also called a modified Newton method, which takes into account second-order derivatives. That is, if the estimates x_0, y_0 are incremented respectively by δx, δy, then second-order approximations of the resulting changes in $f(x, y)$ and $g(x, y)$ are given by the expressions

$$\delta f = f_x \delta x + f_y \delta y + 1/2(f_{xx}\delta x^2 + 2f_{xy}\delta x \cdot \delta y + f_{yy}\delta y^2)$$

$$\delta g = g_x \delta x + g_y \delta y + 1/2(g_{xx}\delta x^2 + 2g_{xy}\delta x \cdot \delta y + g_{yy}\delta y^2) \quad (11.346)$$

A solution of equation (11.346) can be obtained by determining δx and δy such that δf and δg satisfy the constraints

$$\delta f = -f(x_0, y_0)$$
$$\delta g = -g(x_0, y_0) \quad (11.347)$$

Equating equations (11.346) and (11.347) and factoring the right-hand sides of the resulting expressions, we find that

$$-f(x_0, y_0) = \left[f_x + \frac{f_{xx}\delta x}{2} + \frac{f_{xy}\delta y}{2} \right] \delta x$$
$$+ \left[f_y + \frac{f_{yy}\delta y}{2} + \frac{f_{xy}\delta x}{2} \right] \delta y$$
$$-g(x_0, y_0) = \left[g_x + \frac{g_{xx}\delta x}{2} + \frac{g_{xy}\delta y}{2} \right] \delta x$$
$$+ \left[g_y + \frac{g_{yy}\delta y}{2} + \frac{g_{xy}\delta x}{2} \right] \delta y \qquad (11.348)$$

Now if the δx and δy inside the brackets are approximated respectively by $-f/f_x$ and $-f/f_y$ in the first equation and by $-g/g_x$ and $-g/g_y$ in the second equation, the system (11.348) can be written as the following linear equations in δx and δy:

$$-f(x_0, y_0) = \left[f_x - \frac{f_{xx}f}{2f_x} - \frac{f_{xy}f}{2f_y} \right] \delta x$$
$$+ \left[f_y - \frac{f_{yy}f}{2f_y} - \frac{f_{xy}f}{2f_x} \right] \delta y$$
$$-g(x_0, y_0) = \left[g_x - \frac{g_{xx}x}{2g_x} - \frac{g_{xy}g}{2g_y} \right] \delta x$$
$$+ \left[g_y - \frac{g_{yy}g}{2g_y} - \frac{g_{xy}g}{2g_x} \right] \delta y \qquad (11.349)$$

These equations can be directly solved for δx and δy and $(x_0 + \delta x, y_0 + \delta y)$ should be an improved estimate of the solution of $f(x, y) = 0$, $g(x, y) = 0$, depending upon the accuracy of the quadratic Taylor expansions and the approximations of δx and δy by the quotients used in (11.349). If $|\delta x| > \epsilon$ or if $|\delta y| > \epsilon$, the entire process is repeated, after replacing the values of x_0, y_0 by $x_0 + \delta x$ and $y_0 + \delta y$, respectively.

Solution of Equations

Table 11.12 Bailey's Method Example

Iter.	δx	δy	x_{n+1}	y_{n+1}	$f(x, y)$	$g(x, y)$
0	0	0	1	1	-2	1/9
1	0.7961	-0.125	1.7916	0.875	-0.0246	0.1222
2	0.0477	-0.0831	1.8393	0.7919	0.0101	0.0030

Example

The solution of the following set of equations by the Bailey method is shown in Table 11.12.

Solve

$$\begin{aligned} f(x, y) &= x^2 + y^2 - 4 = 0 \\ g(x, y) &= x^2/9 + y^2 - 1 = 0 \end{aligned} \quad (11.350)$$

The convergence of the method is evidently faster than that of the ordinary Newton method.

Greenspan's Modification of Newton's Method

Consider the system of equations

$$f_i(x_1, x_2, x_3, \ldots, x_n) = 0, \quad i = 1, 2, \ldots, n \quad (11.351)$$

Let

$$f_{ij} = \frac{\partial f_i}{\partial x_j} \quad (11.352)$$

and assume that $f_{ij} \neq 0$, $i = 1, 2, \ldots, m$. Then for m, a fixed nonzero constant, and for initial vector

$$x^{(0)} = (x_1^{(0)}, x_2^{(0)}, \ldots, x_m^{(0)}) \quad (11.353)$$

a sequence of vectors

$$x^{(k)} = (x_1^{(k)}, x_2^{(k)}, ..., x_m^{(k)}), \quad k = 1, 2, ... \tag{11.354}$$

is defined by the following iteration process:

$$\begin{aligned} x_1^{(k+1)} &= x_1^{(k)} - m \frac{f_1(x_1^{(k)}, x_2^{(k)}, ...x_m^{(k)})}{f_{11}(x_1^{(k)}, x_2^{(k)}, ...x_m^{(k)})} \\ &\cdots \\ x_m^{(k+1)} &= x_m^{(k)} - m \frac{f_m(x_1^{(k+1)}, x_2^{(k+1)}, ...x_m^{(k+1)})}{f_{mm}(x_1^{(k+1)}, x_2^{(k+1)}, ...x_m^{(k+1)})} \end{aligned} \tag{11.355}$$

for $k = 1, 2, ...$ Then $x^{(k)}$ is a solution of $f(x)$ iff $x^{(k+1)} = x^{(k)}$. If the system of equations (11.355) are linear, the method is identical to the linear overrelaxation method.

A
VECTOR OPERATORS

Rectangular Coordinates:

$$\nabla \psi = \frac{\partial \psi}{\partial x}\hat{a}_x + \frac{\partial \psi}{\partial y}\hat{a}_y + \frac{\partial \psi}{\partial z}\hat{a}_z$$

$$\nabla \cdot \vec{F} = \frac{\partial F_x}{\partial x} + \frac{\partial F_y}{\partial y} + \frac{\partial F_z}{\partial z}$$

$$\nabla \times \vec{F} = \left(\frac{\partial F_z}{\partial y} - \frac{\partial F_y}{\partial z}\right)\hat{a}_x + \left(\frac{\partial F_x}{\partial z} - \frac{\partial F_z}{\partial x}\right)\hat{a}_y + \left(\frac{\partial F_y}{\partial x} - \frac{\partial F_x}{\partial y}\right)\hat{a}_z$$

$$\nabla^2 \psi = \frac{\partial^2 \psi}{\partial x^2} + \frac{\partial^2 \psi}{\partial y^2} + \frac{\partial^2 \psi}{\partial z^2}$$

Appendix A

Cylindrical Coordinates:

$$\nabla \psi = \frac{\partial \psi}{\partial r}\hat{a}_r + \frac{1}{r}\frac{\partial \psi}{\partial \theta}\hat{a}_\theta + \frac{\partial \psi}{\partial z}\hat{a}_z$$

$$\nabla \cdot \vec{F} = \frac{1}{r}\frac{\partial}{\partial r}(rF_r) + \frac{1}{r}\frac{\partial F_\theta}{\partial \theta} + \frac{\partial F_z}{\partial z}$$

$$\nabla \times \vec{F} = \left(\frac{1}{r}\frac{\partial F_z}{\partial \theta} - \frac{\partial F_\theta}{\partial z}\right)\hat{a}_r + \left(\frac{\partial F_r}{\partial z} - \frac{\partial F_z}{\partial r}\right)\hat{a}_\theta + \left(\frac{1}{r}\frac{\partial (rF_\theta)}{\partial r} - \frac{1}{r}\frac{\partial F_r}{\partial \theta}\right)\hat{a}_z$$

$$\nabla^2 \psi = \frac{1}{r}\frac{\partial}{\partial r}\left(r\frac{\partial \psi}{\partial r}\right) + \frac{1}{r^2}\frac{\partial^2 \psi}{\partial \theta^2} + \frac{\partial^2 \phi}{\partial z^2}$$

Spherical Coordinates:

$$\nabla \psi = \frac{\partial \psi}{\partial r}\hat{a}_r + \frac{1}{r}\frac{\partial \psi}{\partial \theta}\hat{a}_\theta + \frac{1}{r\sin\theta}\frac{\partial \psi}{\partial \phi}\hat{a}_\phi$$

$$\nabla \cdot \vec{F} = \frac{1}{r^2}\frac{\partial}{\partial r}(r^2 F_r) + \frac{1}{r\sin\theta}\frac{\partial (\sin\theta F_\theta)}{\partial \theta} + \frac{1}{r\sin\theta}\frac{\partial F_\phi}{\partial \phi}$$

$$\nabla \times \vec{F} = \left(\frac{1}{r\sin\theta}\left(\frac{\partial (\sin\theta F_\phi)}{\partial \theta} - \frac{\partial F_\theta}{\partial \phi}\right)\right)\hat{a}_r + \left(\frac{1}{r\sin\phi}\frac{\partial F_r}{\partial \theta} - \frac{1}{r}\frac{\partial rF_\phi}{\partial r}\right)\hat{a}_\theta$$
$$+ \left(\frac{1}{r}\frac{\partial rF_\theta}{\partial r} - \frac{\partial F_r}{\partial \theta}\right)\hat{a}_\phi$$

$$\nabla^2 \psi = \frac{1}{r^2}\frac{\partial}{\partial r}\left(r^2 \frac{\partial \psi}{\partial r}\right) + \frac{1}{r^2 \sin\theta}\frac{\partial}{\partial \theta}\left(\sin\theta \frac{\partial \psi}{\partial \theta}\right) + \frac{1}{r^2 \sin^2\theta}\frac{\partial^2 \psi}{\partial \phi^2}$$

B

TRIANGLE AREA IN TERMS OF VERTEX COORDINATES

The area of a triangle can be found in terms of its vertex coordinates as follows: Referring to Figure B.1 we see that the triangle area is equal to the area of trapezoid ABEF plus the area of trapezoid AFGC minus the area of trapezoid BEGC. Thus

$$\Delta = \frac{(y_1 + y_2)(x_1 - x_2)}{2} + \frac{(y_1 + y_3)(x_3 - x_1)}{2} - \frac{(y_2 + y_3)(x_3 - x_2)}{2} \quad \text{(B.1)}$$

Performing the multiplications

$$\Delta = \frac{1}{2}(x_1 y_1 + x_1 y_2 - x_2 y_1 - x_2 y_2 + x_3 y_1 + x_3 y_3 - x_1 y_1 \\ - x_1 y_3 - x_3 y_2 - x_3 y_3 + x_2 y_2 + x_2 y_3) \quad \text{(B.2)}$$

or

$$\Delta = \frac{1}{2}[(x_2 y_3 - x_3 y_2) + x_1(y_2 - y_3) + y_1(x_3 - x_1)] \quad \text{(B.3)}$$

so that

$$2\Delta = (x_2 y_3 - x_3 y_2) + x_1(y_2 - y_3) + y_1(x_3 - x_1) \quad \text{(B.4)}$$

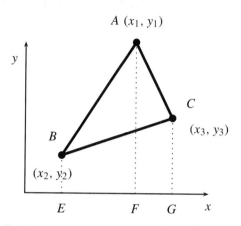

Figure B.1 Triangle Area

C

FOURIER TRANSFORM METHOD

Let us consider a simple L-R circuit and its transient response for a step function. The equation to be solved is

$$v = Ri + L\frac{di}{dt} \tag{C.1}$$

where $v = vu(t)$ and $u(t)$ is the unit step function.

Fourier Transform of (C.1) will yield

$$\hat{i} = \frac{\hat{v}F(u(t))}{R + j\omega L} \tag{C.2}$$

where F is the Fourier transform

$$F(u(t)) = \int_0^\infty u(t)e^{-j\omega t}\, dt \tag{C.3}$$

At $t = 0$, $F(u(t)) = \pi\delta(\omega)$....$\delta(0) = 1$, $\delta(\omega) = 0$ for $\omega > 0$. For $t > 0$,

$$F(u(t)) = \int_0^\infty e^{-j\omega t}\, dt = \frac{1}{j\omega} \tag{C.4}$$

Combining the values of $F(u(t))$ in (C.4) for $t = 0$ and $t > 0$, we have

$$F(u(t)) = \left(\pi\delta(\omega) + \frac{1}{j\omega}\right) \quad \text{(C.5)}$$

Substituting (C.5 into (C.2)

$$\hat{i} = \frac{\hat{v}\left(\pi\delta(\omega) + \frac{1}{j\omega}\right)}{R + j\omega L} \quad \text{(C.6)}$$

The first term on the right-hand side of equation (C.6) is the DC term, and its inverse Fourier transform is obtained as

$$i_{dc} = \frac{1}{2\pi}\int_0^\infty \frac{\hat{v}\pi\delta(\omega)e^{j\omega t}}{R + j\omega L}\,d\omega = \frac{v}{2R} \quad \text{(C.7)}$$

The inverse Fourier transform of the second term on the right-hand side of equation (C.6) can be written as

$$i(t) = \frac{1}{\pi}\int_0^\infty \frac{\hat{v}e^{j\omega t}}{R + j\omega L}\,d\omega \quad \text{(C.8)}$$

Note the $1/\pi$ instead of $1/2\pi$ in (C.8), since this contains sine or cosine terms. Substituting $e^{j\omega t} = \cos\omega t + j\sin\omega t$ into equation (C.8) and using partial fractions

$$i(t) = \frac{v}{\pi}\int_0^\infty \frac{1}{R}\left(\frac{1}{j\omega} - \frac{L}{R + j\omega L}\right)(\cos\omega t + j\sin\omega t)\,d\omega \quad \text{(C.9)}$$

Fourier Transform Method

Since, the time solution must be real, we shall consider only the real part of equation (C.9). Therefore,

$$i(t) = \frac{v}{\pi R} \int_0^\infty \left(\frac{\sin \omega t}{\omega} - \frac{\frac{R}{L} \cos \omega t}{\frac{R^2}{L^2} + \omega^2} - \frac{\omega \sin \omega t}{\frac{R^2}{L^2} + \omega^2} \right) d\omega \qquad (C.10)$$

The definite integrals in equation (C.10) are evaluated as follows:

$$\int_0^\infty \frac{\sin \omega t}{\omega} d\omega = \frac{\pi}{2} \quad \text{for } t > 0$$
$$= 0 \quad \text{for } t = 0$$
$$= -\frac{\pi}{2} \quad \text{for } t < 0 \qquad (C.11)$$

$$\int_0^\infty \frac{\frac{R}{L} \cos \omega t}{\frac{R^2}{L^2} + \omega^2} d\omega = \frac{\pi}{2} e^{-Rt/L} \qquad (C.12)$$

$$\int_0^\infty \frac{\frac{R}{L} \sin \omega t}{\frac{R^2}{L^2} + \omega^2} d\omega = \frac{\pi}{2} e^{-Rt/L} \qquad (C.13)$$

Using equations (C.11) through (C.13) in equation (C.10) for $t > 0$, we have

$$i(t) = \frac{v}{\pi R} \left(\frac{\pi}{2} - \frac{\pi}{2} e^{-Rt/L} - \frac{\pi}{2} e^{-Rt/L} \right)$$

or

$$i(t) = \frac{V}{R} \left(\frac{1}{2} - e^{-Rt/L} \right) \qquad (C.14)$$

Adding the DC term from equation (C.7) to (C.14)

$$i_T(t) - i(t) + i_{dc} = \frac{v}{R}\left(1 - e^{-Rt/L}\right) \tag{C.15}$$

C.1 COMPUTATION OF ELEMENT COEFFICIENT MATRICES AND FORCING FUNCTIONS

Referring to equation (10.127) of Chapter 10, the element coefficient matrix can be written as

$$P(\phi, W) = -\int_v (\sigma + j\omega\epsilon)\nabla W \cdot \nabla\phi \, rdrdz - \int_0^L (\sigma + j\omega\epsilon)\left(C_1 + \frac{C_2 C_3}{j\omega}\right)\frac{\partial W}{\partial z} \cdot \frac{\partial \phi}{\partial z} dz \tag{C.16}$$

where the first integral is over the volume element and the second integral is over the line element.

The weighting or shape functions of the volume and shell elements are chosen to be second order triangular elements and second order line elements respectively for the axisymmetric geometry of the problem being analyzed.

Volume Element Shape Functions:

$$\begin{aligned}\text{Corner Nodes} \quad & W_i = \xi_i(2\xi_i - 1), \text{ where } i \text{ is a triangle vertex} \\ \text{Midside Nodes} \quad & W - i = 4\xi_i \xi_j, \text{ where } i \text{ and } j \text{ are triangle vertices}\end{aligned} \tag{C.17}$$

Fourier Transform Method

Shell Element Shape Functions:

$$W_1 = \frac{\xi}{2}(\xi - 1)$$
$$W_2 = (1 - \xi^2)$$
$$W_3 = \frac{xi}{2}(\xi + 1) \quad \text{(C.18)}$$

where W_i, W_1, W_2 and W_3 are shape functions and ξ_i are non-dimensional area coordinates and ξ non-dimensional line coordinates.

Volume Element

Substituting the values for W from equation (C.17) into (C.16) and integrating the element coefficient matrix for the volume element, one obtains

$$P_v(\psi, W) = (\sigma + j\omega\epsilon)\sigma_{i=1}^3 r_i \sigma_{j=1}^3 \cot\theta_j Q_{ij} \quad \text{(C.19)}$$

where

$$\cot\theta_j = -\frac{1}{2A}(b_{j+1}b_{j-1} + c_{j+1}c_{j-1})$$

$$b_k = z_{k+1} - z_{k-1}$$

$$c_k = r_{k-1} - r_{k+1}$$

$$Q_{ij}\frac{1}{2}(G_{j+1}^2 - G_{j-1}^2)M_i^2(G_{j+1}^2 - G_{j-1}^2)$$

and r_i, z_i are the radial and axial coordinates of the element vertices, A is the triangle area and G_i^2 and M_i^2 are the universal matrices given in reference [117] and are listed below for completeness.

G and M Matrices:

$$G_1^2 = \begin{pmatrix} 3 & 0 & 0 & 0 & 0 & 0 \\ -1 & 4 & 0 & 0 & 0 & 0 \\ -1 & 0 & 4 & 0 & 0 & 0 \end{pmatrix}$$

$$G_2^2 = \begin{pmatrix} 0 & 4 & -1 & 0 & 0 & 0 \\ 0 & 0 & 0 & 3 & 0 & 0 \\ -1 & 0 & 4 & 0 & 0 & 0 \end{pmatrix}$$

$$G_3^2 = \begin{pmatrix} 0 & 0 & 4 & 0 & 0 & -1 \\ 0 & 0 & 0 & 0 & 4 & -1 \\ 0 & 0 & 0 & 0 & 0 & 3 \end{pmatrix}$$

$$M_1^2 = \frac{1}{60} \begin{pmatrix} 6 & 2 & 2 \\ 2 & 2 & 1 \\ 2 & 1 & 2 \end{pmatrix}$$

$$M_2^2 = \frac{1}{60} \begin{pmatrix} 2 & 2 & 1 \\ 2 & 6 & 2 \\ 1 & 2 & 2 \end{pmatrix}$$

$$M_3^2 = \frac{1}{60} \begin{pmatrix} 2 & 1 & 2 \\ 1 & 2 & 2 \\ 2 & 2 & 6 \end{pmatrix}$$

Line Element

The matrix for a second order Lagrange type line element is given by

$$P_{line}(\phi, W) = \frac{(\sigma + j\omega\epsilon)\left(C_1 + \frac{C_2 C_3}{j\omega}\right) a}{3L} \begin{pmatrix} 7 & -8 & 1 \\ -8 & 16 & -8 \\ 1 & -8 & 7 \end{pmatrix} \quad \text{(C.20)}$$

where L is the length of the element, a the inner radius of the tube in Figure 10.12, and the C_1, C_2 and C_3 constants are given in equation (10.122).

Fourier Transform Method

Forcing Function

The forcing function matrix using a second order Lagrangian type line element is obtained by substituting the values of weights W_1, W_2 and W_3 given in equation (C.18) into the second integral on the right-hand side of equation (C.16) and performing the appropriate integrations. Thus the element forcing function is given as

$$R(\phi, W) = -\frac{-(\sigma + j\omega\epsilon)C_2 C_4 a L}{30 j\omega} \begin{pmatrix} 4 & 2 & -1 \\ 2 & 16 & 2 \\ -1 & 2 & 4 \end{pmatrix} \begin{pmatrix} f_1 \\ f_2 \\ f_3 \end{pmatrix} \quad \text{(C.21)}$$

where the C_2 and C_4 constants are given in equation (10.122) and a is the inner radius of the tube.

D

INTEGRALS OF AREA COORDINATES

Table D.1 $I = \frac{1}{\Delta} \int_{area} \xi^\alpha \xi^\beta \xi^\gamma \, dxdy = \frac{P}{Q}$

$\alpha+\beta+\gamma$	α	β	γ	P	Q
0	0	0	0	1	1
1	1	0	0	1	3
2	2	0	0	2	12
2	1	1	0	1	12
3	3	0	0	6	60
3	2	1	0	2	60
3	1	1	1	1	60
4	4	0	0	12	180
4	3	1	0	3	180
4	2	2	0	2	180
4	2	1	1	1	180
5	5	0	0	60	1260
5	4	1	0	12	1260
5	3	2	0	6	1260
5	3	1	1	3	1260
5	2	2	1	2	1260
6	6	0	0	180	5040
6	5	1	0	30	5040
6	4	2	0	12	5040
6	4	1	1	6	5040
6	3	2	1	3	5040
6	2	2	2	1	5040

E

INTEGRALS OF VOLUME COORDINATES

Table E.1 $I = \frac{1}{V} \int_V \xi^\alpha \xi^\beta \xi^\gamma \xi^\theta \, dV = \frac{P}{Q}$

$\alpha + \beta + \gamma$	α	β	γ	θ	P	Q
0	0	0	0	0	1	1
1	1	0	0	0	1	4
2	2	0	0	0	2	20
2	1	1	0	0	1	20
3	3	0	0	0	6	120
3	2	1	0	0	2	120
3	1	1	1	0	1	120
4	4	0	0	0	24	840
4	3	1	0	0	6	840
4	2	2	0	0	4	840
4	2	1	1	0	2	840
4	1	1	1	1	1	840
5	5	0	0	0	60	3360
5	3	2	0	0	6	3360
5	3	1	1	0	3	3360
5	2	2	1	0	2	3360
5	2	1	1	1	2	3360
6	6	0	0	0	360	30240
6	5	1	0	0	60	30240
6	4	2	0	0	24	30240
6	4	1	1	0	12	30240
6	3	3	0	0	18	30240
6	3	2	1	0	6	30240
6	3	1	1	1	3	30240
6	2	2	2	0	4	30240
6	2	2	1	1	2	30240

F

GAUSS–LEGENDRE QUADRATURE FORMULAE, ABSCISSAE, AND WEIGHT COEFFICIENTS

Table F.1 $\int_{-1}^{1} f(x)\,dx = \sum_{j=1}^{n} F_i(a_i) W_i$

n	$\pm a$	W_i
2	0.57735027	1.00000000
3	0.77459667	0.55555555
	0.00000000	0.88888888
4	0.86113631	0.34785485
	0.33998104	0.65214515
5	0.90617985	0.23692689
	0.53846931	0.47862867
	0.00000000	0.56888889
6	0.93246951	0.17132449
	0.66120939	0.36076157
	0.23861919	0.46791393
7	0.94910791	0.12948497
	0.74153119	0.27970539
	0.40584515	0.38183005
	0.00000000	0.41795918
8	0.9602886	0.10122854
	0.79666648	0.22238103
	0.52553241	0.31370665
	0.18343464	0.36268378
9	0.96816024	0.08127439
	0.83603111	0.18064816
	0.61337143	0.26061070
	0.32425342	0.31234708
	0.00000000	0.33023936
10	0.97390653	0.06667134
	0.86506337	0.14945135
	0.67940957	0.21908636
	0.43339539	0.26926672
	0.14887434	0.29552423

G

SHAPE FUNCTIONS FOR 1D FINITE ELEMENTS

Shape Functions for a First-Order Line Element

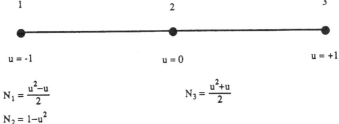

$$N_1 = \frac{1-u}{2} \qquad N_2 = \frac{1+u}{2}$$

Shape Functions for a Second-Order Line Element

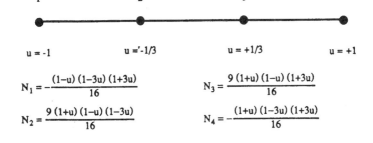

$$N_1 = \frac{u^2 - u}{2} \qquad N_3 = \frac{u^2 + u}{2}$$
$$N_2 = 1 - u^2$$

Shape Functions for a Third-Order Line Element

$$N_1 = -\frac{(1-u)(1-3u)(1+3u)}{16} \qquad N_3 = \frac{9(1+u)(1-u)(1+3u)}{16}$$
$$N_2 = \frac{9(1+u)(1-u)(1-3u)}{16} \qquad N_4 = -\frac{(1+u)(1-3u)(1+3u)}{16}$$

Figure G.1 Line Elements

H

SHAPE FUNCTIONS FOR 2D FINITE ELEMENTS

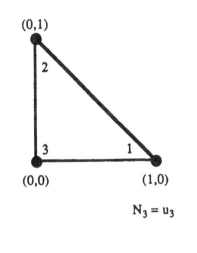

Figure H.1 First-Order Triangular Element

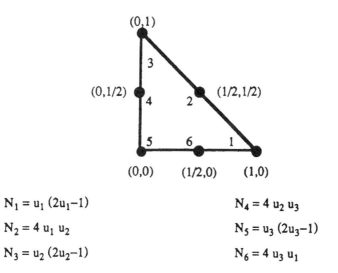

Figure H.2 Shape Functions for a Second-Order Triangular Element

Shape Functions for 2D Finite Elements

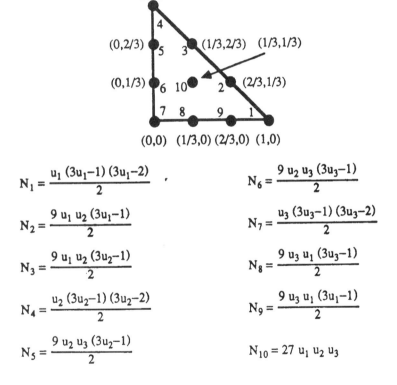

$$N_1 = \frac{u_1 (3u_1-1)(3u_1-2)}{2}$$

$$N_2 = \frac{9 u_1 u_2 (3u_1-1)}{2}$$

$$N_3 = \frac{9 u_1 u_2 (3u_2-1)}{2}$$

$$N_4 = \frac{u_2 (3u_2-1)(3u_2-2)}{2}$$

$$N_5 = \frac{9 u_2 u_3 (3u_2-1)}{2}$$

$$N_6 = \frac{9 u_2 u_3 (3u_3-1)}{2}$$

$$N_7 = \frac{u_3 (3u_3-1)(3u_3-2)}{2}$$

$$N_8 = \frac{9 u_3 u_1 (3u_3-1)}{2}$$

$$N_9 = \frac{9 u_3 u_1 (3u_1-1)}{2}$$

$$N_{10} = 27 u_1 u_2 u_3$$

Figure H.3 Shape Functions for a Third-Order Triangular Element

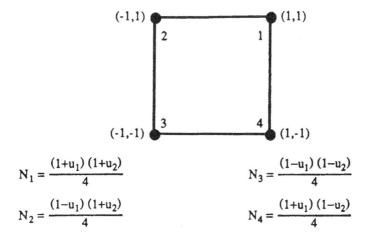

Figure H.4 Shape Functions for a First-Order Quadrilateral Element

Shape Functions for 2D Finite Elements

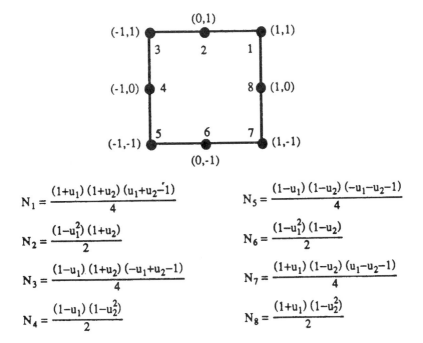

$$N_1 = \frac{(1+u_1)(1+u_2)(u_1+u_2-1)}{4}$$

$$N_2 = \frac{(1-u_1^2)(1+u_2)}{2}$$

$$N_3 = \frac{(1-u_1)(1+u_2)(-u_1+u_2-1)}{4}$$

$$N_4 = \frac{(1-u_1)(1-u_2^2)}{2}$$

$$N_5 = \frac{(1-u_1)(1-u_2)(-u_1-u_2-1)}{4}$$

$$N_6 = \frac{(1-u_1^2)(1-u_2)}{2}$$

$$N_7 = \frac{(1+u_1)(1-u_2)(u_1-u_2-1)}{4}$$

$$N_8 = \frac{(1+u_1)(1-u_2^2)}{2}$$

Figure H.5 Shape Functions for a Second-Order Serendipity Quadrilateral Element

3.2.6. Shape Functions for a Third-Order Quadrilateral Element (Serendipity Type)

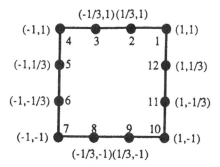

$$N_1 = \frac{(1+u_1)(1+u_2)(9u_1^2+9u_2^2-10)}{32}$$

$$N_2 = \frac{9(1-u_1^2)(1+u_2)(1+3u_1)}{32}$$

$$N_3 = \frac{9(1-u_1^2)(1+u_2)(1-3u_1)}{32}$$

$$N_4 = \frac{(1-u_1)(1+u_2)(9u_1^2+9u_2^2-10)}{32}$$

$$N_5 = \frac{9(1-u_1)(1-u_2^2)(1+3u_2)}{32}$$

$$N_6 = \frac{9(1-u_1)(1-u_2^2)(1-3u_2)}{32}$$

$$N_7 = \frac{(1-u_1)(1-u_2)(9u_1^2+9u_2^2-10)}{32}$$

$$N_8 = \frac{9(1-u_1^2)(1-u_2)(1-3u_1)}{32}$$

$$N_9 = \frac{9(1-u_1^2)(1-u_2)(1+3u_1)}{32}$$

$$N_{10} = \frac{(1+u_1)(1-u_2)(9u_1^2+9u_2^2-10)}{32}$$

$$N_{11} = \frac{9(1+u_1)(1-u_2^2)(1-3u_2)}{32}$$

$$N_{12} = \frac{9(1+u_1)(1-u_2^2)(1+3u_2)}{32}$$

Figure H.6 Shape Functions for a Third-Order Quadrilateral Serendipity Element

Shape Functions for 2D Finite Elements

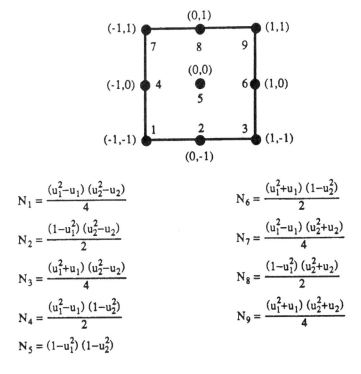

$$N_1 = \frac{(u_1^2-u_1)(u_2^2-u_2)}{4}$$

$$N_2 = \frac{(1-u_1^2)(u_2^2-u_2)}{2}$$

$$N_3 = \frac{(u_1^2+u_1)(u_2^2-u_2)}{4}$$

$$N_4 = \frac{(u_1^2-u_1)(1-u_2^2)}{2}$$

$$N_5 = (1-u_1^2)(1-u_2^2)$$

$$N_6 = \frac{(u_1^2+u_1)(1-u_2^2)}{2}$$

$$N_7 = \frac{(u_1^2-u_1)(u_2^2+u_2)}{4}$$

$$N_8 = \frac{(1-u_1^2)(u_2^2+u_2)}{2}$$

$$N_9 = \frac{(u_1^2+u_1)(u_2^2+u_2)}{4}$$

Figure H.7 Shape Functions for a Second-Order Lagrangian Quadrilateral Element

I

SHAPE FUNCTIONS FOR 3D FINITE ELEMENTS

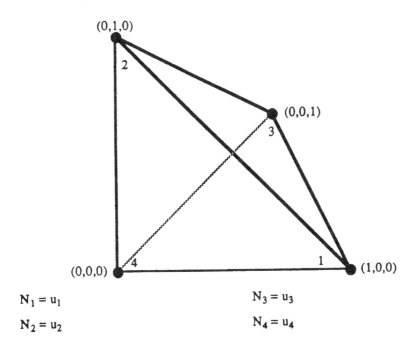

Figure I.1 Shape Functions for a First-Order Tetrahedral Element

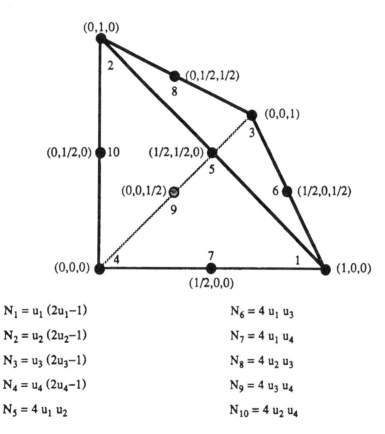

$N_1 = u_1 (2u_1-1)$ $\qquad N_6 = 4 u_1 u_3$

$N_2 = u_2 (2u_2-1)$ $\qquad N_7 = 4 u_1 u_4$

$N_3 = u_3 (2u_3-1)$ $\qquad N_8 = 4 u_2 u_3$

$N_4 = u_4 (2u_4-1)$ $\qquad N_9 = 4 u_3 u_4$

$N_5 = 4 u_1 u_2$ $\qquad N_{10} = 4 u_2 u_4$

Figure I.2 Shape Functions for a Second-Order Tetrahedral Element

Shape Functions for 3D Finite Elements

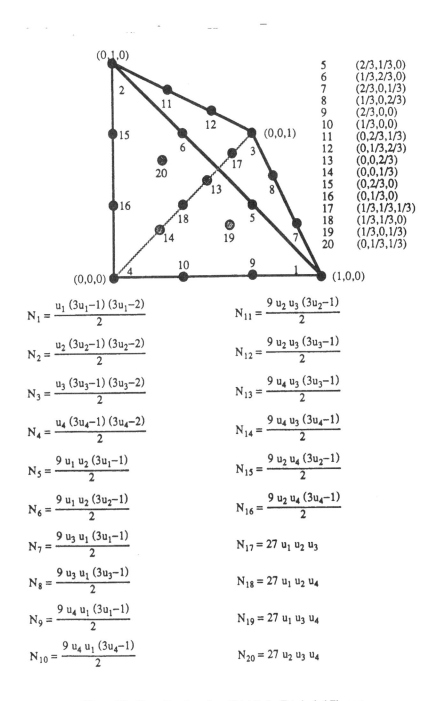

$$N_1 = \frac{u_1 (3u_1-1)(3u_1-2)}{2}$$

$$N_2 = \frac{u_2 (3u_2-1)(3u_2-2)}{2}$$

$$N_3 = \frac{u_3 (3u_3-1)(3u_3-2)}{2}$$

$$N_4 = \frac{u_4 (3u_4-1)(3u_4-2)}{2}$$

$$N_5 = \frac{9 u_1 u_2 (3u_1-1)}{2}$$

$$N_6 = \frac{9 u_1 u_2 (3u_2-1)}{2}$$

$$N_7 = \frac{9 u_3 u_1 (3u_1-1)}{2}$$

$$N_8 = \frac{9 u_3 u_1 (3u_3-1)}{2}$$

$$N_9 = \frac{9 u_4 u_1 (3u_1-1)}{2}$$

$$N_{10} = \frac{9 u_4 u_1 (3u_4-1)}{2}$$

$$N_{11} = \frac{9 u_2 u_3 (3u_2-1)}{2}$$

$$N_{12} = \frac{9 u_2 u_3 (3u_3-1)}{2}$$

$$N_{13} = \frac{9 u_4 u_3 (3u_3-1)}{2}$$

$$N_{14} = \frac{9 u_4 u_3 (3u_4-1)}{2}$$

$$N_{15} = \frac{9 u_2 u_4 (3u_2-1)}{2}$$

$$N_{16} = \frac{9 u_2 u_4 (3u_4-1)}{2}$$

$$N_{17} = 27 u_1 u_2 u_3$$

$$N_{18} = 27 u_1 u_2 u_4$$

$$N_{19} = 27 u_1 u_3 u_4$$

$$N_{20} = 27 u_2 u_3 u_4$$

Figure I.3 Shape Functions for a Third-Order Tetrahedral Element

Appendix I

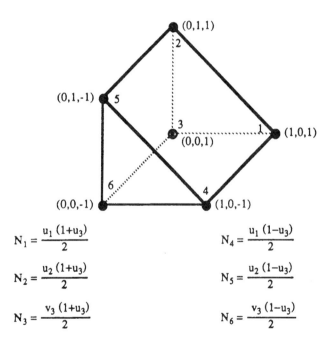

Figure I.4 Shape Functions for a First-Order Triangular Prism Element

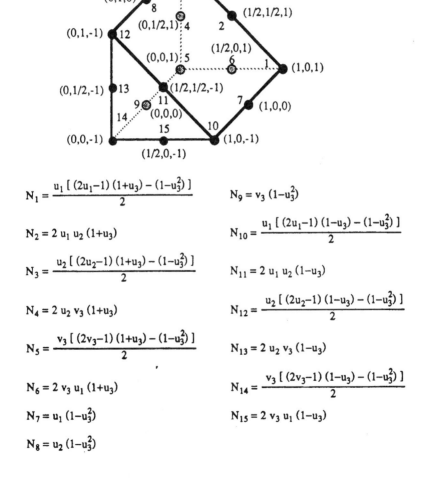

$$N_1 = \frac{u_1 \left[(2u_1-1)(1+u_3) - (1-u_3^2) \right]}{2}$$

$$N_9 = v_3 (1-u_3^2)$$

$$N_2 = 2 u_1 u_2 (1+u_3)$$

$$N_{10} = \frac{u_1 \left[(2u_1-1)(1-u_3) - (1-u_3^2) \right]}{2}$$

$$N_3 = \frac{u_2 \left[(2u_2-1)(1+u_3) - (1-u_3^2) \right]}{2}$$

$$N_{11} = 2 u_1 u_2 (1-u_3)$$

$$N_4 = 2 u_2 v_3 (1+u_3)$$

$$N_{12} = \frac{u_2 \left[(2u_2-1)(1-u_3) - (1-u_3^2) \right]}{2}$$

$$N_5 = \frac{v_3 \left[(2v_3-1)(1+u_3) - (1-u_3^2) \right]}{2}$$

$$N_{13} = 2 u_2 v_3 (1-u_3)$$

$$N_6 = 2 v_3 u_1 (1+u_3)$$

$$N_{14} = \frac{v_3 \left[(2v_3-1)(1-u_3) - (1-u_3^2) \right]}{2}$$

$$N_7 = u_1 (1-u_3^2)$$

$$N_{15} = 2 v_3 u_1 (1-u_3)$$

$$N_8 = u_2 (1-u_3^2)$$

Figure I.5 Shape Functions for a Second-Order Triangular Prism Serendipity Element

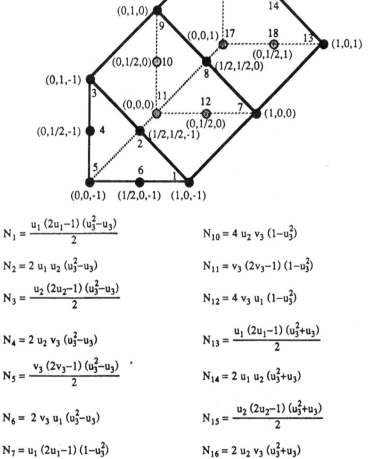

$$N_1 = \frac{u_1(2u_1-1)(u_3^2-u_3)}{2}$$

$$N_2 = 2 u_1 u_2 (u_3^2-u_3)$$

$$N_3 = \frac{u_2(2u_2-1)(u_3^2-u_3)}{2}$$

$$N_4 = 2 u_2 v_3 (u_3^2-u_3)$$

$$N_5 = \frac{v_3(2v_3-1)(u_3^2-u_3)}{2}$$

$$N_6 = 2 v_3 u_1 (u_3^2-u_3)$$

$$N_7 = u_1(2u_1-1)(1-u_3^2)$$

$$N_8 = 4 u_1 u_2 (1-u_3^2)$$

$$N_9 = u_2(2u_2-1)(1-u_3^2)$$

$$N_{10} = 4 u_2 v_3 (1-u_3^2)$$

$$N_{11} = v_3(2v_3-1)(1-u_3^2)$$

$$N_{12} = 4 v_3 u_1 (1-u_3^2)$$

$$N_{13} = \frac{u_1(2u_1-1)(u_3^2+u_3)}{2}$$

$$N_{14} = 2 u_1 u_2 (u_3^2+u_3)$$

$$N_{15} = \frac{u_2(2u_2-1)(u_3^2+u_3)}{2}$$

$$N_{16} = 2 u_2 v_3 (u_3^2+u_3)$$

$$N_{17} = \frac{v_3(2v_3-1)(u_3^2+u_3)}{2}$$

$$N_{18} = 2 v_3 u_1 (u_3^2+u_3)$$

Figure I.6 Shape Functions for a Second-Order Triangular Prism Lagrangian Element

Shape Functions for 3D Finite Elements

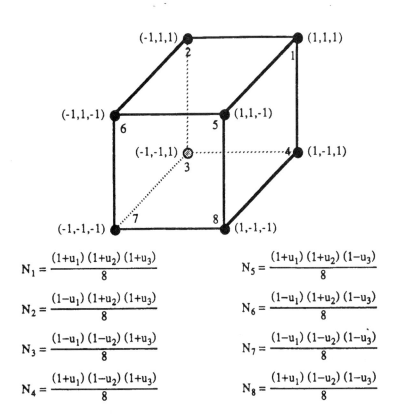

$$N_1 = \frac{(1+u_1)(1+u_2)(1+u_3)}{8} \qquad N_5 = \frac{(1+u_1)(1+u_2)(1-u_3)}{8}$$

$$N_2 = \frac{(1-u_1)(1+u_2)(1+u_3)}{8} \qquad N_6 = \frac{(1-u_1)(1+u_2)(1-u_3)}{8}$$

$$N_3 = \frac{(1-u_1)(1-u_2)(1+u_3)}{8} \qquad N_7 = \frac{(1-u_1)(1-u_2)(1-u_3)}{8}$$

$$N_4 = \frac{(1+u_1)(1-u_2)(1+u_3)}{8} \qquad N_8 = \frac{(1+u_1)(1-u_2)(1-u_3)}{8}$$

Figure I.7 Shape Functions for a First-Order Hexahedral Element

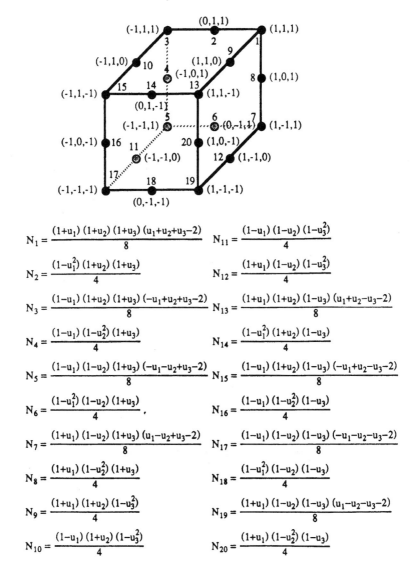

Figure I.8 Shape Functions for a Second-Order Hexahedral Serendipity Element

Shape Functions for 3D Finite Elements

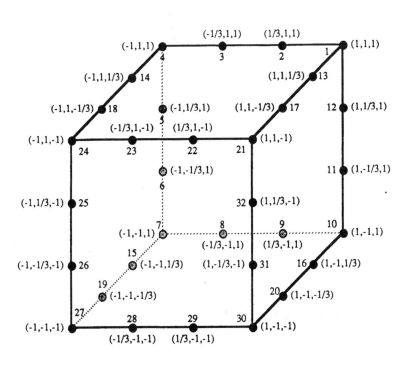

Figure I.9 Shape Functions for a Third-Order Hexahedral Serendipity Element

$$N_1 = \frac{(1+u_1)(1+u_2)(1+u_3)[9(u_1^2+u_2^2+u_3^2)-19]}{64}$$

$$N_{17} = \frac{9(1-u_3^2)(1+u_1)(1+u_2)(1-3u_3)}{64}$$

$$N_2 = \frac{9(1-u_1^2)(1+3u_1)(1+u_2)(1+u_3)}{64}$$

$$N_{18} = \frac{9(1-u_3^2)(1-u_1)(1+u_2)(1-3u_3)}{64}$$

$$N_3 = \frac{9(1-u_1^2)(1-3u_1)(1+u_2)(1+u_3)}{64}$$

$$N_{19} = \frac{9(1-u_3^2)(1-u_1)(1-u_2)(1-3u_3)}{64}$$

$$N_4 = \frac{(1-u_1)(1+u_2)(1+u_3)[9(u_1^2+u_2^2+u_3^2)-19]}{64}$$

$$N_{20} = \frac{9(1-u_3^2)(1+u_1)(1-u_2)(1-3u_3)}{64}$$

$$N_5 = \frac{9(1-u_2^2)(1-u_1)(1+3u_2)(1+u_3)}{64}$$

$$N_{21} = \frac{(1+u_1)(1+u_2)(1-u_3)[9(u_1^2+u_2^2+u_3^2)-19]}{64}$$

$$N_6 = \frac{9(1-u_2^2)(1-u_1)(1-3u_2)(1+u_3)}{64}$$

$$N_{22} = \frac{9(1-u_1^2)(1+3u_1)(1+u_2)(1-u_3)}{64}$$

$$N_7 = \frac{(1-u_1)(1-u_2)(1+u_3)[9(u_1^2+u_2^2+u_3^2)-19]}{64}$$

$$N_{23} = \frac{9(1-u_1^2)(1-3u_1)(1+u_2)(1-u_3)}{64}$$

$$N_8 = \frac{9(1-u_1^2)(1-3u_1)(1-u_2)(1+u_3)}{64}$$

$$N_{24} = \frac{(1-u_1)(1+u_2)(1-u_3)[9(u_1^2+u_2^2+u_3^2)-19]}{64}$$

$$N_9 = \frac{9(1-u_1^2)(1+3u_1)(1-u_2)(1+u_3)}{64}$$

$$N_{25} = \frac{9(1-u_2^2)(1-u_1)(1+3u_2)(1-u_3)}{64}$$

$$N_{10} = \frac{(1+u_1)(1-u_2)(1+u_3)[9(u_1^2+u_2^2+u_3^2)-19]}{64}$$

$$N_{26} = \frac{9(1-u_2^2)(1-u_1)(1-3u_2)(1-u_3)}{64}$$

$$N_{11} = \frac{9(1-u_2^2)(1+u_1)(1-3u_2)(1+u_3)}{64}$$

$$N_{27} = \frac{(1-u_1)(1-u_2)(1-u_3)[9(u_1^2+u_2^2+u_3^2)-19]}{64}$$

$$N_{12} = \frac{9(1-u_2^2)(1+u_1)(1+3u_2)(1+u_3)}{64}$$

$$N_{28} = \frac{9(1-u_1^2)(1-3u_1)(1-u_2)(1-u_3)}{64}$$

$$N_{13} = \frac{9(1-u_3^2)(1+u_1)(1+u_2)(1+3u_3)}{64}$$

$$N_{29} = \frac{9(1-u_1^2)(1+3u_1)(1-u_2)(1-u_3)}{64}$$

$$N_{14} = \frac{9(1-u_3^2)(1-u_1)(1+u_2)(1+3u_3)}{64}$$

$$N_{30} = \frac{(1+u_1)(1-u_2)(1-u_3)[9(u_1^2+u_2^2+u_3^2)-19]}{64}$$

$$N_{15} = \frac{9(1-u_3^2)(1-u_1)(1-u_2)(1+3u_3)}{64}$$

$$N_{31} = \frac{9(1-u_2^2)(1+u_1)(1-3u_2)(1-u_3)}{64}$$

$$N_{16} = \frac{9(1-u_3^2)(1+u_1)(1-u_2)(1+3u_3)}{64}$$

$$N_{32} = \frac{9(1-u_2^2)(1+u_1)(1+3u_2)(1-u_3)}{64}$$

Figure I.10 Shape Functions for a Third-Order Hexahedral Element (continued)

Shape Functions for 3D Finite Elements

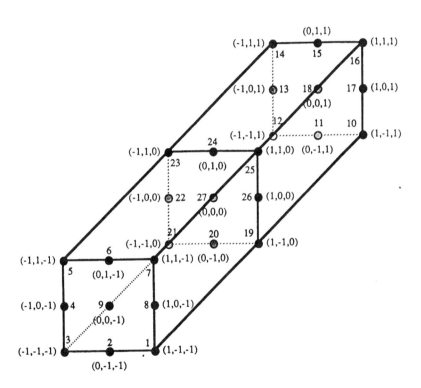

Figure I.11 Shape Functions for a Second-Order Hexahedral Lagrangian Element

$$N_1 = \frac{(u_1^2+u_1)(u_2^2-u_2)(u_3^2-u_3)}{8}$$

$$N_2 = \frac{(1-u_1^2)(u_2^2-u_2)(u_3^2-u_3)}{4}$$

$$N_3 = \frac{(u_1^2-u_1)(u_2^2-u_2)(u_3^2-u_3)}{8}$$

$$N_4 = \frac{(u_1^2-u_1)(1-u_2^2)(u_3^2-u_3)}{4}$$

$$N_5 = \frac{(u_1^2-u_1)(u_2^2+u_2)(u_3^2-u_3)}{8}$$

$$N_6 = \frac{(1-u_1^2)(u_2^2+u_2)(u_3^2-u_3)}{4}$$

$$N_7 = \frac{(u_1^2+u_1)(u_2^2+u_2)(u_3^2-u_3)}{8}$$

$$N_8 = \frac{(u_1^2+u_1)(1-u_2^2)(u_3^2-u_3)}{4}$$

$$N_9 = \frac{(1-u_1^2)(1-u_2^2)(u_3^2-u_3)}{2}$$

$$N_{10} = \frac{(u_1^2+u_1)(u_2^2-u_2)(u_3^2+u_3)}{8}$$

$$N_{11} = \frac{(1-u_1^2)(u_2^2-u_2)(u_3^2+u_3)}{4}$$

$$N_{12} = \frac{(u_1^2-u_1)(u_2^2-u_2)(u_3^2+u_3)}{8}$$

$$N_{13} = \frac{(u_1^2-u_1)(1-u_2^2)(u_3^2+u_3)}{4}$$

$$N_{14} = \frac{(u_1^2-u_1)(u_2^2+u_2)(u_3^2+u_3)}{8}$$

$$N_{15} = \frac{(1-u_1^2)(u_2^2+u_2)(u_3^2+u_3)}{4}$$

$$N_{16} = \frac{(u_1^2+u_1)(u_2^2+u_2)(u_3^2+u_3)}{8}$$

$$N_{17} = \frac{(u_1^2+u_1)(1-u_2^2)(u_3^2+u_3)}{4}$$

$$N_{18} = \frac{(1-u_1^2)(1-u_2^2)(u_3^2+u_3)}{2}$$

$$N_{19} = \frac{(u_1^2+u_1)(u_2^2-u_2)(1-u_3^2)}{4}$$

$$N_{20} = \frac{(1-u_1^2)(u_2^2-u_2)(1-u_3^2)}{2}$$

$$N_{21} = \frac{(u_1^2-u_1)(u_2^2-u_2)(1-u_3^2)}{4}$$

$$N_{22} = \frac{(u_1^2-u_1)(1-u_2^2)(1-u_3^2)}{2}$$

$$N_{23} = \frac{(u_1^2-u_1)(u_2^2+u_2)(1-u_3^2)}{4}$$

$$N_{24} = \frac{(1-u_1^2)(u_2^2+u_2)(1-u_3^2)}{2}$$

$$N_{25} = \frac{(u_1^2+u_1)(u_2^2+u_2)(1-u_3^2)}{4}$$

$$N_{26} = \frac{(u_1^2+u_1)(1-u_2^2)(1-u_3^2)}{2}$$

$$N_{27} = (1-u_1^2)(1-u_2^2)(1-u_3^2)$$

Figure I.12 Shape Functions for a Third-Order Hexahedral Element (continued)

Shape Functions for 3D Finite Elements

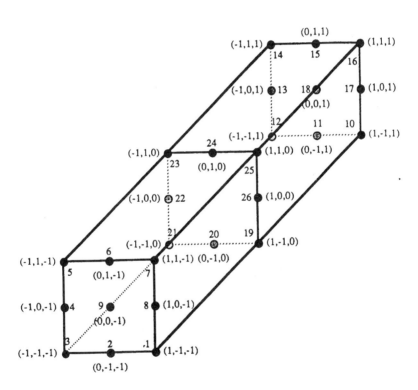

Figure I.13 Shape Functions for a Second-Order Hexahedral Serendipity Element

$$N_1 = \frac{(u_1^2+u_1)(u_2^2-u_2)(u_3^2-u_3) + (1-u_1^2)(1-u_2^2)(1-u_3^2)}{8}$$

$$N_2 = \frac{(u_1^2-1)(u_2-1)(u_2+u_3+1)(u_3-1)}{4} \qquad N_{17} = \frac{(u_1+1)(1-u_2^2)(u_1+u_3-1)(u_3+1)}{4}$$

$$N_3 = \frac{(u_1^2-u_1)(u_2^2-u_2)(u_3^2-u_3) + (1-u_1^2)(1-u_2^2)(1-u_3^2)}{8}$$

$$N_4 = \frac{(u_1-1)(u_2^2-1)(u_1+u_3+1)(u_3-1)}{4} \qquad N_{18} = \frac{(u_1^2-1)(u_2^2-1)(u_3+1)}{2}$$

$$N_5 = \frac{(u_1^2-u_1)(u_2^2+u_2)(u_3^2-u_3) + (1-u_1^2)(1-u_2^2)(1-u_3^2)}{8}$$

$$N_6 = \frac{(u_1^2-1)(u_2+1)(u_2-u_3-1)(u_3-1)}{4} \qquad N_{19} = \frac{(u_1+1)(u_2-1)(u_1-u_2-1)(u_3^2-1)}{4}$$

$$N_7 = \frac{(u_1^2+u_1)(u_2^2+u_2)(u_3^2-u_3) + (1-u_1^2)(1-u_2^2)(1-u_3^2)}{8}$$

$$N_8 = \frac{(u_1+1)(u_2^2-1)(u_1-u_3-1)(u_3-1)}{4} \qquad N_{20} = \frac{(u_1^2-1)(u_2-1)(1-u_3^2)}{2}$$

$$N_9 = \frac{(u_1^2-1)(u_2^2-1)(1-u_3)}{2} \qquad N_{21} = \frac{(u_1-1)(u_2-1)(u_1+u_2+1)(u_3^2-1)}{4}$$

$$N_{10} = \frac{(u_1^2+u_1)(u_2^2-u_2)(u_3^2+u_3) + (1-u_1^2)(1-u_2^2)(1-u_3^2)}{8}$$

$$N_{11} = \frac{(1-u_1^2)(u_2-1)(u_2-u_3+1)(u_3+1)}{4} \qquad N_{22} = \frac{(u_1-1)(u_2^2-1)(1-u_3^2)}{2}$$

$$N_{12} = \frac{(u_1^2-u_1)(u_2^2-u_2)(u_3^2+u_3) + (1-u_1^2)(1-u_2^2)(1-u_3^2)}{8}$$

$$N_{13} = \frac{(1-u_1)(u_2^2-1)(u_1-u_3+1)(u_3+1)}{4} \qquad N_{23} = \frac{(u_1-1)(u_2+1)(u_1-u_2+1)(1-u_3^2)}{4}$$

$$N_{14} = \frac{(u_1^2-u_1)(u_2^2+u_2)(u_3^2+u_3) + (1-u_1^2)(1-u_2^2)(1-u_3^2)}{8}$$

$$N_{15} = \frac{(1-u_1^2)(u_2+1)(u_2+u_3-1)(u_3+1)}{4} \qquad N_{24} = \frac{(u_1^2-1)(u_2+1)(u_3^2-1)}{2}$$

$$N_{16} = \frac{(u_1^2+u_1)(u_2^2+u_2)(u_3^2+u_3) + (1-u_1^2)(1-u_2^2)(1-u_3^2)}{8} \qquad N_{25} = \frac{(u_1+1)(u_2+1)(u_1+u_2-1)(1-u_3^2)}{4}$$

$$N_{26} = \frac{(u_1+1)(u_2^2-1)(u_3^2-1)}{2}$$

Figure I.14 Shape Functions for a Second-Order Hexahedral Serendipity Element (continued)

REFERENCES

[1] J. Van Bladel. *Electromagnetic Fields*. Hemisphere Publishing Corporation, New York, 1985.

[2] M. Abromowitz and I. Stegun. *Handbook of Mathematical Functions*. Dover Publications, New York, 1972.

[3] J. Jackson. *Classical Electrodynamics*. John Wiley and Sons, New York, 1962.

[4] I. D. Mayergoyz. *Mathematical Models of Hysteresis*. Springer-Verlag, New York, 1991.

[5] R. J. Parker and R. J. Studder. *Permanent Magnets and Their Applications*. John Wiley and Sons, NY and London, 1962.

[6] G. W. Carter. *The Electromagnetic Field in Its Engineering Aspect*. Longmans, Green and Co., London, 1954.

[7] O. Biro and K. Preis. Finite element analysis of 3d eddy currents. *IEEE Transactions on Magnetics*, Vol. 26, No. 2:418–423, 1990.

[8] P. M. Morse and H. Feshbach. *Methods of Theoretical Physics: Part I*. McGraw Hill Book Company, New York, 1953.

[9] L. F. Richardson. The free-hand graphic method of determining stream lines and equipotentials. *Phil. Mag. 15*, page 273, 1908.

[10] T. H. Lehmann. Graphical method for determining the tracing lines of force in air. *Lumiere Elec.*, 8, 1909.

[11] K. Kulmann. High tension insulators. *Arch. Electrotech.*, 3:203, 1915.

[12] A. D. Moore. Mapping magnetic and electrostatic fields. *Elect. Journal*, 23:355, 1926.

[13] A. R. Stevenson and R. H. Park. Graphical determination of magnetic fields. *Trans. American Institute of Electrical Engineers*, 46:112, 1927.

[14] B. Hague. *Electromagnetic Problems in Electrical Engineering*. Oxford University Press, London, 1929.

[15] V. L. Bewley. *Two Dimensional Fields in Electrical Engineering*. Macmillan Press, New York, 1948.

[16] K. J. Binns. Predetermination of the no-load magnetization characteristics of large turbogenerators. *Proceeding of the IEE*, Vol.112(No.4), April 1965.

[17] P. Moon and D. E. Spenser. *Field Theory for Engineers*. Van Nostrand, Princeton, N.J., 1960. Chapters 9 & 10.

[18] E. G. Wright and S. Deutscher. *Beama Journal*, 53, January 1946.

[19] G. Kirchoff. Uber den durchgang eines electrischen stromes durch eine ebene. *A. Physics*, 64:947, 1847.

[20] W. G. Adams. On the forms of equipotential curves and surfaces and lines of electric force. *Proc. Royal Society*, 23:280, 1875.

[21] A. E. Kennelly and S. E. Whiting. On the approximate measurement by electrolytic means of the electrostatic capacity between a vertical metallic cylinder and the ground. *Electrical World*, 48:1293, 1906.

[22] L. Dadda. Direct mapping of electric field configuration by means of the electrolytic tank. *Energia Elett*, 26:469, 1949.

[23] K. F. Sander and J.G. Yates. The accurate mapping of electric fields in an electrolytic tank. *Proc. Institute of Electrical Engineers*, 100 (II):p. 167, 1953.

[24] K. F. Raby. Flux plotting on teledeltos paper. *Bull. of Electrical Engineering Education*, Dec. 1958.

[25] K. F. Raby. *Conductive-Sheet Analogues, Chapter 5, Field Analysis: Experimental and Computational Methods, D. Victovitch edn.* Van Nostrand, London, 1966.

[26] G. Liebmann. Solution of partial differential equations with a resistance net analogue. *British Journal of Applied Phys.*, 1, 1950.

[27] E. C. Zachmanoglou and D. W. Thoe. *Introduction to Partial Differential Equations with Applications*. The Williams and Wilkins Company, Baltimore, 1976.

[28] W. F. Ames. *Nonlinear Partial Differential Equations in Engineering*. Academic Press, New York, 1965.

[29] L. Boltzmann. Boltzmann. *Ann.Physik*, Vol.53, 1894.

[30] O. Benedikt. *The Nomographic Computation of Complicated and Highly Saturated Magnetic Circuits*. Pergamon Press, London, 1964.

[31] C. F. Gauss. Brief an gerling. *Werke, Vol.9*, pages 278–281, December 1823.

References

[32] L. Seidel. Uber ein verfahren die gleichungen auf welche die methode der kleinsten quadrate furt, sowie lineare gleichungen uberhaupt durch successive annaherung aufzulasen. *Abhandlungen der Bayerischen Akademie*, Vol. 11:81–108, 1873.

[33] R. V. Southwell. Stress calculation in frameworks by the method of systematic relaxation of constraints. *Proc. Royal. Soc.*, Vol. 151 A:56, 1935.

[34] E. A. Erdelyi and S.V. Ahmed. Flux distribution in saturated d.c. machines. *IEEE Trans. PAS*, Vol. 84, No.5:375–381, May 1965.

[35] J. A. Stratton. *Electromagnetic Theory*. Mcgraw Hill Book Co., New York, 1941.

[36] F. B. Hildebrand. *Advanced Calculus for Applications*. Prentice Hall, Englewood Cliffs, N.J., 1962.

[37] H. C. Martic and G.Carey. *Introduction to Finite Element Analysis – Theory and Applications*. Mcgraw Hill Book Company, New York, 1973.

[38] J. Roberts. Analog treatment of eddy current problems involving two-dimensional fields. *IEE Monograph 341*, 1959.

[39] E. I. King. Equivalent circuits for the magnetic field: I - the static field. *IEEE Transactions on PAS*, Vol. PAS-85, No. 9:927–935, 1966.

[40] A. Thom and C. J. Apelt. *Field Computations in Engineering and Physics*. Van Nostrand, London, 1961.

[41] C. Hirsch. *Numerical Computation of Internal and External Flows*. John Wiley and Sons, Chichester, 1990.

[42] Charles Jordan. *Calculus of Finite Differences*. Chelsea Publishing Company, New York, 1950.

[43] A. Al-Khafaji and J. Tooley. *Numerical Methods in Engineering Practice*. Holt Reinhart and Winston, New York, 1986.

[44] E. I. King. Equivalent circuits for the amgnetic field: Ii - the sinusoidally time varying field. *IEEE Transactions on PAS*, Vol. PAS-85, No. 9:936–945, 1966.

[45] K. S. Yee. Numerical solution of initial boundary value problems involving maxwell's equations in isotropic media. *IEEE Transactions on Antennas and Propagation*, Vol. AP-14:302–307, 1966.

[46] M. N. O. Sadiku. *Numerical Techniques in Electromagnetics*. CRC Press, Boca Raton, Ann Arbor, London, Tokyo, 1992.

[47] G. A. Kriegsmann A. Taflove and K. R. Umashankar. A new formulation for electromagnetic scattering problems. *IEEE Transaction on Antennas and Propagation*, Vol. AP-35:153–161, 1987.

[48] P. P. Silvester and R. L. Ferrari. *Finite Elements for Elecrical Engineers*. Cambridge University Press, Cambridge, U.K., 1982.

[49] Kenneth Huebner. *The Finite Element Method for Engineers*. John Wiley and Sons, New York, 1975.

[50] P. Hammond. *Energy Methods in Electromagnetism*. Clarendon Press, Oxford England, 1981.

[51] C. A. J. Fletcher. *The Galerkin Meethod: An Introduction, Numerical Simulation of Fluid Motion; ed. J. Hoys*. North-Holland Co., 1978.

[52] G. Strang. *Introduction to Applied Mathematics*. Wellesley Cambridge Press, Wellesley, Massachusetts, USA, 1986.

[53] Richard H. Gallacher. *Finite Element Analysis Fundamentals*. Prentice Hall Inc., Engelwood Cliffs, New Jersey, 1975.

[54] C. A. Felippa and R.W. Clough. The finite element method in solid mechanics. *Numerical solution of field problems in Continuum Physics, SIAM-AMS Proceedings American Mathematics Society, Providence, R.I.,*, Vol. 2:210–252, 1970.

[55] E. F. A. Oliveira. Theoretical foundations of the finite element method. *International Journal of Solids and Structures*, Vol. 2:924–952, 1968.

[56] P. Dunne. Complete polynomial displacement fields for the finite element method. *Areonautical Journal*, Vol. 72, 1968.

[57] M. V. K. Chari and P.P. Silvester. *Finite Elements in Electric and Magnetic Fields Problems*. John Wiley and Sons, Chichester England, 1980.

[58] L. Fox. *An Introduction to Numerical Algebra*. Clarendon Press, Oxford, England, 1964.

[59] B. M. Irons. Economical computation techniques for numerically integrating finite elements. *International Journal of Numerical Methods in Engineering*, Vol. 1:20, 1969.

[60] P. Silvester. Tetrahedral polynomial finite elements for the helmholtz equation. *International Journal of Numerical Methods in Engineering*, Vol. 4, No. 2:405–413, 1972.

References

[61] Z. J. Cendes. *The High Order Polynomial Finite Element Method in Electromagnetic Computation — Finite Elements in Electric and Magnetic Fields Problems, ed. MVK Chari and PP Silvester*. John Wiley and Sons, Chichester England, 1980.

[62] P. Silvester. High-order polynomial triangular finite elements for potential problems. *International Journal of Engineering Science*, Vol. 1:849–861, 1969.

[63] J. H. Argyris K.E. Buck I. Fried G. Mareczek and D. W. Schrapf. Some new elements for the matrix displacement method. *Proceedings of the Second Conference on Matrix Methods in Structural Mechanics*, Wright-Patterson Air Force Base, 1968.

[64] J. G. Ergatoudis B. M. Irons and O. C. Zienkiewicz. Curved isoparametric quadrilateral elements for finite element analysis. *International Journal of Solids and Structures*, Vol. 4:31–42, 1968.

[65] Thomas J. R. Hughes. *The Finite Element Method*. Prentice Hall Inc., Englewood Cliffs, New Jersey, 1987.

[66] J. Ergatoudis S. Ahmed O. C. Zienkiewicz, B. M. Irons and F. C. Scott. *Finite Element Methods in Stress Analysis, ed. I. Holland and K. Bell, Isoparametric and Associated Element Families for Two- and Three-Dimensional Analysis*. Tapier Press, Trondheim, Norway, 1968.

[67] A. F. Armor and M. V. K. Chari. Heta flow in the stator core of large turbine generators by the method of three dimensional finite element analysis, part ii: Temperature distribution in the stator iron. *IEEE Transactions on PAS*, PAS - 95, No. 5:1657–1668, 1976.

[68] R. W. Clough and C. A. Filippa. A refined quadrilateral element for anlaysis of plastic bending. *Proceedings of the Second Conference on Matrix Methods in Structural Mechanics*, 1968.

[69] I. C. Taig and R. I. Kerr. *Some Problems in the Discrete Element Representation of Aircraft Structures - (AGARD-ograph 72, B. F. Venbeke (ed))*. Pergamon Press, New York, 1964.

[70] O. C. Zienkiewicz. Isoparametric and allied numerically integrated - a review. *Proceedings of SYmposium on Numerical and Computer Methods in Structural Mechanics*, 1971.

[71] O. C. Zienkiewicz. *The Finite Element Method in Engineering Science*. McGraw Hill, London, 1971.

[72] M. V. K. Chari and P. Silvester. Analysis of turbo-alternator magnetic fields by finite elements. *IEEE Trans.PAS*, PAS-90:454–464, 1971.

[73] A. Konrad and P. P. Silvester. A finite element program package for axisymmetric scalar filed problems. *Computer Physics Communications*, 5:437–455, 1975.

[74] S. J. Salon and J.M. Schneider. The use of a finite element formulation of the electric vector potential in the determination of eddy current losses. *IEEE Transactions on PAS*, PAS-99, No.1:16, 1980.

[75] S. Salon. *Finite Element Analysis of Electrical Machines*. Kluwer Academic Publisher, Boston, 1995.

[76] M. DeBortoli. Personal communication. ., 1995.

[77] J. Jin. *The Finite Element Method in Electromagnetics*. John Wiley and Sons, New York, 1993.

[78] I. G. Petrovsky. *Lectures on the Theory of Integral Equations*. MIR Publishers, Moscow, 1971.

[79] R. F. Harrington. *Field Computation by Moment Methods*. IEEE Press, Piscataway, New Jersey, 1993.

[80] K. Connor and S. Salon. *Supplementary Notes to Fields and Waves I*. Rensselaer Polytechnic Institute, Troy, New York, 1995.

[81] J. P. Peng and S. Salon. A hybrid finite element boundary element formulation of poisson's equation for axisymmteric vector potential problems. *Journal of Aplied Physics*, Vol. 53, No. 11:8420–8422, 1982.

[82] J. Schneider S. Salon and S. Uda. *Boundary Integral Solutions to the Eddy Current Problem; Recent Advances in Boundary Element Methods, ed. C. Brebbia*. Springer Verlag, 1981.

[83] Y. L. Luke. *Integrals of Bessel Functions*. McGraw-Hill, New York, 1962.

[84] S. Salon S. Peaiyoung and I. Mayergoyz. Some techical aspects of implementing boundary element equations. *IEEE Transactions on Magnetics*, Vol. 25, No. 4:2998–3000, 1989.

[85] P. P. Silvester D. A. Lowther C. J. Carpenter and E. A. Wyatt. Exterior finite elements for 2-dimensional field problems with open boundaries. *Proceedings of the IEE*, 124, No. 12:1267–1270, 1977.

[86] H. Hurwitz. Infinitesimal scaling - a new procedure for modeling exterior field problems. *IEEE Transactions on Magnetics*, MAG-20(5), 1984.

[87] B. H. McDonald and A. Wexler. Finite element solution of unbounded field problems. *IEEE Transactions on MTT*, 20, No. 12:841–847, 1972.

References

[88] S. J. Salon and J. D'Angelo. Applications of the hybrid finite element - boundary element method in electromagnetics. *IEEE Transactions on Magnetics*, MAG-24, No. 1:80–85, January 1988.

[89] P. Betess. Infinite elements. *International Journal for Numerical Methods*, Vol. 11:53–64, 1977.

[90] D. A. Lowther C. B. Rajanathan and P. Silvester. Finite element techniques for solving open boundary problems. *IEEE Transactions on Magnetics*, MAG-14, No. 5:467–469, 1978.

[91] O. W. Andersen. Laplacian electrostatic field calculations by finite elements with automatic grid generation. *IEEE Transsaction on PAS*, PAS-92(No. 5), 1973.

[92] John Schneider. *Hybrid Finite Element Boundary Element Solutions to Electromagneic Field Problems*. PhD thesis, Rensselaer Polytechnic Institute, Troy, New York, 1984.

[93] M. A. Jawson and G. T. Symm. Integral equation methods in potential theory - i. *Proceedings of the Royal Society of London*, page 23, 1963.

[94] M. A. Jawson and G. T. Symm. Integral equation methods in potential theory - ii. *Proceedings of the Royal Society of London*, page 33, 1963.

[95] G. Jeng and A. Wexler. Isoparametric finite element variational solution of integral equations for three-dimensional fields. *International Journal of Numerical Mehtods in Engineering*, 11:1455, 1977.

[96] J. Daffe and R. G. Olson. An integral equation method for solving rotationally symmetric electrostatic problems in conducting and dielectric material. *IEEE Transactions on PAS*, PAS-98:1609–1915, 1979.

[97] S. Kagami and I. Fukai. Application of boundary element method to electromagnetic field problems. *IEEE Transactions on MTT*, Vol. 32, No. 4:455–461, 1986.

[98] P. S. Shin. *Hybrid Finite-Boundary Element Analysis of Time Harmonic Electromagnetic Fields*. PhD thesis, Rensselaer Polytechnic Institute, Troy, New York, 1989.

[99] R. E. Collin ad F. J. Zucker. *Antenna Theory - Part I*. McGraw Hill Book Company, New York, 1969.

[100] B. Engquist and A. Majda. Absorbing boundary conditions for the numerical simulation of waves. *Math. of Comp.*, Vol. 31, No. 139:629–651, 1977.

[101] J. J. Bowman and T. B. Senior. *Electromagnetic and Acoustic Scattering by Simple Shapes*. Summa Books, 1987.

[102] Laurent Nicolas. Personal communication. ., 1999.

[103] J. D'Angelo. *Radio Frequency Scattering and Radiation by Finite Element Frequency Domain Methods.* PhD thesis, Rensselaer Polytechnic Institute, Troy, New York, 1994.

[104] Mark J. DeBortoli. *Extensions to the Finite Element Method for the Electromechanical Analysis of Electric Machines.* PhD thesis, Rensselaer Polytechnic Institute, Troy, New York, 1992.

[105] S. J. Salon, M. J. DeBortoli, and R. Palma. Coupling of transient fields, circuits, and motion using finite element analysis. *Journal of Electromagnetic Waves and Applications*, Vol. 4(No. 11), 1990.

[106] E. G. Strangas and K. R. Theis. Shaded pole motor design using coupled field and circuit equations. *IEEE Transactions on Magnetics*, MAG-21, No. 5:1880–1882, 1985.

[107] E. G. Strangas. Coupling the circuit equaion to the non-linear time dependent field solution in inverter driven induction motors. *IEEE Transactions on Magnetics*, MAG-21, No. 6:2408–2411, 1985.

[108] Rodolfo Palma. *Transient Analysis of Induction Machines Using the Finite Element Method.* PhD thesis, Rensselaer Polytechnic Institute, Troy, New York, 1989.

[109] Basim Istfan. *Extensions to the Finite Element Method for Nonlinear Magnetic Field Problems.* PhD thesis, Rensselaer Polytechnic Institute, Troy, New York, 1987.

[110] A. Klinkenberg and J. L. Van der Minne. *Electrostatics in the Petroleum Industry.* Elsevier Publishing Co., Amsterdam, 1958.

[111] J. T. Leonard. Generation of electrostatic charges in fuel handling systems: A literature survey. *NRL Report*, 8484, 1981.

[112] S. Shimzu H. Murata and M. Honda. Electrostatics in power transformers. *IEEE Transactions on PAS*, Vol. PAS-90:1244–1250, 1979.

[113] D. W. Crofts. The static electrification phenonena in power transformers. *Annual Report of IEEE Conference on Electrical Insulation and Dielectric Phenomena (CEIDP)*, pages 222–236, 1986.

[114] J. K. Nelson. Dielectric fluids in motion. *IEEE Electrical Insulation*, Vol. 10:16–28, 1994.

[115] A. D. Moore. *Electrostatics and Its Applications.* John Wiley and Sons, New York, 1973.

[116] J. R. Melcher S. M. Gasworth and M. Zahn. *Electrification Problems Resulting from Liquid Dielectric Flow*. EPRI Report El-4501, Palo Alto, Ca., 1986.

[117] A. Tahani. *Numerical Modeling of Electrification Phenomena and Its Implications for Dielectric Integrity*. PhD thesis, Rensselaer Polytechnic Institute, Troy, New York, 1994.

[118] M. V. K. Chari A. Tahani and S. Salon. Combined finite element and laplace transform method for transient solution of flow-induced fields. *COMPEL*, Vol. 15, No. 1:3–14, 1996.

[119] P. J. Lawrenson and T. J. E. Miller. *Transient Solution to the Diffusion Equation by Discrete Fourier Transformation, Finite Elements in Electrical and Magnetic Field Problems, ed. MVK Chari and PP Silvester, Chapter 8*. New York, 1980.

[120] M. A. Al-shaher. *Numerical Modeling of Static Electrification and its Application for Insulating Structures*. PhD thesis, Rensselaer Polytechnic Institute, Troy, New York, 1998.

[121] J. K. Nelson M. J. Lee and S. Salon. *Electrokinetic Effects in Power Transformers*. EPRI Report TR-101216, Palo Alto, Ca., 1992.

[122] S. Salon E. Plantive and M.V.K. Chari. Advances in axiperiodic magnetostatic analysis of generator end regions. *IEEE Transactions on Magnetics*, 1997.

[123] C. J. Carpenter. Theory and application of magnetic shells. *Proceedings of IEE*, 114(7):995–1000, 1967.

[124] E. Plantive. Personal communication. ., 1998.

[125] W. Tinney and J. Walker. Direct solution of sparse network equations by optimally ordered triangular factorization. *Proceedings of IEEE*, Vol. 55, No. 11, 1967.

[126] E. C. Osbuobiri. Dynamic storage and retrieval in sparsity programming. *IEEE Trans. PAS*, PAS 89:150–155, January 1970.

[127] A. George and J. W. Liu. *Computer Solution of Large Sparse Positive Definite Systems*. Prentice Hall Inc., Englewood Cliffs New Jersey, 1981.

[128] E. Cuthill and J. McKee. Reducing the bandwidth of sparse symmetric matrices. *Proceedings of the 24th National Conference for the Association for Computing Machinery*, ACM Publication P-69:157–172.

[129] P. K. Stockmeyer N.E. Gibbs, W.G. Poole. An algorithm for reducing bandwidth and profile of a sparse matrix. *Society for Industrial and Applied Mathematics Journal on Numerical Analysis*, Vol. 13, No. 2:236–250, .

[130] Hestenes and E. Stiefel. Method of conjugate gradients for solving linear systems. *Report No: 1659, National Bureau of Standards, USA,*, 1952.

[131] G. E. Forsythe. Solving linear algebraic equations can be interesting. *Bull. Amer. Math. Soc.*, 59:299 – 329, 1953.

[132] H. Rutishauser M. Engeli, Th. Ginsburg and E. Stiefel. Refined iterative methods for computation of the solution and the eigenvalues of self-adjoint boundary value problems. *Birkhauser, Basle*, 1959.

[133] T. A. Manteuffel. The shifted incomplete cholesky factorization. *Report No: SANDS 78 - 8226, Sandia Laboratories*, 1978.

[134] Y. Saad and H. Schultz. Gmres, a general minumizal residual algorithm for solving nonsymmetric linear systems. *SIAM. J. Sci. Stat. Comput.*, 7(3):856–869, 1986.

[135] M. Spasojević J. Petrangelo and P. Levin. Iterative solvers and preconditioners for large dense systems. *Worcester Polytechnic Institute*, 1994.

[136] E. Isaacson and H.B. Keller. *Analysis of Numerical Methods*. John Wiley and Sons, New York, 1966.

[137] A. Ostrowski. *Solution of Equations and Systems of Equations*. Academis Press, New York, 1960.

[138] P. Henrici. *Elements of Numerical Analysis*. John Wiley and Sons, New York, 1964.

[139] D. Greenspan. On approximating extremals of functionals. *Bulletin of the International Computation Centre*, Vol. 4:99–120, 1965.

INDEX

Absorbing boundary conditions, 484–487, 508–516
Acceleration equation. See Mechanical acceleration equation
Aitken-Steffenson method, 683–692, 700–703
Alternating relaxation, 82
Amphere's law, 26, 32, 33, 34, 39, 40–43, 71, 72
Area coordinates, integrals of, 719
Axiperiodic analysis
 defining equations, 570–576
 finite element formulation, 579–582
 magnetodynamic T–Ω formulation, 583–589
 MPV formulation to find H0, 576–579
Axisymmetric formulation for eddy current problem using vector potential, 313–320
Axisymmetric integral equations for magnetic vector potential, 389–393
Backward difference formula, 106
Bailey's method, 685–687, 703–705
Ballooning, 427–433
Band storage, 614–621
Bars, 522–523
Bessel functions, 73, 278, 395
Biot-Savart law, 48, 92
Boundary conditions
 absorbing, 484–487
 Dirichlet, 111, 332
 natural, 324
 Neumann, 111–114, 161
 treatment of irregular, 114–115
 wave equations and, 139–141
Boundary element equations for Poisson's equation in two dimensions, 374–381

Boundary element method (BEM), 413
 advantages of, 460
 disadvantages of, 460
 of scalar Poisson equation in three dimensions, 403–408
 transverse electric (TE) polarization, 466–467
 of two-dimensional potential problem, 381–389
 wave applications, 459–467
Boundary integral equations, 440–442
Cauchy-Riemann conditions, 65–66
Central difference formula, 106, 107
Chain rule of differentiation, 248, 303
Charge simulation method (CSM), 370–374
Cholesky decompositions, 545, 605–607
 incomplete, 641–645
Chord method, 302, 667–669
Circuit equations, 522
 time discretization of, 534
Closed form analytical methods, 71–73
Coils, 522–524
Collocation method, 183, 191–195
Computational methods
 closed form analytical methods, 71–73
 conformal mapping, 68
 discrete analytical methods, 74–75
 electroconducting analog, 69–70
 experimental methods, 68–69
 finite difference method, 80–83
 finite element method, 96–104
 graphical methods, 65–68
 historical background, 63–65
 integral equation method, 84–96
 nonlinear magnetic circuit analysis, 79
 resistive analog, 70–71

transformation methods for nonlinear problems, 75–79
Conductors, 11
Conformal mapping, 68
Conjugate gradient method, 627–645
  convergence, 637–645
Connectivity, 49
Conservative fields, 4
Constitutive equation, 25, 40, 43, 44
Continuity equation, 20, 23
Convergence, 666–667
Coulomb, Charles Augustin de, 1
Coulomb gauge, 55
Coulombs, 1
Coulomb's law, 2
Crank-Nicholson method, 131, 132, 531
Curl-curl operator, 51–54, 91, 354, 499
Current density, 522
Current equation, total, 521–522, 526
  formulation of, 529–531
  time discretization of, 532–533
Cuthhill-McKee algorithm and reverse, 618–621
Depth of penetration, 42
Dielectrics, 11, 12–13
Difference equations, 106–108
Differential equations, classification of, 60–61
Differential methods, 359
Diffusion equation, 39–42
  boundary conditions, 139–141
  energy-related functional for, 166–168
  explicit scheme, 126–127
  FDTD for wave equation, 135–141
  functional for steady-state linear time-harmonic, 168–170
  Green function for one-dimensional, 409
  implicit scheme, 131–135
  optimum time step, 130
  time-dependent magnetic, 520–521
  time discretization of, 532
Dipole, potential of, 6–7
Directive gain, 479
Direct methods, 595–598
Dirichlet condition, 111, 332
Discrete analytical methods, 74–75

Discretization, transient analysis and, 566–567
Discretization of time. See Time discretization
Divergence theorem, 17, 47, 48, 54–57, 167
Eddy currents, 39–42
  axisymmetric formulation for, using vector potential, 313–320
  integral equations and two-dimensional, 393–402
Edge elements, finite element method and, 353
  first-order rectangular, 353–356
  triangular, 356–357
Electric fields
  interface conditions, 13–15
  materials and, 11–13
  sinusoidally time-varying, 21–22
  static, 1–2
  three dimensional FEM formulation for, 494–516
  transient, 23–24
Electric flux density, 2
Electric potential, 3–11
  of a dipole, 6–7
  due to a ring of charge, 8–9
  due to point charge, 4–5
  energy density and, 9–11
  logarithmic potential, 5
Electric scalar potential, 10
Electric vector potential, formulation, 59–60
Electrification, flow-induced
  auxiliary variable introduced, 556–557
  computation of charge, potential, and time scale, 558–559
  electric potential calculation, 553–555
  finite element formulation, 561–569
  Laplace transform formulation, 555–556
  numerical versus analytical results, 559–560
  physical model described, 551–553
  solution, 555
Electroconducting analog, 69–70
Electrolytic tank experiment, 68–69
Electromotive force (EMF), 34, 36, 39
Electrostatic integral formulation, 87–90

Electrostatics, defined, 1
Elliptic differential equation, 61
Energy density, potential energy and, 9–11
Energy-related functional
  derivation of, 157–170
  for diffusion equation, 166–168
  nonlinear, for magnetostatic field problem, 164–166
Engquist-Majda absorbing boundary condition, 484, 508
Equivalent circuits, finite difference method and, 115–117, 133–135
Euler-Lagrange equation, 144, 145, 148–149, 150–151, 159
Experimental methods, 68–69
Faraday's law, 33–34, 39–40, 43
Far-field evaluation, 471–480
Field equation, 526
  formulation of, 527–529
  linearization of, 537–539
Field plotting techniques, 63
Fill-ins, 607
Finite difference method, 80–83
  difference equations, 106–108
  diffusion equation, 126–141
  equivalent circuit representation, 115–117, 133–135
  explicit scheme, 126–127
  first-order finite elements and, 320–322
  formulas for high-order schemes, 117–122
  implicit scheme, 131–135
  interfaces between materials, 109–111
  irregular boundaries, treatment of, 114–115
  Neumann boundary conditions, 111–114
  Poisson's equation, 108–109
  symbolic operators and, 122–125
Finite difference time domain (FDTD), formulas for wave equation, 135–141
Finite element equations, 437–439
Finite element method (FEM), 96
  advantages of, 97, 283
  axiperiodic analysis and, 579–582
  axisymmetric formulation for eddy current problem using vector potential, 313–320
  chord method, 302
  discretization and, 566–567
  discretization of time by, 310–312
  edge elements, 353–357
  electrical conduction problem example, 101–104
  finite difference and first-order finite elements, 320–322
  functional minimization and global assembly, 295–301
  Galerkin, 322–326
  hybrid harmonic, 413–417
  interpolatory functions and, 196–201
  Newton-Raphson method, 303–309
  numerical example of matrix formation for isoparametric elements, 342–352
  one-dimensional electrostatic problem and, 284–289
  parallel plate capacitor example, 97–101
  permanent magnets, 338–342
  procedures for implementing, 96–97, 283–284
  three dimensional formulation for electric field, 494–516
  three-element magnetostatic problem, 326–338
  two-dimensional analysis, 289–295
Finite elements for wave equations, high-frequency problems with
  advantages of, 451
  boundary element formulation, 459–467
  disadvantages of, 451
  far-field evaluation, 471–480
  finite element formulation, 452–459
  geometrical theory of diffraction, 472–474
  hybrid element method (HEM) formulation, 467–470, 475–480
  numerical examples, 488–494
  scattering problems, 481–487
  three dimensional FEM formulation for electric field, 494–516
  transverse electric and magnetic polarizations, 452–459
  Wiener-Hopf method, 471
Finite element shapes. See Three-dimensional finite elements

First-order finite elements, finite difference and, 320–322
First order methods, 659–667
First-order rectangular elements, 353–356
Five-point star method, 80
Flux plotting, 65
Forward difference formula, 106
Fourier integral transform method description of, 711–717
  transient analysis by, 562–564, 567–570
Fredholm integral equations, 84–85, 360, 361, 362
Free space, 488
Functional iteration, 663–666
Functional minimization and global assembly, 295–301
Functionals
  description of, 145–152
  energy-related, derivation of, 157–170
  energy-related, for diffusion equation, 166–168
  in more than one space variable and its extremum, 153–157
  nonlinear energy-related, for magnetostatic field problem, 164–166
  Poisson's equation and, 160–161
  Poynting vector method, 161–163
  residual excitation method, 157–159
  Ritz's method, 170–175
  for steady-state linear time-harmonic diffusion equation, 168–170
  variational method, 144–145, 151–152
  wave equation, 176–179
Galerkin finite elements, 322–326
Galerkin formulation, 527–531
Galerkin weighted residual method, 96
  description of, 182–183
  example of, 183–187
  finite element formulation and, 452, 453–459, 484–487
  trial functions, 182 Gauss elimination, 595–598 Gauss' law, 1, 2
  differential form of, 16, 24
  for magnetic fields, 24–25
Gauss-Legendre quadrature formulae, 723

Gauss points, 263–265
  one-point Gaussian quadrature formula, 349–350, 352
  ten-point Gaussian quadrature formula, 456
Generalized method of minimum residuals (GMRES), 645–657
Geometrical theory of diffraction (GTD), 472–474
Geometric isotropy, 205
Givens rotation, 653–657
Global assembly, functional minimization and, 295–301
Global support, 359
Global system of equations
  assembly of, 544
  common solution algorithms, 545–547
  summary of, 542–543
Graphical methods, 65–68
Greenspan's modification, 705–706
Green's theorem/identity/function, 88, 152, 167, 179–181
  half-space, 445–449
  one-dimensional case, 409
  three-dimensional case, 410–411
  two-dimensional case, 410
Hankel function, 410, 463, 465
Hard magnetic material, 27
Helmholtz's (wave) equation. See Wave (Helmholtz's) equation
Hexahedral elements, 251–253
High-frequency problems. See Finite elements for wave equations, high-frequency problems with
High-order schemes, formulas for, 117–122
High-order triangular interpolation functions
  Irons' method, 222–223
  Silvester's method, 224–227
Hodograph transformation, 76, 78–79
Homogeneous Neumann condition, 161
Hybrid element method (HEM) formulation, 467–470, 475–484
Hybrid finite element-boundary element method and, 437–449
Hybrid harmonic finite element method, 413–417

Hyperbolic differential equation, 61
Hysteresis characteristics, intrinsic and normal, 30
Hysteresis loops, 27
Incomplete Cholesky conjugate gradient (ICCG), 545
Infinite element method, 417–426
Infinitesimal scaling, 433–436
Integral equations
  axisymmetric, for magnetic vector potential, 389–393
  basic, 359–362
  BEM solution of scalar Poisson equation in three dimensions, 403–408
  BEM solution of two-dimensional potential problem, 381–389
  boundary, 440–442
  boundary conditions, 86–87
  boundary element equations for Poisson's equation in two dimensions, 374–381
  charge simulation method, 370–374
  eddy currents and, 393–402
  electrostatic integral formulation, 87–90
  example of electric field of a long conductor, 85–86
  Fredholm, 84–85, 360, 361, 362
  Green's functions, 409–411
  magnetostatic integral formulation, 90–96
  method of moments, 183, 362–370
  variational method for, 179–181
Integrals
  of area coordinates, 719
  of volume coordinates, 721
Interface conditions
  for electric fields, 13–15
  finite difference method and, 109–111
  for magnetic fields, 32
Interpolation
  high order triangular, 222–227
  inverse linear, 661–663
  Lagrangian, 211–214
  polynomial, 201–207
  of rectangular elements, 227–234
  of serendipity elements, 234–240
  of three-dimensional finite elements, 241–277
  trigonometric, 279–281
Interpolatory functions, 190–201
Intrinsic and normal hysteresis characteristics, 30
Intrinsic coercive force, 30
Inverse linear interpolation, 661–663
Inverse square law, 2
Isoparametric elements, 254–265
  numerical example of matrix formation for, 342–352
  triangular, 271–272
Jacobians/Jacobian matrix of derivatives, 261–262, 266, 271, 272, 303, 504–507, 513–515
  local to global, 330–332, 336–337
  nonlinear, 336
  is zero, 697–700
Kepler's equation, 659
Kernel function, 84, 361
Kirchhoff transformation, 76, 77, 658
Lagrangian interpolation, 211–214
Laplace's equation, 15–24, 88, 108–111
  Green function for one-dimensional, 409
  Green function for three-dimensional, 410–411
  Green function for two-dimensional, 410
  high-order scheme for, 117
  orthogonal basis functions and, 278–279
Laplace transform formulation, 555–556
Laplacian of a potential, 11
LDL^T decomposition, 602–604
LDU' decomposition, 602
Least squares method, 183, 192
Legendre functions, 278, 723
l'Hôpital's rule, 274
Linearization
  of field equation, 537–539
  of mechanical acceleration equation, 539–541
Linked list structure, 608–609
Local support, 359
Logarithmic potential, 5
LU decomposition, 598–604

Magnetic fields
  energy in, 36–38
  interface conditions, 32
  intrinsic and normal hysteresis characteristics, 30
  materials for, 26–31
  static, 24–36
Magnetic scalar potential, 25
  mixed formulation, 48–51
  second-order elements, use of, 51
  total and reduced, 47–51 Magnetic vector potential (MVP), 33–36
  $\vec{A}^*$ formulation, 58
  A–V formulation, 51–57
  axisymmetric integral equations for, 389–393
  curl-curl equation, 51–54
  diffusion equation for, 39–42
  formulation to find H0, 576–579
Magnetization curve, 27
Magnetodynamic T–Ω formulation, 583–589
Magnetomotive force (MMF), 75, 79
Magnetostatic integral formulation, 90–96
Mapping ratio, 427
Matrix operations and notation, 591–594
Maxwell's equations, 1, 4, 176
  proof of the uniqueness of fields, 45–47
    Mechanical acceleration equation, 526
  linearization of, 539–541
  time discretization of, 536
Mechanical velocity equation, 526
  time discretization of, 536–537
Method of moments (MOM), 183, 362–370
Method of weighted residuals (MWRs) See also Galerkin weighted residual method
  collocation method, 183, 191–195
  least squares method, 183, 192
  method of moments, 183
  subdomain method, 183
Models, use of prototype or scaled, 63
Monopoles, 24
Motion equations, 525
  time discretization of, 536–537 Muller's method, 681–683
Nested dissection reordering, 622–627

Neumann boundary conditions, 111–114, 161
Newton-Raphson method, 303–309, 538, 539, 541, 671–681, 692–706
Node numbering, 617–618
Nonlinear energy-related functional for magnetostatic field problem, 164–166
Nonlinear equations, examples of, 658–659
Nonlinear equations, solution of, 658–706
  Aitken-Steffenson method, 683–692, 700–703
  Bailey's method, 685–687, 703–705
  Chord method/simple iteration, 302, 667–669
  convergence, 666–667
  first order methods, 659–667
  functional iteration, 663–666
  Greenspan's modification, 705–706
  inverse linear interpolation, 661–663
  Muller's method, 681–683
  Newton-Raphson method, 303–309, 538, 539, 541, 671–681, 692–706
  regula falsi, 681
  second order methods, 669–671
  successive bisection, 660–661
  Whittaker's method, 681
Nonlinear magnetic circuit analysis, 79
Nonlinear problems, transformation methods for, 75–79
One-dimensional electrostatic problem, 284–289
One-dimensional Poisson equation, 364–366
Open boundary problems
  ballooning and, 427–433
  hybrid finite element-boundary element method and, 437–449
  hybrid harmonic finite element method and, 413–417
  infinite element method and, 417–426
  infinitesimal scaling and, 433–436
Optimal ordering, 610–614
Orthogonal basis functions, 278–281
Parabolic differential equation, 61
Parallel coil/circuit equation, 522–523, 526
  time discretization of, 535
Parallel connection of coils, 523–524

Partial differential equations (PDEs), 87, 96
    See also Functionals; Variational method
Pascal arraying procedure, 241, 245
Penalty function, 499–503
Permanent magnets, 338–342
Permeability
    recoil, 30
    relative, 27
Permittivity of the material, 13
Pivoting, 597
Point charge, potential due to, 4–5
Point collocation method, 192–195
Pointer method, 609–610
Point-value techniques, 80
Poisson's equation, 11, 15–24, 25, 33, 108–109
    BEM solution of scalar Poisson equation in three dimensions, 403–408
    boundary element equations for, in two dimensions, 374–381
    functional formulation for, 160–161
    one-dimensional, 364–366
Polar coordinates, 73
Polarization, 12–13
Polynomial interpolation, 201–207
Poynting vector method, 161–163
Preconditioned conjugate gradient method, 627–645
    convergence, 637–645
Profile storage, 614–621
Pseudoperipheral node algorithm, 627
QR factorization, 652–653
Quadrilateral elements, 265–267
Recoil permeability, 30
Rectangular elements, first-order, 353–356
Rectangular elements, interpolation of, 227–234
    serendipity elements, 234–240
Recurrence formula, 600
Regula falsi, 681
Relative permeability, 27
Relaxation factors, 302
Relaxation method, 80
    alternating, 82
Residual excitation method, 157–159

Resistive analog, 70–71
Ritz's method, 170–175, 283
Saturation, 27
    curve, 27
Scalar potential, 10
Scattering problems
    in a cylinder, 488–491
    finite element formulation with absorbing boundary conditions, 484–487
    hybrid element method, 481–484
    numerical examples, 488–494
    wedge-shaped, 492–494
Schwartz-Christoffel transformation, 68
Second order methods, 669–671
Serendipity elements, 234–240
Series bar-coil equation, 526
    time discretization of, 534
Series connection of bars to form coils, 522–523
Shape functions
    derivation of, 207–211
    derivation of, for serendipity elements, 234–240
    high order triangular interpolation functions, 222–227
    interpolatory, 190–201
    Lagrangian interpolation, 211–214
    one-dimensional finite elements, 725
    orthogonal basis functions, 278–281
    polynomial interpolation, 201–207
    rectangular elements, 227–234
    three-dimensional finite elements, 241–277, 735–748
    two-dimensional finite elements, 214–222, 727–733
Similarity transformation, 76, 77
Simple iteration, 667–669
Sinusoidally time-varying electric fields, 21–22
Soft magnetic material, 27
Sparse matrix techniques, 607
    band and profile storage, 614–621
    Cuthhill-McKee algorithm and reverse, 618–621
    linked list structure, 608–609

nested dissection reordering, 622–627
node numbering, 617–618
optimal ordering, 610–614
pointer method, 609–610
storage methods, 608–610
Spherical harmonic functions, 73
Squirrel cage induction motor example, 548–550
Static electric fields, 1–2
Static magnetic fields, 24–36
Steady-state linear time-harmonic diffusion equation, functional for, 168–170
Steffenson method, 683–692, 700–703
Stokes' theorem, 37, 71
Streaming electrification
See also Electrification, flow-induced causes of, 550
Subdomain method, 183
Symbolic operators, finite difference method and, 122–125
Symmetric matrix, 597–598
Taylor series, fourth-order formula from, 120–122
Taylor's theorem, 670
Teledeltos paper, use of, 69–70
Tetrahedral elements, 241–246
Three-dimensional finite elements, 241
  curved elements, 268–270
  derivatives and integration procedure, 248–249
  formulation for electric field, 494–516
  hexahedral elements, 251–253
  isoparametric elements, 254–265
  isoparametric elements, triangular, 271–272
  mixed and continuous elements, 275–277
  natural or volume coordinates in, 246–248
  quadrilateral elements, 265–267
  shape functions for, 735–748
  special, 272–275
  tetrahedral elements, 241–246
  triangular prisms, 249–250
Three-dimensional scalar Poisson equation, BEM solution of, 403–408
Three-element magnetostatic problem, finite element method and, 326–338

Time-dependent magnetic diffusion equation, 520–521
Time discretization
  of acceleration equation, 536
  of circuit equations, 534
  of current equation, 532–533
  of diffusion equation, 532
  by finite element method, 310–312
  of mechanical equations, 536
  of parallel circuit equation, 535
  of series bar-coil equation, 534
  of velocity equation, 536–537
Time domain modeling of electromechanical devices, 519
  electromagnetic and mechanical theory, 520–526
  Galerkin formulation, 527–531
  global system of equations, 542–547
  linearization, 537–541
  squirrel cage induction motor example, 548–550
  time discretization, 531–537
Total current equation, 521–522, 526
  time discretization of, 532–533
Total magnetic scalar potential. See Magnetic scalar potential
Transformation methods for nonlinear problems, 75–79
Transient electric fields, 23–24
Transient scalar field problems
  application to electrification problem in dielectric media, 564–570
  finite element formulation and discretization, 566–567
  Fourier integral transform method and, 562–564, 567–570
Transverse electric (TE) polarization, 452, 457–459, 466–467, 483–484, 487, 515–516
Transverse magnetic (TM) polarization, 452–457, 481–483, 515–516
Trial functions, examples of, 182, 189
Triangle area, vertex coordinates and, 709–710
Triangular edge elements, 356–357

Triangular interpolation functions, high order
  Irons' method, 222–223
  Silvester's method, 224–227
Triangular isoparametric elements, 271–272
Triangular prisms, 249–250
Trigonometric interpolation, 279–281
Two-dimensional eddy currents, integral equations and, 393–402
Two dimensional electromagnetic field analysis,
  transverse electric and magnetic polarizations for, 452–459
Two-dimensional finite elements
  analysis, 289–295
  equations, 214–222
  natural coordinates in, 218–222
  shape functions for, 727–733
    Two-dimensional Poisson equation, boundary element equations for, 374–381
Uniqueness theorem, 17–19, 372
Variables, choice of, 44–60
Variational method, 144–145
  accuracy of, 151–152
  integral equations and, 179–181
  Ritz's method, 170–175
  wave equation, 176–179
Vector operators, 707–708

Vector potential
  See also Magnetic vector potential (MVP)
  axisymmetric formulation for eddy current problem using, 313–320
Vector wave equation, 178–179 Velocity equation. See Mechanical velocity equation
Vertex coordinates, triangle area and, 709–710
Volume coordinates, integrals of, 721
Wave (Helmholtz's) equation, 42–44
  See also Finite elements for wave equations, high-frequency problems with finite difference time domain formulas for, 135–141
  Green function for one-dimensional, 409
  Green function for three-dimensional, 410–411
  Green function for two-dimensional, 410
  variational method and, 176–179
  vector, 178–179
Wedge-shaped scattering, 492–494
Weighted residual method. See Galerkin weighted residual method
Whittaker's method, 681
Wiener-Hopf method, 471
z-plane, 68